QUATERNARY ENVIRONMENTAL CHANGE IN SOUTHERN AFRICA

Physical and Human Dimensions

Ongoing climate change necessitates advances in our understanding of the interrelationships between climate, landscape-shaping processes and human activity over long time periods, especially in areas that are already climatically stressed. Southern Africa is a key region in the story of human evolution during the last 2 million years but, until recently, relationships between human evolution, the archaeological record and environmental changes in this region have been poorly understood or connected.

This volume presents new ideas on macroscale landscape evolution; mountain, fluvial and aeolian processes; and environments in southern Africa during the Quaternary. Interdisciplinary in scope, it brings together an international team of experts to synthesise the latest research and understanding of landscape-human relationships in this region. It incorporates ideas and results from the emerging fields of geoarchaeology and cultural landscapes, and utilises the latest data and analytical techniques.

This is a key reference for researchers studying hominid evolution, geoarchaeology and environmental change, and provides a benchmark study of southern African landscape evolution during the Quaternary. It will also appeal to professionals and policymakers with interests in future human-landscape evolution in southern Africa.

JASPER KNIGHT is Professor of Physical Geography at the University of the Witwatersrand. He is a geoscientist with research interests in geomorphology and sediment system responses to climate change during the late Pleistocene and Holocene, specifically on glaciers, rivers, coasts and mountains, in Europe, the USA, Australasia and southern Africa.

STEFAN W. GRAB is Professor of Physical Geography at the University of the Witwatersrand. He is a geomorphologist with a research focus on cold region and mountain geomorphology, Quaternary environmental change in southern Africa, and quantifying climate change during recent and historical times.

QUATERNARY ENVIRONMENTAL CHANGE IN SOUTHERN AFRICA

Physical and Human Dimensions

Edited by

JASPER KNIGHT
University of the Witwatersrand, Johannesburg

STEFAN W. GRAB
University of the Witwatersrand, Johannesburg

CAMBRIDGE
UNIVERSITY PRESS

CAMBRIDGE
UNIVERSITY PRESS

University Printing House, Cambridge CB2 8BS, United Kingdom

Cambridge University Press is part of the University of Cambridge.

It furthers the University's mission by disseminating knowledge in the pursuit of education, learning and research at the highest international levels of excellence.

www.cambridge.org
Information on this title: www.cambridge.org/9781107055797

First published 2016

Printed in the United Kingdom by TJ International Ltd. Padstow Cornwall

A catalogue record for this publication is available from the British Library

Library of Congress Cataloging in Publication Data
Names: Knight, Jasper, editor. | Grab, S. W., editor.
Title: Quaternary environmental change in southern Africa : physical and human dimensions / edited by Jasper Knight, University of the Witwatersrand, Johannesburg, and Stefan W. Grab, University of the Witwatersrand, Johannesburg.
Description: Cambridge : Cambridge University Press, 2016. | Includes bibliographical references and index.
Identifiers: LCCN 2015035034 | ISBN 9781107055797 (Hardback : alk. paper)
Subjects: LCSH: Paleoclimatology–Quaternary. | Paleoclimatology–Africa, Southern. | Climatic changes–Africa, Southern.
Classification: LCC QC884.2.C5 Q83 2016 | DDC 551.7/90968–dc23 LC record available at http://lccn.loc.gov/2015035034

ISBN 978-1-107-05579-7 Hardback

Contents

Contributors

Marion K. Bamford
Evolutionary Studies Institute, University of the Witwatersrand, Private Bag 3, Wits 2050, Johannesburg, South Africa

Mark D. Bateman
Department of Geography, University of Sheffield, Sheffield, S10 2TN, UK

Greg A. Botha
Council for Geoscience, PO Box 900, Pietermaritzburg 3200, South Africa

James Brink
National Museum Bloemfontein, PO Box 266, Bloemfontein 9300, South Africa

Kristian J. Carlson
Institute for Human Evolution, University of the Witwatersrand, Private Bag 3, Wits 2050, Johannesburg, South Africa

Hayley C. Cawthra
Marine Geoscience Unit, Council for Geoscience, 7535 Cape Town, South Africa; *and* Centre for Coastal Palaeoscience, Nelson Mandela Metropolitan University, PO Box 77000, Port Elizabeth 6031, South Africa

John S. Compton
Department of Geological Sciences, Louis Ahrens Building, Library Road, University of Cape Town, Rondebosch 7701, South Africa

J. Andrew G. Cooper
School of Environmental Studies, University of Ulster, Cromore Road, Coleraine, BT52 1SA, UK; *and* Geological Sciences, School of Agriculture, Earth and Environmental Sciences, University of KwaZulu-Natal, Westville Campus, Private Bag X54001, Durban, South Africa

Ander de Lecea
Geological Sciences, School of Agriculture, Earth and Environmental Sciences, University of KwaZulu-Natal, Westville Campus, Private Bag X54001, Durban, South Africa

Sarah Edlund
Department of Anthropology, University of Wisconsin-Madison, 1180 Observatory Drive, Madison, WI 53706, USA

William N. Ellery
Department of Environmental Science, Rhodes University, PO Box 94, Grahamstown 6140, South Africa

Jennifer M. Fitchett
School of Geography, Archaeology and Environmental Studies, University of the Witwatersrand, Private Bag 3, Wits 2050, Johannesburg, South Africa

Stefan W. Grab
School of Geography, Archaeology and Environmental Studies, University of the Witwatersrand, Private Bag 3, Wits 2050, Johannesburg, South Africa

Andrew N. Green
Geological Sciences, School of Agriculture, Earth and Environmental Sciences, University of KwaZulu-Natal, Westville Campus, Private Bag X54001, Durban, South Africa

Michael C. Grenfell
Department of Earth Science, University of the Western Cape, Bellville 7535, South Africa

Suzanne E. Grenfell
Department of Geography and Environmental Studies, University of the Western Cape, Bellville 7535, South Africa

Jason Heaton
Birmingham-Southern College, Campus Box 549022, 900 Arkadelphia Road, Birmingham, AL 35254, USA

Karin Holmgren
Department of Physical Geography and Quaternary Geology, Stockholm University, S-10691 Stockholm, Sweden

Jasper Knight
School of Geography, Archaeology and Environmental Studies, University of the Witwatersrand, Private Bag 3, Wits 2050, Johannesburg, South Africa

Donovan C. Kotze
Centre for Water Resources Research, University of KwaZulu-Natal, Private Bag 01, Scottsville 3209, South Africa

Kathleen Kuman
School of Geography, Archaeology and Environmental Studies, University of the Witwatersrand, Private Bag 3, Wits 2050, Johannesburg, South Africa

Philip M. Marren
Department of Resource Management and Geography, University of Melbourne, Victoria 3010, Australia

Steve McCourt
School of Geological Sciences, University of KwaZulu-Natal, Westville Campus, Private Bag X54001, Durban, 4000, South Africa

Michael E. Meadows
Department of Environmental and Geographical Science, South Lane, Upper Campus, University of Cape Town, Private Bag X3, Rondebosch 7701, South Africa

Peter Mitchell
Institute of Archaeology, University of Oxford, 36 Beaumont Street, Oxford, OX1 2PG, UK; *and* School of Geography, Archaeology and Environmental Studies, University of the Witwatersrand, Private Bag 3, Wits 2050, Johannesburg, South Africa

Frank H. Neumann
Forschungsstelle für Paläobotanik, Westfälische Wilhelms-Universität, Heisenbergstrasse 2, 48149 Münster, Germany; *and* Evolutionary Studies Institute, University of the Witwatersrand, Private Bag 3, Wits 2050, Johannesburg, South Africa

Rebecca Powell
Department of Environmental Science, Rhodes University, PO Box 94, Grahamstown 6140, South Africa

Lynne J. Quick
Department of Environmental and Geographical Science, South Lane, Upper Campus, University of Cape Town, Private Bag X3, Rondebosch 7701, South Africa

Jürgen Runge
Goethe Universität, Institut für Physische Geographie & Zentrum für interdisziplinäre Afrikaforschung (ZIAF), Altenhöferallee 1, 60438 Frankfurt am Main, Germany

Maria H. Schoeman
School of Geography, Archaeology and Environmental Studies, University of the Witwatersrand, Private Bag 3, Wits 2050, Johannesburg, South Africa

Louis Scott
Department of Plant Sciences, Faculty of Natural and Agricultural Sciences, University of the Free State, PO Box 339, 9300 Bloemfontein, South Africa

Paul Shaw
Department of Geography, Faculty of Food and Agriculture, University of the West Indies, St Augustine, Trinidad and Tobago

Rebekah G. Singh
Council for Geoscience, PO Box 900, Pietermaritzburg 3200, South Africa

Dominic Stratford
School of Geography, Archaeology and Environmental Studies, University of the Witwatersrand, Private Bag 3, Wits 2050, Johannesburg, South Africa

Arnaud J.A.M. Temme
Soil Geography and Landscape, WU Environmental Sciences, Wageningen University, PO Box 47, 6700AA Wageningen, The Netherlands

David S.G. Thomas
School of Geography and the Environment, University of Oxford, Oxford, OX1 3QY, UK

Stephen Tooth
Institute of Geography and Earth Science, Aberystwyth University, Aberystwyth, SY23 3DB, UK; *and* School of Geosciences, University of the Witwatersrand, Private Bag 3, Wits 2050, Johannesburg, South Africa

Stephan Woodborne
iThemba Laboratories, Private Bag 11, WITS 2050, Johannesburg, South Africa

Sarah Wurz
Evolutionary Studies Institute *and* School of Geography, Archaeology and Environmental Studies, University of the Witwatersrand, Private Bag 3, Wits 2050, Johannesburg, South Africa; *and* Department of Archaeology, History, Cultural Studies and Religion, University of Bergen, Norway

1

The context of Quaternary environmental change in southern Africa

JASPER KNIGHT AND STEFAN W. GRAB

Abstract

Climate changes and tectonic processes throughout the Cenozoic, and earlier, provide the context for landscape and environmental change in southern Africa during the Quaternary. Changing land surface properties and resource availability, including rock types, topography, soils, ecosystems and drainage patterns, have exerted a strong impact on the processes and patterns of human evolution, technological innovation and behaviour over millennial timescales. The southern African landscape seen today, and the preserved imprint of its past human activities, resulted from the interplay between climate, tectonics and geomorphology over lengthy Cenozoic timescales.

1.1 Introduction

'The contemporary face of southern Africa owes much to events in the Cenozoic' (Partridge and Maud, 2000a, p3). Thus began the last significant book on the topic of Quaternary environmental change in the subcontinent, sixteen years ago. Although this assessment is still broadly true, advances in the fields of palaeoecology and geochronometry in particular have yielded a far more complex pattern of landscape development in southern Africa during the Cenozoic (66 Ma–present), and more widely from the late Cretaceous onwards. Recent research also highlights a degree of spatial and temporal complexity in patterns of climate and environmental change across southern Africa, and the proxies in which such records have been preserved, that has not been previously recognised (Meadows, 2014). This is especially the case for the Quaternary epoch (last 2.58 Ma; Gibbard and Cohen, 2008). These terrestrial records can also be supplemented by data from marine

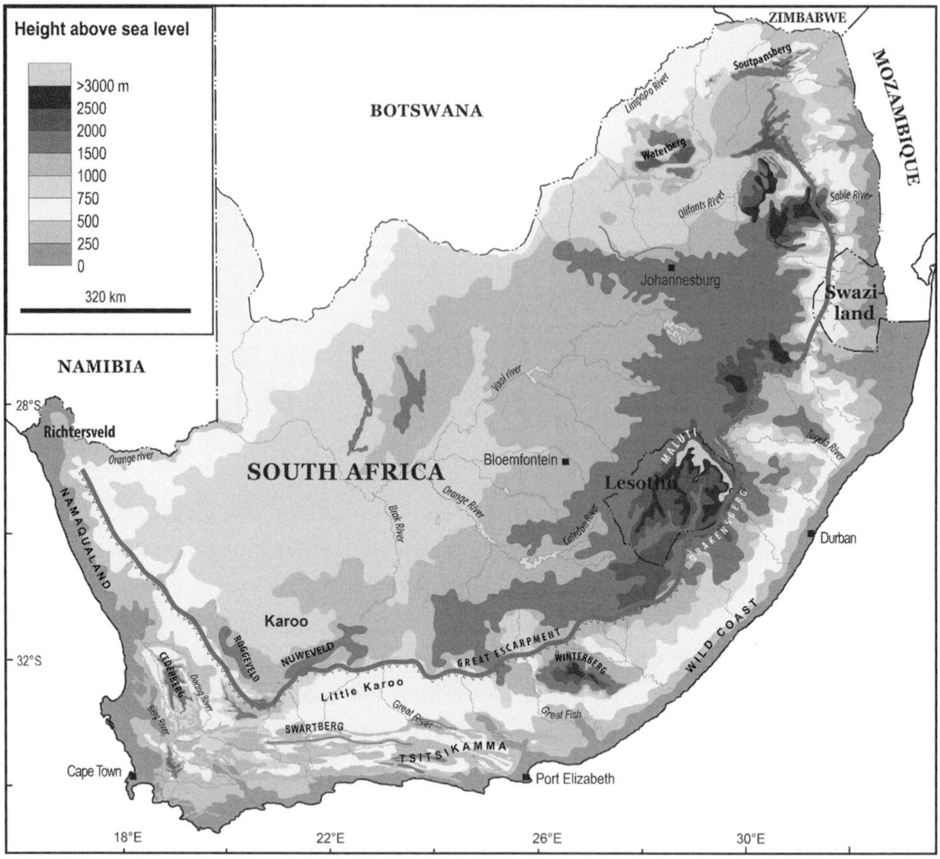

Fig. 1.1. Map of the topography of southern Africa (in m above sea level) and the location of the Great Escarpment (hatched line). Note that other key regional maps are presented elsewhere in this book: regional geology (Fig. 2.1), soils (Fig. 16.2) and ecosystems (Figs. 18.1, 18.2). A black and white version of this figure will appear in some formats. For the colour version, please refer to the plate section.

cores (e.g. *n*-alkanes, $\delta^{18}O$ and ^{13}C records); outputs from the advanced capabilities of climate and land-surface models; and real-time evaluation and modelling of climate and land surface variables on a scale and at a resolution that previously has not been available, based on satellite remote sensing datasets. In addition, there have been significant new palaeo-anthropological finds and modes of data analysis since 2000 that have informed on patterns of human evolution (Beaumont and Vogel, 2006). All of these represent significant advances from Partridge and Maud's (2000b) volume that have increased the power and complexity of Cenozoic datasets. The contemporary topography, climate and environments of southern Africa are remarkably diverse (Figs. 1.1, 1.2), and thus it can be anticipated that climatic conditions in the past, and the geomorphological (landscape-shaping)

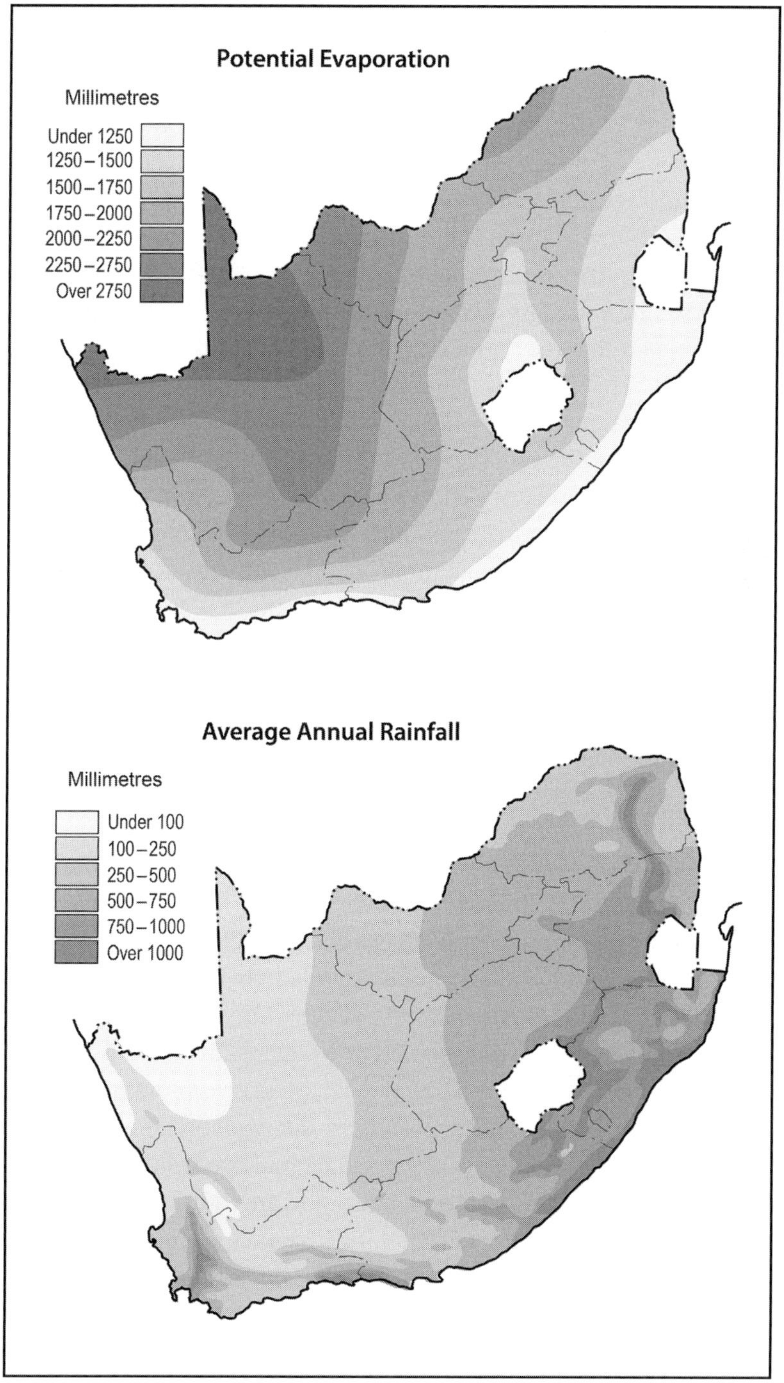

Fig. 1.2. Maps of key climatic features of South Africa, as expressed by potential evaporation (upper panel) and annual average rainfall (lower panel). The contrast between spatial patterns of these parameters shows that northwestern South Africa has a large moisture deficit, which has strongly impacted on geomorphological processes and landforms, ecosystems and the capacity for varied human activities in the region, including agriculture.

processes associated with them, were equally diverse. Many potential palaeoclimatic records in southern Africa remain unexploited, and research is now suggesting that African records, in particular during the Pleistocene, may record climatic events such as Dansgaard–Oeschger cycles and Heinrich events that were previously considered only to be a feature of northern hemisphere records (Itambi et al., 2009). Such results suggest that southern Africa is well placed to record variations in the strength of Southern Ocean overturn, as well as monsoon strength and position of the intertropical convergence zone (Garcin et al., 2007; McGee et al., 2014). These are some of the emerging areas of palaeoclimatic research in the subcontinent.

It can also be said that the 'contemporary face' of southern Africa has changed much since 2000. The impacts of human activity on all parts of the southern African landscape (Binns, 1997), on its geomorphological processes, and on its varied resources (including biodiversity and ecosystem services; soils, agriculture and food security; water security and water quality; waste management) are now recognised as significant issues, not just for environmentalism and landscape conservation and management, but more significantly as a major limitation to achieving sustainable socioeconomic development in the 21st Century (Cole et al., 2014; Ziervogel et al., 2014). Climate assessment reports published by the Intergovernmental Panel on Climate Change (IPCC) also highlight the vulnerability of sub-Saharan Africa to anthropogenic climate change (global warming), and that climate change over coming decades may result in significant, wide-ranging and yet poorly understood impacts on both physical and human environments (Niang et al., 2014).

It is in this wider context that this book is set: there is no more important time than now for developing a better understanding of the Quaternary context of landscape–human relationships in southern Africa. As scientists, we look to the past in both geologic, faunal and cultural records to help inform on the processes and sensitivities of the landscape and its contemporary peoples to future change, and this uniformitarian principle can help us when evaluating the challenges facing southern Africa, its landscapes and peoples, over coming decades (Ziervogel et al., 2014). This chapter considers the wider context of Quaternary environmental change in southern Africa and the nuanced and complex narratives now beginning to emerge from recent inter- and multidisciplinary research into landscape–human relationships. In detail, this chapter (1) discusses the geologic and geomorphic framework whereby such environmental resources are provided; (2) presents models of evolution of the southern African landscape during the Cenozoic, and in particular the Quaternary, that shaped today's landscape; and (3) considers the range of environmental resources offered by the southern African landscape that were (are) of relevance to past (present) patterns of human activity.

1.2 The geologic and geomorphic context of southern Africa

The geological history of southern Africa is exceptionally long, extending from 3600 million years ago to present, and covering a wide range of depositional and tectonic settings (McCarthy and Rubidge, 2005; Hunter et al., 2006). A timeline of critical geological and human evolutionary events over different timescales is given in Figure 1.3. Tectonic processes (continental accretion and rifting, epeirogenic uplift), magmatic processes (intrusion and extrusion of igneous rocks, metamorphism and mineralisation), sedimentary processes (basin formation and infilling), and climate change (land surface weathering and erosion) have been the major regional-scale controls on southern Africa's geological history and resulting rock types. The relationship between underlying geology, topography, geomorphology and climate is complex, and has been examined in several studies (e.g. Bishop, 2007; Partridge et al., 2010).

Cosmogenic dating of exposed and buried rock surfaces can inform on rates of land surface denudation and sediment accumulation, respectively (Decker et al., 2011). Dating studies from different physical environments and locations across southern Africa show different rates of land surface change that vary according to rock type, climate zone and tectonic setting (Portenga and Bierman, 2011), and mesoscale properties such as slope angle, and location relative to hilltops or river valleys (Bierman and Caffee, 2001). The lowest rates of land surface change are found on exposed but isolated summits of mesas, koppies and bornhardts (Bierman and Caffee, 2001), which are preserved largely because of high rock strength and low drainage density (Moon, 1990). Chadwick et al. (2013) demonstrated that landscape-scale weathering and erosion in river catchments of the Kruger National Park (northeastern South Africa) contribute to low but spatially consistent denudation rates of 3–6 m/myr, which are similar to those calculated by independent geochronometric methods. These authors argued that river drainage density and slope gradients within river catchments show a consistent relationship with precipitation, such that climate is the most important control on landscape evolution, rather than geology alone, and that long-term weathering and erosion contributes to downslope sediment supply and sediment yield to river basins, but that this is modified by local geology (Decker et al., 2013). There is a self-limiting relationship between weathering and erosion processes and the evolution of land surface topography. For example, over many footslope locations across southern Africa, an extensive colluvial mantle has developed (Watson et al., 1984), which attests to climatically driven denudation and slope instability from at least the mid-Pleistocene onwards (Dardis, 1990). Valley infilling and decreased slope angle through colluviation reduces slope instability and thus the rate of land surface change. Such a situation, often recorded by large cosmogenic exposure or burial

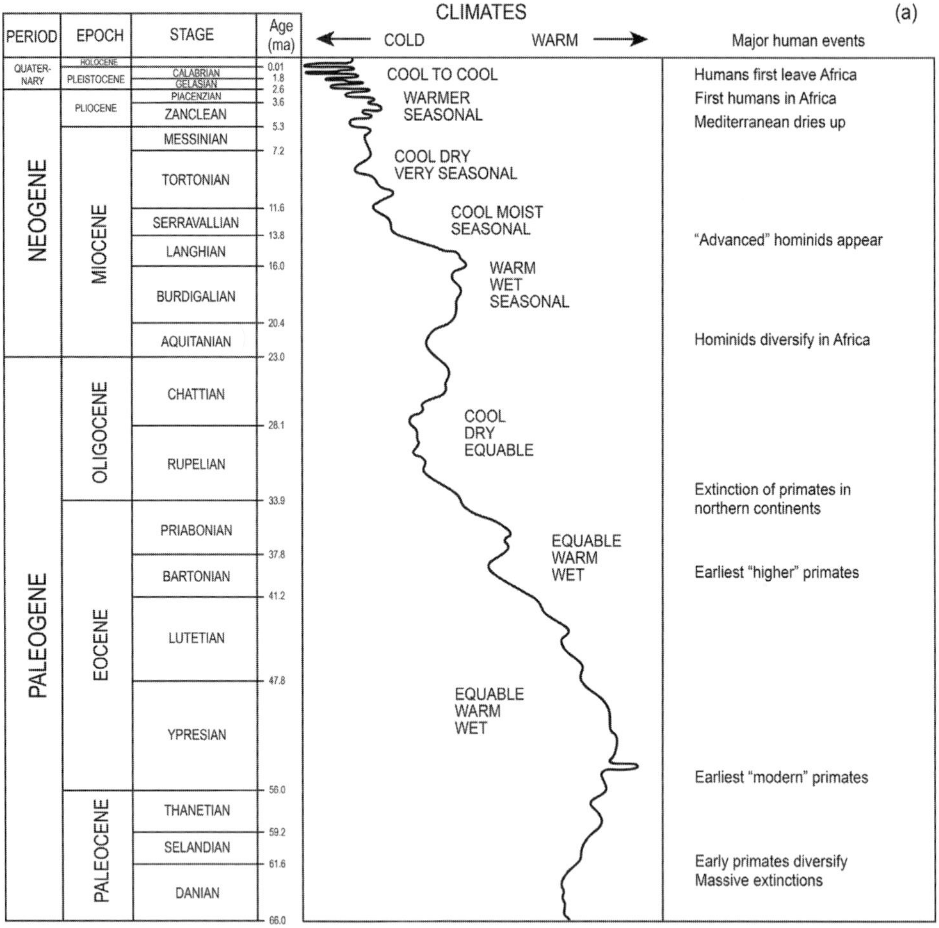

Fig. 1.3. (a) Simplified timeline of the Cenozoic (66 Ma to present), showing the major climatic and human events to have affected southern Africa (adapted from Partridge et al., 1995; geologic timescale from the Geological Society of America; www.geosociety.org/science/timescale/). Note that there is uncertainty on the timing and/or magnitude of climatic and human events. (b) Simplified timeline from the Middle Pleistocene (~460 kyr) to present; showing the δ^{18}O record from SE Atlantic core MD962094 (Stuut et al., 2002); and elemental Fe counts from SE Atlantic core ODP1083 (West et al., 2004), with the major human events over this timescale in southern Africa. LSA: Late Stone Age, MSA: Middle Stone Age. (c) Simplified timeline for the period 25 kyr to present, showing δ^{18}O values from Makapansgat (Limpopo Province, South Africa) speleothem record (Holmgren et al., 2003), where LIA: Little Ice Age, YD/H0: Younger Dryas/Heinrich event 0, H1: Heinrich event 1, H2: Heinrich event 2; principal component analysis scores for moisture variations from three sites (the continuous and dashed lines) at Braamhoek wetland (Free State, South Africa) (Scott et al., 2012), and major human events over this timescale in southern Africa. LSA: Late Stone Age, MSA: Middle Stone Age.

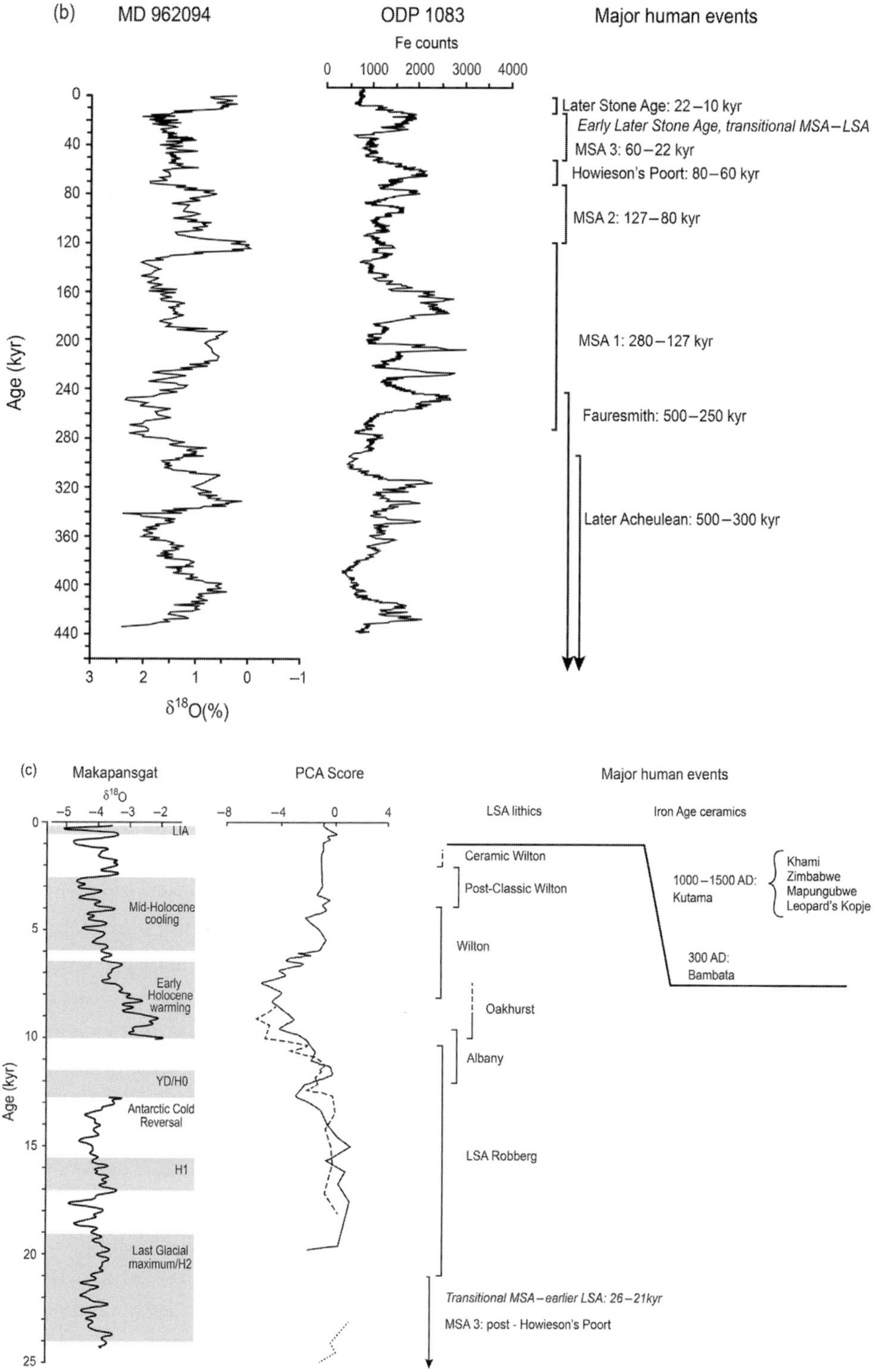
(b)
MD 962094
ODP 1083
Fe counts
0 1000 2000 3000 4000
Major human events
Age (kyr)
0 40 80 120 160 200 240 280 320 360 400 440
3 2 1 0 –1
δ18O(%)
Later Stone Age: 22–10 kyr
Early Later Stone Age, transitional MSA–LSA
MSA 3: 60–22 kyr
Howieson's Poort: 80–60 kyr
MSA 2: 127–80 kyr
MSA 1: 280–127 kyr
Fauresmith: 500–250 kyr
Later Acheulean: 500–300 kyr
(c)
Makapansgat
δ18O
–5 –4 –3 –2
PCA Score
–8 –4 0 4
Major human events
LSA lithics
Iron Age ceramics
LIA
Mid-Holocene cooling
Early Holocene warming
YD/H0
Antarctic Cold Reversal
H1
Last Glacial maximum/H2
Age (kyr)
0 5 10 15 20 25
Ceramic Wilton
Post-Classic Wilton
Wilton
Oakhurst
Albany
LSA Robberg
Transitional MSA–earlier LSA: 26–21kyr
MSA 3: post - Howieson's Poort
1000–1500 AD:
Kutama
Khami
Zimbabwe
Mapungubwe
Leopard's Kopje
300 AD:
Bambata

Fig. 1.3. (*cont.*)

ages (thus low denudation rates), does not reflect land surface stability at all, but rather low accommodation space in upland source areas. Cosmogenic dating, although critical for evaluating land surface evolution in southern Africa, has limited applicability where not all physical environments have been dated, or are capable of being dated, and where interpretation of cosmogenic dates is based on limited understanding of their geomorphological and sedimentary context.

However, this viewpoint of 'climatically controlled geomorphology' is but one approach. Other studies have argued that spatial and temporal patterns of land surface denudation in southern Africa can be explained by tectonic processes. For example, based on modelling, thermochronometric and cosmogenic dating evidence, episodes of tectonic uplift from the late Cretaceous onwards are followed by episodes of accelerated rates of land surface denudation (e.g. Braun et al., 2014). However, there are relatively large errors associated with the use of such evidence to calculate denudation rates, and many land surface feedbacks through the accumulation of weathering mantles, ecosystem and carbon cycle responses, and river system responses, are difficult to evaluate (de Wit, 2007).

1.3 Climate and environmental changes in southern Africa prior to the Quaternary

Climate and environmental changes of the late Cenozoic (Neogene; 23.0–2.6 Ma) in southern Africa are set within the wider context of ongoing tectonic evolution, speciation and ecosystem changes, and hominin evolution during this time period (Partridge et al., 1995). These linked changes are often difficult to distinguish in terms of cause and effect, within dating resolution, hence the causal relationships between these components are often still unclear or contentious in many cases (Lavier et al., 2001; de Wit, 2007; Jung et al., 2014). This poses an interpretive challenge with respect to identifying the relative roles of tectonics or climate in macroscale land surface evolution. For example, Cenozoic and pre-Cenozoic volcanism in southern Africa, associated with emplacement of the Karoo and Etendeka igneous provinces, led to considerable topographic land surface changes (Partridge et al., 1995; Phillips et al., 2000). In turn, this resulted in enhanced subsequent land surface weathering and erosion, and river water and geochemical export into surrounding oceans, affecting water chemistry, biotic processes, and hemispheric climate feedbacks (e.g. Roelandt et al., 2011).

Most evidence for Neogene environments and climates comes from isolated onshore and offshore depocentres where mineral and organic sediments have accumulated and been preserved (Fig. 1.4). Marine cores in the South Atlantic show palynological and sedimentary evidence for changes in zonal climate and atmosphere–ocean circulation patterns. For example, a drying trend in the Middle

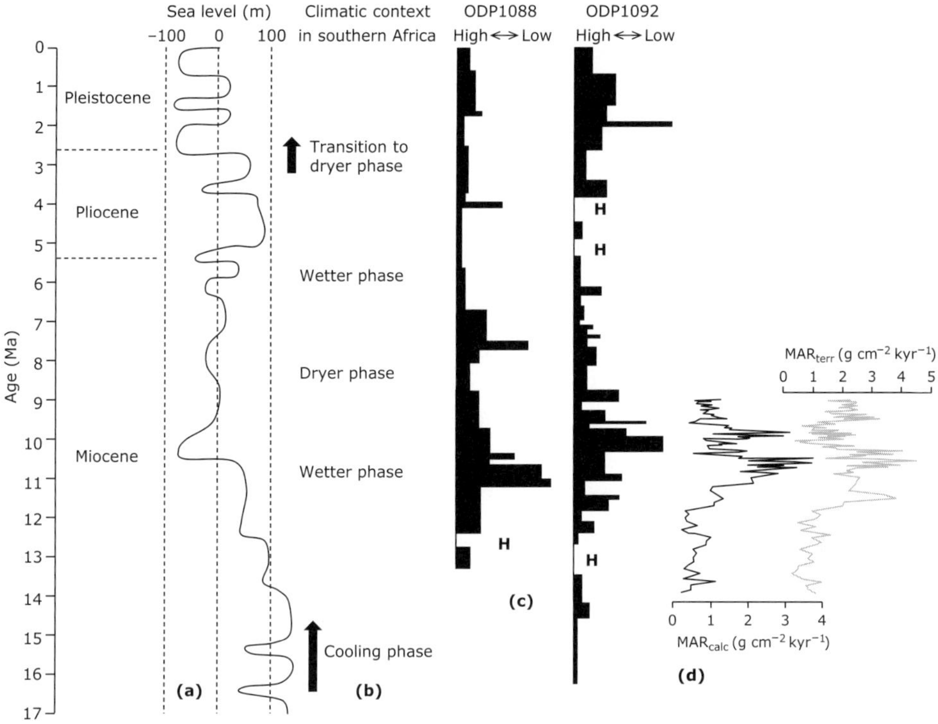

Fig. 1.4. Environmental changes during the late Cenozoic, with a particular focus on southern Africa. (a) Global sea level changes relative to present (Haq et al., 1987); (b) key regional climatic events (from various sources named in the text); (c) relative sediment accumulation rates at South Atlantic ODP sites 1088 and 1092 (Diekmann et al., 2003). H = hiatuses; (d) variations in marine accumulation rates (MAR) for calcium carbonate (calc) and terrigeneous (terr) debris (grey line), ODP1085A, offshore Orange River, South Atlantic (Roters and Henrich, 2010).

Miocene is marked by increased aeolian contribution to ocean sediments from the developing Namib Desert (Senut et al., 2009; Roters and Henrich, 2010), and land surface aridity in southwest Africa (Pickford et al., 2014). This period (around 14–11 Ma) matches with continental glaciation of Antarctica, the spread of cold Antarctic Deep Water, invigorated Benguela Current upwelling, and zonal atmospheric circulation (Jung et al., 2014). In the late Miocene and Pliocene, changes in Polar Front position over the Southern Ocean influenced rainfall seasonality in Namibia, including moisture source (Dupont et al., 2013). Increased winter rainfall and resulting Renosterveld vegetation displaced xerophytic savannah in Namibia as the Polar Front migrated northwards, increasing cold water advection and thus rainfall into the region (Dupont, 2006). This also had an impact on the distribution of C_3/C_4 flora (Franz-Odendaal et al., 2002; Ségalen et al., 2006) and on mammalian fossil assemblages found at key human evolution sites such as Swartkrans (Avery, 2001). Periods of aridity during the Miocene are marked by the development of silcrete and calcrete weathering profiles found on the African Surfaces, discussed later (Blumel and Eitel, 1994; Marker et al., 2002).

1.4 Models of landscape evolution in southern Africa

Cosmogenic dating has invigorated debate on land surface evolution models. These are of particular interest in southern Africa through the ground-breaking work of du Toit (1954) and King (1963). Initially, du Toit (1954, pp. 573–575) invoked Davisian ideas of cyclic episodes of landscape evolution, comprising Tertiary and Quaternary tectonic uplift and/or sea-level change around coastal fringes, followed by formation of peneplains at different levels, which were later incised. Later, King (1963, pp. 49–55) expanded on these ideas and discussed a 'landscape cycle' of pediplanation for the evolution of the southern African landscape. This 'pediplanation' involved a period of river incision followed by one of parallel scarp retreat. Using the Davisian evolutionary stages of youth, maturity and old age, King argued that the youth stage is dominated by river incision, and as this stage progresses, stream density decreases as rivers become larger. The maturity stage is dominated by scarp retreat as river valleys widen and flat land surfaces, inherited from the youth stage, reduce in size. In the old-age stage, opposing scarps meet each other and the resulting ridges gradually decrease in height, forming a low-relief landscape. King mentions some specific examples of these evolutionary stages from southern Africa, arguing that these stages can coexist in different areas.

The persistence of flat-lying surfaces in the landscape, equivalent to King's 'pediplains', has been long noted (Partridge and Maud, 1987, their Appendix A; Partridge, 1998). In broad age-order, these are the African Surface (Cretaceous age), Post-African I (mid-Tertiary), and Post-African II (late Tertiary) Surfaces (Fig. 1.5). Preserved remnants of earlier and later surfaces have also been identified. These surfaces have been used as important markers in the landscape evolution story of southern Africa, but the notion of time-equivalent and accordant surfaces is inherently problematic and involves substantial circular reasoning based on untestable hypotheses. For example, all accordant surfaces of the same altitude are argued to be of the same age and thus affected by the same spatially uniform tectonic regime. Surfaces located at different altitudes must therefore be of different ages or affected by different tectonic regimes. Surfaces are assumed to be of a single 'age' that represents a Davisian-style peneplain that is at equilibrium. The extent of land surface incision has also been used as an indicator of relative age (Partridge and Maud, 1987, 2000a). There are also implications for understanding

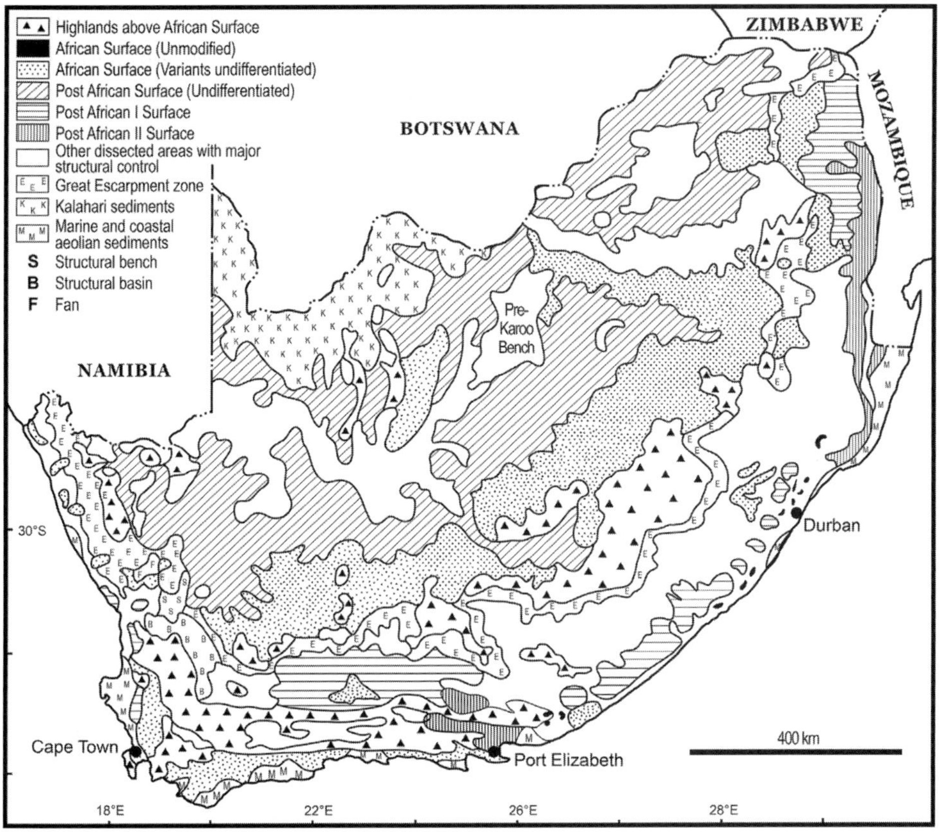

Fig. 1.5. Map of South Africa showing assumed and geochronometrically dated land surface ages, and assumed correlatives. The African Surface ages are identified largely on lithostratigraphic principles (redrawn from Moon and Dardis, 1988).

the landscape context of hominin evolution, which is necessary in order to correctly evaluate the environmental controls on this evolution (Dirks and Berger, 2013).

Two lines of evidence now clearly show that such ideas of macroscale land surfaces are untenable. (1) Cosmogenic dates from equivalent land surfaces, discussed earlier, show highly variable ages (in the range 10^5–10^6 years) and thus variable resulting rates of land surface denudation. They show that land surfaces are not passive relicts but are actively being dissected and denuded with high spatial and temporal variability (Fleming et al., 1999; Tinker et al., 2008; Moore et al., 2009; Erlanger et al., 2012). The notion of landscape palimpsests, now commonly discussed in areas undergoing differential areal denudation (e.g. Stroeven et al., 2002), provides a geomorphological interpretive framework for these 'non-surfaces'. (2) Some recent studies using different radiometric dating methods, including apatite fission track thermochronology of igneous rocks that are now exposed on the surface, show that very substantial land surface uplift, and thus denudation, has taken place since rock formation. For example, Tinker et al., (2008) showed that around 1 km thickness of rock from the Cape Fold Belt has been eroded since the late Cretaceous. This is highly inconsistent with the static equilibrium and preservation of isochronous surfaces. The view presented by Partridge and Maud (2000a) on the macroscale landscape evolution of southern Africa is therefore very dated and reflects a Davisian-style approach to landscape dynamics. New digital satellite datasets of the land surface, including spectral reflectance, altimetry and gravity, also enable a better understanding of spatial and temporal variations of the land surface and its attributes.

1.5 South African landscapes and environmental resources

The environmental resources (rocks/minerals, soil, water and ecosystems) provided in varied southern African landscapes have made these landscapes particularly favourable for the development of the *Homo* species and human activities over long time scales (Vrba et al., 1995; Reed, 1997). The general environmental settings favourable for human evolution and spread, in particular during the fluctuating climates of the Cenozoic, are well met in southern Africa. The presence of generally flat grasslands (velds) and isolated hilltops (koppies) favoured the movement and occupation of human social groups, respectively, which likely mirrored the evolution and spread of other animal species, and thus facilitated development of the high faunal and floral endemism recorded in many areas. Prominent landscape positions provided by local geomorphology, such as caves and rock shelters, acted as foci for a range of human domestic, territorial and ritual activities. Freshwater resources in the form of rivers and ephemeral lakes (pans) in the dry interior of the region were fundamental to the spread inland and then occupation of the landscape by hominid groups, as well as other species. Rivers and pans not only attracted large animals and thus hunter populations, but also provided fertile areas for food and fibre resources, and facilitated the spread of species along river valleys. Weathering and soil development in footslope locations and where rainfall is sufficient, favoured the development of settled agriculture. More marginal and drier areas favoured hunter-gathering activities, in which populations utilised different faunal and floral resources. Likewise, the availability of specific geological resources such as materials suitable for working, influenced both the locations and impacts of small-scale knapping activities and, later, the large-scale working of areas by mining. In turn, these economic activities strongly influenced hominid/human settlements, trading routes, and the types and spread of different cultural activities, including pottery and metalworking, traditions and beliefs. The coincidence of locations where a range of environmental resources is provided, and known archaeological sites, shows the importance of these resources in influencing the trajectory and manifestation of human evolution (both morphological and behavioural) throughout the Quaternary.

1.6 About this book

This book covers a range of topics concerning the physical and human environments of southern Africa during the Quaternary. Chapters in this book include contributions by 39 authors from 24 institutions in 9 different countries; the Quaternary of southern Africa is therefore of global research concern. In detail, this book is organised into three sections. **Section I** deals largely with the pre-Quaternary background of southern Africa, including geology, tectonics, climate and long-term hominid evolution. **Section II** deals with the Quaternary and Holocene environmental and climatic records of different southern African physical environments, and focuses in more depth on human evolution and development in the context of landscape and environmental change. **Section III** aims to draw together the physical and human dimensions of Quaternary environmental change in southern Africa, discussing the interconnections between the physical and human environments, and identifying the critical research strands for future decades. We argue that both physical and human Quaternary environmental research in southern Africa can profit from a mutual understanding, and that their intersection in the fields of cultural anthropology, behavioural archaeology, landscape archaeology and geoarchaeology, and in geo- and landscape-heritage conservation, is a critical future research area.

Some recent books that include discussion on southern African geomorphology include those by Moon and Dardis (1988), Partridge and Maud (2000b), Holmes and Meadows (2012), and Grab and Knight (2015). This volume differs from these previous books in that it explicitly identifies and discusses the relationships between climate and environmental change, geomorphological processes, and human evolution and activities. As such, it represents a significant change in the ways in which the physical and human dimensions of the southern African landscape can be conceptualised and drawn together in a coherent and interdisciplinary way, to address the research needs of the 21st Century, and to apply modern methods to investigate the past.

1.7 Summary

Climatic and tectonic events over the last 66 Ma have impacted the macroscale geomorphology of southern Africa and, in turn, the geomorphological processes operating on the land surface. These processes are important on long time scales, not only for shaping the nature of today's land surface, but also for providing the environmental context within which human evolution has taken place during the Quaternary. Furthermore, the close interconnections in the southern African landscape between physical (environmental) and human factors can help identify and value cultural landscape resources and geoheritage.

Acknowledgements

In compiling this book we thank Susan Francis and Zoe Pruce at Cambridge University Press, Wendy Phillips (Wits University) for drawing many diagrams, the National Research Foundation (South Africa) for independently funding our research, and the many (40+) international scientists from many different countries who kindly peer-reviewed these chapters. Dominic Stratford and Peter Holmes are thanked for reviewing this introduction chapter.

References

Avery, D. M. (2001). The Plio-Pleistocene vegetation and climate of Sterkfontein and Swartkrans, South Africa, based on micromammals. *Journal of Human Evolution*, 41, 113–132.

Beaumont, P. B. and Vogel, J. C. (2006). On a timescale for the past million years of human history in central South Africa. *South African Journal of Science*, 102, 217–229.

Bierman, P. R. and Caffee, M. (2001). Slow rates of rock surface erosion and sediment production across the Namib Desert and escarpment, southern Africa. *American Journal of Science*, 301, 326–358.

Binns, J. A. (1997). People, environment and development in Africa. *South African Geographical Journal*, 79, 13–18.

Bishop, P. (2007). Long-term landscape evolution: Linking tectonics and surface processes. *Earth Surface Processes and Landforms*, 32, 329–365.

Blumel, W. D. and Eitel, B. (1994). Tertiary calcic sediment covers and calcretes in Namibia – origin and geomorphic significance. *Zeitschrift für Geomorphologie, N.F.*, 38, 385–403.

Braun, J., Guillocheau, F., Robin, C., Baby, G. and Jelsma, H. (2014). Rapid erosion of the Southern African Plateau as it climbs over a mantle superswell. *Journal of Geophysical Research*, 119, 6093–6112, doi:10.1002/2014JB010998.

Chadwick, O. A., Roering, J. J., Heinsath, A. M., Levick, S. R., Asner, G. P. and Khomo, L. (2013). Shaping post-orogenic landscapes by climate and chemical weathering. *Geology*, 41, 1171–1174.

Cole, M. J., Bailey, R. M. and New, M. G. (2014). Tracking sustainable development with a national barometer for South Africa using a downscaled "safe and just space" framework. *Proceedings of the National Academy of Sciences*, 111, E4399–E4408.

Dardis, G. F. (1990). Late Holocene erosion and colluvium deposition in Swaziland. *Geology*, 18, 934–937.

Decker, J. E., Niedermann, S. and de Wit, M. J. (2011). Climatically influenced denudation rates of the southern African plateau: Clues to solving a geomorphic paradox. *Geomorphology*, 190, 48–60.

Decker, J. E., Niedermann, S. and de Wit, M. J. (2011). Soil erosion rates in South Africa compared with cosmogenic ^{3}He-based rates of soil production. *South African Journal of Geology*, 114, 475–488.

de Wit, M. (2007). The Kalahari Epeirogeny and climate change: Differentiating cause and effect from core to space. *South African Journal of Geology*, 110, 367–392.

Diekmann, B., Fälker, M. and Kuhn, G. (2003). Environmental history of the south-eastern South Atlantic since the Middle Miocene: Evidence from the sedimentological records of ODP Sites 1088 and 1092. *Sedimentology*, 50, 511–529.

Dirks, P. H. G. M. and Berger, L. R. (2013). Hominin-bearing caves and landscape dynamics in the Cradle of Humankind, South Africa. *Journal of African Earth Sciences*, 78, 109–131.

du Toit, A. L. (1954). *Geology of South Africa*, 3rd edition. Edinburgh: Oliver and Boyd, 611pp.

Dupont, L. M. (2006). Late Pliocene vegetation and climate in Namibia (southern Africa) derived from palynology of ODP Site 1082. *Geochemistry, Geophysics, Geosystems*, 7, Q05007, doi:10.1029/2005GC001208.

Dupont, L. M., Rommerskirchen, F., Mollenhauer, G. and Schefuß, E. (2013). Miocene to Pliocene changes in South African hydrology and vegetation in relation to the expansion of C_4 plants. *Earth and Planetary Science Letters*, 375, 408–417.

Erlanger, E. D., Granger, D. E. and Gibbon, R. J. (2012). Rock uplift rates in South Africa from isochron burial dating of fluvial and marine terraces. *Geology*, 40, 1019–1022.

Fleming, A., Summerfield, M. A., Stone, J. O., Fifield, L. K. and Cresswell, R. G. (1999). Denudation rates for the southern Drakensberg escarpment, SE Africa, derived from in-situ-produced cosmogenic ^{36}Cl: Initial results. *Journal of the Geological Society*, 156, 209–212.

Franz-Odendaal, T. A., Lee-Thorp, J. A. and Chinsamy, A. (2002). New evidence for the lack of C_4 grassland expansions during the early Pliocene at Langebaanweg, South Africa. *Paleobiology*, 28, 378–388.

Garcin, Y., Vincens, A., Williamson, D., Buchet, G. and Guiot, J. (2007). Abrupt resumption of the African Monsoon at the Younger Dryas-Holocene climatic transition. *Quaternary Science Reviews*, 26, 690–704.

Gibbard, P. and Cohen, K. M. (2008). Global chronostratigraphical correlation table for the last 2.7 million years. *Episodes*, 31, 243–247.

Grab, S. W. and Knight, J. (eds) (2015). *Landscapes and Landforms of South Africa*. Switzerland: Springer, 187pp.

Haq, B. U., Hardenbol, J. and Vail, P. R. (1987). Chronology of fluctuating sea levels since the Triassic. *Science*, 235, 1156–1167.

Holmes, P. J. and Meadows, M. E. (eds) (2012). *Southern African Geomorphology: Recent Trends and New Directions*. Bloemfontein: Sun Press, 431pp.

Holmgren, K., Lee-Thorp, J. A., Cooper, G. R. J., Lundblad, K., Partridge, T. C., Scott, L., Sithaldeen, R., Talma, A. S. and Tyson, P. D. (2003). Persistent millennial-scale climatic variability over the past 25,000 years in Southern Africa. *Quaternary Science Reviews*, 22, 2311–2326.

Hunter, D. R., Johnson, M. R., Anhaeusser, C. R. and Thomas, R. J. (2006). Introduction. In *The Geology of South Africa*, ed. M. R. Johnson, C. R. Anhaeusser and R. J. Thomas. Johannesburg/Pretoria: GSSA/Council for Geoscience, pp. 1–7.

Itambi, A. C., von Dobeneck, T., Mulitza, S., Bickert, T. and Heslop, D. (2009). Millennial-scale northwest African droughts related to Heinrich events and Dansgaard-Oeschger cycles: Evidence in marine sediments from offshore Senegal. *Paleoceanography*, 24, PA1205, doi:10.1029/2007PA001570.

Jung, G., Prange, M. and Schulz, M. (2014). Uplift of Africa as a potential cause for Neogene intensification of the Benguela upwelling system. *Nature Geoscience*, 7, 741–747.

King, L. C. (1963). *South African Scenery: A Textbook of Geomorphology*. Edinburgh: Oliver & Boyd, 308pp.

Lavier, L. L., Steckler, M. S. and Brigaud, F. (2001). Climatic and tectonic control on the Cenozoic evolution of the West African margin. *Marine Geology*, 178, 63–80.

McCarthy, T. and Rubidge, B. (2005). *The Story of Earth & Life: A Southern African Perspective on a 4.6-billion-year Journey*. Cape Town: Struik Publishers, 333pp.

McGee, D., Donohoe, A., Marshall, J. and Ferreira, D. (2014). Changes in ITCZ location and cross-equatorial heat transport at the Last Glacial Maximum, Heinrich Stadial 1, and the mid-Holocene. *Earth and Planetary Science Letters*, 390, 69–79.

Meadows, M. E. (2014). Recent methodological advances in Quaternary palaeoecological proxies. *Progress in Physical Geography*, 38, 807–817.

Moon, B. P. (1990). Rock mass strength as a control on slope development: Evidence from southern Africa. *South African Geographical Journal*, 72, 50–53.

Moon, B. P. and Dardis, G. F. (eds) (1998). *The Geomorphology of Southern Africa*. Johannesburg: Southern Book Publishers, 320pp.

Moore, A., Blenkinsop, T. and Cotterill, F. (2009). Southern African topography and erosion history: Plumes or plate tectonics? *Terra Nova*, 21, 310–315.

Niang, I., Ruppel, O. C., Abdrabo, M. A., Essel, A., Lennard, C., Padgham, J. and Urquhart, P. (2014). Africa. In *Climate Change 2014: Impacts, Adaptation, and Vulnerability. Part B: Regional Aspects. Contribution of Working Group II to the Fifth Assessment Report of the Intergovernmental Panel on Climate Change*, ed. V. R. Barros, C. B. Field, D. J. Dokken, M. D. Mastrandrea, K. J. Mach, T. E. Bilir, M. Chatterjee, K. L. Ebi, Y. O. Estrada, R. C. Genova, B. Girma, E. S. Kissel, A. N. Levy, S. MacCracken, P. R. Mastrandrea and L. L. White. Cambridge: Cambridge University Press, pp. 1199–1265.

Partridge, T. C. (1998). Of diamonds, dinosaurs and diastrophism: 150 million years of landscape evolution in southern Africa. *South African Journal of Geology*, 101, 167–184.

Partridge, T. C., Bond, G. C., Hartnady, C. J. H., deMenocal, P. B. and Ruddiman, W. F. (1995). Climatic effects of Late Neogene tectonism and volcanism. In *Paleoclimate and Evolution, with Emphasis on Human Origins*, ed. E. S. Vrba, G. H. Denton, T. C. Partridge and L. H. Burckle. New Haven, CT: Yale University Press, pp. 8–23.

Partridge, T. C., Dollar, E. S. J., Moolman, J. and Dollar, L. H. (2010). The geomorphic provinces of South Africa, Lesotho and Swaziland: A physiographic subdivision for earth and environmental scientists. *Transactions of the Royal Society of South Africa*, 65, 1–47.

Partridge, T. C. and Maud, R. R. (1987). Geomorphic evolution of southern Africa since the Mesozoic. *South African Journal of Geology*, 90, 179–208.

Partridge, T. C. and Maud, R. R. (2000a). Macro-scale geomorphic evolution of southern Africa. In *The Cenozoic of Southern Africa*, ed T. C. Partridge and R. R. Maud. Oxford: Oxford University Press, pp. 3–18.

Partridge, T. C. and Maud, R. R. (2000b). *The Cenozoic of Southern Africa*. Oxford: Oxford University Press, 406pp.

Phillips, D., Kiniets, G. B., Biddulph, M. G. and Madav, M. K. (2000). Cenozoic volcanism. In *The Cenozoic of Southern Africa*, ed. T. C. Partridge and R. R. Maud. Oxford: Oxford University Press, pp. 182–197.

Pickford, M., Senut, B., Mocke, H., Mourer-Chauviré, C., Rage, J.-C. and Mein, P. (2014). Eocene aridity in southwestern Africa: Timing of onset and biological consequences, *Transactions of the Royal Society of South Africa*, 69, 139–144.

Portenga, E. W. and Bierman, P. R. (2011). Understanding Earth's eroding surface with ^{10}Be. *GSA Today*, 21 (8), 4–10.

Reed. K. E. (1997). Early hominid evolution and ecological change through the African Plio-Pleistocene. *Journal of Human Evolution*, 32, 289–322.

Roelandt, C., Godderis, Y., Bonnet, M. P. and Sondag, F. (2011). Coupled modeling of biospheric and chemical weathering processes at the continental scale. *Global Biogeochemical Cycles*, 24, GB2004, doi:10.1029/2008GB003420.

Roters, B. and Henrich, R. (2010). The middle to late Miocene climatic development of Southwest Africa derived from the sedimentological record of ODP Site 1085A. *International Journal of Earth Science (Geol Rundsch)*, 99, 459–471.

Scott, L., Neumann, F. H., Brook, G. A., Bousman, C. B., Norström, E. and Metwally, A. A. (2012). Terrestrial fossil-pollen evidence of climate change during the last 26 thousand years in Southern Africa. *Quaternary Science Reviews*, 32, 100–118.

Ségalen, L., Renard, M., Lee-Thorp, J. A., Emmanuel, L., Le Callonnec, L., de Rafélis, M., Senut, B., Pickford, M. and Melice, J.-L. (2006). Neogene climate change and emergence of C_4 grasses in the Namib, southwestern Africa, as reflected in ratite ^{13}C and ^{18}O. *Earth and Planetary Science Letters*, 244, 725–734.

Senut, B., Pickford, M. and Ségalen, L. (2009). Neogene desertification of Africa. *Comptes Rendus Geoscience*, 341, 591–602.

Stroeven, A. P., Fabel, D., Hättestrand, C. and Harbor, J. (2002). A relict landscape in the centre of Fennoscandian glaciation: Cosmogenic radionuclide evidence of tors preserved through multiple glacial cycles. *Geomorphology*, 44, 145–154.

Stuut, J.-B. W., Prins, M. A., Schneider, R. R., Weltje, G. J., Jansen, J. H. F. and Postma, G. (2002). A 300 kyr record of aridity and wind strength in Southwestern Africa: Inferences from grain-size distributions of sediments on Walvis Ridge, SE Atlantic Ocean. *Marine Geology*, 180, 221–233.

Tinker, J., de Wit, M. and Brown, R. (2008). Mesozoic exhumation of the southern Cape, South Africa, quantified using apatite fission track thermochronology. *Tectonophysics*, 455, 77–93.

Vrba, E. S., Denton, G. H., Partridge, T. C. and Burckle, L. H. (eds) (1995). *Paleoclimate and Evolution, with Emphasis on Human Origins*. New Haven, CT: Yale University Press, 547pp.

Watson, A., Price-Williams, D. and Goudie, A. S. (1984). The palaeoenviromental interpretation of colluvial sediments and palaeosols of the Late Pleistocene Hypothermal in southern Africa. *Palaeogeography, Palaeoclimatology, Palaeoecology*, 45, 225–249.

West, S., Jansen, J. H. F. and Stuub, J.-B. (2004). Surface water conditions in the Northern Benguela Region (SE Atlantic) during the last 450 ky reconstructed from assemblages of planktonic foraminifera. *Marine Micropaleontology*, 51, 321–344.

Ziervogel, G., New, M., Archer van Garderen, E., Midgley, G., Taylor, A., Hamann, R., Stuart-Hill, S., Myers, J. and Warburton, M. (2014). Climate change impacts and adaptation in South Africa. *WIREs Climate Change*, 5, 605–620.

2

A brief geological history of southern Africa

STEVE MCCOURT

Abstract

The geological history of southern Africa extends back to 3650 Ma with the great majority of the resultant rock record formed in the Precambrian period (>545 Ma). This history is described chronologically as a series of events starting with the formation of the Kaapvaal Craton in the Mesoarchaean and ending with the formation of a series of Mesozoic basins between 135 and 115 Ma during the break-up of Gondwana. The tropical climate of the Cretaceous period facilitated deep weathering and rapid erosion, with the bulk of eroded sediments deposited on the continental shelf. The Cenozoic was characterised by tectonic uplift and a cooling climate which controlled the processes responsible for shaping southern Africa's current topography. The geologic history of southern Africa is reflected in present-day patterns of landforms, soils and ecosystems, which have affected resource availability and thus patterns of human activity.

2.1 Introduction

The geological history of the continental crust forming southern Africa dates back to the Eoarchaean (>3.6 Ga) with the oldest dated rock unit (part of the Ancient Gneiss Complex) having a U-Pb age of 3644±4 Ma (Compston and Kröner, 1988). This history is preserved as a series of events in both the Precambrian (Archaean and Proterozoic Eons of geological time) and the Phanerozoic (<545 Ma) that can be interpreted within the context of current plate tectonic theory. The rocks produced during this long and complex history are preserved as a mosaic of crustal fragments with each fragment providing evidence for both the various

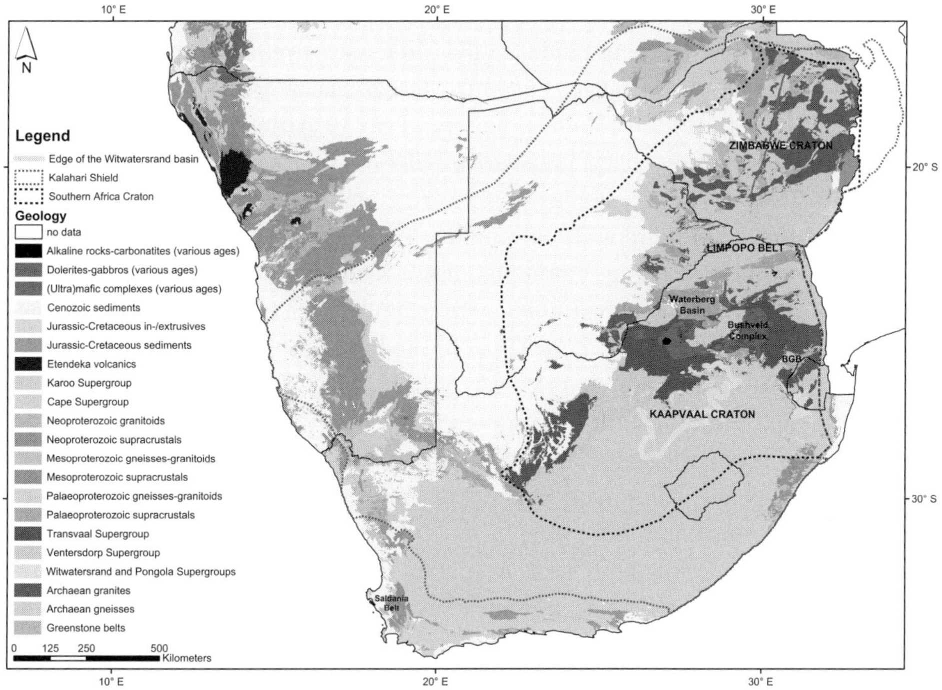

Fig. 2.1. Simplified geology of southern Africa, showing components named in the text (map compiled by Dr Hielke Jelsma). A black and white version of this figure will appear in some formats. For the colour version, please refer to the plate section.

stages involved in the geological evolution of that fragment, and the geological events that characterise each of the stages recognised. In the sections that follow, this rich history will be summarised chronologically with reference to the geological map included as Figure 2.1. For a detailed description of the lithostratigraphy of the region, the reader is referred to the summary by McCarthy and Rubidge (2005), and individual chapters within Johnson et al. (2006a).

2.2 The Archaean Eon (3.85–2.65 Ga)

The Archaean Eon is the period of geological time between ~3850 and 2650 Ma and within South Africa as elsewhere, the early part of the Archaean (the Eoarchaean) is characterised by the formation of the first fragments of continental crust and the growth of these fragments into continents. During the succeeding Mesoarchean, these early continents acted as basement for the accumulation of sedimentary and volcanic rocks (e.g. the rocks of the Witwatersrand Supergroup) in some of the oldest sedimentary basins yet to be recognised. The processes by which these fragments of ancient continental crust formed are still a matter of debate but there is increasing evidence that processes similar to present-day plate tectonics have been in operation since the Mesoarchaean (3.2–2.8 Ga).

The earliest part of the geological history of South Africa is represented by the granitoid rocks and associated greenstone belts of the Kaapvaal Craton covering an area of 1.2 x 10^6 km^2 within South Africa and adjacent parts of Botswana. Age data from single zircon U-Pb studies (e.g. Poujol et al., 2003; Eglington and Armstrong, 2004) on the granitoids of the Kaapvaal Craton indicate multiple magmatic events over more than 1 Ga duration, with Na-rich tonalite–trondhjemite–granodiorite rocks dominant between 3.65 and 3.1 Ga and potassic granitoids thereafter. The greenstone belts are typically linear in plan with widths of 10–50 km and strike lengths of 100–300 km (Brandl et al., 2006), and are dominated by lavas of basaltic composition. Sedimentary rocks are predominantly clastic and typically dominate the upper part of the supracrustal sequence. Granitoid-greenstone terranes occur across the Kaapvaal Craton with the crust decreasing in age north and west from the Barberton terrane (Fig. 2.1). The Barberton granitoid-greenstone terrane comprises the supracrustal rocks of the Barberton Supergroup which defines the Barberton Greenstone Belt, together with granitoids that range in age from the Eoarchaean (>3.6 Ga) through to the early Mesoarchaean.

The Barberton Greenstone Belt comprises a 120 x 50 km ENE-trending remnant of strongly deformed Palaeoarchaean (3.6–3.2 Ga) supracrustal rocks (the Barberton Supergroup; SACS 1980) that define a lithotectonic complex subdivided into Northern and Southern Terranes separated by the strike-parallel Inyoka–Saddleback fault system (Brandl et al., 2006, and references therein). The oldest rocks of the Barberton Supergroup comprise a tectonic stack of ultramafic and mafic volcanic rocks some 12–15 km thick and termed the Onverwacht Group. The lower part of the Onverwacht Group comprises Mg-rich ultramafic volcanic rocks (komatiites) that may represent Archaean oceanic crust. The stratigraphy of the Onverwacht Group is a matter of some debate although there is general agreement that it represents a succession disrupted and duplicated by folding and faulting rather than an original sequence.

The sedimentary rocks of the Barberton Greenstone Belt are subdivided into the older Fig Tree Group which unconformably overlies the Onverwacht Group and comprises deep- to shallow-water shale, greywacke, jaspilitic banded iron formation and carbonaceous chert (Hofmann, 2005); and the younger Moodies Group of conglomerate, coarse, cross-bedded, quartz-rich sandstones and shales that were deposited in tidally influenced basins (Eriksson, 1980). The rocks of the Moodies Group are unconformable with respect to the Fig Tree Group and represent a molasse-type sequence deposited in synorogenic basins (Heubeck et al., 2013).

The oldest granitoid rocks of the Barberton terrane (e.g. the Ancient Gneiss Complex, the 3.51 Ga Steynsdorp Pluton, and the 3.44 Ga Theespruit Pluton) were emplaced as diapirs which contributed to the deformation seen in the supracrustal rocks. The younger K-granitoids are sheet-like intrusions and include the 3.10 Ga Nelspruit and Mpuluzi batholiths (Robb et al., 2006, and references therein). An important result of this magmatism was the progressive stabilisation of the developing continental crust, such that by 3.1 Ga a continental fragment of definable size had formed. de Wit et al. (1992) refer to this fragment as the Kaapvaal Shield and identify it as the micro-continent on which the Mesoarchaean cover sequences were deposited, and which are now preserved as the Dominion Group and Witwatersrand Supergroup in the central part of the Kaapvaal Craton, and the Pongola Supergroup in the southeast of the craton.

The rocks of the Dominion Group comprise quartzite with minor conglomerate overlain by volcanic rocks that include both basalt and rhyolite, a sequence that suggests the Dominion Group accumulated in a rift basin. A minimum age for this rifting event is 3.07 Ga, the age of the rhyolite lava in the upper part of the Dominion Group (Armstrong et al., 1991). The Nsuze Group forms the lower part of the Pongola Supergroup and, like the Dominion Group, is characterized by a ~3.0 Ga rift-related volcano-sedimentary sequence, thereby allowing for correlation between units of the Nsuze Group and the Dominion Group (Gold, 2006). In the central part of the craton, the Dominion Group is overlain by the sedimentary rocks of the Witwatersrand Supergroup, whereas to the southeast the Nsuze Group is overlain by the largely sedimentary Mozaan Group which defines the upper part of the Pongola Supergroup. The rocks of the Witwatersrand Supergroup are subdivided into the lower West Rand Group and the upper Central Rand Group. Deposition of the West Rand Group started around 2.97 Ga (Robb and Meyer, 1995) and the Crown Lava towards the top of the Group was erupted at 2.91 Ga. The rocks of the West Rand Group are dominated by quartzite and shale in roughly equal proportions, and are subdivided into three subgroups (Hospital Hill, Government and Jeppestown) and sixteen formations (McCarthy, 2006). The Central Rand Group consists mainly of quartzite and conglomerate including the renowned auriferous units (gold reefs), and was deposited between 2.91 Ga and the eruption of the Klipriviersberg Group (see below) at 2.714 Ga. The rocks of the Central Rand Group are subdivided into two subgroups (Johannesburg and Turffontein) and nine formations (McCarthy, 2006). Correlation between the cover rocks of the central and southeastern parts of the Kaapvaal Craton has been extended into the Witwatersrand Supergroup, with the Mozaan Group (Pongola Supergroup) regarded as a correlative of the West Rand Group (Beukes and Cairncross, 1991).

Sedimentation to produce the Witwatersrand Supergroup was replaced by a volcanic event that produced the thick pile of lava preserved as the Ventersdorp

Supergroup. This volcanic event started with the outpouring of the basaltic Klipriviersberg Group followed unconformably by the, initially sedimentary, Platberg Group in which felsic volcanics of the Makwassie Formation have been dated to 2.708 Ga (Armstrong et al., 1991). The Platberg Group is overlain with a pronounced unconformity by the Bothaville Formation, and the sequence completed by mafic to intermediate lava of the ~2.68 Ga Allanridge Formation.

The Limpopo Orogeny was the event that produced the Limpopo Belt, a zone of granulite facies gneisses between the Kaapvaal and Zimbabwe cratons. The geology of the Limpopo Belt is very complex, being the product of high-grade metamorphism, extreme ductile deformation and widespread crustal melting (van Reenen et al., 2011). The details of the processes involved in its formation are still a matter of debate, but there is general agreement that the high-grade gneisses represent the root zone of an ancient mountain belt formed by plate tectonic processes similar to those responsible for the Himalaya (see Kramers et al., 2011, for discussion). The collision of the Kaapvaal and Zimbabwe cratons to produce the Limpopo Belt occurred between ~2.7 and 2.65 Ga and the resultant crustal fragment, the core of the continental crust of southern Africa, has been referred to as the Azanian Craton (McCourt et al., 2004).

2.3 The Neoarchaean and Palaeoproterozoic Eras

Following the Limpopo Orogeny, or perhaps in response to it, a new rift developed in the Kaapvaal section of the Azanian Craton in which the rocks of the 2.66 Ga Wolkberg Group and associated sequences accumulated. This rifting was followed by thermal subsidence of the majority of the Kaapvaal Craton to form a large shallow continental shelf basin on which the sedimentary rocks of the Transvaal Supergroup were deposited. The remnants of this sequence are preserved in three structural basins with the same basic succession recognised in each, *viz*: a basal clastic unit overlain by a thick sequence of chemical sedimentary rocks which are in turn separated by an unconformity from an upper clastic unit. The hiatus represented by this unconformity is poorly constrained but in the case of the Transvaal Basin it may be as much as 150 Ma (Moore et al., 2012). Age constraints indicate that the basal unit was deposited around 2.64 Ga, the carbonate rocks around 2.58 Ga and the overlying sequence between 2.5 and 2.43 Ga. Within the Transvaal Basin, the minimum age of the upper clastic sequence (the Pretoria Group) is provided by the emplacement age of the world's largest mafic layered intrusion, the Bushveld Complex.

The Bushveld Complex comprises three different groups of igneous rocks. The oldest component of the Complex is the 2.061 Ga Rooiberg Group, a succession of predominantly rhyolitic lavas that erupted along an unconformity surface at the top of the Pretoria Group. The second component is the 2.054 Ga Rustenburg Layered Suite (RLS), a sill-like intrusion of mafic and ultramafic magma into the rocks of the Pretoria Group and host to enormous reserves of platinum-group elements, chromite and vanadium-bearing magnetite. The youngest component of the Bushveld Complex is the Lebowa Granite Suite produced by partial melting of the lower crust during the emplacement of the RLS.

At around 2.15 Ga, rifting occurred along the northwest margin of the Azanian Craton (now northwest Zimbabwe) to accommodate the development of the Magondi Basin and the deposition of the rocks of the Magondi Supergroup. By about 1.95 Ga, this rift system had produced a continental shelf along the western margin of the former Kaapvaal Craton on which shallow-water marine sediments began to accumulate. Today this shelf sequence is represented by the quartzite, shale and local carbonate of the Olifantshoek Supergroup, the age of which is constrained by the 1.93 Ga lava of the Hartley Formation towards the base of the sequence (Cornell et al., 1998). Palaeoproterozoic extension also affected the interior of the Kaapvaal Craton and produced basins in which the rocks of the Waterberg Group and Soutpansberg Group were deposited. The clastic rocks in both these sequences are characterised by a strong red colour, indicating deposition in an atmosphere that contained free oxygen and, as such, record an important change in the Earth's environment. These 'Red Bed' successions were deposited between 2.05 and 1.87 Ga.

2.4 Rodinia and Gondwana (1.6–0.5 Ga)

The geological history of southern Africa during the Mesoproterozoic Era (1.6–1.0 Ga) records the formation of island arcs to the south and west of the Kaapvaal Craton followed by accretion of these arc complexes to the craton margin to form the Namaqua–Natal Metamorphic Province. The rocks defining the Namaqua–Natal Metamorphic Province can be traced from southern Namibia, southeast through the Northern Cape Province, to the coastline of KwaZulu–Natal (Fig. 2.1). Within each sector of the orogen a number of terranes have been recognised (McCourt et al., 2006, and references therein). The orogenic event responsible for the Namaqua–Natal Metamorphic Province is linked to the assembly of the supercontinent Rodinia and, within southern Africa, led to the growth of the Azanian Craton into the Kalahari Craton.

The geology formed during the Neoproterozoic Era (1000–524 Ma) is represented by lithostratigraphic units that characterise the Gariep Belt to the north of 32°S (Richtersveld Suite, Gariep Supergroup, Nama Group and Vanrhynsdorp Group) and the Saldania Belt to the south (Gresse et al., 2006). The sedimentary rocks of the Gariep Supergroup were deposited in a shallow-water marine environment and then deformed and metamorphosed to produce the Gariep Belt. The rocks of the Nama and Vanrhynsdorp Groups represent foreland basin deposits inboard of the belt. Within the Saldania Belt, the main occurrence of supracrustal rocks lies to the north and northeast of Cape Town (Malmesbury Group) with smaller occurrences farther east in the Oudtshoorn (Cango Caves and Kansa Groups), George (Kaaimans Group) and Port Elizabeth (Gamtoos Group) areas (Gresse et al., 2006). The metamorphosed rocks of the Malmesbury and Kaaimans Groups are intruded by the 550–507 Ma granitoids of the Cape Granite Suite and are well exposed at the Sea Point locality in Cape Town. Globally, the Gariep and Saldania belts are related to the assembly of the southwest part of Gondwana and, together with the Kaoko and Damara belts in Namibia and the Mozambique Belt, define the southern African sector of the Pan-African orogenic system.

2.5 The Phanerozoic Eon

Subsequent to the Saldanian Orogeny and the intrusion of the Cape Granite Suite, southern Africa lay close to the southern margin of Gondwana, and it is within this context that the rocks which record the final chapter in the geological history of continental southern Africa formed. Today these rocks of Gondwana (McCarthy and Rubidge, 2005) are preserved as the Cape Supergroup, the Natal Group, and the Karoo Supergroup. Collectively these Palaeozoic to Mesozoic rocks cover more than 60% of the surface of South Africa, are integral to the landscape of the region, and preserve a unique record of the early evolution of fishes, amphibians, reptiles and mammals (McCarthy and Rubidge, 2005). The Cape Supergroup was deposited between ~500 and 330 Ma in a basin that developed along the southern margin of South Africa from present-day Clanwilliam to Port Alfred. The rocks of the Cape Supergroup are Palaeozoic in age and subdivided into three lithologically distinctive units called the Table Mountain Group (Ordovician to early Devonian sandstones), the Bokkeveld Group (early to middle Devonian shale and sandstone containing a rich and diverse marine invertebrate fossil record), and the Witteberg Group (late Devonian to early Carboniferous sandstone and mudrock with a rich invertebrate, fish and plant fossil record). The Natal Group (Marshall, 2006) is preserved in a northeast-trending, coast-parallel basin that has been interpreted as part of the system that accommodated the deposition of the Cape Supergroup (McCarthy and Rubidge, 2005), and as a foreland basin to the Pan African Mozambique Belt (Marshall, 2006). Deposition of the reddish-grey sandstone-dominated Natal Group started in the early Ordovician (~490 Ma; Thomas et al., 1992), but although similar in age, there is no evidence to support a correlation, on lithological grounds, with the Table Mountain Group.

Rocks of the Karoo Supergroup range in age from late Carboniferous to middle Jurassic (310–180 Ma) and represent the last major supracrustal sequence to be deposited on the continental crust of southern Africa. The bulk of the dominantly sedimentary succession occupies the erosional remnant of the main basin which covers some 700,000 km^2 of south-central South Africa (Johnson et al., 2006b) with smaller but significant deposits present in the Springbok Flats, Ellisras, Tshipise and Tuli Basins farther north. The succession in the Main Karoo Basin is strongly asymmetric in cross section, being significantly thicker in the south against the mountains of the Cape Fold Belt and thinning northwards across the Kaapvaal Craton. Based on age and lithological character, Johnston et al. (2006b) subdivide the sedimentary rocks in the Main Basin into the basal Dwyka Group (glacial deposits), the Ecca Group (deltaic deposits including coal), Beaufort Group (floodplain deposits), Molteno Formation, Elliot Formation and Clarens Formation (desert conditions), reflecting a significant change in climate due to the northward drift of Gondwana from polar to warmer latitudes. The youngest unit of the Karoo Supergroup is the Drakensberg Group comprising the basaltic lava and associated dolerite intrusions that identify the start of Gondwana breakup at some 180 Ma (Duncan and Marsh, 2006; Watkeys, 2006), an event that was to have an important effect on the present-day landscape of South Africa.

The Mesozoic (Jurassic and Cretaceous) clastic sedimentary rocks that are preserved in isolated fault-bound onshore basins from northern KwaZulu–Natal to Worcester in the Western Cape accumulated in basins related to extension of the newly-formed continental margin during the break-up of Gondwana (Shone, 2006). The Algoa Basin to the north of Port Elizabeth is the largest of the basins along the southern margin of South Africa, and accommodated the deposition of the late Jurassic to early Cretaceous Uitenhage Group, comprising the Enon, Kirkwood and Sundays River Formations. The Cretaceous deposits of the Zululand Group in northern KwaZulu–Natal are subdivided into the Makatini, Mzinene and St Lucia Formations. Exposure is confined to a narrow belt from Mthubathuba north to the border with Mozambique, but drilling on the coastal plain has confirmed up to 2000 m of Cretaceous rocks under the Cenozoic cover (Shone, 2006).

Mesozoic offshore basins are located along the western, southern and eastern margins of continental South Africa and are presently important exploration targets for oil and gas. The basins developed in response to the separation of East and West Gondwana (e.g. Watkeys, 2006). Rifting along the western margin led to the development of the Orange Basin, the fill of which comprises two synrift phases and five drift intervals (Broad et al., 2006). The sedimentary rocks are siliclastic and predominantly deltaic with the oldest being early Aptian (112 Ma) rocks ranging from continental red beds to marine sandstones and shales (Broad et al.,

2006). The Outeniqua Basin is situated off the south coast of South Africa and comprises a series of southeast-trending sub-basins named, from west to east, the Bredasdorp, Pletmos, Gamtoos and Algoa sub-basins (Broad et al., 2006). The Bredasdorp sub-basin is approximately 200 km long and 80 km wide and the site of the majority of oil and gas exploration activities, thus the geology is well documented (Broad et al., 2006). The east coast of South Africa has two Mesozoic offshore basins, namely the Durban and Zululand Basins located between Port Shepstone and the Mozambique border, with the Zululand Basin being the southern component of the Mozambique Basin (Broad et al., 2006). Exploration activity is limited, thus the geology of these basins is poorly understood. In the Durban Basin, synrift rocks have been correlated with continental red beds at the base of the Zululand Group onshore (Broad et al., 2006).

The formation and stratigraphy of the Cenozoic deposits of southern Africa can be directly linked to the tectonic and geomorphic events that followed the breakup of Gondwana (Partridge et al., 2006). At this time the average elevation of southern Africa was around 2000 m (Partridge and Maud, 1987). Rifting of this elevated region created a narrow coastal plain flanked by a high escarpment. The climate was warm and humid which, together with the high elevation, promoted rapid and effective erosion such that by the end of the Cretaceous (66 Ma) when the Earth's climate started to cool, the escarpment had been eroded back to a position 120 km inland from the southeastern coast and 50 km from the west coast (Partridge et al., 2006). The result of this post-rifting period of erosion and denudation was the formation of an interior plateau separated from surrounding lower-lying regions by the Great Escarpment. The coeval erosion surfaces above and below the Great Escarpment define the African Surface (Partridge and Maud, 1987, and references therein) on which pedogenic duricrusts formed above deeply kaolinised weathering profiles (Botha, 2000) and Cenozoic sediments were deposited in both interior (Kalahari Group; Partridge et al., 2006) and coastal (Maputland, Algoa and Bredasdorp Groups; Roberts et al., 2006) basins. At ~30 Ma, the African plate stopped its northward drift and began to bulge upwards (Burke, 1996, and references therein) with important periods of uplift documented in the early Miocene (~20 Ma) and particularly the Pliocene (~5 Ma). These periods of uplift would have increased erosion and facilitated the processes responsible for shaping southern Africa's current topography. This recent geologic history is discussed by Knight and Grab (this volume).

2.6 Summary

The long geological history of southern Africa is testimony to the interplay between tectonic, sedimentary and climatic processes, which have driven periods of uplift, rifting and volcanism; metamorphism and formation of ore deposits; the formation and closure of sedimentary basins; and landscape denudation. The geological relationship to the present-day landscape is shown in patterns of landforms, soils and ecosystems, and provides the context for later human occupation of the landscape and exploitation of its resources.

References

Armstrong, R. A., Compston, W., Retief, E. A., Williams, I. S. and Welke, H. J. (1991). Zircon ion microprobe studies bearing on the age and evolution of the Witwatersrand triad. *Precambrian Research*, 53, 243–266.

Beukes, N. J. and Cairncross, B. (1991). A lithostratigraphic-sedimentological reference profile for the late Archaean Mozaan Group, Pongola Sequence: Application to sequence stratigraphy and correlation with the Witwatersrand Supergroup. *South African Journal of Geology*, 94, 44–69.

Botha, G. A. (2000). Paleosols and Duricrusts. In *The Cenozoic of Southern Africa*, ed. T. C. Partridge and R. R. Maud. New York: Oxford University Press, pp. 131–144.

Brandl, G., Cloete, M. and Anhauesser, C. R. (2006). Archaean Greenstone Belts. In *The Geology of South Africa*, ed. M. R. Johnson, C. R. Anhaeusser and R. J. Thomas. Pretoria: Geological Society of South Africa/Council for Geoscience, pp. 9–56.

Broad, D. S., Jungslager, E. H. A., McLachlan, I. R. and Roux, J. (2006). Offshore Mesozoic Basins. In *The Geology of South Africa*, ed. M. R. Johnson, C. R. Anhaeusser and R. J. Thomas. Pretoria: Geological Society of South Africa/Council for Geoscience, pp. 553–571.

Burke, K. (1996). The African Plate: The 24th Alex du Toit Memorial Lecture. *South African Journal of Geology*, 99, 341–409.

Compston, W. and Kröner, A. (1988). Multiple zircon growth within early Archaean tonalitic gneiss from the Ancient Gneiss Complex, Swaziland. *Earth and Planetary Science Letters*, 87, 13–28.

Cornell, D. H., Armstrong, R. A. and Walraven, F. (1998). Geochronology of the Proterozoic Hartley Formation, South Africa: Constraints on the Kheis tectogenesis and the Kaapvaal Craton's earliest Wilson Cycle. *Journal of African Earth Sciences*, 26, 5–27.

de Wit, M. J., de Ronde, C. E. J., Tredoux, M., Roering, C., Hart, R. J., Armstrong, R. A., Green, R. W. E., Peberdy, E. and Hart, R. A. (1992). Formation of an Archaean continent. *Nature*, 357, 553–562.

Duncan, A. R. and Marsh, J. S. (2006). The Karoo Igneous Province. In *The Geology of South Africa*, ed. M. R. Johnson, C. R. Anhaeusser and R. J. Thomas. Pretoria: Geological Society of South Africa/Council for Geoscience, pp. 501–520.

Eglington, B. M. and Armstrong, R. A. (2004). The Kaapvaal Craton and adjacent orogens, southern Africa: A geochronological database and overview of the geological development of the craton. *South African Journal of Geology*, 107, 13–32.

Eriksson, K. A. (1980). Transitional sedimentation styles in the Moodies and Fig Tree Groups, Barberton Mountain Land, South Africa: Evidence favouring an Archaean continental margin. *Precambrian Research*, 12, 141–160.

Gold, D. J. C. (2006). The Pongola Supergroup. In *The Geology of South Africa*, ed. M. R. Johnson, C. R. Anhaeusser and R. J. Thomas. Pretoria: Geological Society of South Africa/Council for Geoscience, pp. 135–147.

Gresse, P. G., von Veh, M. W. and Frimmel, H. E. (2006). Namibian (Neoproterozoic) to early Cambrian successions. In *The Geology of South Africa*, ed. M. R. Johnson, C. R. Anhaeusser and R. J. Thomas. Pretoria: Geological Society of South Africa/Council for Geoscience, pp. 395–420.

Heubeck, C., Engelhardt, J., Byerly, G. R., Zeh, A., Sell, B., Luber, T. and Lowe D. R. (2013). Timing of deposition and deformation of the Moodies Group (Barberton Greenstone, Belt, South Africa): Very-high-resolution of Archaean surface processes. *Precambrian Research*, 231, 236–262.

Hofmann, A. (2005). The geochemistry of sedimentary rocks from the Fig Tree Group, Barberton greenstone belt: Implications for tectonic, hydrothermal and surface processes during mid-Archaean times. *Precambrian Research*, 143, 23–49.

Johnson, M. R., Anhaeusser, C. R. and Thomas, R. J. (eds) (2006a). *The Geology of South Africa*. Pretoria: Geological Society of South Africa/Council for Geoscience, 691pp.

Johnson, M. R., van Vuuren, C. J., Visser, J. N. J., Cole, D. I., de V. Wickens, H., Christie, A. D. M., Roberts, D. L. and Brandl, G. (2006b). Sedimentary rocks of the Karoo Supergroup. In *The Geology of South Africa*, ed. M. R. Johnson, C. R. Anhaeusser and R. J. Thomas. Pretoria: Geological Society of South Africa/Council for Geoscience, pp. 461–499.

Kramers, J. D., McCourt, S., Roering, C., Smit, C. A. and van Reenen, D. D. (2011). Tectonic models proposed for the Limpopo Complex: mutual compatibilities and constraints. In *Origin and evolution of Precambrian high-grade gneiss terranes, with special emphasis on the Limpopo Complex of southern Africa*, ed. D. D. van Reenen, J. D. Kramers, S. McCourt and L. L. Perchuk. Boulder, CO: Geological Society of America, Memoir 207, pp. 311–324.

Marshall, C. G. A. (2006). The Natal Group. In *The Geology of South Africa*, ed. M. R. Johnson, C. R. Anhaeusser and R. J. Thomas. Pretoria: Geological Society of South Africa/Council for Geoscience, pp. 433–441.

McCarthy, T. S. (2006). The Witwatersrand Supergroup. In *The Geology of South Africa*, ed. M. R. Johnson, C. R. Anhaeusser and R. J. Thomas. Pretoria: Geological Society of South Africa/Council for Geoscience, pp. 155–186.

McCarthy, T. and Rubidge, B. (2005). *The story of Earth & Life: A southern African perspective on a 4.6-billion-year journey*. Cape Town: Struik Publishers, 333pp.

McCourt, S., Kampunzu, A. B., Bagai, Z. and Armstrong, R. A. (2004). The crustal architecture of Archaean terranes in Northeastern Botswana. *South African Journal of Geology*, 107, 147–158.

McCourt, S., Armstrong, R. A., Grantham, G. H. and Thomas, R. J. (2006). Geology and evolution of the Natal Belt, South Africa. *Journal of African Earth Sciences*, 46, 71–92.

Moore, J. M., Polteau, S., Armstrong, R.A., Corfu, F. and Tsikos, H. (2012). The age and correlation of the Postmasburg Group, southern Africa: Constraints from detrital zircon grains. *Journal of African Earth Sciences*, 64, 9–19.

Partridge, T. C. and Maud, R. R. (1987). Geomorphic evolution of southern Africa since the Mesozoic. *South African Journal of Geology*, 90, 179–208.

Partridge, T. C., Botha, G. A. and Haddon, I. G. (2006). Cenozoic deposits of the interior. In *The Geology of South Africa*, ed. M. R. Johnson, C. R. Anhaeusser and R. J. Thomas. Pretoria: Geological Society of South Africa/Council for Geoscience, pp. 585–604.

Poujol, M., Robb, L. J., Anhaeusser, C. R. and Gericke, B. (2003). A review of the geochronological constraints on the evolution of the Kaapvaal Craton, South Africa. *Precambrian Research*, 127, 181–213.

Robb, L. J., Brandl, G., Anhaeusser, C. R. and Poujol, M. (2006). Archaean Granitoid Intrusions. In *The Geology of South Africa*, ed. M. R. Johnson, C. R. Anhaeusser and R. J. Thomas. Pretoria: Geological Society of South Africa/Council for Geoscience, pp. 57–94.

Robb, L. J. and Meyer, F. M. (1995). The Witwatersrand Basin, South Africa: Geological framework and mineralization processes. *Ore Geology Reviews*, 10, 67–94.

Roberts, D. L., Botha, G. A., Maud. R. R. and Pether, J. (2006). Coastal Cenozoic Deposits. In *The Geology of South Africa*, ed. M. R. Johnson, C. R. Anhaeusser and R. J. Thomas. Pretoria: Geological Society of South Africa/Council for Geoscience, pp. 605–628.

Shone, R. W. (2006). Onshore post-Karoo Mesozoic deposits. In *The Geology of South Africa*, ed. M. R. Johnson, C. R. Anhaeusser and R. J. Thomas. Pretoria: Geological Society of South Africa/Council for Geoscience, pp. 541–552.

Thomas, R. J., Marshall, C. G. A., Watkeys, M. K., Fitch, F. J. and Miller, J. A. (1992) K-Ar and $^{40}Ar/^{39}Ar$ dating of the Natal group, Southeast Africa: A post Pan-African molasse? *Journal of African Earth Sciences*, 15, 453–471.

van Reenen, D. D., Kramers, J. D., McCourt, S. and Perchuk, L. L. (eds) (2011). *Origin and Evolution of Precambrian High-Grade Gneiss Terranes, with Special Emphasis on the Limpopo Complex of Southern Africa*. Boulder, CO: Geological Society of America, Memoir 207, 324pp.

Watkeys, M. K. W. (2006). Gondwana break-up: A South African perspective In *The Geology of South Africa*, ed. M. R. Johnson, C. R. Anhaeusser and R. J. Thomas. Pretoria: Geological Society of South Africa/Council for Geoscience, pp. 531–539.

3

A continental-scale perspective on landscape evolution in southern Africa during the Cenozoic

JASPER KNIGHT AND STEFAN W. GRAB

Abstract

The evolution of southern Africa during the Cenozoic (last 66 Ma) results from the interplay between regional-scale tectonic and climatic processes, and complex geomorphic feedbacks and responses that give rise to variations in preservation and denudation of the land surface. As such, this history of landscape evolution reflects a classic forcing–response model that is typical of many different geomorphological systems. In detail, however, the timescales and feedbacks are poorly known, and the palimpsest nature of land surface features (supported by evidence from radiometric dating) shows that the operation of these processes across southern Africa is not spatially uniform, which has not been previously discussed. The climatic and land surface feedbacks associated with mantle swells and periods of Cenozoic rifting and earlier Mesozoic volcanism are also uncertain. These are important future research challenges.

3.1 Introduction

Tectonic and climatic changes during the Cenozoic (66 Ma to present), both on global and regional (southern Africa) scales, have profoundly affected land surface processes, the development of different landforms, and the extent to which such signatures are preserved or destroyed in the landscape. An understanding of these issues frames discussion of landscape evolution in southern Africa during the Cenozoic, and thus the reasons why today's landscapes exist. As such, Cenozoic landscape evolution can also inform on the evolution of land ecosystems (Bamford et al., this volume), large herbivores (Brink, this volume), and hominids (Heaton, this volume; Carlson and Edlund, this volume).

The southern African landscape provides a good opportunity to examine the interplay between different drivers of landscape change, including tectonics and climate (e.g. Partridge and Maud, 1987, 2000; Botha and de Wit, 1996; Séranne and Anka, 2005; Moore et al., 2009; Cotterill and de Wit, 2011; Kounov et al., 2013). Different studies have taken various viewpoints as to which driver is most dominant (e.g. Champagnac et al., 2014), but there are two observations to make based on these viewpoints. First, it is likely that the relative importance of these drivers varied over time and space, and thus that one viewpoint is unlikely to fully describe the evolving nature of the land surface over geological time, or adequately explain the way a given landscape appears today. Second, there are feedback loops between tectonic and climatic processes that (i) are important to consider in evaluating their interrelationships, and (ii) are little understood in southern Africa compared to other areas worldwide. These issues are discussed in detail in Section 3.5 with respect to developing an understanding of landscape evolution in southern Africa during the Cenozoic.

This chapter discusses the tectonic and climatic contexts of landscape evolution in southern Africa on regional to continental spatial scales and over Cenozoic timescales. As such, it is complementary to the shorter time frame discussed by Knight and Grab (this volume).

3.2 Africa's tectonic and geologic setting

Africa as a whole represents an intracratonic setting, as evidenced by the relatively stable and ancient land surface and its old Precambrian basement of the interior. The tectonic history of southern Africa has been recently summarised by Watkeys (2006). This history focuses on the breakup of Gondwana during the period ~180–90 Ma, but has wider implications for changes in continental-scale climate and land surface processes as a result of continental drift and development of ocean circulation patterns in emerging ocean basins (Reyment and Dingle, 1987). The continental interior of southern Africa, comprising the Kaapvaal craton, was not strongly affected by tectonic deformation at this time, although sheared and mountain-building zones including the Lebombo, Cape Fold and Gariep belts were formed around craton peripheries prior to this period (James et al., 2001). Tectonic subsidence may have contributed to the formation of accommodation space on the Kalahari and Congo craton surface, which was overlain by volcanic and sedimentary sequences such as of the Witwatersrand, Pongola, Ventersdorp, Transvaal, Olifantshoek and Karoo Supergroups which were formed during the Mesozoic and Cenozoic (Bumby and Guiraud, 2005; Said et al., 2015).

Over long (10^6–10^8 year) timescales, the interplay of several factors has contributed to the evolution of the African land surface (Ashwal and Burke, 1989).

Climatic changes during the Cenozoic and earlier represent one such factor (outlined in Knight and Grab, this volume), but climatic feedbacks arising from tectonic and volcanic factors are also significant (e.g. Sepulchre et al., 2006). Mantle swells and continental rifting are important tectonic and volcanic controls on the African land surface (Ashwal and Burke, 1989; Moucha and Forte, 2011), which have further implications for erosion and sedimentation, in particular along continental margins (Lavier et al., 2001).

3.2.1 Swells

Mantle swells likely form as a result of increased mantle heating and buoyancy of mantle plumes, and/or convergent mantle circulation systems, and can be identified from large scale deep seismic and gravity surveys (Gurnis et al., 2000; James et al., 2001), including identifying large low shear velocity provinces (LLSVPs) (Burke et al., 2008; Austermann et al., 2014). The net result of these processes is domal lithospheric uplift, which can likely be persistent over 10^7–10^8 years. The 'African Superswell' located beneath eastern and southern Africa (Fig. 3.1) has elevated the land surface to average values of ~1000 m above sea level (asl), compared to typical cratonic values of 400–500 m asl (Lithgow-Bertelloni and Silver, 1998). What is unclear is the detailed spatial pattern of mantle swells beneath Africa, their relative sizes, and their patterns of temporal evolution. This is because they may change in size and position due to changes in mantle processes (Moucha and Forte, 2011), and unroofing by increased land surface erosion over topographic highs can lead to isostatic rebound effects that can encourage differential uplift. The proposed stability of LLSVPs (Mulyukova et al., 2015) suggests that the spatial distribution of swells beneath Africa is an outcome of plate movement over LLSVP and associated mantle plumes (Steinberger and Torsvik, 2012). This in turn has implications for the formation of Large Igneous Provinces (LIPs), including the Karoo volcanics, that are associated with the activity of these plumes (Torsvik and Cocks, 2013). Formation of the Karoo LIP during the Toarcian (early Jurassic) is thought to have led to significant impacts on global climate, biodiversity, marine anoxia and geochemical cycling (Sell et al., 2014), and more regional impacts on weathering patterns and long-term sediment yield across southern Africa (Burke and Gunnell, 2008). Moreover, the presence of acid Karoo volcanic rocks has exerted an impact on soil, vegetation and hydrological properties throughout the subsequent Cenozoic.

Several studies have examined the impacts of differential land surface uplift along the axes of adjacent swells, which are also related to phases of rifting and volcanism. These impacts are well expressed in the migration of major drainage divides, which can be evaluated through changes in sediment provenance (Moore et al., 2009). The relationship between uplift of the land surface by mantle swells

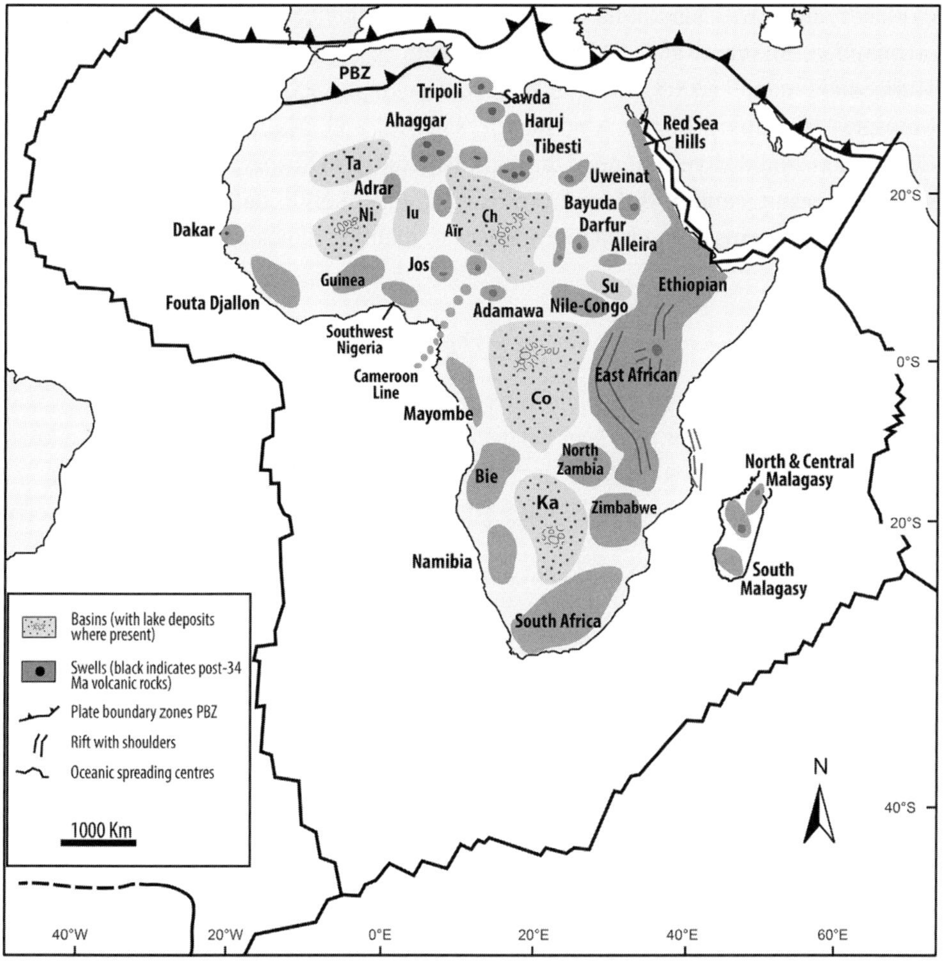

Fig. 3.1. Map of Africa's major mantle swells (dark shading), rifts, and sedimentary basins (light shading). Basin abbreviations are: Taoudeni (Ta), Niger (Ni), Iullemeden (Iu), Chad (Ch), Sudd (Su), Congo (Co), Kalahari (Ka) (adapted from Burke and Gunnell, 2008; reproduced with permission of the Geological Society of America).

and development of the erosional African Surfaces was discussed by Burke and Gunnell (2008). They argued that a low-lying and weathered early Cenozoic African land surface was uplifted by the Afar (Ethiopian) mantle plume at ~31 Ma. Subsequently, differential weathering and erosion have partly preserved and partly removed this surface, leaving bauxite and laterite deposits.

Tinker et al. (2008) present thermochronometric data from the southern Cape that suggest high denudation rates during the mid to late Cretaceous (80–100 Ma) compared to low rates during the Cenozoic. The Cretaceous denudation episode may have been triggered by epeirogenic uplift associated with emplacement of LIPs. This evidence, along with similar thermochronometric evidence elsewhere across southern Africa (e.g. Brown et al., 2002) contrasts with the viewpoint of Burke (1996) and Burke and Gunnell (2008) that much of the topographic evolution of southern Africa postdates the timing of the Afar mantle swell at ~31 Ma. Although land surface denudation would have continued throughout geological history, the available data are spatially sparse, capture different time slices, and likely hide spatial and temporal complexity in terms of landscape responses to mantle plumes and swells.

3.2.2 Rifts

Rifting arises as a result of plate movement, shearing, continental breakup and causes the development of ocean basins (Paton, 2006). Stages of rifting have been summarised from across Africa by Bumby and Guiraud (2005) with significant African rifts shown in Figure 3.2. Marginal upwarping is a characteristic property

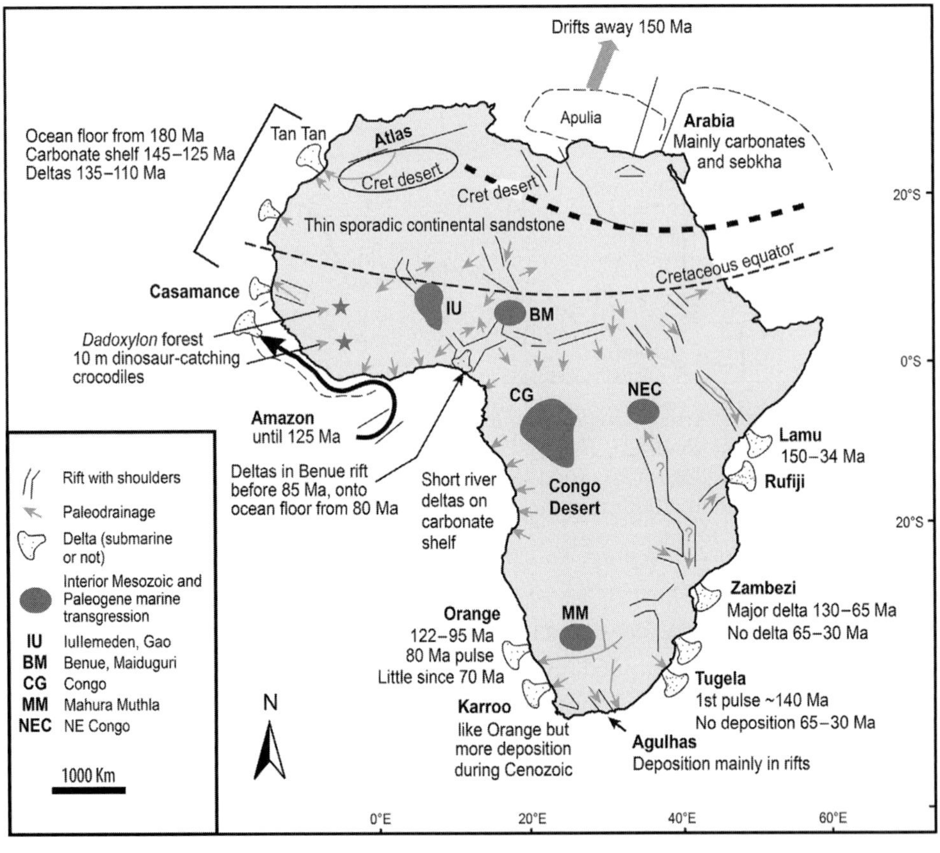

Fig. 3.2. Schematic relationship between rifts, drainage patterns, and deltaic deposition in adjacent oceans (from Burke and Gunnell, 2008; reproduced with permission of the Geological Society of America).

of rifting; in southern Africa this is estimated to have contributed to ~600 m of uplift compared to the continental interior (Gilchrist and Summerfield, 1990). This differential uplift is an isostatic response to differential denudation rates, and fits with wider discussion of forcing–response relationships to land surface uplift. For example, Matmon et al. (2002) argued that uplift of rifted continental margins takes place only during initial stages of ocean basin development. In this model, the Great Escarpment of southern Africa does not experience significant and uniform parallel slope retreat over time, but the escarpment edge experiences an increase in sinuosity (indentedness of the scarp line position in plan view) (e.g. van der Beek et al., 2002). The denudation response of creation of relief by rifting can be seen in the adjacent sediment record: thick sequences (<600 m thickness) of synrift sediments covering the last 14.5 Ma are preserved in Uganda, recording both rifting and climatic phases along the southern Albert Rift system (Roller et al., 2010). Foster and Gleadow (1992) argue a direct causal relationship between climate and denudation rates of rift-margin mountains. However, this does not take into account changes in accommodation space within rift valleys. Periods of low accommodation space and/or arid climatic conditions are shown by the development of chemical weathering products (iron, carbonate and gypsum deposits) within palaeosols in these rift valley sequences.

3.3 Dating the African land surface

Several different approaches have been used for dating the age and dynamic behaviour, such as denudation history, of the African land surface (see Woodborne, this volume). Most commonly, cosmogenic exposure or burial age dating, or thermochronometric dating, has been used. Cosmogenic radionuclide dating can inform on the timing of exposure of rock surfaces following erosion, or the timing of burial of rock surfaces by eroded sediments. These related applications of the cosmogenic method can be used to calculate point-specific and averaged rates of land surface denudation and aggradation over (generally) long timescales, commonly based upon one single radiometric age with relatively wide errors. The validity of such calculated rates has not been discussed in the literature, but regional-scale models of land surface denudation history have been based on such very limited data (e.g. Flemming et al., 1999; Kounov et al., 2007). Generally, these cosmogenic dating approaches suggest that denudation rates are both climatically and geologically dependent, with areas of high weathering rates promoting high sediment production and slope sediment transport (Kounov et al., 2007; Chadwick et al., 2013; Decker et al., 2013). The thermochronometric approach, using apatite fission track dating, is based on calculated palaeotemperature and thus depth of emplacement of host igneous rocks, and reconstructed thermal history. In turn, this can be used to calculate the depth of land surface erosion. However, this method is complicated by variations in epeirogenic uplift and mantle buoyancy (Tinker et al., 2008; Flowers and Schoene, 2010).

The aggradational products of land surface weathering and erosion, including slope, fluvial and aeolian sediments, generally accumulate more episodically and under specific environmental conditions, when compared to the weathering processes that create them. This means that there is a mismatch between the timing and sediment yield of weathering, and the timing and sedimentation rate of aggradation. Some studies have suggested that this mismatch may be related to transient sediment storage within river basins (e.g. Coulthard and Van de Wiel, 2013). It is also likely that sediment volumes within subsiding sedimentary basins or in the offshore zone are poorly estimated, which makes it difficult to backstrip these sediment volumes in order to calculate past basin accumulation rates or fluvial sediment yield.

3.4 Landscape responses to Cenozoic climate changes

Climate variability during the Cenozoic of southern Africa has been described in several studies (e.g. Partridge et al., 1995; Senut et al., 2009; Roters and Henrick, 2010). A potential problem with these studies is the extent to which site-scale data reflect climates across a wider region, and whether significant climatic gradients existed in the past as they do today. Irrespective of Cenozoic climatic conditions, it is likely that the land surface experienced spatially and temporally variable responses to this climatic forcing which cannot be adequately resolved from the existing dating controls on denudation and deposition (e.g. Champagnac et al., 2014).

A range of land surface responses takes place as a result of climate forcing. Within any one landscape, however, different geomorphological elements may exhibit different spatial and temporal scales of response, and be affected by different time lags, feedbacks and geomorphological thresholds (Phillips, 2009; von Elderfeldt, 2012; Knight and Harrison, 2013). This means that climate forcing across any one landscape can lead to spatial and temporal variations in geomorphic responses. The most useful ways to measure these geomorphic responses are through (1) variations in land surface radiometric ages and averaged denudation rates, and/or (2) variations in sediment yield on slopes and within river basins. Both these approaches have been used in order to evaluate the geomorphic impacts of Cenozoic climate changes in southern Africa.

3.4.1 Erosional responses: the African Surfaces

The different African Surfaces, formed as a result of peneplanation following tectonic uplift, are presented and discussed by Knight and Grab (this volume).

Land surfaces that have been previously correlated to the same African Surface level show a range of ages and altitudes (Partridge and Maud, 1987), thus their grouping is somewhat arbitrary. For example, uplift during the Pliocene was associated with the formation of the Post-African I Surface, but the uplift ranged from 100 m in the west to 900 m in the east of southern Africa (Partridge and Maud, 1987, 2000; Moon and Dardis, 1988). In turn, this differential uplift resulted in very different land surface responses with respect to river incision and sediment yield, giving rise to major deposition in the Kalahari basin at this time. Much emphasis has been given to river system responses to uplift, with studies focusing on drainage basin evolution and migration (e.g. Moore and Larkin, 2001; Moore and Blenkinsop, 2006) and changes in river long profiles (Partridge and Maud, 1987, 2000). These viewpoints focus mainly on erosional responses, yet river deposition/aggradational responses and slope instability responses are equally important, but have not really been considered within the context of climate- or tectonics-driven sediment supply (Wintle et al., 1995; Botha and Partridge, 2000).

3.4.2 Residual deposits

The relationship of residual (weathered) deposits to long-term landscape evolution and climate change in Africa has been described by several authors (e.g. Marker and Holmes, 1999; Pickford et al., 1999; Marker et al., 2002). The relationship between the development (in thickness and spatial extent) of residual deposits and their impacts on land surface denudation is poorly known and has not been adequately resolved in any model (see Faniran and Jeje, 1983). For example, chemical weathering under the typically hot, humid conditions found in many parts of central and southern Africa results in kaolinisation, leaching of silica and bases and enrichment of iron and aluminium. Such *in situ* weathering followed by chemical translocation may lead to negative feedbacks such as the development of iron hardpans that progressively inhibit translocation and interflow. Likewise, development of silcrete and calcrete at or just beneath the land surface can progressively decrease net denudation rates over time. For example, Gunnell (2003) suggested that bauxitisation in West Africa took place around 45–50 Ma and one of several phases of lateritisation around 24–25 Ma. These estimates were based on calculated denudation rates of ~2 m/Ma on lateritic plateaus and 7–13 m/Ma on non-lateritic surfaces. Here, the presence of a laterite cap is suggested to decrease denudation by a factor of 4–6, but it is likely that such differences in denudation rates are time-dependent. Thus, such denudation rates are not appropriate for the entire time period over which they are calculated. Several studies argue that weathering-limited denudation is the correct theoretical approach to understand landscape responses to Cenozoic climate changes (e.g. Decker et al., 2011, 2013; Chadwick et al., 2013), but none has considered the temporal evolution of weathering and denudation rates. Enrichment of surface minerals by lateritisation has been argued to have taken place preferentially on the African Surfaces that were subject to long periods of weathering (Beukes et al., 2004; Prendergast, 2013).

3.4.3 Depositional responses: fans

Footslope deposits of alluvial and colluvial fans are key elements in the landscape history of southern Africa, because they can often be dated, and may occur on different scales from microscale footslope wedges to megafans (Tooth and McCarthy, 2007). The smaller systems (<1 ha) form wetlands in valley confluence or piedmont settings with high river water and sediment supply, and are late Pleistocene or Holocene in age (e.g. Grenfell et al., 2009). The larger systems were in nearly all cases formed during the mid-Cenozoic when pluvial conditions led to the formation of large lakes in southern Africa, including Palaeolake Etosha and Palaeolake Makgadikgadi (Namibia/Botswana). These lakes offered shallow water conditions with migrating channels, islands and wetlands, and formed as endorheic basins fed by rivers which had their catchments in nearby highland regions. The Etosha and Cubango megafans formed under these conditions (Lindenmaier et al., 2014), and now have a high preservation potential underneath a younger and relatively inactive surface, where aridification during the Neogene and Quaternary resulted in abandonment and preservation of the megafan form (Miller et al., 2010) and partial drying out of fans such as the Okavango Delta.

The contemporary sedimentary system of the Okavango Delta (Botswana) is the uppermost part of a much larger relict, buried megafan (McCarthy, 2013). Seismic profiling data show that this is around 150 m thick and was formed in the very large (>90,000 km^2) Palaeolake Makgadikgadi (Podgorski et al., 2013, 2015). Sediments progressively infilled this sag basin during the Cenozoic (Haddon and McCarthy, 2005). Internally, Okavango sediments comprise mainly sand, silt and mud interbeds, typical of a shifting sandy braidplain delta with seasonal variations in water fluxes.

Larger fan systems that are much older and more subdued in surface relief are also found, often in piedmont settings where sediments accumulate due to a decrease in flow energy. In the Witteberg of the Cape Fold Belt, elements of the land surface of a relict alluvial fan were dated by cosmogenic methods to ~2 Ma (Kounov et al., 2014). Many other relict alluvial fans are observed in the Richtersveld and Cederberg (Western and Northern Cape), although these have not been studied in detail.

3.4.4 Marine deposition

The activity of sediment transport systems inland often contributes to sediment supply to marine basins or shelf break canyons. For example, climatic changes from the Cretaceous to Cenozoic resulted in phases of marine sedimentation driven by erosional cycles inland (Fig. 3.2). This hints at, first, the longevity of many river systems (Moore et al., 2012) and, second, the macroscale connections between different parts of integrated sediment systems that operate on a continental scale (Macgregor, 2010). Modelling shows that sediment fluxes are positively enhanced during periods of tectonic uplift and tilting inland (Braun et al., 2014), but changing land surface topography can result in migrating watersheds and alterations in the sediment balance within individual catchments (Macgregor, 2010). For example, the Congo basin appears to have received sediment from a number of different sources throughout its history (Giresse, 2005). Offshore and onshore patterns of basinal sedimentation may record slightly different things related to both tectonics and climate, as a result of shifting drainage divides, changing accommodation and endorheic sag basins on land, and rifting and basin subsidence in the ocean (Lavier et al., 2001; Guillocheau et al., 2012). The sediment record of marine basins may therefore have a more complex and longer history than previously considered. Over shorter Quaternary timescales, sediment dynamics of the Orange River shelf responded to glacial–interglacial changes in sea level and turbidite activity (Compton and Wiltshire, 2009).

3.5 Discussion

Most of the existing landscape evolution models for southern Africa focus on tectonic forcing followed by steady weathering/geomorphic responses (Cotterill and de Wit, 2011; Kounov et al., 2013; Champagnac et al., 2014). These models are theoretically similar to the model presented by du Toit (1933) some 80 years earlier.

The events described earlier, however, represent regional- to continental-scale drivers of landscape change over specific timescales, and should be qualified in two ways. First, tectonic events prior to the Cenozoic, including linked patterns of mantle plume migration, land surface uplift and erosional responses, are known to have taken place during the late Cretaceous (Braun et al., 2014). These are very likely to have influenced both lithospheric relaxation and topographic evolution over 10^6-year timescales, and thus well into the Cenozoic. Swell-driven uplift, subsidence and migration cannot be easily reconciled with simple and deterministic patterns of tectonic uplift and lithospheric relaxation (e.g. Erlanger et al., 2012; Kounov et al., 2013; Bierman et al., 2014). Second, several studies have directly

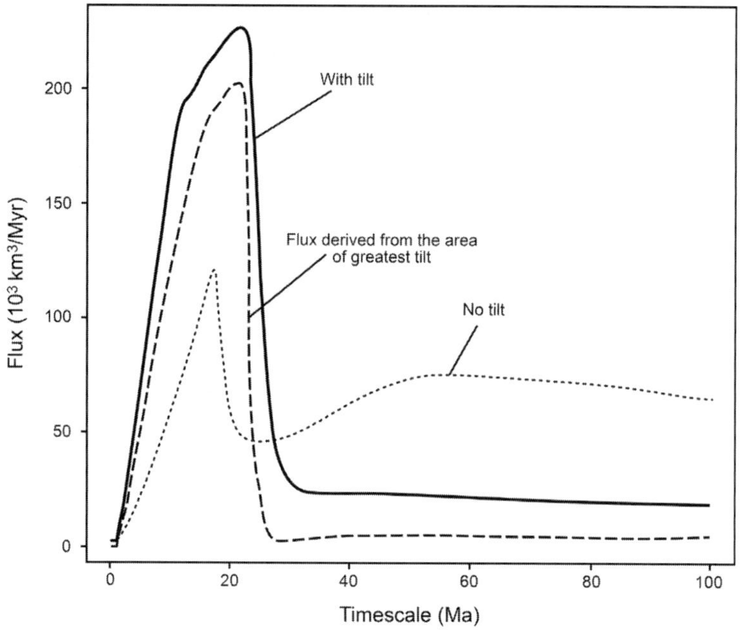

Fig. 3.3. Illustration of the modelled relationship between mantle uplift and land surface tilting, and associated response through predicted sediment flux (modified from Braun et al., 2014). Note that tilted land surfaces provide higher initial sediment yields, but that these values are exceeded by sediment yields from non-tilted land surfaces after around 30 Ma.

examined the relationships between tectonics, climate, and land surface responses (Fig. 3.3), mainly through examining topographic (relief) and elevational (land surface height) changes, using different dating and analytical methods (e.g. Decker et al., 2013; Kounov et al., 2013; Champagnac et al., 2014), geologic patterns of drainage evolution (Moore, 1999; Moore et al., 2012), or climate modelling approaches (Sepulchre et al., 2006; Jung et al., 2014). These studies are somewhat supported by dating evidence which suggests that the southern African land surface is not of uniform age (Bierman et al., 2014). However, more widely it illustrates the potential of the southern African landscape as a palimpsest comprised of different landscape elements of different ages (Cotterill and de Wit, 2011), and exhibiting different sensitivities to climatic and associated geomorphic forcing (Knight and Grab, this volume).

A number of future research strands can be developed to better evaluate the role of tectonic and climatic events on Cenozoic landscape evolution in southern Africa. These may include: (1) using geophysical methods to evaluate weathering depth, (2) using geochemical and radiometric methods to evaluate sediment source areas and dynamics within river and slope systems, (3) intensive cosmogenic dating of land surfaces in order to evaluate the range of land surface ages within any one area, and (4) testing the concept of African Surfaces through digital hypsometry.

3.6 Summary

The southern African landscape shows complex patterns of evolution during the Cenozoic, comprising tectonic, climatic and geomorphic interactions and feedbacks. Landscape evolution models and radiometric dating of the land surface and its weathering products provide some control on the timings of different events, but no coherent spatial or temporal patterns have yet emerged. Integrated tectonic, geophysical, geochemical, geomorphological and dating studies may illuminate on some of these issues, but this requires considerable multidisciplinary research effort.

Acknowledgements

We thank Grant Bybee, Zubair Jinnah and Bruce Rubidge for commenting on a previous version of this chapter.

References

Ashwal, L. D. and Burke, K. (1989). African lithospheric structure, volcanism, and topography. *Earth and Planetary Science Letters*, 96, 8–14.

Austermann, J., Kaye, B. T., Mitrovica, J. X. and Huybers, P. (2014). A statistical analysis of the correlation between large igneous provinces and lower mantle seismic structure. *Geophysical Journal International* 197, 1–9.

Beukes, N. J., van Niekerk, H. S. and Gutzmer, J. (2004). Post Gondwana African land surface and pedogenetic ferromanganese deposits on the Witwatersrand at the West Wits gold mine, South Africa. *South African Journal of Geology*, 102, 65–82.

Bierman, P. R., Coppersmith, R., Hanson, K., Neveling, J., Portenga, E. W. and Rood, D. H. (2014). A cosmogenic view of erosion, relief generation, and the age of faulting in southern Africa. *GSA Today*, 24 (9), 4–11.

Botha, G. A. and de Wit, M. J. C. (1996). Post-Gondwanan continental sedimentation, Limpopo region, southeastern Africa. *Journal of African Earth Sciences*, 23, 163–187.

Botha, G. A. and Partridge, T. C. (2000). Colluvial deposits. In *The Cenozoic of Southern Africa*, ed. T. C. Partridge and R. R. Maud. Oxford: Oxford University Press, pp. 88–99.

Braun, J., Guillocheau, F., Robin, C., Baby, G. and Jelsma, H. (2014). Rapid erosion of the Southern African Plateau as it climbs over a mantle superswell. *Journal of Geophysical Research: Solid Earth*, 119, 6093–6112.

Brown, R. W., Summerfield, M. A. and Gleadow, A. J. W. (2002). Denudational history along a transect across the Drakensberg Escarpment of southern Africa derived from apatite fission track thermochronology. *Journal of Geophysical Research–Solid Earth*, 107, 2350, doi:10.1029/2001JB000745.

Bumby, A. J. and Guiraud, R. (2005). The geodynamic setting of the Phanerozoic basins of Africa. *Journal of African Earth Sciences*, 43, 1–12.

Burke, K. (1996). The African Plate. *South African Journal of Geology*, 99, 341–409.

Burke, K. and Gunnell, Y. (2008). *The African surface: A continental-scale synthesis of geomorphology, tectonics, and environmental change over the past 180 million years*. Boulder, CO: Geological Society of America, Memoir 201, 66pp.

Burke, K., Steinberger, B., Torsvik, T. H. and Smethurst, M. A. (2008). Plume Generation Zones at the margins of Large Low Shear Velocity Provinces on the core–mantle boundary. *Earth and Planetary Science Letters* 265, 49–60.

Chadwick, O. A., Roering, J. J., Heimsath, A. M., Levick, S. R., Asner, G. P. and Khomo, L. (2013). Shaping post-orogenic landscapes by climate and chemical weathering. *Geology*, 41, 1171–1174.

Champagnac, J.-D., Valla, P. G. and Herman, F. (2014). Late-Cenozoic relief evolution under evolving climate: A review. *Tectonophysics*, 614, 44–65.

Compton, J. S. and Wiltshire, J. G. (2009). Terrigenous sediment export from the western margin of South Africa on glacial to interglacial cycles. *Marine Geology*, 266, 212–222.

Cotterill, F. P. D. and de Wit, M. (2011). Geoecodynamics and the Kalahari epeirogeny: Linking its genomic record, tree of life and palimpsest into a unified narrative of landscape evolution. *South African Journal of Geology*, 114, 489–514.

Coulthard, T. J. and Van de Wiel, M. J. (2013). Climate, tectonics or morphology: What signals can we see in drainage basin sediment yields? *Earth Surface Dynamics*, 1, 13–27.

Decker, J. E., Niedermann, S. and de Wit, M. J. (2011). Soil erosion rates in South Africa compared with cosmogenic ^{3}He-based rates of soil production. *South African Journal of Geology*, 114, 475–488.

Decker, J. E., Niedermann, S. and de Wit, M. J. (2013). Climatically influenced denudation rates of the southern African plateau: Clues to solving a geomorphic paradox. *Geomorphology*, 190, 48–60.

du Toit, A. L. (1933). Crustal movements as a factor in the geographical evolution of South Africa. *South African Geographical Journal*, 16, 1–33.

Erlanger, E. D., Granger, D. E. and Gibbon, R. J. (2012). Rock uplift rates in South Africa from isochron burial dating of fluvial and marine terraces. *Geology*, 40, 1019–1022.

Faniran, A. and Jeje, L. K. (1983). *Humid Tropical Geomorphology*. New York: Longman, 414pp.

Fleming, A., Summerfield, M. A., Stone, J. O., Fifield, L. K. and Cresswell, R. G. (1999). Denudation rates for the southern Drakensberg escarpment, SE Africa, derived from in-situ-produced cosmogenic ^{36}Cl: initial results. *Journal of the Geological Society*, 156, 209–212.

Flowers, R. M. and Schoene B. (2010). (U-Th)/He thermochronometry constraints on unroofing of the eastern Kaapvaal craton and significance for uplift of the southern African Plateau. *Geology*, 38, 827–830.

Foster, D. A. and Gleadow, A. J. W. (1992). The morphotectonic evolution of rift-margin mountains in central Kenya: Constraints from apatite fission-track thermochronology. *Earth and Planetary Science Letters*, 113, 157–171.

Gilchrist, A. R. and Summerfield, M. A. (1990). Differential denudation and flexural isostasy in formation of rifted-margin upwarps. *Nature*, 346, 739–742.

Giresse, P. (2005). Mesozoic-Cenozoic history of the Congo Basin. *Journal of African Earth Science*, 43, 301–315.

Grenfell, M. C., Ellery, W. N. and Grenfell, S. E. (2009). Valley morphology and sediment cascades within a wetland system in the KwaZulu-Natal Drakensberg Foothills, Eastern South Africa. *Catena*, 78, 20–35.

Guillocheau, F., Rouby, D., Robin, C., Helm, C., Rolland, N., Le Carlier de Veslud, C. and Braun, J. (2012). Quantification and causes of the terrigeneous sediment budget at the scale of a continental margin: A new method applied to the Namibia–South Africa margin. *Basin Research*, 24, 3–30.

Gunnell, Y. (2003). Radiometric ages of laterites and constraints on long-term denudation rates in West Africa. *Geology*, 31, 131–134.

Gurnis, M., Mitrovica, J. X., Ritsema, J. and van Heijst, H. J. (2000). Constraining mantle density structure using geological evidence of surface uplift rates: The case of the African superplume. *Geochemistry, Geophysics, Geosystems*, 1, 1020, doi:10.1029/1999GC000035.

Haddon, I. G. and McCarthy, T. S. (2005). The Mesozoic-Cenozoic interior sag basins of Central Africa: The Late-Cretaceous-Cenozoic Kalahari and Okavango basins. *Journal of African Earth Sciences*, 43, 316–333.

James, D. E., Fouch, M. J., VanDecar, J. C., van der Lee, S. and Kaapvaal Seismic Group. (2001). Tectospheric structure beneath southern Africa. *Geophysical Research Letters*, 28, 2485–2488.

Jung, G., Prange, M. and Schulz, M. (2014). Uplift of Africa as a potential cause from Neogene intensification of the Benguela upwelling system. *Nature Geoscience*, 7, 741–747.

Knight, J. and Harrison, S. (2013). The impacts of climate change on terrestrial Earth surface systems. *Nature Climate Change*, 3, 24–29.

Kounov, A., Nierdermann, S., de Wit, M. J., Codilean, A. T., Viola, G., Andreoli, M. and Christl, M. (2014). Cosmogenic ^{21}Ne and ^{10}Be reveal a more than 2 Ma alluvial fan flanking the Cape Mountains, South Africa. *South African Journal of Geology*, 118, 129–144.

Kounov, A., Niedermann, S., de Wit, M.J., Viola, G., Andreoli, M. and Erzinger, J. (2007). Present denudation rates at selected sections of the South African escarpment and the elevated continental interior based on cosmogenic ^{3}He and ^{21}Ne. *South African Journal of Geology*, 110, 235–248.

Kounov, A., Viola, G., Dunkl, I. and Frimmel, H. E. (2013). South African perspectives on the long-term morpho-tectonic evolution of cratonic interiors. *Tectonophysics*, 601, 177–191.

Lavier, L.L., Steckler, M.S. and Brigaud, F. (2001). Climatic and tectonic control on the Cenozoic evolution of the West African margin. *Marine Geology*, 178, 63–80.

Lindenmaier, F., Miller, R., Fenner, J., Christelis, G., Dill, H. G., Himmelsbach, T., Kaufhold, S., Lohe, C., Quinger, M., Schildknecht, F., Symons, G., Walzer, A. and van Wyk, B. (2014). Structure and genesis of the Cubango Megafan in northern Namibia: Implications for its hydrogeology. *Hydrogeology Journal*, 22, 1307–1328.

Lithgow-Bertelloni, C. and Silver, P. G. (1998). Dynamic topography, plate driving forces and the African superswell. *Nature*, 395, 269–272.

Macgregor, D. S. (2010). Understanding African and Brazilian margin climate, topography and drainage systems, implications for predicting deepwater reservoirs and source rock burial history. *AAPG, Search and Discovery*, Article 10270, 38pp.

Marker, M. E. and Holmes, P. J. (1999). Laterisation on limestones of the Teritary Wankoe Formation and its relationship to the African Surface, Southern Cape, South Africa. *Catena*, 38, 1–21.

Marker, M. E., McFarlane, M. J. and Wormald, R. J. (2002). A laterite profile near Albertinia, Southern Cape, South Africa: Its significance in the evolution of the African Surface. *South African Journal of Geology*, 105, 67–74.

Matmon, A., Bierman, P. and Enzel, Y. (2002). Pattern and tempo of great escarpment erosion. *Geology*, 30, 1135–1138.

McCarthy, T. S. (2013). The Okavango delta and its place in the geomorphological evolution of Southern Africa. *South African Journal of Geology*, 116, 1–54.

Miller, R. McG., Pickford, M. and Senut, B. (2010). The geology, palaeontology and evolution of the Etosha pan, Namibia: Implications for terminal Kalahari deposition. *South African Journal of Geology*, 113, 307–334.

Moon, B. P. and Dardis, G. F. (1988). Introduction. In *The Geomorphology of Southern Africa*, ed. B. P. Moon and G. F. Dardis. Johannesburg: Southern Book Publishers, pp. 1–11.

Moore, A. E. (1999). A reappraisal of epeirogenic flexure axes in southern Africa. *South African Journal of Geology*, 102, 363–376.

Moore, A. and Blenkinsop, T. (2006). Scarp retreat versus pinned drainage divide in the formation of the Drakensberg escarpment, southern Africa. *South African Journal of Geology*, 109, 599–610.

Moore, A., Blenkinsop, T. and Cotterill, F. (2009). Southern African topography and erosion history: Plumes or plate tectonics? *Terra Nova*, 21, 310–315.

Moore, A., Blenkinsop, T. and Cotterill, F. P. D. (2012). Dynamic evolution of the Zambezi-Limpopo watershed, central Zimbabwe. *South African Journal of Geology*, 115, 551–560.

Moore, A. E. and Larkin, P. A. (2001). Drainage evolution in south-central Africa since the breakup of Gondwana. *South African Journal of Geology*, 104, 47–68.

Moucha, R. and Forte, A. M. (2011). Changes in African topography driven by mantle convection. *Nature Geoscience*, 4, 707–712.

Mulyukova, E., Steinberger, B., Dabrowski, M. and Sobolev, S. V. (2015). Survival of LLSVPs for billions of years in a vigorously convecting mantle: Replenishment and destruction of chemical anomaly. *Journal of Geophysical Research–Solid Earth,* 120, 3824–3847.

Partridge, T. C., Bond, G. C., Hartnady, C. J. H., deMenocal, P. B. and Ruddiman, W. F. (1995). Climatic effects of Late Neogene tectonism and volcanism. In *Paleoclimate and Evolution, with Emphasis on Human Origins*, ed. E. S. Vrba, G. H. Denton, T. C. Partridge and L. H. Burckle. New Haven, CT: Yale University Press, pp. 8–23.

Partridge, T. C. and Maud, R. R. (1987). Geomorphic evolution of southern Africa since the Mesozoic. *South African Journal of Geology*, 90, 179–208.

Partridge, T. C. and Maud, R. R. (2000). Macro-scale geomorphic evolution of southern Africa. In *The Cenozoic of Southern Africa*, ed T. C. Partridge and R. R. Maud. Oxford: Oxford University Press, pp. 3–18.

Paton, D. A. (2006). Influence of crustal heterogeneity on normal fault dimensions and evolution: Southern South Africa extensional system. *Journal of Structural Geology*, 28, 868–886.

Phillips, J. D. (2009). Changes, perturbations, and responses in geomorphic systems. *Progress in Physical Geography*, 33, 17–30.

Pickford, M., Eisenmann, V. and Senut, B. (1999). Timing of landscape development and calcrete genesis in northern Namaqualand, South Africa. *South African Journal of Science*, 95, 357–359.

Podgorski, J. E., Green, A. G., Kgotlhang, L., Kinzelbach, W. K. H., Kalscheuer, T., Auken E. and Ngwisanyi, T. (2013). Paleo-megalake and paleo-megafan in southern Africa. *Geology*, 41, 1155–1158.

Podgorski, J. E., Green, A. G., Kalscheuer, T., Kinzelbach, W. K. H., Horstmeyer, H., Maurer, H., Rabenstein, L., Doetsch, J., Auken, E., Ngwisanyi, T., Tshoso, G., Jaba, B. C., Ntibinyane, O. and Laletsang, K. (2015). Integrated interpretation of helicopter and ground-based geophysical data recorded within the Okavango Delta, Botswana. *Journal of Applied Geophysics*, 114, 52–67.

Prendergast, M. D. (2013). Landscape evolution, regolith formation and nickel laterite development in the northern part of the Great Dyke, Zimbabwe. *South African Journal of Geology*, 116, 219–240.

Reyment, R. A. and Dingle, R. V. (1987). Palaeogeography of Africa during the Cretaceous period. *Palaeogeography, Palaeoclimatology, Palaeoecology*, 59, 93–116.

Roller, S., Hornung, J., Hinderer, M. and Ssemmanda, I. (2010). Middle Miocene to Pleistocene sedimentary record of rift evolution in the southern Albert Rift (Uganda). *International Journal of Earth Science*, 99, 1643–1661.

Roters, B. and Henrich, R. (2010). The middle to late Miocene climatic development of Southwest Africa derived from the sedimentological record of ODP Site 1085A. *International Journal of Earth Science (Geol Rundsch)*, 99, 459–471.

Said, A., Moder, C., Clark, S. and Ghorbal, B. (2015). Cretaceous–Cenozoic sedimentary budgets of the Southern Mozambique Basin: Implications for uplift history of the South African plateau. *Journal of African Earth Sciences*, 109, 1–10.

Sell, B., Ovtcharova, M., Guex, J., Bartolini, A., Jourdan, F., Spangenberg, J. E., Vicente, J.-C. and Schaltegger, U. (2014). Evaluating the temporal link between the Karoo LIP and climatic-biologic events of the Toarcian Stage with high-precision U-Pb geochronology. *Earth and Planetary Science Letters* 408, 48–56.

Sepulchre, P., Ramstein, G., Fluteau, F., Schuster, M., Tiercelin, J.-J. and Brunet, M. (2006). Tectonic uplift and eastern Africa aridification. *Science*, 313, 1419–1423.

Senut, B., Pickford, M. and Ségalen, L. (2009). Neogene desertification of Africa. *Comptes Rendus Geoscience*, 341, 591–602.

Séranne, M. and Anka, Z. (2005). South Atlantic continental margins of Africa: A comparison of the tectonic vs climate interplay on the evolution of equatorial west Africa and SW Africa margins. *Journal of African Earth Sciences*, 43, 283–300.

Steinberger, B. and Torsvik, T. H. (2012). A geodynamic model of plumes from the margins of Large Low Shear Velocity Provinces. *Geochemistry, Geophysics, Geosystems* 13, Q01W09, doi:10.1029/2011GC003808.

Tinker, J., de Wit, M. and Brown, R. (2008). Mesozoic exhumation of the southern Cape, South Africa, quantified using apatite fission track thermochronology. *Tectonophysics*, 455, 77–93.

Tooth, S. and McCarthy, T. S. (2007). Wetlands in drylands: Geomorphological and sedimentological characteristics, with emphasis on examples from southern Africa. *Progress in Physical Geography*, 31, 3–41.

Torsvik, T. H. and Cocks, L. R. M. (2013). Gondwana from top to base in space and time. *Gondwana Research* 24, 999–1030.

van der Beek, P., Summerfield, M. A., Braun, J., Brown, R. W. and Fleming, A. (2002). Modeling postbreakup landscape development and denudational history across the southeast African (Drakensberg Escarpment) margin. *Journal of Geophysical Research–Solid Earth*, 107, 2351, doi:10.1029/2001JB000744.

von Elderfeldt, K. (2012). *System theory in geomorphology. Challenges, Epistemological Consequences and Practical Implications*. Dordrecht: Springer Theses, 139pp.

Watkeys, M. K. W. (2006). Gondwana break-up: A South African perspective. In *The Geology of South Africa*, ed. M. R. Johnson, C. R. Anhaeusser and R. J. Thomas. Pretoria: Geological Society of South Africa/Council for Geoscience, pp. 531–539.

Wintle, A. G., Botha, G. A., Li, S. H. and Vogel, J. C. (1995). Chronological framework for colluviation during the last 110 ka in Natal. *South African Journal of Science*, 91, 134–139.

4

Hominin origins and evolution during the Neogene

JASON L. HEATON

Abstract

The record of early hominin evolution derives mainly from central and eastern African fossils, with basal forms appearing in the late Miocene (~8.00–5.33 Ma). Recent research suggests that a key distinguishing hominin feature, bipedalism, may have first emerged primitively in wooded environments. Miocene hominins are divided into three genera, *Sahelanthropus, Orrorin* and *Ardipithecus*. These groups may represent anagenesis during the late Miocene and early Pliocene. With the appearance of *Australopithecus* in the early Pliocene, hominins arguably fully commit to bipedality. At present, data indicate that an archaic form of *Australopithecus*, *A. afarensis*, was the first eurytopic hominin and the first to expand into southern Africa. By the end of the Neogene, separate hominin species can be recognised in eastern and southern Africa. At present, the earliest South African *Australopithecus* fossils derive from Sterkfontein, Member 2 and Makapansgat, Member 3.

4.1 Introduction

During the Neogene, the earliest, or basal, hominins appear in the record of central and eastern African fossil deposits (Fig. 4.1). Changes in late Miocene environments may have provided the selective force for terrestrial bipedality, a key distinguishing feature of this group. At present, the interplay between morphological (evolutionary) and environmental change remains uncertain. Previous hypotheses suggested that savannas, or a mosaic of savannas and forests, played a significant role in evolution (e.g. Dart, 1925; Ardrey, 1961; Jolly, 1970; Lovejoy, 1981; Coppens, 1994; Elton, 2008). More recently, there has been debate on the use

Fig. 4.1. Map of major Mio-Pliocene hominin sites in eastern and southern Africa.

of the term 'savanna' (see Domínguez-Rodrigo, 2014) and detailed analysis of the role of this ecosystem in early hominin evolution (WoldeGabriel et al., 1994, 2009; Cerling et al., 2010, 2011; White et al., 2010).

Research at basal hominin sites suggests that bipedalism may have first emerged in more wooded rather than open environments (e.g. White et al., 1994, 2009a; Pickford and Senut, 2001). For example, fauna associated with the hominin *Orrorin tugenensis* suggest a gallery forest fringed with streams, shifting to open woodland away from the water sources (Pickford and Senut, 2001). Associated with *Ardipithecus kadabba* fossil remains, the environment varied from closed and open woodland to floodplain grasslands (Su et al., 2009). At Aramis (Ethiopia), slightly younger *Ardipithecus* fossils suggest less ecological variation. Fossil wood, seeds and associated fauna recovered alongside *A. ramidus* (WoldeGabriel et al., 2009), combined with carbon isotopes from hominin teeth, suggest a woodland-to-forest habitat (White et al., 2009b). In their analysis of isotopic data from 1300 paleosols in the Awash and Omo-Turkana Valleys (Kenya), Cerling et al. (2011) argue that *A. ramidus* lived in grasslands and wooded grasslands during the late Miocene–early Pliocene (cf. White et al., 2010). These reconstructions are vital to understanding the hominin shift to bipedality. The environment may also help explain why these species appear to be stenotopic and geographically restricted.

In the early Pliocene, the appearance of *Australopithecus* signifies a major commitment to bipedalism. Evidence suggests that *Australopithecus* was tolerant, possibly more so than its predecessors, to variations in its environment. This may explain how the genus successfully expanded its range into southern Africa near the early to late Pliocene transition. Presently, the earliest South African *Australopithecus* fossils derive from Sterkfontein, Member 2 (>3.0 Ma; Bruxelles et al., 2014), and Makapansgat, Member 3.

4.2 Origin of the basal hominins

Three genera, *Sahelanthropus*, *Orrorin* and *Ardipithecus,* have been proposed as primitive representatives of our lineage (Fig. 4.2). Recognition of these basal hominins depends upon a suite of characteristics that, taken together, differentiates them from other groups such as Miocene or modern apes. The basal hominins exhibit less prognathism, possess small canines lacking honing wear, and demonstrate varying adaptations to bipedality. Basal hominins have been recovered in Chad (*Sahelanthropus*) and the East African Rift in Kenya (*Orrorin*) and Ethiopia (*Ardipithecus*). Recently, Haile-Selassie et al. (2004) suggested that all three basal hominins may represent a single species.

The oldest of the basal, bipedal group, *Sahelanthropus tchadensis,* was dated biostratigraphically to 7–6 Ma (Vignaud et al., 2002), but ^{10}Be cosmogenic nuclide dating on the sediments indicates that the fossils may date to the upper end of that range (7.2–6.8 Ma; Lebatard et al., 2008). These dates closely approximate estimates from genetic research (Bradley, 2008; Fabre et al., 2009) indicating a divergence between *Pan* and the hominin lineage near 8–6 Ma. The *Sahelanthropus* cranium (TM 266-01-060-1) possesses traits such as a small (360–370 ml), low braincase and a relatively vertical face, a short canine with apical wear, implying a lack of a canine-premolar honing complex, and moderately thick enamel, suggestive of a transitional form between the Miocene apes and later hominins (Brunet et al., 2002, 2005). Complicating its phylogenetic assessment, *Sahelanthropus* is only

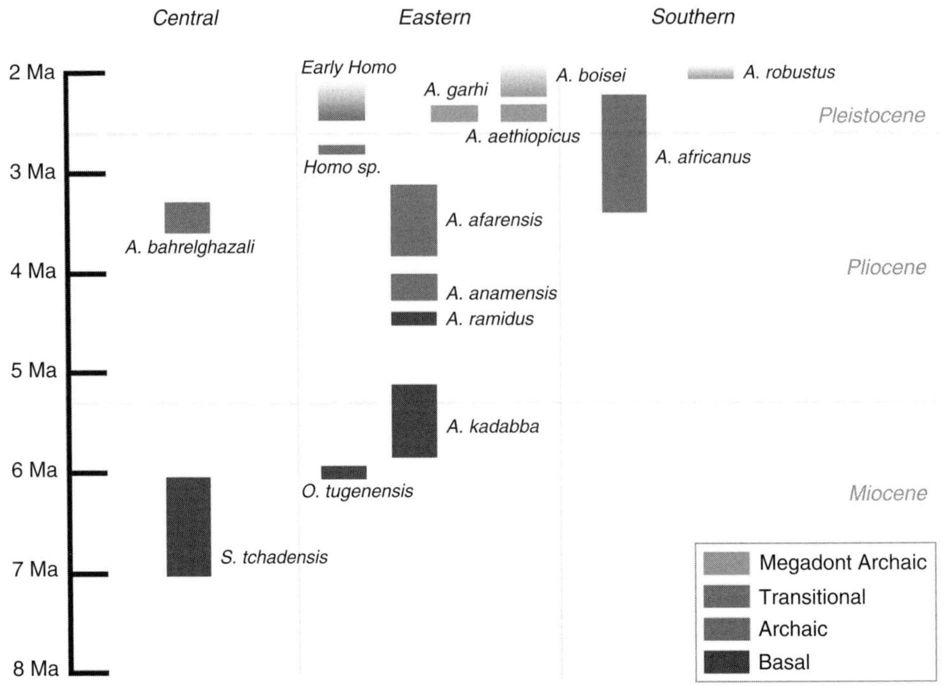

Fig. 4.2. Spatiotemporal distribution of Neogene hominins (based upon Wood, 2010, his fig. 1).

known from craniodental remains. Other researchers have suggested that the cranium belongs to a female ape and not a hominin (Wolpoff et al., 2002, 2006). Nevertheless, virtual reconstruction of the heavily distorted cranium recently suggests that *Sahelanthropus* was a biped (Zollikofer et al., 2005) due to its inferiorly directed nuchal plane and anteriorly placed foramen magnum.

Another basal hominin, *Orrorin tugenensis* (~6.0 Ma), retains more ape-like dental morphology than the older *Sahelanthropus*. For example, *Orrorin* possesses a robust ape-like I^1 and triangular, pointed maxillary canines with mesial grooves, similar to that found in male monkeys and apes (Senut et al., 2001). In contrast, traits such as the long femoral neck, an obturator externus groove, and a shallow trochanteric fossa in the *Orrorin* femur (BAR 1002'00) strongly imply bipedality (Pickford et al., 2002) and support its status as an early hominin. Recent morphometric analysis (Richmond and Jungers, 2008) argues that *Orrorin*'s femur was most similar to the morphology, and perhaps gait, observed in *Australopithecus* and *Paranthropus,* rather than *Homo*, as previously claimed (Senut et al., 2001; Pickford et al., 2002). Almécija et al. (2013), however, concluded that *Orrorin* femoral morphology is not intermediate between extant great apes and humans, but rather intermediate between plesiomorphic Miocene apes and Plio-Pleistocene hominins. Additionally, they argue that extant apes are derived relative to the ancestral hominoid morphology and exhibit higher diversity in proximal femoral morphology. Moreover, similarities among the extant great apes appear to be homoplastic (Almécija et al., 2013).

The last of the Mio-Pliocene basal hominin groups is represented by *Ardipithecus*, which includes two temporally separated species, *Ardipithecus kadabba* (5.8–5.2 Ma; Haile-Selassie, 2001; Haile-Selassie et al., 2004) and *A. ramidus* (4.4 Ma; WoldeGabriel et al., 2009). The Miocene species *Ardipithecus kadabba* is known from the Asa Koma (5.54–5.77 Ma) and Kuseralee (~5.2 Ma) sites in the Middle Awash, Ethiopia (Haile-Selassie, 2001; WoldeGabriel et al., 2001; Haile-Selassie et al., 2004; Renne et al., 2009). Initially designated as a sub-species of *A. ramidus* (Haile-Selassie, 2001), the potential presence of a primitive C/P_3 honing complex (C^1=ASK-VP-3/400, C_1=STD-VP-2/61 and P_3=ASK-VP-3/403) resulted in elevation to its own species (Haile-Selassie et al., 2004). Some post-cranial elements, such as the proximal pedal phalanx (AME-VP-1/71), are suggestive of bipedality (Haile-Selassie et al., 2009). It exhibits a dorsally canted proximal articular surface, a feature associated with toeing-off in bipeds (Latimer and Lovejoy, 1990).

Ardipithecus ramidus, the better known of the two species, is represented by 110 specimens including a partial female skeleton (White et al., 2009b). In initial descriptions, White et al. (1994) noted thin enamel, a lower premolar (P_3) with a single cusp, and a small posterior dentition – ape-like traits suggesting closeness to the last common ancestor of chimps and humans. Recent studies (White et al., 1994, 2009b; Semaw et al., 2005, Lovejoy et al., 2009a–d; Suwa et al., 2009a, b) have revealed characteristics linking *A. ramidus* to later hominins. These include: small canines lacking evidence of a C/P_3 honing complex, an anteriorly positioned foramen magnum, implied lumbar lordosis, an anterior iliac spine, and a greater sciatic notch. Additionally, very little sexual dimorphism was observed among the Aramis *A. ramidus* individuals. However, *A. ramidus* also exhibits primitive characters, such as the absence of a longitudinal arch in the foot and relatively equal fore- to hindlimb lengths (Lovejoy et al., 2009a, c). Furthermore, post-cranial evidence reveals adaptations for terrestrial bipedalism and arboreal clambering (Lovejoy et al., 2009c) with a uniquely competent (and primitive) grasping hallux (Lovejoy et al., 2009a). The pelvis of *A. ramidus* (Lovejoy et al., 2009d) displays a mixture of palmigrade clambering and terrestrial bipedal features – a more primitive form of bipedality than later *Australopithecus* (White et al., 2009b).

4.3 Diversification of the archaic hominins

The term archaic refers to the taxa between the basal hominins and *Homo* (Wood and Lonergan, 2008), including the genera *Australopithecus* and *Kenyanthropus*. Much of the australopith anatomy suggests an increasing commitment to bipedality, with current evidence (e.g. Leakey et al., 1995) suggesting that *Australopithecus* originated in eastern Africa. When compared to *A. ramidus* (White et al., 2009b), the spatial and temporal distribution of *Australopithecus* localities indicate habitats from closed woodland and forest to open grassland and shrub (Behrensmeyer and Reed, 2013; Brown et al., 2013). However, one should be cautious of assigning individual species to stenotopic or eurytopic categories (Bobe et al., 2002). The oldest species (4.2–3.9 Ma), *Australopithecus anamensis*, is preserved at sites such as Kanapoi and Allia Bay, Kenya (Leakey et al., 1995; Ward et al., 2001, 2013), Asa Issie, Ethiopia (White et al., 2006) and possibly Fejej, southern Ethiopia (Fleagle et al., 1991). The transition from *Ardipithecus* to *Australopithecus* around 4.4–4.2 Ma was rapid. However, the mode of speciation is unresolved (White et al., 2006) although changes in local to regional environments may have been significant. Existing data suggest a more wooded biome for the Ethiopian material (White et al., 2006), contrasting with the mosaic wooded/savanna reconstructions for the Kenyan *A. anamensis* sites (Coffing et al., 1994; Leakey et al., 1995; Schoeninger et al., 2003).

Craniodentally, *A. anamensis* lacked derived traits seen in later *Australopithecus*. It possessed parallel post-canine toothrows and a distinctive canine-premolar complex with mesiodistally longer honing teeth and large canine roots, especially at Kanapoi (Ward et al., 2013). While similar, hypodigm maxillae from Kanapoi (KNM-KP 29283) and the Middle Awash (ARA-VP-14/1) vary considerably in molar size and canine shape (White et al., 2006). Molar crown dimensions of some specimens (e.g. ASI-VP-2/334) from Asa Issie are at the upper end, or above, the previously known ranges for *A. anamensis*. One canine from Asa Issie (ASI-VP-2/334) is the largest hominin canine yet found (White et al., 2006). As a group, *A. anamensis* canines are still smaller than extant apes. All three canines from Asa Issie (ASI-VP-2/2, ASI-VP-2/334 and ASI-VP-2/367) possessed well-developed mesiolingual ridges, intermediate in form between *Ardipithecus* and *Australopithecus afarensis* (White et al., 2006). Additionally, the mandibular symphysis, as seen in the holotype KNM-KP 29281A, reaches as far back as the first molar (M_1). Furthermore, it is also associated with a fragmentary temporal bone (KNM-KP 29281B), a mandibular fossa lacking an articular eminence (Ward et al., 2001). Research (Teaford and Ungar, 2000) suggests that *A. anamensis* was the first hominin to exhibit thicker enamel when compared to earlier species, such as *Ardipithecus* (Suwa et al., 2009a, b). *Australopithecus* was better adapted to a more abrasive diet requiring greater mastication (Teaford and Ungar, 2000; Ward et al., 2001). This dietary shift, however, does not appear to be connected to any global climatic change (Alemseged, 2013).

Postcranially, *A. anamensis* demonstrates a continuing adaptation to bipedalism whilst retaining many primitive and ape-like features in the upper skeleton (for a review, see Alemseged, 2013). For instance, an *A. anamensis* proximal tibia (KNM-KP 29285) illustrates conformation to a bipedal pattern with an extended knee adapted to habitual bipedality (Ward et al., 2001). The upper limb, however, still retained significant primitive characters, such as a long radius and a mildly curved manual phalanx. Ward et al. (2001), however, caution against over-interpretation of these features. Arguments vary on the importance of arboreal traits in *Australopithecus*, especially with regard to a later hominin (*A. afarensis*), as reflections of locomotor behaviour (Senut and Tardieu, 1979; Stern and Susman, 1983; Stern, 2000; Alemseged et al., 2006; Green and Alemseged, 2012) or retentions from a common ancestor (Lovejoy, 1981; Latimer, 1991; Ward, 2013).

As first proposed by Kimbel et al. (2006), new evidence from Woranso-Mille (Ethiopia) supports an ancestor-descendant relationship between *A. anamensis* and *A. afarensis* (Haile-Selassie et al., 2010b). There is also debate as to whether *A. afarensis* should subsume *Kenyanthropus platyops* (Leakey et al., 2001; White, 2003). Biogeographically, *A. afarensis* ranges as far south as Laetoli, Tanzania (White, 2002), with possibly its earliest occurrence (~3.8 Ma) at Belohdelie, Ethiopia (Asfaw, 1987). The species' last occurrence is well-documented at Hadar (Ethiopia) in deposits dated between 3.4 and 2.9 Ma. In addition, the Hadar *A. afarensis* sample preserves the strongest evidence for australopith eurytopy, where it existed through fluctuations in its environment (Bonnefille et al., 2004; Campisano and Feibel, 2007; Reed, 2008). Younger deposits in the Omo basin (Brown and de Heinzelin, 1983) portray an important transitional period (2.9–2.7 Ma), but Suwa et al. (1996) note that the hominin remains are too fragmentary to be diagnostic.

A recent review estimated that cranial capacities for *A. afarensis* were within and slightly above (380–550 ml) the chimpanzee range (Holloway et al., 2004). Relative to earlier hominins, the face was slightly reduced, but still moderately prognathic, possessing an ape-like nasoalveolar clivus (e.g. AL 417-1). In the species, the primitive lower face is paired with a more derived mandible. Although mandibular morphology is highly variable (Kimbel et al., 2004), a novel feature, an upright mandibular symphysis, is observed for the first time in some specimens (e.g. AL 417-1, AL 444-2) (Kimbel and Delezene, 2009). Lockwood et al. (2000) recognised that late-occurring *A. afarensis* sites, such as Hadar, hosted the largest overall mandibles. When compared to the earlier *A. anamensis*, the maxillary (I^1) and mandibular (I_1) incisors of *A. afarensis* are smaller (Ward et al., 2001).

Conversely, canines were not appreciably different between *A. anamensis*, *A. afarensis* and *A. africanus* (Kimbel and Delezene, 2009). Canine attrition was intermediate (e.g. AL 200-1) between ape-like honing wear and the apical form of later hominins. Unlike the basal hominins and *A. anamensis*, the P_3 of *A. afarensis* possessed a more symmetrical single cusp with frequent development of a second cusp (Kimbel, 2007). However, Alemseged (2013) notes that a temporal trend (in P_3 morphology) cannot be discerned in the *A. afarensis* sample. Even juvenile specimens (Alemseged et al., 2006) exhibited a moralisation of the premolars (e.g. dp_3). Further advancing megadontia, first noted in *A. anamensis*, and relative to its body size, *A. afarensis* displayed large premolars and molars. However, overall molar size between *A. anamensis* and *A. afarensis* was similar (Ward et al., 2001).

Possibly related to dietary changes, *A. afarensis* developed weak to moderate articular eminences and more derived mandibular fossae. Kimbel et al. (2004) note that some specimens (e.g. AL 444-2 and AL 822-1) fall between the previous ranges (for mandibular fossae) of *A. afarensis* and the more South African derived *A. africanus*. In many cases, posteriorly oriented sagittal crests or closely approximated temporal lines can be observed on *A. afarensis* crania, possibly suggesting changes in chewing stresses (Kimbel and Delezene, 2009).

Post-cranially, *A. afarensis* exhibited many ape-like features in its wrist, shoulder, primitive limb proportions and long curved phalanges (Susman and Stern, 1991; Stern, 2000; Alemseged et al., 2006; Arias-Martorell et al., 2015). Regarding juvenile scapulae from Dikika, Ethiopia (DIK-1-1), Green and Alemseged (2012) argue that *A. afarensis* exhibited substantial suspensory behaviour. The benefit of cranially oriented glenoid fossae rests in the ability to transfer the weight more directly from the vertebral column to the forelimb when climbing, or hanging from one arm (Hunt, 1991). But a recently recovered adult scapula from Woranso-Mille (KSD-VP-1/1g) appears more *Homo*-like, questioning the ape-like morphology previously attributed to *A. afarensis*. Haile-Selassie et al. (2010a) remark that the scapula does not offer evidence for a suspensory history in *A. afarensis*. When applied to the fossils, a new methodology (Green et al., 2015) may further reveal the functional nuances. Kimbel and Delezene (2009) note that the humerofemoral index in *A. afarensis* was intermediate (~85) relative to small-bodied *Homo sapiens* (~74) and *P. paniscus* (~98). The relatively long upper limb may reflect sleeping and/or feeding activities in the trees, much like their great ape relatives (Sept, 1992). However, the intermediate proportions result from a relatively short (ape-like) femur, and not a long (ape-like) humerus (Jungers, 1982; Jungers and Stern, 1983).

While there is debate on the exact form of locomotion exhibited by *A. afarensis* (see Stern, 2000; Ward, 2013, for discussion), data strongly suggest terrestrial bipedality. Hypotheses range from modern, human-like extended knee locomotion (e.g. Latimer and Lovejoy, 1990) to a bent knee and bent hip gait (e.g. Stern and

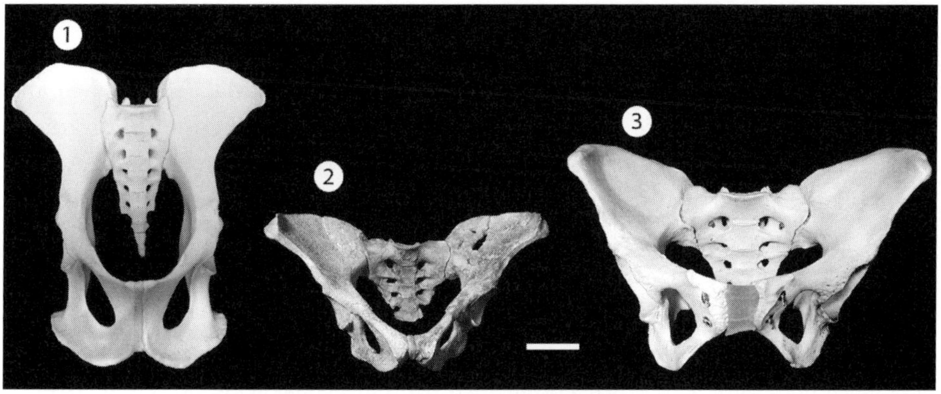

Fig. 4.3. Pelvis comparisons. Shown are the complete pelvic girdles of three hominid species: (1) *P. troglodytes,* (2) *A. afarensis* (AL 288-1, cast) and (3) *H. sapiens. Pan* exhibits narrow, elongated ilia whilst in *Australopithecus* and *Homo,* they are shortened and flared. Scale = 5 cm (photo: Jason Heaton).

Susman, 1983). The pelvic evidence reveals changes in the orientation of the iliac blades which are more medially rotated, and dramatically shortened, when compared to extant apes (Fig. 4.3). Because *A. afarensis* (e.g. AL 288) possessed valgus femora, the feet were placed directly beneath the pelvis. Additionally, footprints preserved in volcanic ash at Laetoli (~3.6 Ma) indicate the presence of longitudinal arches, as well as an adducted hallux, or great toe (Raichlen et al., 2010). Some, however, disagree with the level of hallucial adduction postulated for *A. afarensis* (Stern and Susman, 1983; Clarke and Tobias, 1995). Newly recovered material, a 4th metatarsal (AL 333-160), reveals transverse and longitudinal arches in a foot with no midtarsal break (Ward et al., 2011). The question is thus not whether *A. afarensis* was bipedal when on the ground (cf. Sarmiento, 1998), but whether its arboreal features reflected a significant contribution to its locomotor behaviour (Stern, 2000).

Reconstructions of sexual dimorphism in *A. afarensis* vary from *Gorilla-* and *Pongo*-like (Gordon et al., 2008) to modern human-like (Reno et al., 2003, 2010). Ape-like dimorphism would signal substantial male–male competition for reproductive access (Plavcan and van Schaik, 1997), but a reduction in dimorphism could indicate (Reno et al., 2003, 2010) male provisioning of females (Lovejoy, 1981). Thus, based upon present evidence, potential dimorphism is unresolved (Gordon, 2013).

4.4 Dispersal of hominins into South Africa

Hominin populations do not appear to have expanded southwards until the Pliocene. At Langebaanweg, a South African Mio-Pliocene (~5 Ma) site, a diverse array of vertebrates is preserved (Hendey, 1984), but no hominin has been recovered. Current evidence (see Ségalen et al., 2007) suggests that hominin migrations were timed with the expansion of C_4 grasses. Around 4.0–3.0 Ma, African mammal migrations (Strait and Wood, 1999) coincide with the first occurrence of grasses and their associated fauna at Makapansgat, the northernmost South African hominin site (Reed, 1997; Sponheimer et al., 1999; Hopley et al., 2007). Farther south, palaeoenvironmental reconstructions from the Pliocene deposits at Sterkfontein (Vrba, 1974; McKee, 1991; Bamford, 1999; Pickering et al., 2004) and Taung (Berger and Clarke, 1995) suggest moister, wooded habitats.

In South Africa, three species were initially identified: *Australopithecus africanus* at Taung (Dart, 1925), *Plesianthropus transvaalensis* at Sterkfontein (Broom, 1938) and *Australopithecus prometheus* at Makapansgat (Dart, 1948). Each possessed features that were thought to distinguish them, but Robinson (1954) grouped all three into *A. africanus*. Overall, the body of *A. africanus* was similar to *A. afarensis* with a few notable exceptions (Ward, 2013). An analysis (Clarke and Tobias, 1995) of the foot of StW 573, the oldest South African hominin (Bruxelles et al., 2014), suggests a greater range of motion than its predecessor (cf. McHenry and Jones, 2006). Additionally, the limbs of *A. africanus*, and their proportions, are suggestive of increased arboreality (Green et al., 2007). These may reflect autapomorphic traits resulting from local adaptations to the more wooded environment of the Sterkfontein Valley.

In *A. africanus*, the cranium (e.g. Sts 5) was more globular and less ape-like than in previous hominins. Compared to modern humans, however, the braincase remained relatively small (~428–515 ml) and the face is bordered by a large brow ridge and sloping forehead. Between the brow and maxilla, the face remains prognathic and chimp-like, but with a flattened nose and nasoalveolar clivus. Anteriorly, the palate is more parabolic, trending toward *Homo* in morphology. Unlike extant apes and most early hominins, the mandibular fossa is deepened with a distinct articular eminence. Compared to *A. afarensis,* the incisors and canines are reduced whilst the posterior dentition is slightly enlarged (Kimbel and Delezene, 2009). Adult dental morphology is nearly devoid of primitive features, though the Taung child, a juvenile *A. africanus* (Fig. 4.4), retains an ape-like deciduous third premolar (dp_3).

A potential problem for the single species hypothesis, as it relates to South African Pliocene hominins, is the high degree of morphological variation, especially at Sterkfontein. Some researchers (Aguirre, 1970; Clarke, 1985, 1988) have suggested that two distinct species are present. At Makapansgat, Tobias (1967) noted the presence of 'robust' features, suggesting that they were more similar to *Australopithecus robustus*, a Pleistocene megadont hominin, than the Sterkfontein sample. Several analyses (Kimbel and Rak, 1993; Ahern, 1998; Lockwood and Tobias,

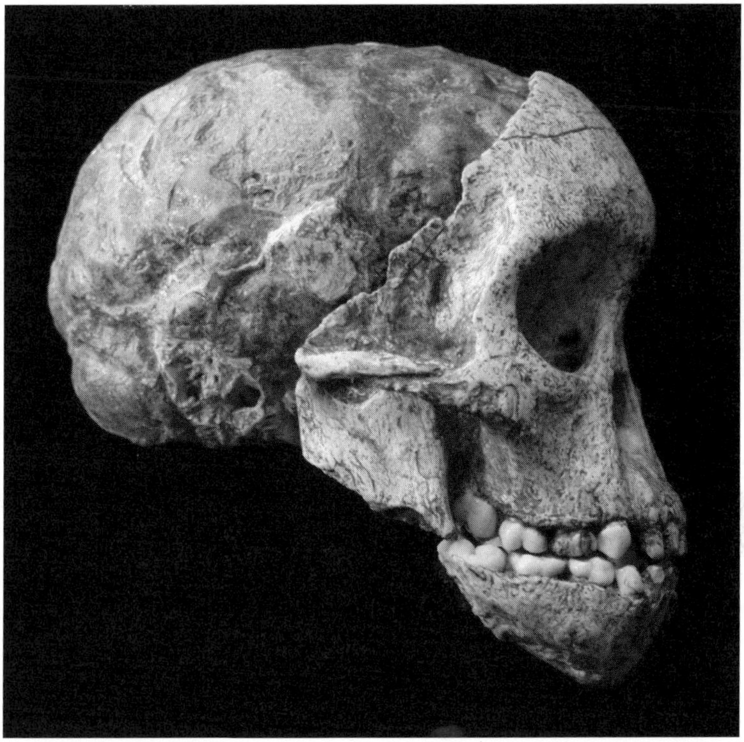

Fig. 4.4. Taung Child. An example of a juvenile of the South African species, *A. africanus*. The lower deciduous third premolar (dp_3) is ape-like in morphology. In later hominins this is replaced by the modern molariform type (photo: Jason Heaton).

2005) have attempted to address the range of variability, with little agreement (see review by Grine, 2013). Kimbel and White (1988) suggested that the Sterkfontein crania could also be separated by the degree of prognathism, with Sts 5 and Sts 71 representing the two groups (see Fig. 4.5 for comparison). Recently, Clarke (2013) assigned two groups at Sterkfontein to separate species, *A. africanus* and *A. prometheus* (i.e. the second species). Illustrating his case, a mandible (StW 384) and maxillae (Sts 1 and StW 183) of *A. prometheus* exhibit characteristics of later robust australopith groups, such as extreme bunodonty and comparatively large molars.

Preliminary analyses of new postcranial elements, as discussed in Clarke (2013), provide additional evidence of two morphs (e.g. StW 562 and StW 595), the latter of which is more ape-like. The presence of two species at Sterkfontein would seemingly reject monophyly among the South African (*Australopithecus robustus*) and East African (*A. aethiopicus* and *A. boisei*) 'robust' forms.

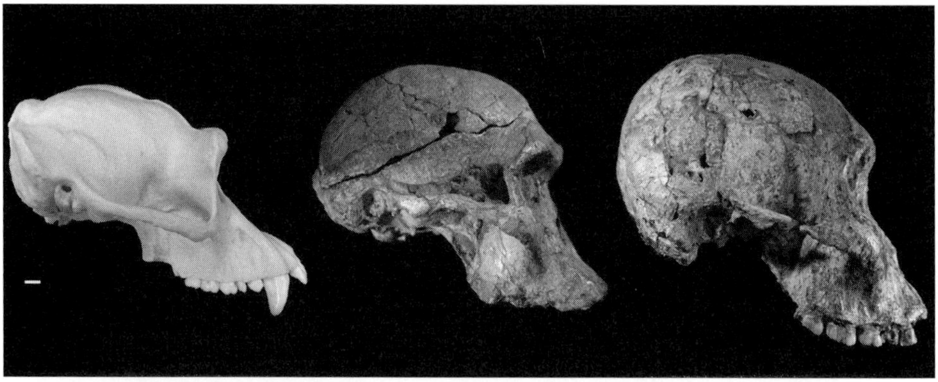

Fig. 4.5. Cranial comparisons. *Pan troglodytes* (left) exhibits a smaller and less globular braincase when compared to *A. africanus* (Sts 5-Ms Ples [middle] and Sts 71 [right]). Prognathism is also reduced in *Australopithecus*. Note that the dentition is missing in Sts 5. Scale = 1 cm (photo: Jason Heaton).

These species possibly originated through cladogenesis in the late Pliocene (3.6–2.6 Ma), but the nature of that event is debated (Grine, 1988). Dating of the Sterkfontein deposits is also important to questions of evolutionary change, as Member 4 precedes and possibly bridges the Plio-Pleistocene boundary (Herries et al., 2013).

What is clear is that the Neogene record was dominated by bipedal hominins exhibiting varying degrees of arboreality. Near the close of the Pliocene (~2.6 Ma), a dramatic shift in behaviour occurred, as populations of hominins began to produce stone tools (Semaw et al., 2003). Unfortunately, there are currently no directly associated hominin remains. As a result, a late variant of australopith, *A. garhi*, is currently the best contender for the earliest toolmaker (Semaw, 2000). This behavioural shift set the stage for subsequent cultural evolution during the Pleistocene and its dramatic impact on the origins and evolution of the genus *Homo*. Now, evidence suggests that the first occurrence of *Homo* may have extended into the Pliocene (Villmoare et al., 2015).

4.5 Summary

During the Neogene, the hominin fossil record suggests a probable evolutionary path from the basal hominins (*Sahelanthropus, Orrorin* and *Ardipithecus*) to early forms of *Australopithecus*. Debate remains on the role that wooded environments had on the origins of bipedalism, a defining hominin trait. Undoubtedly, variations in woody cover played a significant role during early hominin evolution, and adaptations developed as a result of local responses, rather than coinciding with global climatic events. In reaction, the hominin lineage arguably becomes wholly committed to bipedality, which made possible later development of the complex cultural behaviours observed in the Quaternary. Near this Neogene–Quaternary boundary, cladogenesis in archaic hominins produced two lineages: one leading to 'robust', megadont forms of *Australopithecus,* and the other leading to *Homo.*

Acknowledgements

I am indebted to the editors, Jasper Knight and Stefan Grab, for the opportunity to contribute to this volume. Jasper's patience is also deeply appreciated. I would like to thank Tim White and three anonymous reviewers for comments on the initial draft which greatly improved its scope and content. However, any errors, or omissions are solely my own. This work is dedicated to my late mother, Anne Heaton.

References

Aguirre, E. (1970). Identificacion de "*Paranthropus*" en Makapansgat. *Crónica del XI Congreso Nacional de Arqueologia. Merida*, 1969, 98–124.

Ahern, J. C. (1998). Underestimating intraspecific variation: The problem with excluding Sts 19 from *Australopithecus africanus*. *American Journal of Physical Anthropology*, 105, 461–480.

Alemseged, Z. (2013). *Australopithecus* in Ethiopia. In *The Paleobiology of Australopithecus*, ed. K. E. Reed, J. G. Fleagle and R. E. Leakey. Dordrecht: Springer, pp. 63–71.

Alemseged, Z., Spoor, F., Kimbel, W. H., Bobe, R., Geraads, D., Reed, D. and Wynn, J. G. (2006). A juvenile early hominin skeleton from Dikika, Ethiopia. *Nature*, 443, 296–301.

Almécija, S., Tallman, M., Alba, D. M., Pina, M., Moyà-Solà, S. and Jungers, W. L. (2013). The femur of *Orrorin tugenensis* exhibits morphometric affinities with both Miocene apes and later hominins. *Nature Communications*, 4, 2888, doi:10.1038/ncomms3888.

Ardrey, R. (1961). *African Genesis: A personal Investigation into the Animal Origins and Nature of Man*. London: Bantam, 384pp.

Arias-Martorell, J., Potau, J. M., Bello-Hellegouarch, G. and Pérez-Pérez, A. (2015). Like father, like son: Assessment of the morphological affinities of A.L. 288–1 (*A. afarensis*), Sts 7 (*A. africanus*) and Omo 119–73–2718 (*Australopithecus*) through a three-dimensional shape analysis of the shoulder joint. *PLoS ONE,* 10, e0117408, doi:10.1371/journal.pone.0117408.

Asfaw, B. (1987). The Belohdelie frontal: New evidence of early hominid cranial morphology from the Afar of Ethiopia. *Journal of Human Evolution*, 16, 611–624.

Bamford, M. K. (1999). Pliocene fossil woods from an early hominid cave deposit, Sterkfontein, South Africa. *South African Journal of Science*, 95, 231–237.

Behrensmeyer, A. K. and Reed, K. E. (2013). Reconstructing the habitats of *Australopithecus*: Paleoenvironments, site taphonomy and faunas. In *The Paleobiology of Australopithecus*, ed. K. E. Reed, J. G. Fleagle and R. E. Leakey. Dordrecht: Springer, pp. 41–60.

Berger, L. R. and Clarke, R. J. (1995). Eagle involvement in the accumulation of the Taung child fauna. *Journal of Human Evolution*, 29, 275–299.

Bobe, R., Behrensmeyer, A. K. and Chapman, R. E. (2002). Faunal change, environmental variability and late Pliocene hominin evolution. *Journal of Human Evolution*, 42, 475–497.

Bonnefille, R., Potts, R., Chalié, F., Jolly, D. and Peyron, O. (2004). High-resolution vegetation and climate change associated with Pliocene *Australopithecus afarensis. Proceedings of the National Academy of Sciences,* 101, 12125–12129.

Bradley, B. J. (2008). Reconstructing phylogenies and phenotypes: A molecular view of human evolution. *Journal of Anatomy*, 212, 337–353.

Broom, R. (1938). The Pleistocene anthropoid apes of South Africa. *Nature*, 142, 377–379.

Brown, F. H. and de Heinzelin, J. (1983). The lower Omo basin. In *The Omo Group: Archives of the interal Omor Research Expedition, Annales des Sciences Geologies*, ed. L. de Heinzelin. Tervuren: Musee Royal de L'Afrique Centrale, pp. 7–24.

Brown, F. H., McDougall, I. and Gathogo, P. N. (2013). Age ranges of *Australopithecus* species, Kenya, Ethiopia and Tanzania. In *The Paleobiology of Australopithecus*, ed. K. E. Reed, J. G. Fleagle and R. E. Leakey. Dordrecht: Springer, pp. 7–20.

Brunet, M., Guy, F., Pilbeam, D., Lieberman, D. E., Likius, A., Mackaye, H. T., Ponce de Leon, M. S., Zollikofer, C. P. E. and Vignaud, P. (2005). New material of the earliest hominid from the Upper Miocene of Chad. *Nature*, 434, 752–755.

Brunet, M., Guy, F., Pilbeam, D., Mackaye, H. T., Likius, A., Djimdoumalbaye, A., Beauvilain, A., Blondel, C., Bocherens, H., Boisserie, J.-R., De Bonis, L., Coppens, Y., Dejax, J., Denys, C., Duringer, P., Eisenmann, V., Fanone, G., Fronty, P., Geraads, D., Lehmann, T., Lihoreau, F., Louchart, A., Mahamat, A., Merceron, G., Mouchelin, G., Otero, O., Pelaez Campomanes, P., Ponce De Leon, M., Rage, J.-C., Sapanet, M., Schuster, M., Sudre, J., Tassy, P., Valentin, X., Vignaud, P., Viriot, L., Zazzo, A. and Zollikofer, C. (2002). A new hominid from the Upper Miocene of Chad, Central Africa. *Nature*, 418, 145–151.

Bruxelles, L., Clarke, R. J., Maire, R., Ortega, R. and Stratford, D. (2014). Stratigraphic analysis of the Sterkfontein StW 573 *Australopithecus* skeleton and implications for its age. *Journal of Human Evolution*, 70, 36–48.

Campisano, C. J. and Feibel, C. S. (2007). Connecting local environmental sequences to global climate patterns: Evidence from the hominin-bearing Hadar Formation, Ethiopia. *Journal of Human Evolution*, 53, 515–527.

Cerling, T. E., Levin, N. E., Quade, J., Wynn, J. G., Fox, D. L., Kingston, J. D., Klein, R. G. and Brown, F. H. (2010). Comment on the Paleoenvironment of *Ardipithecus ramidus*. *Science*, 328, 1105.

Cerling, T. E., Wynn, J. G., Andanje, S. A., Bird, M. I., Korir, D. K., Levin, N. E., Mace, W., Macharia, A. N., Quade, J. and Remien, C. H. (2011). Woody cover and hominin environments in the past 6 million years. *Nature*, 476, 51–56.

Clarke, R. J. (1985). *Australopithecus* and early *Homo* in southern Africa. In *Ancestors: The hard evidence*, ed. E. Delson. New York: Alan R. Liss, pp. 171–177.

Clarke, R. J. (1988). A new *Australopithecus* cranium from Sterkfontein and its bearing on the ancestry of *Paranthropus*. In *Evolutionary History of the 'Robust' Australopithecines*, ed. F. E. Grine. New York: Aldine de Gruyter, pp. 285–292.

Clarke, R. J. (2013). *Australopithecus* from Sterkfontein Caves, South Africa. In *The Paleobiology of Australopithecus*, ed. K. E. Reed, J. G. Fleagle and R. E. Leakey. Dordrecht: Springer, pp. 105–123.

Clarke, R. J. and Tobias, P. V. (1995). Sterkfontein member 2 foot bones of the oldest South African hominid. *Science*, 269, 521–524.

Coffing, K., Feibel, C., Leakey, M. and Walker, A. (1994). Four-million-year-old hominids from East Lake Turkana, Kenya. *American Journal of Physical Anthropology*, 93, 55–65.

Coppens, Y. (1994). East side story: The origin of humankind. *Scientific American*, 270, 88–95.

Dart, R. A. (1925). *Australopithecus* africanus: The Man-Ape of South Africa. *Nature*, 115, 195–199.

Dart, R. A. (1948). The Makapansgat protohuman *Australopithecus prometheus*. *American Journal of Physical Anthropology*, 6, 259–283.

Domínguez-Rodrigo, M. (2014). Is the "Savanna Hypothesis" a dead concept for explaining the emergence of the earliest hominins? *Current Anthropology*, 55, 59–81.

Elton, S. (2008). The environmental context of human evolutionary history in Eurasia and Africa. *Journal of Anatomy*, 212, 377–393.

Fabre, P.-H., Rodrigues, A. and Douzery, E. J. P. (2009). Patterns of macroevolution among Primates inferred from a supermatrix of mitochondrial and nuclear DNA. *Molecular Phylogenetics and Evolution*, 53, 808–825.

Fleagle, J. G., Rasmussen, D. T., Yirga, S., Bown, T. M. and Grine, F. E. (1991). New hominid fossils from Fejej, Southern Ethiopia. *Journal of Human Evolution*, 21, 145–152.

Gordon, A. D. (2013). Sexual size dimorphism in *Australopithecus*: Current understanding and new directions. In *The Paleobiology of Australopithecus*, ed. K. E. Reed, J. G. Fleagle and R. E. Leakey. Dordrecht: Springer, pp. 195–212.

Gordon, A. D., Green, D. J. and Richmond, B. G. (2008). Strong postcranial size dimorphism in *Australopithecus afarensis*: Results from two new resampling methods for multivariate data sets with missing data. *American Journal of Physical Anthropology*, 135, 311–328.

Green, D. J. and Alemseged, Z. (2012). *Australopithecus afarensis* scapular ontogeny, function, and the role of climbing in human evolution. *Science*, 338, 514–517.

Green, D. J., Gordon, A. D. and Richmond, B. G. (2007). Limb-size proportions in *Australopithecus afarensis* and *Australopithecus africanus*. *Journal of Human Evolution*, 52, 187–200.

Green, D. J., Serrins, J. D., Seitelman, B., Martiny, A. R. and Gunz. P. (2015). Geometric morphometrics of hominoid infraspinous fossa shape. *The Anatomical Record*, 298, 180–194.

Grine, F. E. (ed) (1988). *Evolutionary History of the "Robust" Australopithecines*. New York: A. de Gruyter, 550pp.

Grine, F. E. (2013). The alpha taxonomy of *Australopithecus africanus*. In *The Paleobiology of Australopithecus*, ed. K. E. Reed, J. G. Fleagle and R. E. Leakey. Dordrecht: Springer, pp. 73–104.

Haile-Selassie, Y. (2001). Late Miocene hominids from the Middle Awash, Ethiopia. *Nature*, 412, 178–181.

Haile-Sellassie, Y., Latimer, B. M., Alene, M., Deino, A. L., Gilbert, L., Meililo, S. M., Saylor, B. Z., Scott, G. R. and Lovejoy, C. O. (2010a). An early *Australopithecus afarensis* postcranium from Woranso-Mille, Ethiopia. *Proceedings of the National Academy of Sciences,* 107, 12121–12126.

Haile-Selassie, Y., Saylor, B. Z., Deino, A., Alene, M. and Latimer, B. M. (2010b). New hominid fossils from Woranso-Mille (Central Afar, Ethiopia) and taxonomy of early *Australopithecus*. *American Journal of Physical Anthropology*, 141, 406–417.

Haile-Selassie, Y., Suwa, G. and White, T. D. (2004). Late Miocene teeth from Middle Awash, Ethiopia, and early hominid dental evolution. *Science*, 303, 1503–1505.

Haile-Selassie, Y., Suwa, G. and White, T. D. (2009). Hominidae. In *Ardipithecus kadabba: Late Miocene Evidence from the Middle Awash, Ethiopia*, ed. Y. Haile-Selassie and G. WoldeGabriel. Oakland, CA: University of California Press, pp. 159–236.

Hendey, Q. B. (1984). Southern African Late Tertiary vertebrates. In *Southern African Prehistory and Paleoenvironments*, ed. R. G. Klein. Boston: Balkema, pp. 81–106.

Herries, A. I. R., Pickering, R., Adams, J., Curnoe, D., Warr, G., Latham, A. and Shaw, J. (2013). A multi-disciplinary perspective on the age of *Australopithecus* in Southern Africa. In *The Paleobiology of Australopithecus*, ed. K. E. Reed, J.G. Fleagle and R. E. Leakey. Dordrecht: Springer, pp. 21–40.

Holloway, R. L., Broadfield, D. C. and Yuan, M. S. (2004). *The Human Fossil Record: Vol. 3 Brain Endocasts: The Paleoneurological Evidence*. New York: Wiley-Liss, 316pp.

Hopley, P. J., Marshall, J. D., Weedon, G. P., Latham, A. G., Herries, A. I. R. and Kuykendall, K. L. (2007). Orbital forcing and the spread of C_4 grasses in the late Neogene: Stable isotope evidence from South African speleothems. *Journal of Human Evolution*, 53, 620–634.

Hunt, K. D. (1991). Mechanical implications of chimpanzee positional behavior. *American Journal of Physical Anthropology*, 86, 521–536.

Jolly, C. J. (1970). The seed-eaters: A new model of hominid differentiation based on the baboon analogy. *Man*, 5, 5–26.

Jungers, W. L. (1982). Lucy's limbs: Skeletal allometry and locomotion in *Australopithecus afarensis*. *Nature*, 297, 676–678.

Jungers, W. L. and Stern, J. T. (1983). Body proportions, skeletal allometry and locomotion in the Hadar hominids: A reply to Wolpoff. *Journal of Human Evolution*, 12, 673–684.

Kimbel, W. H. (2007). The species and diversity of Australopiths. In *Handbook of Paleoanthropology*, ed. W. Henke and I. Tattersall. New York: Springer, pp. 1539–1573.

Kimbel, W. H. and Delezene, L. K. (2009). "Lucy" redux: A review of research on *Australopithecus afarensis*. *Yearbook of Physical Anthropology*, 52, 2–48.

Kimbel, W. H., Lockwood, C. A., Ward, C. V., Leakey, M. G., Rak, Y. and Johanson, D. C. (2006). Was *Australopithecus anamensis* ancestral to *A. afarensis*? A case of anagenesis in the hominin fossil record. *Journal of Human Evolution*, 51, 134–152.

Kimbel, W. H. and Rak, Y. (1993). The importance of species taxa in paleoanthropology and an argument for the phylogenetic concept of the species category. In *Species, Species Concepts, and Primate Evolution*, ed. W. H. Kimbel and L. B. Martin. New York: Plenum Press, pp. 461–484.

Kimbel, W. H., Rak, Y. and Johanson, D. C. (2004). *The Skull of Australopithecus afarensis*. New York: Oxford University Press, 272pp.

Kimbel, W. H. and White, T. D. (1988). Variation, sexual dimorphism and the taxonomy of *Australopithecus*. In *Evolutionary History of the 'Robust' Australopithecines*, ed. F. E. Grine. New York: Aldine de Gruyter, pp. 175–192.

Latimer, B. (1991). Locomotor adaptations in *Australopithecus afarensis*: The issue of arboreality. In *Origine(s) de la Bipedie chez les Hominides*, ed. Y. Coppens and B. Senut. Paris: Centre Nationale del Researche Scientifique, pp. 169–176.

Latimer, B. and Lovejoy, C. O. (1990). Hallucal tarsometatarsal joint in *Australopithecus afarensis*. *American Journal of Physical Anthropology*, 82, 125–133.

Leakey, M. G., Fiebel, C., McDougall, I. and Walker, A. (1995). New four-million-year-old hominid species from Kanapoi and Allia Bay, Kenya. *Nature*, 376, 565–571.

Leakey, M. G., Spoor, F., Brown, F. H., Gathogo, P. N., Kiarie, C., Leakey, L. N. and McDougall, I. (2001). New hominin genus from eastern Africa shows diverse middle Pliocene lineages. *Nature*, 410, 433–440.

Lebatard, A.-E., Bourlès, D. L., Duringer, P., Jolivet, M., Braucher, R., Carcaillet, J., Schuster, M., Arnaud, N., Monié, P., Lihoreau, F., Likius, A., Mackaye, H. T., Vignaud, P. and Brunet, M. (2008). Cosmogenic nuclide dating of *Sahelanthropus tchadensis* and *Australopithecus bahrelghazali*: Mio-Pliocene hominids from Chad. *Proceedings of the National Academy of Sciences,* 105, 3226–3231.

Lockwood, C. A., Kimbel, W. H. and Johanson, D. C. (2000). Temporal trends and metric variation in the mandibles and dentition of *Australopithecus afarensis*. *Journal of Human Evolution*, 39, 23–55.

Lockwood, C. A. and Tobias, P. V. (2005). Morphology and affinities of new hominin cranial remains from Member 4 of the Sterkfontein Formation, Gauteng Province, South Africa. *Journal of Human Evolution*, 42, 389–450.

Lovejoy, C. O. (1981). The origin of man. *Science*, 211, 341–350.

Lovejoy, C. O., Latimer, B., Suwa, G., Asfaw, B. and White, T. D. (2009a). Combining prehension and propulsion: The foot of *Ardipithecus ramidus*. *Science*, 326, 72e1–72e8.

Lovejoy, C. O., Simpson, S. W., White, T. D., Asfaw, B. and Suwa, G. (2009b). Careful climbing in the Miocene: The forelimbs of *Ardipithecus ramidus* and humans are primitive. *Science*, 326, 70e1–70e8.

Lovejoy, C. O., Suwa, G., Simpson, S. W., Matternes, J. H. and White, T. D. (2009c). The great divides: *Ardipithecus ramidus* reveals the postcrania of our last common ancestors with African apes. *Science*, 326, 100–106.

Lovejoy, C. O., Suwa, G., Spurlock, L., Asfaw, B. and White, T. D. (2009d). The pelvis and femur of *Ardipithecus ramidus*: The emergence of upright walking. *Science*, 326, 71e1–71e6.

McHenry, H. M. and Jones, A. L. (2006). Hallucial convergence in early hominids. *Journal of Human Evolution*, 50, 534–539.

McKee, J. K. (1991). Palaeo-ecology of the Sterkfontein hominids: A review and synthesis. *Palaeontologia Africana*, 28, 41–51.

Pickering, T. R., Clarke, R. J. and Heaton, J. L. (2004). The context of Stw 573, an early hominid skull and skeleton from Sterkfontein Member 2: Taphonomy and paleoenvironment. *Journal of Human Evolution*, 46, 279–297.

Pickford, M. and Senut, B. (2001). The geological and faunal context of Late Miocene hominid remains from Lukeino, Kenya. *Comptes Rendus de l'Academie des Sciences Serie II A Sciences de la Terre et des Planetes*, 332, 145–152.

Pickford, M., Senut, B., Gommery, D. and Treil, J. (2002). Bipedalism in *Orrorin tugenensis* revealed by its femora. *Comptes Rendus Palevol*, 1, 191–203.

Plavcan, J. M. and van Schaik, C. P. (1997). Interpreting hominid behavior on the basis of sexual dimorphism. *Journal of Human Evolution*, 32, 345–374.

Raichlen, D. A., Gordon, A. D., Harcourt-Smith, W. E. H., Foster, A. D. and Haas Jr, W. R. (2010). Laetoli footprints preserve earliest direct evidence of human-like bipedal biomechanics. *PLoS One*, 5, e9769, doi:10.1371/journal.pone.0009769.

Reed, K. E. (1997). Early hominid evolution and ecological change through the African Plio-Pleistocene. *Journal of Human Evolution*, 32, 289–322.

Reed, K. E. (2008). Paleoecological patterns at the Hadar hominin site, Afar Regional State, Ethiopia. *Journal of Human Evolution*, 54, 743–768.

Renne, P. R., Morgan, L. E., WoldeGabriel, G., Hart, W. K. and Haile-Selassie, Y. (2009). Geochronology. In *Ardipithecus kadabba: Late Miocene Evidence from the Middle Awash, Ethiopia*, ed. Y. Haile-Selassie and G. WoldeGabriel. Oakland, CA: University of California Press, pp. 93–103.

Reno, P. L., McCollum, M. A., Meindl, R. S. and Lovejoy, C. O. (2010). An enlarged postcranial sample confirms *Australopithecus afarensis* dimorphism was similar to modern humans. *Philosophical Transactions of the Royal Society of London, Series B*, 365, 3355–3363.

Reno, P. L., Meindl, R. S., McCollum, M. A. and Lovejoy, C. O. (2003). Sexual dimorphism in *Australopithecus afarensis* was similar to that of modern humans. *Proceedings of the National Academy of Sciences,* 100, 9404–9409.

Richmond, B. G. and Jungers, W. L. (2008). *Orrorin tugenensis* femoral morphology and the evolution of hominin bipedalism. *Science*, 319, 1662–1665.

Robinson, J. T. (1954). The genera and species of the Australopithecinae. *American Journal of Physical Anthropology*, 12, 181–200.

Sarmiento, E. E. (1998). Generalized quadrupeds, commited bipeds, and the shift to open habitats: An evolutionary model of hominid divergence. *American Museum novitates*, 3250, 78pp.

Schoeninger, M. J., Reeser, H. and Hallin, K. (2003). Paleoenvironment of *Australopithecus anamensis* at Allia Bay, East Turkana, Kenya: Evidence from mammalian herbivore enamel stable isotopes. *Journal of Anthropological Archaeology*, 22, 200–207.

Ségalen, L., Lee-Thorp, J. A. and Cerling, T. (2007). Timing of C_4 grass expansion across sub-Saharan Africa. *Journal of Human Evolution*, 53, 549–559.

Semaw, S. (2000). The world's oldest stone artefacts from Gona, Ethiopia: Their implications for understanding stone technology and patterns of human evolution between 2.6–1.5 million years ago. *Journal of Archaeological Science*, 27, 1197–1214.

Semaw, S., Rogers, M. J., Quade, J., Renne, P. R., Butler, R. F., Dominguez-Rodrigo, M., Stout, D., Hart, W. S., Pickering, T. and Simpson, S. W. (2003). 2.6-Million-year-old stone tools and associated bones from OGS-6 and OGS-7, Gona, Afar, Ethiopia. *Journal of Human Evolution*, 45, 169–177.

Semaw, S., Simpson, S. W., Quade, J., Renne, P. R., Butler, R. F., McIntosh, W. C., Levin, N., Dominguez-Rodrigo, M. and Rogers, M. J. (2005). Early Pliocene hominids from Gona, Ethiopia. *Nature*, 433, 301–305.

Senut, B., Pickford, M., Gommery, D., Mein, P., Cheboi, K. and Coppens, Y. (2001). First hominid from the Miocene (Lukeino Formation, Kenya). *Comptes Rendus Académie des Sciences, Serie II-A*, 332, 137–144.

Senut, B. and Tardieu, C. (1979). Functional aspects of Plio-Pleistocene hominid limb bones: Implications for taxonomy and phylogeny. In *Ancestors: The hard evidence*, ed. E. Delson. New York: A.R. Liss, pp. 193–201.

Sept, J. M. (1992). Was there no place like home?: A new perspective on early hominid archaeological sites from the mapping of chimpanzee nests. *Current Anthropology*, 33, 187–207.

Sponheimer, M., Reed, K. E. and Lee-Thorp, J. A. (1999). Combining isotopic and ecomorphological data to refine bovid paleodietary reconstruction: A case study from the Makapansgat Limeworks hominin locality. *Journal of Human Evolution*, 36, 705–718.

Stern, J. T. (2000). Climbing to the top: A personal memoir of *Australopithecus afarensis*. *Evolutionary Anthropology*, 9, 113–133.

Stern, J. T. and Susman, R. L. (1983). The locomotor anatomy of *Australopithecus afarensis*. *American Journal of Physical Anthropology*, 60, 279–317.

Strait, D. S. and Wood, B. A. (1999). Early hominid biogeography. *Proceedings of the National Academy of Sciences,* 96, 9196–9200.

Su, D., Ambrose, S. H., Degusta, D. and Haile-Selassie, Y. (2009). Paleoenvironment. In *Ardipithecus kadabba: Late Miocene Evidence from the Middle Awash, Ethiopia*, ed. Y. Haile-Selassie and G. WoldeGabriel. Oakland, CA: University of California Press, pp. 521–548.

Susman, R. L. and Stern, J. T. (1991). Locomotor behavior of early hominids: Epistemology and fossil evidence. In *Origine(s) de la Bipedie chez les Hominides*, ed. Y. Coppens and B. Senut. Paris: Centre Nationale del Researche Scientifique, pp. 121–131.

Suwa, G., Asfaw, B., Kono, R. T., Kubo, D., Lovejoy, C. O. and White, T. D. (2009a). The *Ardipithecus ramidus* skull and its implications for hominid origins. *Science*, 326, 68e1–68e7.

Suwa, G., Kono, R. T., Simpson, S. W., Asfaw, B., Lovejoy, C. O. and White, T. D. (2009b). Paleobiological implications of the *Ardipithecus ramidus* dentition. *Science*, 326, 94–99.

Suwa, G., White, T. D. and Howell, F. C. (1996). Mandibular postcanine dentition from the Shungura Formation, Ethiopia: Crown morphology, taxonomic allocations, and Plio-Pleistocene hominid evolution. *American Journal of Physical Anthropology*, 101, 247–282.

Teaford, M. F. and Ungar, P. S. (2000). Diet and the evolution of the earliest human ancestors. *Proceedings of the National Academy of Sciences,* 97, 13506–13511.

Tobias, P. V. (1967). *Olduvai Gorge, volume 2: The cranium and maxillary dentition of Australopithecus (Zinjanthrous) boisei*. Cambridge: Cambridge University Press, 264pp.

Vignaud, P., Duringer, P., Mackaye, H. T., Likius, A., Blondel, C., Boisserie, J.-R., de Bonis, L., Eisenmann, V., Etienne, M.-E., Geraads, D., Guy, F., Lehmann, T., Lihoreau, F., Lopez-Martinez, N., Mourer-Chauviré, C., Otero, O., Rage, J.-C., Schuster, M., Viriot, L., Zazzo, A. and Brunet, M. (2002). Geology and palaeontology of the upper Miocene Toros-Menalla hominid locality, Chad. *Nature*, 418, 152–155.

Villmoare, B., Kimbel, W. H., Seyoum, C., Campisano, C. J., DiMaggio, E. N., Rowan, J., Braun, D. R., Arrowsmith, J. R. and Reed, K. E. (2015). Early *Homo* at 2.8 Ma from Ledi-Geraru, Afar, Ethiopia. *Science*, 347, 1352–1355.

Vrba, E. S. (1974). Chronological and ecological implications of the fossil Bovidae at the Sterkfontein *Australopithecine* site. *Nature*, 250, 19–23.

Ward, C. V. (2013). Postural and locomotor adaptations of *Australopithecus* species. In *The Paleobiology of Australopithecus*, ed. K. E. Reed, J. G. Fleagle and R. E. Leakey. Dordrecht: Springer, pp. 235–245.

Ward, C. V., Leakey, M. G. and Walker, A. (2001). Morphology of *Australopithecus anamensis* from Kanapoi and Allia Bay, Kenya. *Journal of Human Evolution*, 41, 255–368.

Ward, C. V., Manthi, F. K. and Plavcan, J. M. (2013). New fossils of *Australopithecus anamensis* from Kanapoi, West Turkana, Kenya (2003-2008). *Journal of Human Evolution*, 65, 501–524.

White, T. D. (2002). Earliest Hominids. In *The Primate Fossil Record*, ed. W. C. Hartwig. Cambridge: Cambridge University Press, pp. 407–417.

White, T. D. (2003). Early Hominids–Diversity or Distortion? *Science*, 299, 1994–1997.

White, T. D., Ambrose, S. H., Suwa, G., Su, D. F., DeGusta, D., Bernor, R. L., Boisserie, J.-R., Brunet, M., Delson, E., Frost, S., Garcia, N., Giaourtsakis, I. X., Haile-Selassie, Y., Howell, F. C., Lehmann, T., Likius, A., Pehlevan, C., Saegusa, H., Semprebon, G., Teaford, M. and Vrba, E. (2009a). Macrovertebrate paleontology and the Pliocene habitat of *Ardipithecus ramidus*. *Science*, 326, 87–93.

White, T. D., Ambrose, S. H., Suwa, G. and WoldeGabriel, G. (2010). Response to comment on the paleoenvironment of *Ardipithecus ramidus*. *Science*, 328, 1105.

White, T. D., Asfaw, B., Beyene, Y., Haile-Selassie, Y., Lovejoy, C. O., Suwa, G. and WoldeGabriel, G. (2009b). *Ardipithecus ramidus* and the paleobiology of early hominids. *Science*, 326, 75–86.

White, T. D., Suwa, G. and Asfaw, B. (1994). *Australopithecus ramidus*, a new species of early hominid from Aramis, Ethiopia. *Nature*, 371, 306–312.

White, T. D., WoldeGabriel, G., Asfaw, B., Ambrose, S., Beyene, Y., Bernor, R. L., Boisserie, J.-R., Currie, B., Gilbert, H., Haile-Selassie, Y., Hart, W. K., Hlusko, L. J., Howell, F. C., Kono, R. T., Lehmann, T., Louchart, A., Lovejoy, C. O., Renne, P. R., Saegusa, H., Vrba, E. S., Wesselman, H. and Suwa, G. (2006). Asa Issie, Aramis and the origin of *Australopithecus*. *Nature*, 440, 883–889.

WoldeGabriel, G., Ambrose, S. H., Barboni, D., Bonnefille, R., Bremond, L., Currie, B., DeGusta, D., Hart, W. K., Murray, A. M., Renne, P. R., Jolly-Saad, M. C., Stewart, K. M. and White, T. D. (2009). The geological, isotopic, botanical, invertebrate, and lower vertebrate surroundings of *Ardipithecus ramidus*. *Science*, 326, 65e1–65e5.

WoldeGabriel, G., Haile-Selassie, Y., Renne, P. R., Hart, W. K., Ambrose, S. H., Asfaw, B., Heiken, G. and White, T. (2001). Geology and palaeontology of the Late Miocene Middle Awash valley, Afar rift, Ethiopia. *Nature*, 412, 175–178.

WoldeGabriel, G., White, T. D., Suwa, G., Renne, P., de Heinzelin, J., Hart, W. K. and Heiken, G. (1994). Ecological and temporal placement of early Pliocene hominids at Aramis, Ethiopia. *Nature*, 371, 330–333.

Wolpoff, M. H., Hawks, J., Senut, B., Pickford, M. and Ahern, J. (2006). An ape or the ape: Is the Toumaï Cranium TM 266 a Hominid? *PaleoAnthropology*, 2006, 36–50.

Wolpoff, M. H., Senut, B., Pickford, M. and Hawks, J. (2002). Palaeoanthropology (communication arising): *Sahelanthropus* or 'Sahelpithecus'? *Nature*, 419, 581–582.

Wood, B. (2010). Reconstructing human evolution: Achievements challenges, and opportunities. *Proceedings of the National Academy of Sciences,* 107, 8902–8909.

Wood, B. and Lonergan, N. (2008). The hominin fossil record: Taxa, grades and clades. *Journal of Anatomy*, 212, 354–376.

Zollikofer, C. P. E., Ponce de Leon, M. S., Lieberman, D. E., Guy, F., Pilbeam, D., Likius, A., Mackaye, H. T., Vignaud, P. and Brunet, M. (2005). Virtual cranial reconstruction of *Sahelanthropus tchadensis*. *Nature*, 434, 755–759.

5

Hominin evolution in Africa during the Quaternary

KRISTIAN J. CARLSON AND SARAH EDLUND

Abstract

During the Quaternary, the hominin evolutionary tree can be best characterised as bushy. Fossil discoveries in recent decades have shown that different hominin taxa co-existed more than was previously thought when *Homo habilis* was first deemed the earliest member of genus *Homo*. Such phylogenetic complexity at the origin of *Homo* is indicative of an adaptive radiation, in this case often attributed to a palaeoclimatic drying trend in Africa. Establishing evolutionary relationships amongst different taxa at the origins of *Homo*, and ultimately with the descendent species of one of them, *Homo erectus*, is made difficult by their mosaic of primitive and derived morphological characteristics. Linking these morphologies with coeval environmental change is made difficult by conflicting signals between regional palaeoclimate indicators. By the end of the Quaternary, hominin behavioural and cognitive changes arguably become more evident in the 'evolving' archaeological record of southern Africa than in its fossil record. New discoveries and new types of morphological analyses are clearly needed for sharpening phylogenetic resolution, particularly early in the Quaternary.

5.1 Introduction

The Quaternary (2.58 Ma until present; Gibbard and Cohen, 2008) witnessed a hominin evolutionary tree that became substantially 'bushier' than it had been during the preceding Neogene. Here, the term hominin is used to refer to the tribe Hominini, or the clade containing modern humans, all extinct human species, and all immediate ancestors of humans (Wood, 2010). This is used in place of the term hominid, which refers to the family Hominidae that includes living apes and their fossil ancestors, as well as members of the tribe Hominini. Globally, climates began cooling and drying from about 3 Ma onwards, while climatic variability increased (Behrensmeyer, 2006). Local palaeoenvironmental records in the Awash Valley and the Omo–Turkana Basin (Ethiopia and Kenya) indicate a reduction in woody cover and an expansion of more open grasslands between 3.6 and 1.4 Ma (Cerling et al., 2011a). While there are attempts to link global climatic drying trends with shrinking African forests and expanding open grasslands, and in turn implicate these as a causal factor for habitat-specific adaptive radiations in the hominin lineage (e.g. Vrba, 1988; Stanley, 1992; Reed, 1997; Potts, 1998; Cerling et al., 2011a), these should be made with caution because regional and global climatic signals may not always be the same (Behrensmeyer, 2006). What is clear is that, whether driven by climate change and/or additional factors (e.g. material culture technology), the co-existence of multiple hominin taxa became the norm from the start of the Quaternary.

Some noteworthy events marking the Quaternary fossil hominin and archaeological records include the emergences of genus *Homo* (Kimbel et al., 1997) and *Paranthropus* (Walker et al., 1986), a tripling of hominin endocranial volume (Holloway et al., 2004), the initial hominin migration out of Africa (Lordkipanidze et al., 2007), the earliest evidence of stone tools (Semaw et al., 1997), the first widely accepted evidence of cut-marked bone (de Heinzelin et al., 1999; Semaw, 2000), and the earliest evidence for controlled use of fire (Brain and Sillen, 1988; Berna et al., 2012). Recent discoveries in Ethiopia, however, suggest that the origin of *Homo* may extend into the Neogene as early as 2.8 Ma (Villmoare et al., 2015), and that controversial stone tool cut-marks may have appeared as early as 3.39 Ma (McPherron et al., 2010; but see Domínguez-Rodrigo et al., 2011).

An increasing number of candidates has been put forward as intermediate taxa between either *Australopithecus* (*Australopithecus afarensis* or *Australopithecus africanus*) or *Kenyanthropus platyops*, and *H. erectus* during the Quaternary. These taxa, which contribute to the bushiness of the hominin evolutionary tree, include *Australopithecus garhi*, *Australopithecus sediba*, *Homo rudolfensis*, and *H. habilis*. Each of these exhibit different primitive and derived craniodental and/or postcranial features (Johanson et al., 1987; Wood, 1991; Asfaw et al., 1999; Leakey et al., 2001; Berger et al., 2010; Spoor et al., 2010). *H. erectus* (*sensu lato*) has traditionally been considered the first hominin to migrate out of Africa (Klein, 2009), possibly facilitated by ecological changes such as grassland expansion (e.g. Antón et al., 2002). However, the 1.77 Ma date for Dmanisi (Georgia) *H. erectus* material (Lordkipanidze et al., 2007) implies that at least one of these intermediate taxa may have been the first to leave Africa unless, as some believe (Lordkipanidze et al., 2013), the intermediate taxa also all represent *H. erectus*. Near the end of the Quaternary at least one additional out-of-Africa migration occurred, this time involving modern *Homo sapiens* (Henn et al., 2011), likely as a result of gradually accumulating behavioural and cognitive changes (McBrearty and Brooks, 2000).

A number of hominin fossils from the Quaternary of southern Africa has contributed to clarifying hominin evolution. Here, evidence from several sites is placed in the broader context of hominin evolution (Fig. 5.1). Some of these fossils have been attributed to *H. habilis* (e.g. partial crania: SK 27, StW 53; see Fig. 5.2) or *H. erectus* (e.g. partial mandibles: SK 15, SK 45, StW 80; partial cranium: SK 847; see Fig. 5.3), although these attributions are debated (Grine et al., 1996). Others simply attribute some of these same specimens to *Australopithecus* or *Homo* spp. (Kuman and Clarke, 2000; Grine, 2005; Clarke, 2013; Dusseldorp et al., 2013). This attribution highlights the differences between southern and eastern African fossils (Grine et al., 1996). Other researchers designate a new species altogether for the southern African collection – *H. gautengensis* (Curnoe, 2010). The fossils from southern Africa assigned to early *Homo* (e.g. StW 53) are predominantly derived from Sterkfontein Cave, either Member 5 (StW 53 Infill: Hughes and Tobias, 1977; Partridge, 2000) or Member 4 (Kuman and Clarke, 2000), and Swartkrans Cave, Members 1–3 (Susman et al., 2001; Grine, 2005). Recently, dental material from Drimolen has been attributed to *Homo* (Keyser et al., 2000; Grine et al., 2009; Moggi-Cecchi et al., 2010). Comparatively speaking, most Neogene hominin material from southern Africa (predominantly Sterkfontein Member 4: Clarke, 2013) appears to represent *A. africanus* or another species of *Australopithecus* altogether, whereas most Quaternary hominin material represents *Paranthropus robustus* (e.g. Cooper's Cave: de Ruiter et al., 2009; Drimolen: Keyser et al., 2000; Kromdraai: Thackeray et al., 2001; Swartkrans Cave, Members 1–4: Sutton et al., 2009) or *Homo*, where *A. sediba* fossils represent a notable exception (Berger et al., 2010). Apart from material representing >1 Ma in human evolutionary history, the fossil record of southern Africa is relatively sparse (Dusseldorp et al., 2013) until the appearance of archaic *H. sapiens* (e.g. Florisbad) at ~0.259 Ma (Grün et al., 1996) and eventually modern *H. sapiens* (e.g. at Border Cave and Klasies River Cave), with a notable exception being the partial hominin cranium from Elandsfontein (Drennen, 1953; Singer, 1954) that could be as old as 1 Ma (Klein et al., 2006), or considerably younger when based on associated fauna (Klein and Cruz-Uribe, 1991). The more geochronologically recent hominin material from southern Africa continues to play a substantial role in debates concerning the origins of modern humans (Bräuer, 2008; Pearson, 2008).

5.2 The emergence of genus *Homo*

Traditionally, genus *Homo* is thought to have emerged from an australopithecine ancestor such as *Australopithecus anamensis*, *A. afarensis*, *A. africanus*, or *A. sediba*

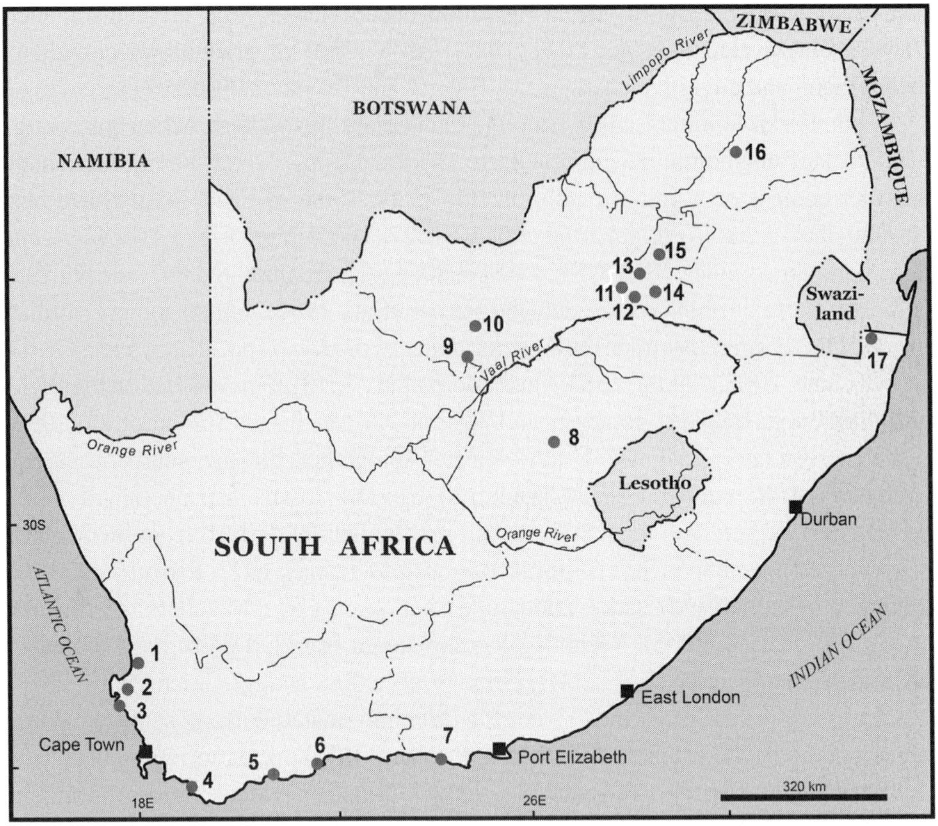

Fig. 5.1. Map of relevant hominin-bearing sites in southern Africa. Numbered sites are 1: Diepkloof Rock shelter, 2: Elandsfontein, 3: Sea Harvest, 4: Die Kelders Cave, 5: Blombos, 6: Pinnacle Point, 7: Klasies River Cave, 8: Florisbad, 9: Equus Cave, 10: Witkrans Cave, 11: Swartkrans, 12: Sterkfontein, 13: Drimolen, 14: Kromdraai, 15: Malapa, 16: Cave of Hearths, 17: Border Cave.

(Klein, 2009; Antón et al., 2014), or possibly the more poorly known taxon *K. platyops* (Leakey et al., 2001). Adaptive transformations that took place (see Fleagle and Grine, 2014) include (1) enlarged body size; (2) increased encephalisation, probably facilitated by greater exploitation of high value food resources such as meat (e.g. increased emphasis on hunting behaviour; Aiello and Wheeler, 1995); (3) reduced masticatory apparatus (e.g. smaller tooth dimensions, less mandibular reinforcement and lower facial prognathism); and (4) expanded ranging behaviour in the context of drying climates and increasingly open grassland habitats, which selected for elongating lower limbs to improve locomotor efficiency (Bramble and Lieberman, 2004; Pontzer, 2012). Overall body and limb proportions also changed towards more modern morphologies (Holliday, 2012). The relative order and timing of these changes, however, is less certain and remain important research questions (Wood, 2014).

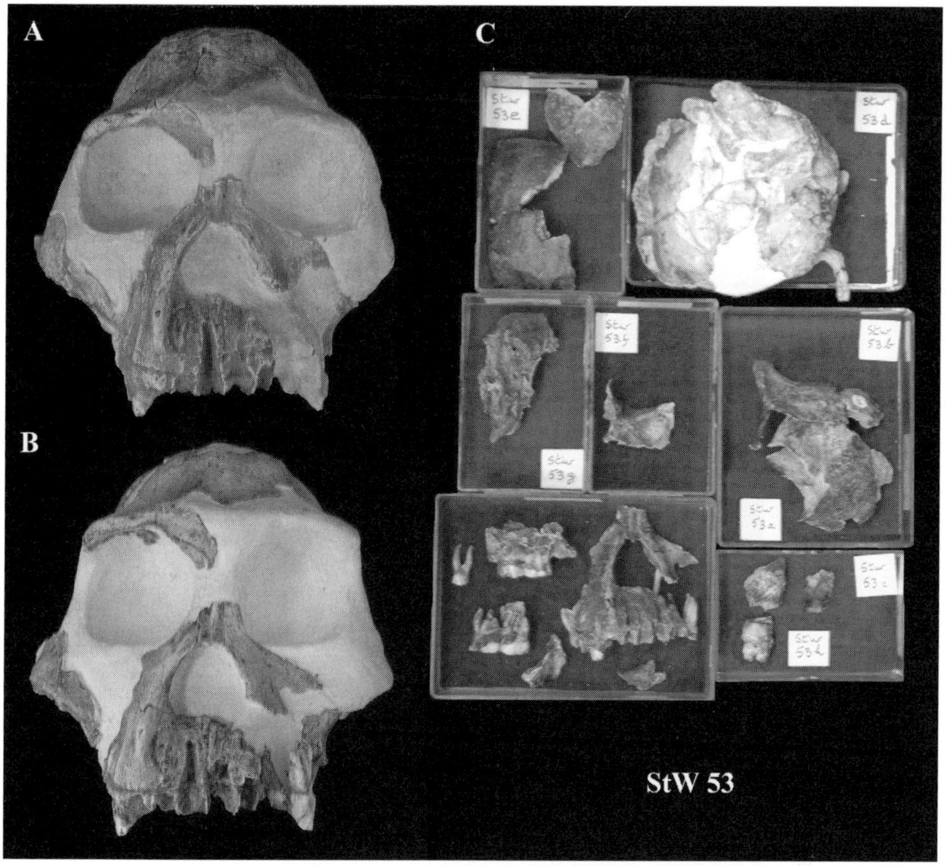

Fig. 5.2. The StW 53 cranium from Sterkfontein has been attributed to *Australopithecus* or *Homo* by various authors. (A) Frontal view of the original reconstruction by Ron Clarke; (B) frontal view of the subsequent reconstruction by Curnoe and Tobias (2006), which was reconstructed with a larger cranial capacity; (C) the partial cranium illustrated in its entirety (photo: Kristian J. Carlson).

H. habilis was initially put forth as an intermediate taxon between *Australopithecus* and *H. erectus,* from Olduvai Gorge, Tanzania (Leakey et al., 1964). Leakey et al. (1964) defined the new intermediate taxon *H. habilis* based on a collection of craniodental materials (parietals, mandible and dentition) and hand bones from a juvenile known as Olduvai Hominid (OH) 7, a partial skull with dentition from a second individual (OH 13), craniodental material assigned to a third individual (OH 6), a partial mandible with dentition assigned to a fourth individual (OH 4), and foot bones assigned to fifth individual (OH 8). Others have suggested that the hand and foot material from OH 7 and OH 8, respectively, belong to a single individual (Susman and Stern, 1982). According to

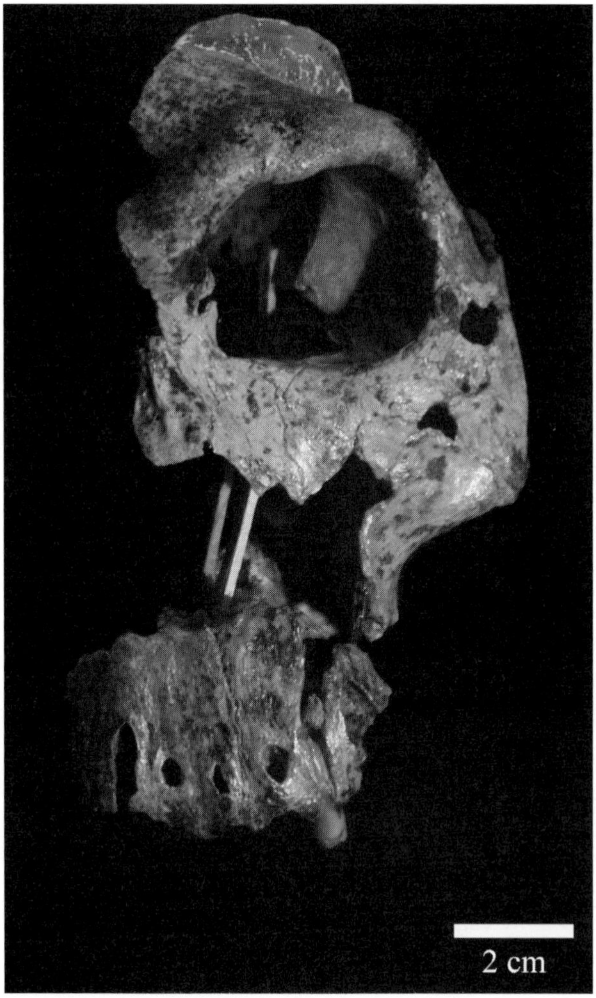

Fig. 5.3. The SK 847 cranium from Swartkrans that is often attributed to *Homo erectus* (photo reproduced by permission of Jason Heaton).

Leakey et al. (1964), the overall morphology of *H. habilis* did not fit into the range of variation characterising either *Australopithecus*, predominately represented at the time by southern African fossils (i.e. *A. africanus*), or the temporally later *H. erectus*, predominately represented at the time by Asian fossils. Not long before the *H. habilis* discovery, hominin fossils from Swartkrans in South Africa that had originally been assigned to *Telanthropus capensis* (Broom and Robinson, 1949; Robinson, 1953) were reassigned to *H. erectus* (Robinson, 1961). The fossils from Swartkrans, however, were considered at the time to be too fragmentary for establishing definitive relationships with the new *H. habilis* fossils from Olduvai, although similarities were noted between them, particularly in the dentition (see Tobias, 1991a). In describing *H. habilis*, Leakey et al. (1964) emphasised its maxillary and mandibular reduction and reduced buccolingual breadth of the postcanine dentition to within *H. erectus* and later *Homo* levels, but they still noted larger anterior dentition in *H. habilis* than was exhibited in either *Australopithecus* or *H. erectus*. The cranial morphology (e.g. muscular ridge form) and facial prognathism of *H. habilis* were described as intermediate between *Australopithecus* and *H. erectus*. In other anatomical regions, Leakey et al. (1964) noted some morphological resemblance to modern humans. Specifically, the hand bones of *H. habilis* exhibited broadened apical tufts of terminal manual phalanges as well as similar forms of metacarpophalangeal joint surfaces, compared to those of modern human hands. However, *H. habilis* hand bones also exhibited more robusticity and more phalangeal dorsal curvature, with stronger flexor sheath ridges, compared to modern human hand bones. Finally, Leakey et al. (1964) pointed to similarities between *H. habilis* and modern human foot bones as evidence of similar bipedal adaptations, including an enlarged and adducted hallux and well-marked longitudinal and transverse arches.

Importantly, the original naming of *H. habilis* required lowering the minimum endocranial volume threshold for the genus *Homo* to 600 cc (Leakey et al., 1964; Tobias, 1991b) from a previous minimum of 700–800 cc. Recent discovery of the small-bodied and small-brained (417 cc) *Homo floresiensis* from the late Quaternary of Flores (Indonesia) (Brown et al., 2004; Falk et al., 2009) has challenged this threshold, particularly because the postcranial features of *H. floresiensis* firmly place it within *Homo* (Jungers et al., 2009). Further re-examination is also warranted by the similarly sized *A. sediba* MH 1 endocast (420 cc) with its accompanying prefrontal reorganisation, which is more in keeping with *H. sapiens* than earlier *A. africanus* endocasts (e.g. Sts 5, Sts 60) (Carlson et al., 2011).

Subsequently, more hominin fossils have been attributed to *H. habilis*, including craniodental (e.g. OH 24: Leakey et al., 1971; KNM-ER 1813: Leakey, 1974; StW 53: Hughes and Tobias, 1977; SK 27: Curnoe and Tobias, 2006; KNM-ER 42703: Spoor et al., 2007; OH 65: Clarke, 2012), dental (Suwa et al., 1996), and postcranial specimens (e.g. OH 35: Susman and Stern, 1982; KNM-ER 3735: Leakey and Walker, 1985; Leakey et al., 1989; OH 62: Johanson et al., 1987), although these attributions are widely debated, as are those of specimens excluded from this taxon (Blumenschine et al., 2003; Curnoe and Tobias, 2006; Collard and Wood, 2007; Antón et al., 2014; Wood, 2014). Amongst these excluded specimens, the partial cranium from Sterkfontein, StW 53 (Hughes and Tobias, 1977), has substantial significance (see Fig. 5.2). Depending on whether this fossil originates from Member 4 (Kuman and Clarke, 2000) or Member 5 (Hughes and Tobias, 1977), stone tools found only in the latter Member could be attributed to this taxon. Also, if StW 53,

SK 27, or other material from South Africa ultimately represents *H. habilis*, it would extend the geographic distribution of this taxon to South Africa.

Reconstructing postcranial adaptations of *H. habilis* has benefited from the recovery of subsequent finds that include associated fossils and partial skeletons. The original material attributed to the taxon (OH 7 and OH 8) exhibited a powerful grasping hand characterised by overall robust and curved middle phalanges with marked flexor sheath ridges (i.e. climbing features), but more human-like distal phalanges (e.g. broadened apical tufts), suggesting human-like manual manipulatory capabilities, a thumb with mixed ape-like and human-like features, and the foot of a dedicated biped (Susman and Stern, 1982). Additional partial skeletons (e.g. OH 62, KNM-ER 3735) also contribute to the picture of postcranial adaptations. While determining the *Homo*-like or ape-like limb proportions of *H. habilis* partial skeletons OH 62 and KNM-ER 3735 has been hampered by the incomplete nature of their diaphyses (Richmond et al., 2002; Haeusler and McHenry, 2004), relative limb strength of the OH 62 partial skeleton has shown compelling similarities to chimpanzee-like limb strength proportions (Ruff, 2009). Thus, the first appearance in the fossil record of the modern human bauplan – an obligate biped – occurs in *H. erectus* (e.g. represented by the partial skeleton KNM-WT 15000, as well as the less complete KNM-ER 803 and KNM-ER 1808 skeletons), which is ~0.5 Ma or more after the initial emergence of genus *Homo* (Bramble and Lieberman, 2004; Klein, 2009).

Following some of the additional fossil discoveries (e.g. the partial cranium KNM-ER 1470 from Koobi Fora, Kenya: Leakey, 1973), it was perhaps unsurprising that other taxa challenged the role of *H. habilis* as the only potential intermediate taxon in the early Quaternary. Overall size and facial configurations of KNM-ER 1813 and KNM-ER 1470 were shown to be too different to encompass both within a single species (Wood, 1992, 2010; Rightmire, 1993; Kramer et al., 1995; Antón et al., 2014). Wood (1992) proposed two coexisting hominin lineages, one being smaller brained (*H. habilis*) with a less robust mandible and smaller teeth (e.g. KNM-ER 1813), the other larger brained (*H. rudolfensis*) with a more robust mandible and larger teeth (e.g. KNM-ER 1470). Unfortunately, these comparisons have been constrained by the absence of postcranial material reliably attributed to *H. rudolfensis* (Wood, 2010).

Additional fossil discoveries pre-dating *H. habilis* (*sensu stricto*), but still attributed to genus *Homo*, now extend this genus earlier into the Plio-Pleistocene. For example, a first right lower molar of a juvenile individual (KNM-WT 42718) has been dated to 2.34 Ma (Prat et al., 2005). The 2.33 Ma maxilla with partial dentition (A.L. 666-1) from Hadar (Ethiopia) shares several features (e.g. palate shape and thin postcanine enamel) with later *Homo* fossils, to the exclusion of *Australopithecus* (Kimbel et al., 1997). Kimbel et al. (1997) assigned the partial maxilla to *Homo* aff. *H. habilis* because they argued that these characters are apomorphic in the *Homo* clade, meaning they would not be useful for distinguishing a different species within that genus. A partial hominin mandible from the Chiwondo Beds in the Malawi Rift (UR 501) was biochronologically dated to 2.4 Ma. The Malawa mandible was attributed to *H. rudolfensis* based on the similarity of overall corpus dimensions and strength, and morphology in the mental region, to *Homo* from Koobi Fora (Bromage et al., 1995). An additional fossil, a partial right temporal (KNM-BC1) from 2.4 Ma deposits in the Chemeron Formation of Kenya, was attributed to *Homo* based on its medially positioned mandibular fossa and sharp petrous crest (Hill et al., 1992). Most recently, a partial mandible (LD 350-1) attributed to *Homo* sp. *indet.* and dating to 2.8 Ma has been recovered from the Ledi-Geraru research area in the Afar Regional State, Ethiopia (Villmoare et al., 2015).

Additional fossils from southern Africa that have had important contributions to the debate on genus *Homo* origins include those from the Malapa site dated to 1.977 Ma and attributed to *A. sediba* (Berger et al., 2010) (Fig. 5.1). Comparative studies of *A. sediba* craniodental material, the MH 1 endocast, and upper limb, pelvis, and lower limb anatomy from two partial skeletons (MH 1 and MH 2) (Berger et al., 2010; Carlson et al., 2011; Kibii et al., 2011; Churchill et al., 2013; DeSilva et al., 2013) suggest that *A. sediba* may have been directly ancestral to *H. erectus*. Potential incipient prefrontal reorganisation of the MH 1 endocast (Carlson et al., 2011), specifically in areas of the orbitofrontal cortex (e.g. Broadmann's areas 10 and 47), that recently have been linked in modern human brains to successful statistical/probabilistic modelling of environmental events (O'Reilly et al., 2013), may have signalled improved fitness via better hunting/scavenging skills. Ultimately, if the lineage represented by *A. sediba* was ancestral to *H. erectus*, then contemporary taxa such as *H. rudolfensis* or *H. habilis* would be precluded from being ancestral to *H. erectus*, eliminating the possibility that these or other *Homo* taxa would have made a genetic contribution to *H. sapiens*.

5.3 *Paranthropus* and the 'robust' australopithecine hypodigm

The initial fossil attributed to the 'robust' australopithecine morph (TM 1517) was found by Broom (1938) at Kromdraai, South Africa. Decades later the discovery of a cranium at Olduvai Gorge, Tanzania (OH 5) with tiny incisors, large postcanine teeth, and a prominent sagittal crest, initially called *Zinjanthropus boisei* at the time by Louis Leakey (1959), geographically extended the *Paranthropus* hypodigm to eastern Africa. *Paranthropus* was pushed back geochronologically to 2.5 Ma when a hyper-robust cranium, KNM-WT 17000, was discovered in the West Turkana area of Kenya (Walker et al., 1986). Both the eastern (*P. boisei*) and southern African (*P. robustus*) representatives of *Paranthropus* are thought to have arisen from *P. aethiopicus*, but doubt about the monophyly of *Paranthropus* remains (Constantino and Wood, 2007). While questions also abound regarding the ancestor of *Paranthropus*, it is clear that it ultimately disappeared from the fossil record approximately 1.0 Ma in southern Africa, and as early as 1.34 Ma in eastern Africa (Grine, 1988; Constantino and Wood, 2007; Domínguez-Rodrigo et al., 2013).

The *Paranthropus* hypodigm is constructed predominantly from craniodental remains, with emphasis placed on megadontia, sagittal crests, and small endocranial volumes (~500 cc), as well as substantial sexual dimorphism (Susman et al., 2001; Constantino and Wood, 2007). The genus is well-represented by fossils chiefly from Olduvai, Koobi Fora, Omo, and West Turkana in eastern Africa; and Drimolen, Gondolin, Kromdraai, and Swartkrans in southern Africa. The existence of *Paranthropus* in both eastern and southern Africa alongside other hominins, including *Australopithecus* or *Homo*, is largely due to niche separation. Palaeoclimatic drying trends in Africa over the course of the Quaternary indicate that woodlands diminished as open grasslands returned (Cerling et al., 2011a), which supports the idea that niche specialisation is reflected in the extreme craniodental feeding adaptations of *Paranthropus* (Rak, 1983; Grine, 1986). Recent analyses of dental microwear and stable carbon isotope ratios (Ungar et al., 2008; Cerling et al., 2011b, 2013), however, suggest that *Paranthropus* emphasised low-quality, fibrous foods in its diet, irrespective of the dietary adaptations suggested by its specialised craniofacial hard tissue anatomy. Thus, there is evidence that *Paranthropus* likely had a broad enough diet that its eventual extinction should not be attributed solely to dietary overspecialisation (Wood and Strait, 2003).

Few postcranial fossils have been attributed to *P. boisei*, primarily because of the challenges inherent in finding associated (and diagnostic) craniodental material. Postcranial remains attributed to *P. robustus* (e.g. partial femora from Swartkrans: SK 82, SK 97, SK 3121, and SKW 19) demonstrate the unambiguous locomotor adaptations of a biped, albeit of a possibly altered form of gait compared to modern humans (Ruff et al., 1999; Susman et al., 2001). Thus far, the only partial skeleton attributed to *P. boisei* (OH 80) supports such interpretations (Domínguez-Rodrigo et al., 2013). *Paranthropus* has not been precluded from the list of tool-makers or tool-users by some workers (e.g. Susman, 1988; Semaw, 2000). The capability of *P. robustus* for performing tool behaviour (e.g. precision gripping) is suggested by its hand skeletal morphology, specifically a well-defined insertion of *m. flexor pollicis longus* and a broad apical tuft on the pollical distal phalanx (Susman, 1988). Moreover, similarities between bone tools from the sites of Swartkrans (Member 3) and Drimolen, and suggestive associations of these tools with *P. robustus* fossils from both sites (Backwell and d'Errico, 2003, 2008;

d'Errico and Backwell, 2009), lend support to this notion. While it is important to note that there remains no firm association between *Paranthropus* fossils and stone/bone tools, the possibility that *Paranthropus* was a tool-user or -maker suggests that *Homo* may not have been the exclusive example at Swartkrans, and that it would not be prudent to attribute the extinction of *P. robustus* to a lack of tool behaviour (Susman, 1988).

5.4 Emergence of modern humans

The oldest fossil evidence attributed to modern *H. sapiens* comes from Ethiopia: two crania from Member 1 of the Omo–Kibish region (approximately 0.195 Ma; McDougall et al., 2005), and several crania from Herto (approximately 0.16 Ma) in the Middle Awash region (White et al., 2003). Subsequent to fossil records of early *Homo* and *Paranthropus* in the Cradle of Humankind, the fossil record of southern Africa contributes to this debate with material postdating ~0.2 Ma, but with intermittent gaps still appearing throughout. A notable exception is the partial hominin cranium from Elandsfontein (Drennen, 1953; Singer, 1954) that could be as old as 1.0 Ma (Klein et al., 2006), or as young as 0.4 Ma based on associated fauna (Klein and Cruz-Uribe, 1991).

Fossils attributed to modern *H. sapiens* in southern Africa are usually associated with Middle Stone Age (MSA) technologies. Modern *H. sapiens* material from MSA levels at Klasies River Mouth (Singer and Wymer, 1982; Rightmire and Deacon, 1991) has played a crucial role in the debate on modern human origins (Stringer, 2002; Rightmire, 2009; Dusseldorp et al., 2013). Fossils from the site of Border Cave (de Villiers, 1973) have also been used to characterise early modern *H. sapiens* cranial (Smith et al., 1989) and postcranial morphologies (Pearson and Grine, 1996), but these are subject to dating problems (Bird et al., 2003; Grün et al., 2003). The Hofmeyr skull dated to 36 kyr BP shares affinities with Upper Paleolithic Eurasian rather than southern African populations, suggesting that the former can be traced back to a sub-Saharan Africa ancestor during the Late Pleistocene (Grine et al., 2007). Additional sites in southern Africa with MSA levels, typically along the coast, contribute small numbers of modern *H. sapiens* fossils (mostly teeth). These sites include Blombos (Grine et al., 2000), Cave of Hearths (Tobias, 1971), Die Kelders Cave 1 (Grine, 2000), Diepkloof Rock shelter (Verna et al., 2013), Equus Cave (Grine and Klein, 1985), Pinnacle Point (Marean et al., 2004), Sea Harvest (Grine and Klein, 1993), and Witkrans Cave (McCrossin, 1992) (Fig. 5.1).

Fossils from the MSA of southern Africa that contribute to documenting postcranial differences between modern *H. sapiens* and their hominin precursors *H. erectus, H. antecessor* and *H. heidelbergensis* are relatively sparse, but these fossils reflect improvements in locomotor (bipedalism) efficiency rather than substantial differences in locomotor activities or repertoires (Klein, 2009). A reduction in postcranial robusticity, but not to present day levels of gracility, is also evident in modern *H. sapiens* MSA material compared to earlier hominins, but this trend is mixed with one of retained primitive features (Churchill et al., 1996; Pearson and Grine, 1996). Of particular note is the massive proximal femur from Berg Aukas (Namibia) of unknown geochronological age. The specimen exhibits external morphology such as a head size relative to proximal shaft breadth recalling that of recent humans, but other features such as neck shaft angle appear to be unique (Grine et al., 1995). In terms of its shaft and overall morphology, the enigmatic Berg Aukas specimen most closely resembles archaic *H. sapiens* (Grine et al., 1995). This has led some to suggest that hominin postcranial fossils associated with MSA levels were still behaviourally archaic compared to Later Stone Age (LSA) hominins (Churchill et al., 1996).

Compared to the origins of *Homo* or *Paranthropus*, the origins of modern humans benefits from corroboratory or complementary evidence from non-fossil genetics and material culture sources. Studies of genetic diversity amongst world-wide populations have documented the greatest diversity in Africa, suggesting that modern *H. sapiens* originated there (Conrad et al., 2006; Campbell and Tishkoff, 2009), and particularly in southern Africa where genetic diversity in modern Khoisan-speaking hunter-gatherers is particularly high compared to populations elsewhere in the world (Henn et al., 2011). In terms of material culture, McBrearty and Brooks (2000) refuted the idea that an abrupt behavioural revolution took place at the origin of modern humans, instead suggesting gradual development of behavioural complexity over the duration of the African MSA, with material culture and symbolic behaviour considered proxies of cognitive complexity. Wurz (this volume) updates the case to be made in the MSA of southern Africa for the emergence and gradual expression of complex behaviours, and thus changing cognition.

Towards the end of the Quaternary, emigration of modern *H. sapiens* out of Africa, whether modelled as a biological population leaving Africa and replacing existing local populations of hominins around the world by sharing genes or not (i.e. variations of a replacement model), or whether modelled as local biological populations of hominins contemporaneously evolving whilst sharing genes (i.e. a multiregional model), has long been debated (Smith et al., 1989). The number of emigration events and their timings is complex (Atkinson et al., 2008), but the latter is estimated to be ~200–100 kyr BP (Lambert and Tishkoff, 2009). Genetic diversity is related to both population size and its mobility. A population bottle-neck refers to a temporary period of low population size, such as where genetic diversity is constrained due to a massive die-off, often attributed to a global environmental event. A population release refers to the period following the bottleneck where rapid population increase takes place. It appears from genetic evidence that a global population bottleneck occurred at ~50 kyr, after emigration events but before a population release (Atkinson et al., 2008). Reasons for the bottleneck are not well understood, but the subsequent population release may be linked to technological improvements accompanying the transition to the LSA at ~50 kyr BP (Villa et al., 2012). An alternative reason may be stabilisation of global climates after the eruption of Mt Toba around 73–71 kyr BP (Ambrose, 1998). Irrespective of the ultimate causes of the population bottleneck and release, global population growth rates increased rapidly everywhere after ~50 kyr BP, although the timing of these increased rates differed in different regions.

5.5 Summary

The hominin evolutionary tree became substantially more complex during the Quaternary compared to the preceding Neogene, possibly from adaptive radiation driven by a climatic drying trend that took place during much of the Quaternary. Noteworthy events marking its fossil record include the emergence of *Homo* and its sister taxon *Paranthropus* in Africa, a tripling of hominin endocranial volume, and the initial hominin migrations of *Homo* out of Africa. While a number of potential evolutionary links between *A. afarensis* (or *A. africanus*) and *H. erectus* have been proposed, the relationship between these taxa ultimately remains uncertain. Contemporaneous events in the Quaternary archaeological record include arguably the earliest evidence of stone tools, the first evidence of cut-marked bone, and the earliest evidence for controlled use of fire. By the end of the Quaternary, modern *H. sapiens* had emerged from Africa and spread throughout the globe. At this time, distinct behavioural and cognitive changes in *H. sapiens* are evident in the archaeological record of southern Africa, more than is evident in its fossil record.

Acknowledgements

We would like to thank the editors, Jasper Knight and Stefan Grab, for the opportunity to contribute to this volume. Their patience has been greatly appreciated. We also appreciate the constructive suggestions of Jasper Knight, John Fleagle and Sandrine Prat that immeasurably improved this chapter. We note, however, that all errors within it are solely our responsibility. Ultimately, the topic of this chapter excites us and we only hope that our efforts have effectively transferred the same excitement to the reader.

References

Aiello, L. C. and Wheeler, P. (1995). The expensive tissue hypothesis: The brain and digestive system in human and primate evolution. *Current Anthropology*, 36, 199–221.

Ambrose, S. H. (1998). Late Pleistocene human population bottlenecks, volcanic winter, and differentiation of modern humans. *Journal of Human Evolution*, 34, 623–651.

Antón, S. C., Leonard, W. R. and Robertson, M. L. (2002). An ecomorphological model of the initial hominid dispersal from Africa. *Journal of Human Evolution*, 43, 773–785.

Antón, S. C., Potts, R. and Aiello, L. C. (2014). Evolution of early *Homo*: An integrated biological perspective. *Science*, 345, 45.

Asfaw, B., White, T., Lovejoy, O., Latimer, B., Simpson, S. and Suwa, G. (1999). *Australopithecus garhi*: A new species of early hominid from Ethiopia. *Science*, 284, 629–635.

Atkinson, Q. D., Gray, R. D. and Drummond, A. J. (2008). mtDNA variation predicts population size in humans and reveals a major southern Asian chapter in human prehistory. *Molecular Biology and Evolution*, 25, 468–474.

Backwell, L. R. and d'Errico, F. (2003). Additional evidence on the early hominid bone tools from Swartkrans with reference to spatial distribution of lithic and organic artefacts. *South African Journal of Science*, 99, 259–267.

Backwell, L. R. and d'Errico, F. (2008). Early hominid bone tools from Drimolen, South Africa. *Journal of Archaeological Science*, 35, 2880–2894.

Behrensmeyer, A. K. (2006). Climate change and human evolution. *Science*, 311, 476–478.

Berger, L. R., de Ruiter D. J., Churchill, S. E., Schmid, P., Carlson, K. J., Dirks, P. H. G. M. and Kibii, J. M. (2010). *Australopithecus sediba*: A new species of *Homo*-like australopithecine from South Africa. *Science*, 328, 195–204.

Berna, F., Goldberg, P., Horwitz, L. K., Brink, J., Holt, S., Bamford, M. and Chazan, M. (2012). Microstratigraphic evidence of in situ fire in the Acheulean strata of Wonderwerk Cave, Northern Cape province, South Africa. *Proceedings of the National Academy of Sciences*, 109, E1215–E1220.

Bird, M. I., Fifield, L. K., Santos, G. M., Beaumont, P. B., Zhou, Y., Di Tada, M. I., Hausladen, P. A. (2003). Radiocarbon dating from 40–60 ka BP at Border Cave, South Africa. *Quaternary Science Reviews*, 22, 943–947.

Blumenschine, R. J., Peters, C. R., Masao, F. T., Clarke, R. J., Deino, A. L., Hay, R. L., Swisher, C. C., Stanistreet, I. G., Ashley, G. M., McHenry, L. J., Sikes, N. E., van der Merwe, N. J., Tactikos, J. C., Cushing, A. E., Deocampo, D. M., Njau, J. K. and Ebert, J. I. (2003). Late Pliocene *Homo* and hominid land use from western Olduvai Gorge, Tanzania. *Science*, 299, 1217–1221.

Brain, C. K. and Sillen, A. (1988). Evidence from the Swartkrans Cave for the earliest use of fire. *Nature*, 336, 464–466.

Bramble, D. M. and Lieberman, D. E. (2004). Endurance running and the evolution of *Homo*. *Nature*, 432, 345–352.

Bräuer, G. (2008). The origin of modern anatomy: By speciation or intraspecific evolution? *Evolutionary Anthropology*, 17, 22–37.

Bromage, T. G., Schrenk, F. and Zonneveld, F. W. (1995). Paleoanthropology of the Malawi Rift: An early hominid mandible from the Chiwondo Beds, northern Malawi. *Journal of Human Evolution*, 28, 71–108.

Broom, R. (1938). The Pleistocene anthropoid apes of South Africa. *Nature*, 142, 377–379.

Broom, R. and Robinson, J. T. (1949). A new type of fossil man. *Nature*, 164, 322–323.

Brown, P., Sutikna, T., Morwood, M. J., Socjono, R. P., Jatmiko, Wayhu Saptomo, E. and Awe Due, R. (2004). A new small-bodied hominin from the late Pleistocene of Flores, Indonesia. *Nature*, 431, 1055–1061.

Campbell, M. C. and Tishkoff, S. A. (2009). African genetic diversity: Implications for human demographic history, modern human origins, and complex disease mapping. *Annual Review of Genomics and Human Genetics*, 9, 403–433.

Carlson, K. J., Stout, D., Jashashvili, T., de Ruiter, D. J., Tafforeau, P., Carlson, K. and Berger, L. R. (2011). The Endocast of MH1, *Australopithecus sediba. Science*, 333, 1402–1407.

Cerling, T. E., Manthi, F. K., Mbua, E. N., Leakey, L. N., Leakey, M. G., Leakey, R. E., Brown, F. H., Grine, F. E., Hart, J. A., Kaleme, P., Roche, H., Uno, K. T. and Wood, B. A. (2013). Stable isotope-based diet reconstructions of Turkana Basin hominins. *Proceedings of the National Academy of Sciences*, 110, 10501–10506.

Cerling, T. E., Mbua, E., Kirera, F. M., Manthi, F. K., Grine, F. E., Leakey, M. G., Sponheimer, M. and Uno, K. T. (2011b). Diet of *Paranthropus boisei* in the early Pleistocene of East Africa. *Proceedings of the National Academy of Sciences*, 108, 9337–9341.

Cerling, T. E., Wynn, J. G., Andanje, S. A., Bird, M. I., Korir, D. K., Levin, N. E., Mace, W., Macharia, A. N., Quade, J. and Remien, C. H. (2011a). Woody cover and hominin environments in the past 6 million years. *Nature*, 476, 51–56.

Churchill, S. E., Holliday, T. W., Carlson, K. J., Jashashvili, T., Macias, M. E., Mathews, S., Sparling, T. L., Schmid, P., de Ruiter, D. J. and Berger, L. R. (2013). The upper limb of *Australopithecus sediba*. *Science*, 340, 1233477-1–1233477-6.

Churchill, S. E., Pearson, O. M., Grine, F. E., Trinkaus, E. and Holliday, T. W. (1996). Morphological affinities of the proximal ulna from Klasies River Mouth Main Site: Archaic or modern? *Journal of Human Evolution*, 31, 213–237.

Clarke, R. J. (1985). *Australopithecus* and early *Homo* in southern Africa. In *Ancestors: The Hard Evidence*, ed. E. Delson. New York: Alan R. Liss, pp. 171–177.

Clarke, R. J. (2012). A *Homo habilis* maxilla and other newly-discovered hominid fossils from Olduvai Gorge, Tanzania. *Journal of Human Evolution*, 63, 418–428.

Clarke, R. J. (2013). *Australopithecus* from Sterkfontein Caves, South Africa. In *The Palaeobiology of Australopithecus*, ed. K. E. Reed, J. G. Fleagle and R. E. Leakey. Amsterdam: Springer, pp. 105–123.

Collard, M. and Wood, B. A. (2007). Defining the genus *Homo*. In *Handbook of Paleoanthropology, Vol. 3 of Phylogeny of hominids*, ed. W. Henke and I. Tattersall. Berlin: Springer, pp. 1575–1611.

Conrad, D. F., Jakobsson, M., Coop, G., Wen, X., Wall, J. D., Rosenberg, N. A. and Pritchard, J. K. (2006). A worldwide survey of haplotype variation and linkage disequilibrium in the human genome. *Nature Genetics*, 38, 1251–1260.

Constantino, P. and Wood, B. (2007). The evolution of *Zinjanthropus boisei. Evolutionary Anthropology*, 16, 49–62.

Curnoe, D. (2010). A review of early *Homo* in southern Africa focusing on cranial, mandibular and dental remains, with the description of a new species (*Homo gautengensis* sp. nov.). *HOMO – Journal of Comparative Human Biology*, 61, 151–177.

Curnoe, D. and Tobias, P. V. (2006). Description, new reconstruction, comparative anatomy, and classification of the Sterkfontein Stw 53 cranium, with discussions about the taxonomy of other southern African early *Homo* remains. *Journal of Human Evolution*, 50, 36–77.

de Heinzelin, J., Clark, J. D., White, T. W., Hart, W., Renne, P., WoldeGabriel, G., Beyene, Y. and Vrba, E. (1999). Environment and behavior of 2.5-million-year-old Bouri hominids. *Science*, 284, 625–629.

de Ruiter, D. J., Pickering, R., Steininger, C. M., Kramers, J. D., Hancox, P. J., Churchill, S. E., Berger, L. R. and Backwell, L. (2009). New *Australopithecus robustus* fossils and associated U-Pb dates from Cooper's Cave (Gauteng, South Africa). *Journal of Human Evolution*, 56, 497–513.

d'Errico, F. and Backwell, L. R. (2009). Assessing the function of early hominin bone tools. *Journal of Archaeological Science*, 36, 1764–1773.

DeSilva, J. M., Holt, K. G., Churchill, S. E., Carlson, K. J., Walker, C. S., Zipfel, B. and Berger, L. R. (2013). The lower limb and mechanics of walking in *Australopithecus sediba*. *Science*, 340, 1232999-1–1232999-5.

de Villiers, H. (1973). Human skeletal remains from Border Cave, Ingwavumu District, KwaZulu, South Africa. *Annals of the Transvaal Museum*, 28, 229–256.

Domínguez-Rodrigo, M., Pickering, T. R., Baquedano, E., Mabulla, A., Mark, D. F., Musiba, C., Bunn, H. T., Uribelarrea, D., Smith, V., Diez-Martin, F., Pérez-González, A., Sánchez, P., Santonja, M., Barboni, D., Gidna, A., Ashley, G., Yravedra, J., Heaton, J. L. and Arriaza, M. C. (2013). First partial skeleton of a 1.34-million-year-old *Paranthropus boisei* from Bed II, Olduvai Gorge, Tanzania. *PLoS One*, 8 (12): e80347, doi:10.1371/journal.pone.0080347.

Domínguez-Rodrigo, M., Pickering, T. R. and Bunn, H. T. (2011). Reply to McPherron et al.: Doubting Dikika is about data, not paradigms. *Proceedings of the National Academy of Sciences*, 108, E117.

Drennen, M. R. D. (1953). A preliminary note on the Saldanha skull. *South African Journal of Science*, 50, 7–11.

Dusseldorp, G., Lombard, M. and Wurz, S. (2013). Pleistocene *Homo* and the updated Stone Age sequence of South Africa. *South African Journal of Science*, 109, Art. #0042, 7pp, doi:10.1590/sajs.2013/20120042.

Falk, D., Hildebolt, C., Smith, K., Morwood, M. J., Sutikna, T., Jatmiko, Saptomo, E. W. and Prior, F. (2009). LB1's virtual endocast, microcephaly, and hominin brain evolution. *Journal of Human Evolution*, 57, 597–607.

Fleagle, J. G. and Grine, F. E. (2014). The genus *Homo* in Africa. In *The Cambridge World Prehistory – Africa, South and Southeast Asia, and the Pacific, Vol. 1*, ed. C. Renfrew and P. Bahn. Cambridge: Cambridge University Press, pp. 85–105.

Gibbard, P. and Cohen, K. M. (2008). Global chronostratigraphical correlation table for the last 2.7 million years. *Episodes*, 31, 243–247.

Grine, F. E. (1986). Dental evidence for dietary differences in *Australopithecus* and *Paranthropus*: A quantitative analysis of permanent molar microwear. *Journal of Human Evolution*, 15, 783–822.

Grine, F. E. (Ed.) (1988). *Evolutionary History of the 'robust' Australopithecines*. New York: Aldine de Gruyter, 546pp.

Grine, F. E. (2000). Middle Stone Age human fossils from Die Kelders Cave 1, Western Cape Province, South Africa. *Journal of Human Evolution*, 38, 129–145.

Grine, F. E. (2005). Early *Homo* at Swartkrans, South Africa: A review of the evidence and an evaluation of recently proposed morphs. *South African Journal of Science*, 101, 43–52.

Grine, F. E., Bailey, R. M., Harvati, K., Nathan, R. P., Morris, A. G., Henderson, G. M., Ribot, I. and Pike, A. W. G. (2007). Late Pleistocene human skull from Hofmeyr, South Africa, and modern human origins. *Science*, 315, 226–229.

Grine, F. E., Henshilwood, C. S. and Sealy, J. C. (2000). Human remains from Blombos Cave, South Africa: (1997–1998 excavations). *Journal of Human Evolution*, 38, 755–765.

Grine, F. E., Jungers, W. L. and Schultz, J. (1996). Phenetic affinities among early *Homo* crania from East and South Africa. *Journal of Human Evolution*, 30, 189–225.

Grine, F. E., Jungers, W. L., Tobias, P. V. and Pearson, O. M. (1995). Fossil *Homo* femur from Berg Aukas, Northern Namibia. *American Journal of Physical Anthropology*, 97, 151–185.

Grine, F. E. and Klein, R. G. (1985). Pleistocene and Holocene human remains from Equus cave, South Africa. *Anthropology*, 8, 55–98.

Grine, F. E. and Klein, R. G. (1993). Late Pleistocene human remains from the Sea Harvest site, Saldanha Bay, South Africa. *South African Journal of Science*, 89, 145–152.

Grine, F. E., Smith, H. F., Heesy, C. P. and Smith, E. J. (2009). Phenetic affinities of Plio-Pleistocene *Homo* fossils from South Africa: Molar cusp proportions. In *The First Humans: Origins and Early Evolution of the Genus Homo*, ed. F. E. Grine, J. G. Fleagle and R. E. Leakey. Berlin: Springer, pp. 49–62.

Grün, R., Beaumont, P., Tobias, P. V. and Eggins, S. (2003). On the age of Border Cave 5 human mandible. *Journal of Human Evolution*, 45, 155–167.

Grün, R., Brink, J. S., Spooner, N. A., Taylor, L., Stringer, C. B., Franciscus, R. G. and Murray, A. S. (1996). Direct dating of Florisbad hominid. *Nature*, 382, 500–501.

Haeusler, M. and McHenry, H. M. (2004). Body proportions of *Homo habilis* reviewed. *Journal of Human Evolution*, 46, 433–465.

Hartwig-Scherer, S. and Martin, R. D. (1991). Was "Lucy" more human than her "child"? Observations on early hominid postcranial skeletons. *Journal of Human Evolution*, 21, 439–449.

Henn, B. M., Gignoux, C. R., Jobin, M., Granka, J. M., Macpherson, J. M., Kidd, J. M., Rodríguez-Botigué, L., Ramachandran, S., Hon, L., Brisbin, A., Lin, A. A., Underhill, P. A., Comas, D., Kidd, K. K., Norman, P. J., Parham, P., Bustamante, C. D., Mountain, J. L. and Feldman, M. W. (2011). Hunter-gatherer genomic diversity suggests a southern African origin for modern humans. *Proceedings of the National Academy of Sciences*, 108, 5154–5162.

Hill, A., Ward, S., Deino, A., Curtis, G. and Drake, R. (1992). Earliest *Homo*. *Nature*, 355, 719–722.

Holliday, T. W. (2012). Body size, body shape, and the circumscription of the Genus *Homo*. *Current Anthropology*, 53 (supplement 6), S330–S345.

Holloway, R. L., Broadfield, D. C. and Yuan, M. S. (2004). *The Human Fossil Record, Vol. 3: Brain Endocasts – the Paleoneurological Evidence*. Hoboken NJ, Wiley-Liss, 315pp.

Hughes, A. R. and Tobias, P. V. (1977). A fossil skull probably of the genus *Homo* from Sterkfontein, Transvaal. *Nature*, 265, 310–312.

Johanson, D., Masao, F., Eck, G., White, T., Walter, R., Kimbel, W., Asfaw, B., Manega, P., Ndessokia, P. and Suwa, G. (1987). New partial skeleton of *Homo habilis* from Olduvai Gorge, Tanzania. *Nature*, 327, 205–209.

Jungers, W. L., Larson, S. G., Harcourt-Smith, W., Morwood, M. J., Sutikna, T., Awe Due, R. and Djubiantono, T. (2009). Descriptions of the lower limb skeleton of *Homo floresiensis*. *Journal of Human Evolution*, 57, 538–554.

Keyser, A. W., Menter, C. G., Moggi-Cecchi, J., Pickering, T. R. and Berger, L. R. (2000). Drimolen: A new hominid-bearing site in Gauteng, South Africa. *South African Journal of Science*, 96, 193–197.

Kibii, J. M., Churchill, S. E., Schmid, P., Carlson, K. J., Reed, N. D., de Ruiter, D. J. and Berger, L. R. (2011). A partial pelvis of *Australopithecus sediba*. *Science*, 333, 1407–1411.

Kimbel, W. H., Johanson, D. C. and Rak, Y. (1997). Systematic assessment of a maxilla of *Homo* from Hadar, Ethiopia. *American Journal of Physical Anthropology*, 103, 235–262.

Klein, R. G. (2009). *The Human Career – Human Biological and Cultural Origins*. Chicago, IL: University of Chicago Press, 840pp.

Klein, R. G., Avery, G., Cruz-Uribe, K. and Steele, T. E. (2006). The mammalian fauna associated with an archaic hominin skullcap and later Acheulean artifacts from Elandsfontein, Western Cape Province, South Africa. *Journal of Human Evolution*, 52, 164–186.

Klein, R. G. and Cruz-Uribe, K. (1991). The bovids from Elandsfontein, South Africa, and their implications of the age, palaeoenvironment and origins of the site. *African Archaeological Review*, 9, 21–79.

Kramer, A., Donnelly, S. M., Kidder, S. D., Ousley, S. D. and Olah, S. M. (1995). Craniometric variation in large-bodied hominoids: Testing the single-species hypothesis for *Homo habilis*. *Journal of Human Evolution*, 29, 443–462.

Kuman, K. and Clarke, R. J. (2000). Stratigraphy, artefact industries and hominid associations for Sterkfontein, Member 5. *Journal of Human Evolution*, 38, 827–847.

Lambert, C. A. and Tishkoff, S. A. (2009). Genetic structure in African populations: Implications for human demographic history. *Cold Spring Harbor Symposium on Quantitative Biology*, 74, 395–402.

Leakey, L. S. B. (1959). A new fossil skull from Olduvai. *Nature*, 184, 491–493.

Leakey, L. S. B., Tobias, P. V. and Napier, J. R. (1964). A new species of genus *Homo* from Olduvai Gorge. *Nature*, 202, 7–9.

Leakey, M. D., Clarke, R. J. and Leakey, L. S. B. (1971). New hominid skull from Bed I, Olduvai Gorge, Tanzania. *Nature*, 232, 308–312.

Leakey, M. G., Spoor, F., Brown, F. H., Gathogo, P. N., Kiarie, C., Leakey, L. N. and McDougall, I. (2001). New hominin genus from eastern Africa shows diverse middle Pliocene lineages. *Nature*, 410, 433–440.

Leakey, R. E. F. (1973). Evidence of an advanced Plio-Pleistocene hominid from East Rudolf, Kenya. *Nature*, 242, 447–450.

Leakey, R. E. F. (1974). Further evidence of Lower Pleistocene hominids from East Rudolf, North Kenya, 1973. *Nature*, 248, 653–656.

Leakey, R. E. F. and Walker, A. C. (1985). Further hominids from the Plio-Pleistocene of Koobi Fora, Kenya. *American Journal of Physical Anthropology*, 67, 135–163.

Leakey, R. E., Walker, A., Ward, C. V. and Grausz, H. M. (1989). A partial skeleton of a gracile hominid from the Upper Burgi Member of the Koobi Fora formation, East Lake Turkana, Kenya. In *Hominidae: Proceedings of the 2nd International Congress of Human Paleontology*, ed. G. Giacobini. Milan: Editoriale Jaca Book, pp. 167–173.

Lordkipanidze, D., Jashashvili, T., Vekua, A., Ponce de León, M. S., Zollikofer, C. P. E., Rightmire, G. P., Pontzer, H., Ferring, R., Oms, O., Tappen, M., Bukhsianidze, M., Agusti, J., Kahlke, R., Kiladze, G., Martinez-Navarro, B., Mouskhelishvili, A., Nioradze, M. and Rook, L. (2007). Postcranial evidence from early *Homo* from Dmanisi, Georgia. *Nature*, 449, 305–310.

Lordkipanidze, D., Ponce de León, M. S., Margvelashvili, A., Rak, Y., Rightmire, G. P., Vekua, A. and Zollikofer, C. P. E. (2013). A complete skull from Dmanisi, Georgia, and the evolutionary biology of early *Homo*. *Science*, 342, 326–331.

Marean, C. W., Nilssen, P. J., Brown, K., Jerardino, A. and Stynder, D. (2004). Paleoanthropological investigations of Middle Stone Age sites at Pinnacle Point, Mossel Bay (South Africa): Archaeology and hominid remains from the 2000 field season. *Paleoanthropology*, 1, 14–83.

McBrearty, S. and Brooks, A. S. (2000). The revolution that wasn't: a new interpretation of the origin of modern human behavior. *Journal of Human Evolution*, 39, 453–563.

McCrossin, M. L. (1992). Human molars from Later Pleistocene deposits of Witkrans Cave, Gaap Escarpment, Kalahari margin. *Human Evolution*, 7, 1–10.

McDougall, I., Brown, F. H. and Fleagle, J. G. (2005). Stratigraphic placement and age of modern humans from Kibish, Ethiopia. *Nature*, 433, 733–736.

McPherron, S. P., Alemseged, Z., Marean, C. W., Wynn, J. G., Reed, D., Geraads, D., Bobe, R. and Béarat, H. A. (2010). Evidence for stone-tool-assisted consumption of animal tissues before 3.39 million years ago at Dikika, Ethiopia. *Nature*, 466, 857–860.

Moggi-Cecchi, J., Menter, C., Boccone, S. and Keyser, A. (2010). Early hominin dental remains from the Plio-Pleistocene of Drimolen, South Africa. *Journal of Human Evolution*, 58, 374–405.

O'Reilly, J. X., Jbabdi, S., Rushworth, M. F. S. and Behrens, T. E. J. (2013). Brain systems for probabilistic and dynamic prediction: Computational specificity and integration. *PLoS Biology*, 11(9), e1001662, doi:10.1371/journal.pbio.1001662.

Pearson, O. M. (2008). Statistical and biological definition of "anatomically modern" humans: Suggestions for a unified approach to modern morphology. *Evolutionary Anthropology*, 17, 38–48.

Pearson, O. M. and Grine, F. E. (1996). Morphology of the Border Cave hominid ulna and humerus. *South African Journal of Science*, 92, 231–236.

Pontzer, H. (2012). Ecological energetics in early *Homo*. *Current Anthropology*, 53 (supplement 6), S346–S358.

Potts, R. (1998). Environmental hypotheses of hominin evolution. *Yearbook of Physical Anthropology*, 41, 93–136.

Prat, S., Brugal, J.-P., Tiercelin, J.-J., Barrat, J.-A., Bohn, M., Delagnes, A., Harmand, S., Kimeu, K., Kibunjia, M., Texier, P.-J. and Roche, H. (2005). First occurrence of early *Homo* in the Nachukui Formation (West Turkana, Kenya) at 2.3–2.4 Myr. *Journal of Human Evolution*, 49, 230–240.

Rak, Y. (1983). *The Australopithecine Face*. New York: Academic Press, 250pp.

Reed, K. E. (1997). Early hominid evolution and ecological change through the African Plio-Pleistocene. *Journal of Human Evolution*, 32, 289–322.

Richmond, B. G., Aiello, L. C. and Wood, B. A. (2002). Early hominin limb proportions. *Journal of Human Evolution*, 43, 529–548.

Rightmire, G. P. (1993). Variation among early *Homo* crania from Olduvai Gorge and the Koobi Fora region. *American Journal of Physical Anthropology*, 90, 1–33.

Rightmire, G. P. (2009). Middle and later Pleistocene hominins in Africa and Southwest Asia. *Proceedings of the National Academy of Sciences*, 106, 16046–16050.

Rightmire, G. P. and Deacon, H. J. (1991). Comparative studies of Late Pleistocene human remains from Klasies River Mouth, South Africa. *Journal of Human Evolution*, 20, 131–156.

Robinson, J. T. (1953). *Telanthropus* and its phylogenetic significance. *American Journal of Physical Anthropology*, 11, 445–501.

Robinson, J. T. (1961). The australopithecines and their bearing on the origin of man and of stone tool-making. *South African Journal of Science*, 57, 3–13.

Ruff, C. B. (2009). Relative limb strength and locomotion in *Homo habilis*. *American Journal of Physical Anthropology*, 138, 90–100.

Ruff, C. B., McHenry, H. M. and Thackeray, J. F. (1999). Cross-sectional morphology of the SK 82 and 97 proximal femora. *American Journal of Physical Anthropology*, 109, 509–521.

Semaw, S. (2000). The world's oldest stone artefacts from Gona, Ethiopia: Their implications for understanding stone technology and patterns of human evolution between 2.6–1.5 million years ago. *Journal of Archaeological Science*, 27, 1197–1214.

Semaw, S., Renne, P., Harris, J. W. K., Feibel, C. S., Bernor, R. L., Fesseha, N. and Mowbray, K. (1997). 2.5-million-year-old stone tools from Gona, Ethiopia. *Nature*, 385, 333–336.

Singer, R. (1954). The Saldanha skull from Hopefield, South Africa. *American Journal of Physical Anthropology*, 12, 345–362.

Singer, R. and Wymar, J. (1982). *The Middle Stone Age at Klasies River Mouth in South Africa*. Chicago IL: University of Chicago Press, 312pp.

Smith, F. H., Falsetti, A. B. and Donnelly, S. M. (1989). Modern human origins. *Yearbook of Physical Anthropology*, 32, 35–68.

Spoor, F., Leakey, M. G., Gathogo, P. N., Brown, F. H., Anton, S. C., McDougall, I., Kiarie, C., Manthi, F. K. and Leakey, L. N. (2007). Implications of new early *Homo* fossils from Ileret, east of Lake Turkana, Kenya. *Nature*, 448, 688–691.

Spoor, F., Leakey, M. G. and Leakey, L. N. (2010). Hominin diversity in the Middle Pleistocene of eastern Africa: The maxilla of KNM-WT 40000. *Philosophical Transactions of the Royal Society of London, Series B*, 365, 3377–3388.

Stanley, S. M. (1992). An ecological theory for the origin of *Homo*. *Paleobiology*, 18, 237–257.

Stringer, C. (2002). Modern human origins: Progress and prospects. *Philosophical Transactions of the Royal Society of London, Series B*, 357, 563–579.

Susman, R. L. (1988). Hand of *Paranthropus robustus* from Member 1. Swartkrans: fossil evidence for tool behaviour. *Science*, 240, 781–784.

Susman, R. L., de Ruiter, D. and Brain, C. K. (2001). Recently identified postcranial remains of *Paranthropus* and early *Homo* from Swartkrans Cave, South Africa. *Journal of Human Evolution*, 41, 607–629.

Susman, R. L. and Stern, J. T. Jr. (1982). Understanding stone technology and patterns of human evolution between 2.6 – 1.5 million years ago. Functional morphology of *Homo habilis*. *Science*, 217, 931–933.

Sutton, M. B., Pickering, T. R., Pickering, R., Brain, C. K., Clarke, R. J., Heaton, J. L. and Kuman, K. (2009). Newly discovered fossil- and artifact-bearing deposits, uranium-series ages, and Plio-Pleistocene hominids at Swartkrans Cave, South Africa. *Journal of Human Evolution*, 57, 688–696.

Suwa, G., White, T. D. and Howell, F. C. (1996). Mandibular postcanine dentition from the Shungura formation, Ethiopia: Crown morphology, taxonomic allocations, and Plio-Pleistocene hominid evolution. *American Journal of Physical Anthropology*, 101, 247–282.

Thackeray, J. F., de Ruiter, D. J., Berger, L. and van der Merwe, N. (2001). Hominid fossils from Kromdraai: A revised list of specimens discovered since 1938. *Annals of the Transvaal Museum*, 38, 43–56.

Tobias, P. V. (1971). Human skeletal remains from the Cave of Hearths, Makapansgat, Northern Transvaal. *American Journal of Physical Anthropology*, 34, 335–368.

Tobias, P. V. (1991a). The species *Homo habilis*: Example of a premature discovery. *Annales Zoologici Fennici*, 28, 371–380.

Tobias, P. V. (1991b). *The Skulls and Endocasts of Homo habilis. Olduvai Gorge, Vol. 4.* Cambridge: Cambridge University Press, 921pp.

Ungar, P. S., Grine, F. E. and Teaford, M. F. (2008). Dental microwear and diet of the Plio-Pleistocene hominin *Paranthropus boisei. PLoS One*, 3, e2044, doi:10.1371/journal.pone.0002044.

Verna, C., Texier, P.-J., Rigaud, J.-P., Poggenpoel, C. and Parkington, J. (2013). The Middle Stone Age human remains from Diepkloof Rock Shelter (Western Cape, South Africa). *Journal of Archaeological Science*, 40, 3532–3541.

Villa, P., Soriano, S., Tsanova, T., Degano, I., Higham, T. F. G., d'Errico, F., Backwell, L., Lucejko, J. J., Colombini, M. P. and Beaumont, P. B. (2012). Border Cave and the beginning of the Later Stone Age in South Africa. *Proceedings of the National Academy of Science*, 109, 13208–13213.

Villmoare, B., Kimbel, W. H., Seyoum, C., Campisano, C. J., DiMaggio, E., Rowan, J., Braun, D. R., Arrowsmith, J. R. and Reed, K. E. (2015). Early *Homo* at 2.8 Ma from Ledi-Geraru, Afar, Ethiopia. *Science,* 347, 1352–1355.

Vrba, E. S. (1988). Late Pliocene climatic events and hominid evolution. In *Evolutionary History of the 'robust' Australopithecines*, ed. F. E. Grine. New York: Aldine de Gruyter, pp. 405–426.

Walker, A., Leakey, R. E., Harris, J. M. and Brown, F. H. (1986). 2.5-Myr *Australopithecus boisei* from west of Lake Turkana, Kenya. *Nature*, 322, 517–522.

White, T. D., Asfaw, B., DeGusta, D., Gilbert, H., Richards, G. D., Suwa, G. and Howell, F. C. (2003). Pleistocene *Homo sapiens* from Middle Awash, Ethiopia. *Nature*, 423, 742–747.

Wood, B. A. (1991). *Hominid Cranial Remains, Koobi Fora Research Project.* Vol. 4. Oxford: Clarendon Press, 496pp.

Wood, B. A. (1992). Origin and evolution of the genus *Homo*. *Nature*, 355, 783–790.

Wood, B. A. (2010). Reconstructing human evolution: Achievements, challenges, and opportunities. *Proceedings of the National Academy of Sciences*, 107, 8902–8909.

Wood, B. A. (2014). Fifty years after *Homo habilis*. *Nature*, 508, 31–33.

Wood, B. A. and Strait, D. (2003). Patterns of resource use in early *Homo* and *Paranthropus. Journal of Human Evolution*, 46, 119–162.

6

Quaternary environmental change on the southern African coastal plain

JOHN S. COMPTON

Abstract

The low-lying South African coastal plain, bounded by the sea and by high-relief terrain of the continental interior, forms an important habitat for many animals, including humans. The present-day coastal plain is generally narrow, but expanded in width when sea level was lowered by between 75 and 130 m during Quaternary glacial periods. During these periods there was moderate expansion of the western and eastern coastal plains, whereas the southern coastal plain expanded by up to a factor of five onto the adjacent Agulhas Bank to form a continuous coastal plain. The offshore marine geology in this region allows for extrapolation of modern vegetation biomes onto comparable substrates exposed on the glacial-age coastal plain. With the exception of cemented dunes (aeolianite), Quaternary deposits on the exposed shelf are generally highly reworked and condensed. However, meandering rivers, wetlands and lakes on the low-lying southern coastal plain can be inferred, and may have provided a refuge from the dry interior during glacial periods. The southern coastal plain possibly served as a geographical region of origin by periodically isolating populations over glacial to interglacial cycles, promoting human evolution.

6.1 Introduction

Southern Africa is atypical amongst passive continents because it has an extensive continental interior plateau rising to elevations in excess of 1000 m (Fig. 6.1) and it has remained tectonically active long after the Cretaceous breakup of Gondwana. The elevated interior plateau, or African superswell, and tectonic activity are

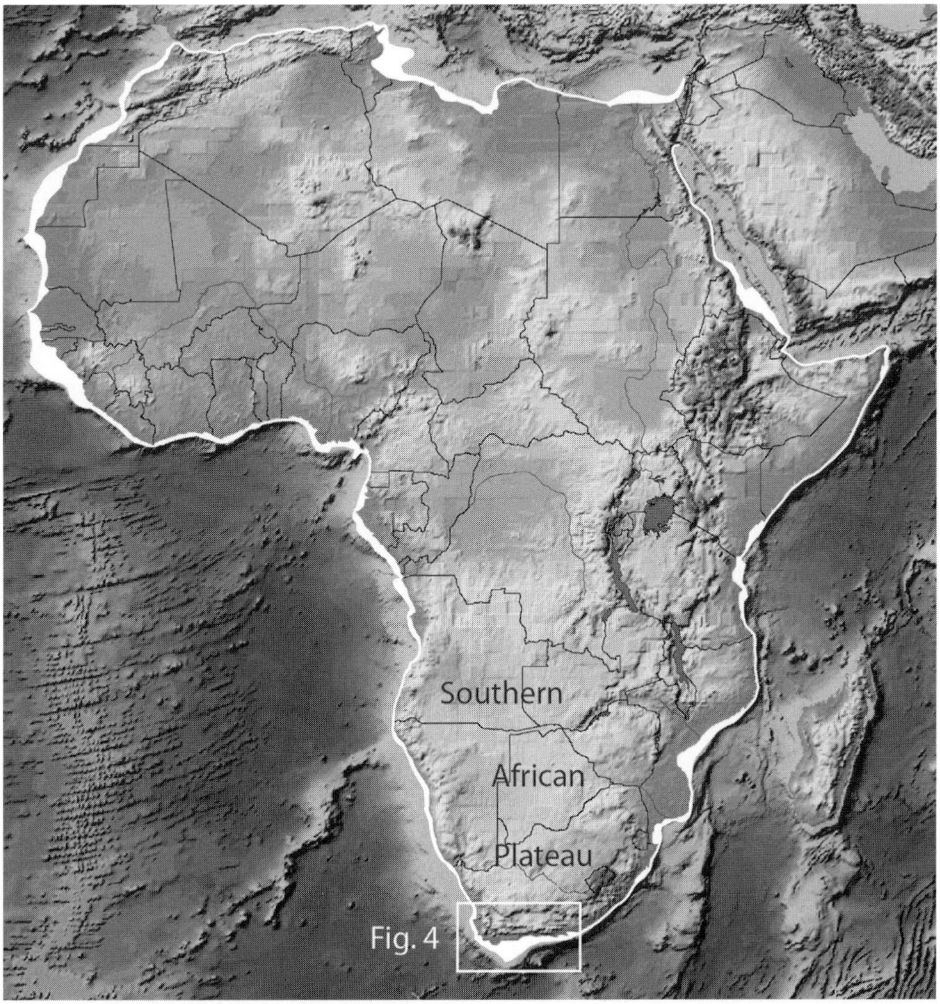

Fig. 6.1. Elevation and bathymetry of southern Africa showing the terrain of the southern African plateau and generally narrow coastal plain (data from National Geophysical Data Center, NOAA). The expansion of the coastal plain during glacial periods is shown in white.

considered expressions of dynamic topography sustained by the flow of a large mantle plume below (Lithgow-Bertelloni and Silver, 1998; Al-Hajri et al., 2014). Fringing the elevated continental interior is the Great Escarpment that along parts of the eastern and western margin marks the transition to the coastal plain. Along the southwestern and southern margins, however, the coastal plain is defined by the Cape Fold Belt (CFB), which is a series of resistant sandstone mountains that are erosional remnants of Gondwana orogenic belts (de Wit and Ransome, 1992). Today, the coastal plain ranges from non-existent where ocean waves crash against cliff faces of the CFB along the south coast, to relatively wide stretches of flat to gently rolling hilly terrain underlain by shale bedrock (Fig. 6.2). Much of the present-day coastal plain was established by the end of the Cretaceous, following the Gondwana breakup (Tinker et al., 2008), with additional topographic features established from Neogene uplift (Partridge and Maud, 1987, 2000). Neogene uplift combined with high-amplitude Quaternary sea-level fluctuations has resulted in a largely erosional coastal plain on which few Cenozoic deposits are preserved.

The width of the coastal plain has varied in the past because of fluctuations in sea level, with particularly high-amplitude (from 130 m below to 10 m above present-day sea level) and high-frequency (5–100 kyr) fluctuations over the last 0.9 million years. Global (eustatic) sea-level variations primarily reflect changes in global glacial to interglacial ice volumes (Fig. 6.3). In addition, regional sea-level variations depend on tectonic uplift/subsidence and sediment supply. Although southern Africa has remained tectonically active, the net amount of uplift at the coast, including glacioisostatic adjustments, is generally thought to have been small compared to the amplitude of glacioeustatic sea-level fluctuations during the Quaternary (Compton and Wiltshire, 2009).

Fig. 6.2. Features of the coastal plain are vertically exaggerated in this composite Landsat and radar topography image looking east from the Cape Peninsula, across the Cape Flats and False Bay to the mountains of the Cape Fold Belt. Beyond the CFB are the undulating hills of the Overberg with erosional outliers of the CFB that extend as rocky headlands separating sandy beaches along the coast (data from NASA/JPL-Caltech/NIMA). A black and white version of this figure will appear in some formats. For the colour version, please refer to the plate section.

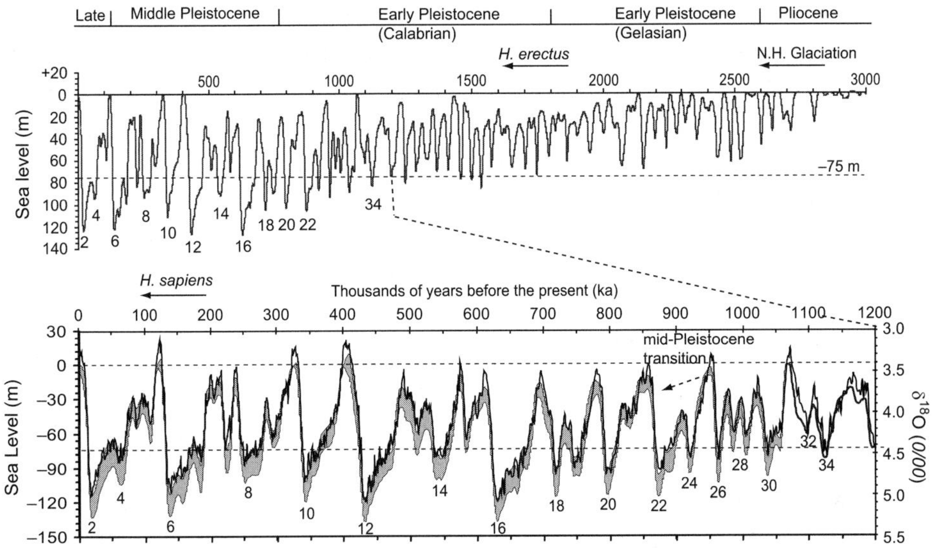

Fig. 6.3. Sea-level fluctuations derived from the composite marine oxygen isotope record of glacial to interglacial climate cycles (Lisiecki and Raymo, 2005) since 3 Ma (top) (Bintanja and van de Wal, 2008) and since 1.2 Ma (bottom) (Bintanja et al., 2005). Glacial periods are indicated by even-numbered marine isotope stages (MIS). Dashed lines mark 0 m and -75 m sea levels (after Compton, 2011).

The southern African shelf is generally sediment starved (Compton et al., 2008), but in some places sediment supply was sufficient to influence local sea level (Compton, 2006). During Quaternary interglacial periods, sea level was generally as high or up to 25 m higher than today, and the coastal plain was somewhat reduced in width. However, for most of the Quaternary, sea level was lower than today and the coastal plain was, in places, significantly wider (Fig. 6.4), depending primarily on shelf bathymetry (Fig. 6.1). The eastern South African margin has a narrow shelf that only broadens across the Natal Bight, north of Durban (Green, 2009; Green and Garlick, 2011). The western South African margin has a broad shelf but most of it is too deep to be exposed during glacials (de Wet, 2013). It is only along the southern margin that the coastal plain underwent considerable expansion during glacials by virtue of the adjacent shallow-water Agulhas Bank (Fig. 6.4).

The southern coastal plain expands approximately 10% when sea level is lowered by 45 m and 32% when sea level is lowered by 75 m, and coincides with several prominent wave-cut platforms (marine terraces) which were formed at these depths when sea level was lowered between 2.6 and 0.9 Ma (Fig. 6.3). Knickpoints up to 130 m water depth were cut during glacial maxima since 0.9 Ma when the amplitude of sea-level fluctuations increased. The width of the southern

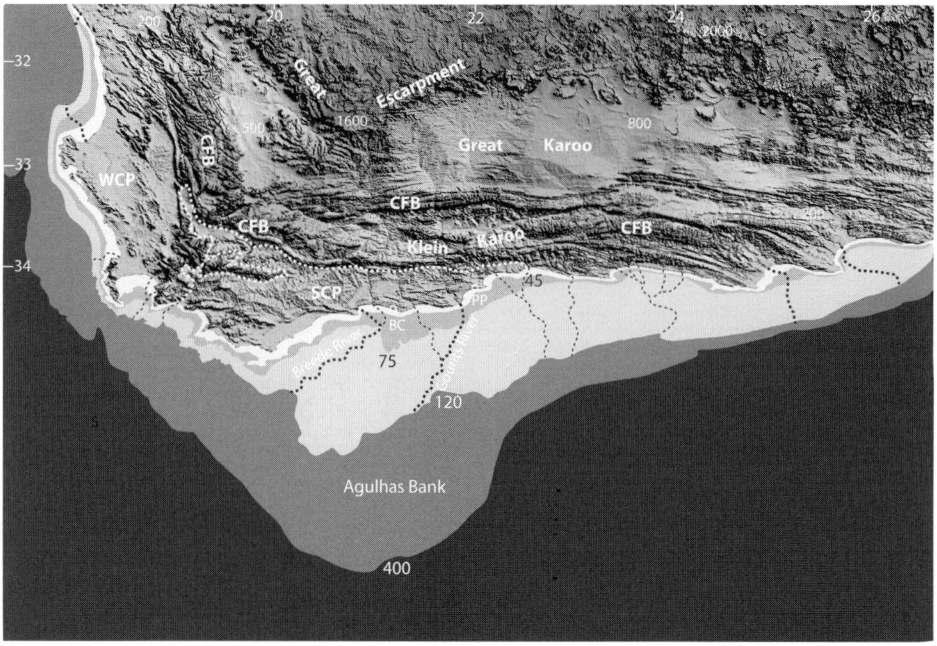

Fig. 6.4. Topography and bathymetry of the southern margin showing the Great Escarpment, Cape Fold Belt (CFB) and expansion of the western coastal plain (WCP) and southern coastal plain (SCP) when sea level is lowered by 45, 75 and 120 m (Compton, 2011). River courses are projected onto the glacial coastal plain, BC: Blombos Cave, PP: Pinnacle Point. A black and white version of this figure will appear in some formats. For the colour version, please refer to the plate section.

coastal plain expands by up to a factor of five when sea level falls from 75 to 130 m during glacial maxima (Fig. 6.5) to form one continuous southern African coastal plain (Figs. 6.1, 6.4).

In general, the coastal plain forms an erosional surface underlain by Precambrian, Palaeozoic and Mesozoic bedrock. Cenozoic deposits are dominated by aeolianite and beach deposits proximal to the present-day shoreline (Roberts et al., 2006), and represented by drowned lowstand deposits offshore with a number of prominent linear aeolianite dune ridges on the eastern and southern shelf (Martin and Flemming, 1987; Cawthra et al., 2012; de Wet, 2013). The extent and thickness of sand deposits onshore related to barrier dunes and coastal embayments would have influenced local relative sea level, and are susceptible to redistribution during sea-level cycles as the strandline migrates across the shelf (Cawthra et al., 2014). Unlike sand, mud deposits on the shelf do not appear to survive sea-level fluctuations (Compton and Wiltshire, 2009). The erosion of highstand nutrient-rich muds in the transitions to Quaternary glacial periods may have enhanced productivity in downstream upwelling regions (Compton et al., 2008).

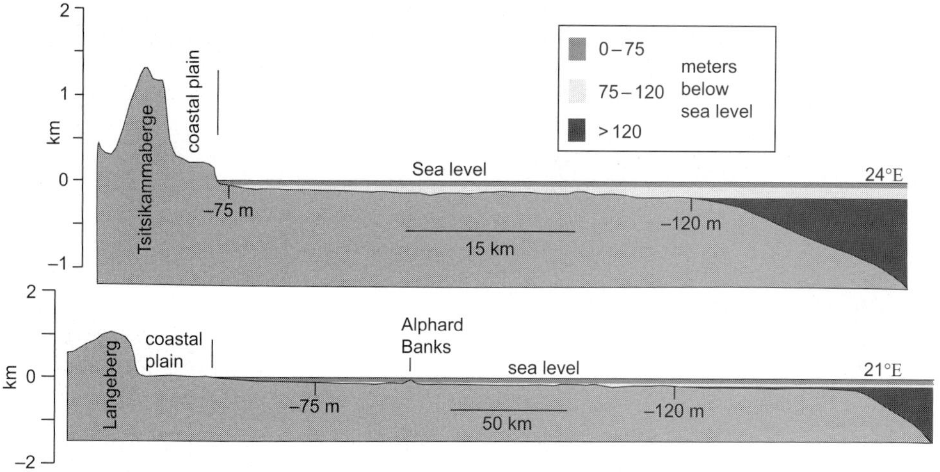

Fig. 6.5. Margin profiles at 21°E and 24°E (shown by dotted lines in Fig. 6.6), showing -75 m wave-cut knickpoints (after Compton, 2011).

6.2 Palaeoenvironments of the coastal plain

The incomplete and highly reworked deposits have made it difficult to reconstruct Quaternary palaeoenvironments of the glacial lowstand coastal plain. Past coastal systems likely consisted of a similar mix of rocky and sandy coasts as is found today. Migration of the high-energy strandline likely eroded down to a hard rock substrate in areas offshore, where limestone and phosphorite hardgrounds are overlain by only a thin veneer of Quaternary sand (Bremner and Rogers, 1990). It is more difficult to establish the palaeoenvironments inland of beach and dune coastal systems because preserved fluvial, lake and wetland records are commonly absent prior to the present interglacial (Holocene) as a result of repeated strandline migration. An alternative approach is to extrapolate today's known climates and vegetation biomes (Mucina and Rutherford, 2006) onto the glacial coastal plain (Fig. 6.6). Extrapolation of modern onshore environments onto modest shelf expansion when sea level was up to 75 m lower is reasonably constrained by known geological substrates offshore. The problem with such an extrapolation is that glacial climates were significantly different from those today (Chase and Meadows, 2007), and the substrate may have varied depending on the amount and type of sediment cover retained on the emergent shelf (Dingle and Rogers, 1972).

During the last glacial maximum when sea level fell by up to 130 m, a small expansion of modern day vegetation was likely on the east and west coasts of South Africa because the continental shelf was limited in extent. On the south coast, where the shelf is wider, it is likely that rocky headlands extended seawards and constrained vegetation expansion. For example, in the Western Cape the

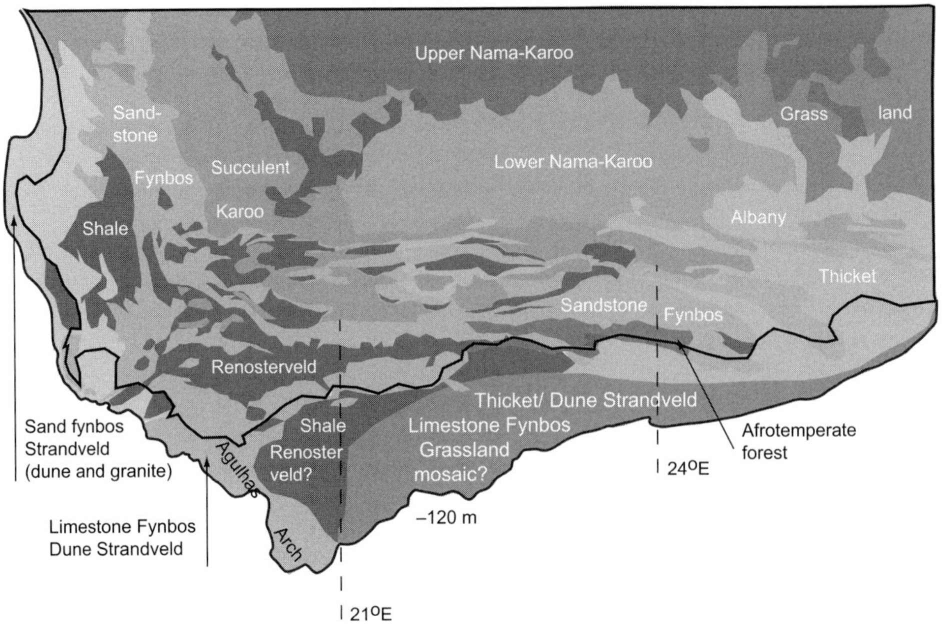

Fig. 6.6. Distribution of major vegetation biomes, simplified from Mucina and Rutherford (2006), and their proposed extrapolation onto the exposed coastal plain during periods of lowered sea level (after Compton, 2011). A black and white version of this figure will appear in some formats. For the colour version, please refer to the plate section.

resistant Table Mountain Group sandstone can be followed offshore as more subdued but distinct submarine ridges, which are separated by low-lying areas underlain by less resistant shales of the Malmesbury and Bokkeveld groups. As is the case today, fynbos biomes likely dominated sandstone and granite elevated substrates, and the Renosterveld biome in low-lying valleys underlain by shale (Fig. 6.6).

Extrapolation of vegetation biomes offshore in the area east of Cape Agulhas is more difficult because of the extensive area of the exposed coastal plain, and given that the offshore geology is less well known. Only a thin veneer of unconsolidated sediment covers much of the Agulhas Bank (Rogers, 1971) and it is likely that migration of the high-energy shoreline would have exposed the underlying Cretaceous to Neogene bedrock. However, some areas likely remained covered by aeolianites, river channel fill or shelf sand sheets not completely eroded away by strandline migration (Cawthra et al., 2014). Therefore, the exposed Agulhas Bank was probably a complex mosaic of different rock and sediment substrates. The offshore Cretaceous shale probably supported vegetation similar to Renosterveld and Albany thicket in the western and eastern areas, respectively, comparable to that observed today (Mucina and Rutherford, 2006) (Fig. 6.6). The vegetation of the more distal Paleogene and Neogene limestones with variable sand and aeolianite cover was perhaps a mix of shrub, grassland and wetland vegetation.

Glacial palaeolandscapes of the exposed coastal plain would have been far more subdued than the present-day coastal plain, reflecting the low topographic gradient of the exposed shelf (Figs. 6.2, 6.5). The exposed landscape varied from gently rolling hills and modest rocky ridges immediately offshore to expansive, low-lying terrain (flatlands) farther offshore. The flatlands were likely similar to, but more extensive than, the present-day Agulhas coastal plain and the Cape Flats. The Breede River valley was defined by a gentle, broad ridge to the east (Alphard Rise) without many topographic features besides the exceptional pinnacles formed by volcanic plugs (Alphard Banks). Rivers east of the Breede River, including the Gouritz River and many other small perennial rivers, were shallowly incised and meandering across the exposed coastal plain but did not form major incised valleys (Cawthra et al., 2014). The low topographic gradient would have resulted in extensive wetlands, lakes, estuaries and lagoons, particularly where these coastal plains had significant aeolianite dune deposits damming water into lakes, similar to the modern Wilderness Lakes area (Bateman et al., 2011; Cawthra et al., 2014). Therefore, the emergent coastal plain flatlands were likely wind-swept shrublands with dispersed wetlands or lakes landward of dune ridges and a mix of C_3 and C_4 grasses, herbaceous and riparian vegetation.

How wet the glacial-period coastal plain was would depend on the amount and seasonality of rainfall. Low-lying areas on the western coastal plain, such as the Cape Flats, form seasonal wetlands today, but low-lying areas on the southern coastal plain receive rainfall all year round and may have had perennial wetlands. Cooler sea surface temperatures are likely to have resulted in less summer rainfall, but shifting of the westerlies equatorward may have increased winter rainfall (Chase and Meadows, 2007). The speleothem record from Crevice Cave at Pinnacle Point (Western Cape) indicates that rainfall was highly variable on the southern coastal plain between 90 and 50 kyr BP (Bar-Matthews et al., 2010). The cause of variability in the Crevice Cave record is uncertain, but superimposed on the long-term glacial–interglacial climate cycles were short-term climate variations such as Heinrich events and Dansgaard–Oeschger cycles that may have had an impact on southern African climate (Zeigler et al., 2013).

6.3 Life on the edge: The coastal plain and human evolution

The major changes outlined here in the extent and palaeoenvironments of the southern African coastal plain during the Quaternary would have impacted the animals, including humans, living there. The onset of northern hemisphere glaciation around 2.6 Ma reflects the 41 kyr obliquity cycle of global climate variation, during which the build-up of continental ice lowered global sea level by up to 80 m (Fig. 6.3). The climate rhythm then shifted at the mid-Pleistocene transition from the modest 41 kyr cycle to the much larger 80–120 kyr glacial–interglacial cycle (Maslin et al., 2001). It was following the mid-Pleistocene transition that the coastal plain in southern Africa underwent periodic major expansions and contractions, responding to global glacial–interglacial cycles (Fig. 6.3).

There is increasing evidence from the archaeological record of coastal cave sites, such as at Blombos and Pinnacle Point, indicating that southern Africa was an important geographical region where hominid evolution took place (Marean, 2010). Owing to the limited expansion of the coastal plain when sea level fell by up to 75 m, these coastal caves remained within reach of the coast and its rich resources (van Andel, 1989). In addition to marine resources, the meandering rivers, wetlands and lakes of the glacial-period coastal plain would have provided a refuge for animals and humans in search of water and food resources when the interior of the continent became dry. These bodies of water would have enhanced survival during periodic droughts by providing a rich source of food when game animals were reduced farther inland. The riparian vegetation along meandering river banks and wetlands would have provided a rich source of aquatic foods as well as a local source of freshwater for those exploiting nearby intertidal marine resources.

During interglacial periods, human populations previously scattered across an extended coastal plain, would have become isolated or trapped as the coastal plain contracted, constrained by the CFB mountains (Compton, 2011). Expansion of the coastal plain during glacial periods would have facilitated access and movement, including migration to new areas, and the spread of new technologies and gene flow. Over glacial–interglacial cycles, the periodic expansion and contraction of the coastal plain may have alternately isolated and connected populations living along the coast with those in the continental interior. Relatively small groups isolated for periods of several tens of thousands of years, adapting to the unique conditions of the coastal plain habitat, may have diverged significantly from other groups in Africa and promoted rapid human evolution by allopatric speciation (Compton, 2011).

6.4 Summary

The present-day coastal plain of South Africa is limited in extent by the elevated terrain of the southern African plateau and the CFB. However, the width of the coastal plain has varied considerably over Quaternary glacial–interglacial cycles. Owing to the large, shallow-water Agulhas Bank, the southern coastal plain expanded dramatically during glacial lowstands to form a continuous and fully connected coastal plain that wrapped around southern Africa. This made circumventing the mountains of the CFB possible and its low-lying wetlands and rivers likely served as a refuge to the dry interior. The opening and closing of the expanded southern coastal plain alternately isolated and then reconnected populations to provide a potential geographical region of origin for human populations.

References

Al-Hajri, Y., White, N. and Fishwick, S. (2014). Scales of transient convective support beneath Africa. *Geology*, 37, 883–887.

Bar-Matthews, M., Marean, C. W., Jacobs, Z., Karkanas, P., Fisher, E. C., Herries, A. I. R., Brown, K., Williams, H. M., Bernatchez, J., Ayalon, A. and Nilssen, P. J. (2010). A high resolution and continuous isotopic speleothem record of paleoclimate and paleoenvironment from 90 to 53 ka from Pinnacle Point on the south coast of South Africa. *Quaternary Science Reviews*, 29, 2131–2145.

Bateman, M. D., Carr, A. S., Dunajko, A. C., Holmes, P. J., Roberts, D. L., McLaren, S. J., Bryant, R. G., Marker, M. E. and Murray-Wallace, C. V. (2011). The evolution of coastal barrier systems: A case study of the Middle–Late Pleistocene Wilderness barriers, South Africa. *Quaternary Science Reviews*, 30, 63–81.

Bintanja, R. and van de Wal, R. S. W. (2008). North American ice-sheet dynamics and the onset of 100,000-year glacial cycles. *Nature*, 454, 869–872.

Bintanja, R., van de Wal, R. S. W. and Oerlemans, J. (2005). Modelled atmospheric temperatures and global sea levels over the past million years. *Nature*, 437, 125–128.

Bremner, J. M. and Rogers, J. (1990). Phosphorite deposits on the Namibian Continental Shelf. In *Phosphate Deposits of the World, Vol. 3: Neogene to Recent Phosphorites*, eds. W. C. Burnett and S. R. Riggs. Cambridge: Cambridge University Press, pp. 143–152.

Cawthra, H. C., Bateman, M. D., Carr, A. S., Compton, J. S. and Holmes, P. J. (2014). Understanding Late Quaternary change at the land–ocean interface: A synthesis of the evolution of the Wilderness coastline, South Africa. *Quaternary Science Reviews*, 99, 210–223.

Cawthra, H. C., Neumann, F. H., Uken, R., Smith, A. M., Guastella, L. A. and Yates, A. (2012). Sedimentation on the narrow (8 km wide), oceanic current-influenced continental shelf off Durban, Kwazulu-Natal, South Africa. *Marine Geology*, 323–325, 107–122.

Chase, B. M. and Meadows, M. E. (2007). Late Quaternary dynamics of southern Africa's winter rainfall zone. *Earth-Science Reviews*, 84, 103–138.

Compton, J. S. (2006). The mid-Holocene sea-level highstand at Bogenfels Pan on the southwest coast of Namibia. *Quaternary Research*, 66, 303–310.

Compton, J. S. (2011). Pleistocene sea-level fluctuations and human evolution on the southern coastal plain of South Africa. *Quaternary Science Reviews*, 30, 506–527.

Compton, J. S., Herbert, C. and Schneider, R. (2008). Holocene organic-rich terrigenous mud on the western margin of South Africa: Nutrient source to the Southern Ocean? *Global Biogeochemical Cycles*, 23, GB4030; doi:10.1029/2008GB003427.

Compton, J. S. and Wiltshire, J. G. (2009). Terrigenous sediment export from the western margin of South Africa on glacial to interglacial cycles. *Marine Geology*, 266, 212–222.

de Wet, W. (2013). *Bathymetry of the South African continental shelf.* Unpublished MSc thesis, University of Cape Town.

de Wit, M. J. and Ransome, I. G. D. (1992). Regional inversion tectonics along the southern margin of Gondwana. In *Inversion Tectonics of the Cape Fold Belt, Karoo and Cretaceous Basins of Southern Africa*, eds. M. J. de Wit and I. G. D. Ransome. Rotterdam: A.A. Balkema, pp. 15–21.

Dingle, R. V. and Rogers, J. (1972). Pleistocene palaeogeography of the Agulhas Bank. *Transactions of the Royal Society South Africa*, 40, 155–165.

Green, A. (2009). Sediment dynamics on the narrow, canyon-incised and current-swept shelf of the northern KwaZulu-Natal continental shelf, South Africa. *Geo-Marine Letters*, 29, 201–219.

Green, A. and Garlick, G. L. (2011). A sequence stratigraphic framework for a narrow, current-swept continental shelf: The Durban Bight, central KwaZulu-Natal, South Africa. *Journal of African Earth Sciences*, 60, 303–314.

Lisiecki, L. E. and Raymo, M. E. (2005). A Pliocene-Pleistocene stack of 57 globally distributed benthic $\delta^{18}O$ records. *Paleoceanography*, 20, doi:10.1029/2004PA001071.

Lithgow-Bertelloni, C. and Silver, P. G. (1998). Dynamic topography, plate driving forces and the African superswell. *Nature*, 395, 269–272.

Marean, C. W. (2010). Pinnacle Point Cave 13B (Western Cape Province, South Africa) in context: The Cape Floral Kingdom, shellfish, and modern human origins. *Journal of Human Evolution*, 59, 425–443.

Martin, A. K. and Flemming, B. W. (1987). Aeolianites of the South African coastal zone and continental shelf as sea-level indicators. *South African Journal of Science*, 83, 507–508.

Maslin, M., Seidov, D. and Lowe, J. (2001). Synthesis of the nature and causes of rapid climate transitions during the Quaternary. In *The Oceans and Rapid Climate Change: Past, Present, and Future*, eds. D. Seidov, B. J. Haupt and M. Maslin. Washington, D.C.: American Geophysical Union, Geophysical Monograph, 126, pp. 9–52.

Mucina, L. and Rutherford, M. C. (2006). *The Vegetation of South Africa, Lesotho and Swaziland. Strelitzia 19.* Pretoria: South African National Biodiversity Institute, 807pp.

Partridge, T. C. and Maud, R. R. (1987). Geomorphic evolution of South Africa since the Mesozoic. *South African Journal of Geology*, 90, 179–208.

Partridge, T. C. and Maud, R. R. (2000). Macro-scale geomorphic evolution of Southern Africa. In *The Cenozoic of Southern Africa*, eds. T. C. Partridge and R. R. Maud. New York: Oxford University Press, pp. 3–18.

Roberts, D. L., Botha, G. A., Maud, R. R. and Pether, J. (2006). Coastal Cenozoic deposits. In *The Geology of South Africa*, eds. M. R. Johnson, C. R. Anhaeusser and R. J. Thomas. Pretoria: Council for Geoscience, pp. 585–604.

Rogers, J. (1971). *Sedimentology of Quaternary deposits on the Agulhas Bank.* Bulletin of the Joint Geological Survey/University of Cape Town Marine Geoscience Group, 1, 1–117.

Tinker, J., de Wit, M. and Brown, R. (2008). Mesozoic exhumation of the southern Cape, South Africa, quantified using apatite fission track thermochronology. *Tectonophysics*, 455, 77–93.

Van Andel, T. H. (1989). Late Pleistocene sea levels and the human exploitation of the shore and shelf of southern South Africa. *Journal of Field Archaeology*, 16, 133–155.

Ziegler, M., Simon, M. H., Hall, I. R., Barker, S., Stringer, C. and Zahn, R. (2013). Development of Middle Stone Age innovation linked to rapid climate change. *Nature Communications*, 4, 1905, doi:10.1038/ncomms2897.

7

Dating the southern African landscape

STEPHAN WOODBORNE

Abstract

The southern African landscape has been subject to denudation, sediment mobilisation, and deposition over the last 5 Ma. Dating the geological, geomorphological and archaeological evidence of these processes has been achieved through the application of different dating techniques (cosmogenic, luminescence, radiocarbon, electron spin resonance and uranium series disequilibrium). Dating evidence can address issues such as tectonic versus isostatic uplift, chemical versus physical weathering, water processes in the landscape, the age of hominid remains, and the archaeological trajectory of humans from early modernity to the present. Although dating methods are improving, many problems lie in user errors linked to inadequate understanding of the dating context and limitations of the dating methods. Confidence in chronologies is enhanced through reproducibility, stratigraphic consistency and cross-referencing between different techniques.

7.1 Introduction

Geochronology is a critical aspect of understanding geological evidence of past events and processes, and is the means by which events that took place in one location may be correlated with those elsewhere. A range of dating techniques has been developed that allows many landscape features to be dated, and in southern Africa a coherent multi-technique chronology that spans the last 5 Ma is emerging. The coherence nurtures confidence in the dating techniques, but dating is neither simple nor is it a static entity. Established dating methods are continuously improving and new techniques are emerging. An objective of this chapter is to outline the dating techniques that have been used to investigate climate and environmental changes in the southern African landscape, and equivalent changes in the archaeological record, and to provide insight into the confidence that should be placed in dating results.

It is not tenable to review all dating initiatives in southern Africa, so this chapter discusses the three main issues to which dating of the southern African landscape has been applied. (1) The first is that of erosion. The problem with erosive surfaces is that they seldom retain evidence for the processes that brought them to that state. Dating what is not preserved is a unique methodological challenge. (2) The second is the movement of sediments from one place to another, which may include aeolian, fluvial and in some cases cryogenic and gravity-driven sediment transport processes. (3) The third issue is the chronology of sediment aggradation (basin infilling) on local to sub-continental scales related to changing climates and environments. These sequences provide evidence for landscape evolution, but are also repositories of archaeological remains. Southern Africa has an almost complete record of fossil remains of early hominids through to the development of modern human behaviour. Dating techniques are thus applied to a diverse range of problems and scientific contexts.

7.2 The most commonly used dating techniques

Surface rock exposures can be dated using cosmogenic 'surface exposure' dating. This is based on the accumulation of nuclei formed from cosmic radiation interacting with the earth surface (Ivy-Ochs and Kober, 2008; Dunai, 2010). These are known as 'cosmogenic' nuclides and those most commonly used in dating rock surfaces include ^{10}Be, ^{26}Al, ^{14}C and the noble gas cosmogenic nuclides ^{38}Ar, ^{3}He and ^{21}Ne. The method is useful for dating events such as landslides that expose deep rocks to cosmic radiation. The related cosmogenic 'burial dating' method is based on the radioactive decay of cosmogenic nuclides when, after a period of surface exposure during which cosmogenic nuclides accumulate in the rock, the surface is buried beyond the influence of cosmic radiation. The age of burial is determined from the radioactive decay of these nuclides (Granger and Muzikar, 2001). Geological settings with a complex history of cosmogenic nuclide inheritance associated with cyclic burial and exposure can be dated by comparing ratios of different nuclides with different production and decay rates (Lal, 1991; Gosse and Phillips, 2001). The cosmogenic techniques can date in the range of 100 years up to 5 Ma and are applied to surface rock exposures, sediment inclusions in caves, as well as colluvial and alluvial deposits.

An alternative dating technique for colluvial and alluvial sediments is luminescence dating. This is a set of techniques that includes thermoluminescence (TL), infrared stimulated luminescence (IRSL) and optically stimulated luminescence (OSL) that are based on the accumulation of radiation damage in crystalline substances. Mineral grains of feldspars and quartz are commonly found in sediments, and exposure to sunlight during their mobilisation and deposition is often sufficient to 'bleach' or 'zero' the radiation damage. By measuring the radiation damage that has accumulated in the mineral grains subsequent to burial, and by measuring the radiation dose rate at the site of burial, it is possible to calculate the time since the sunlight exposure event (see Wintle (2010) for a detailed description of luminescence dating). Part of the versatility of the technique is the development of methods to date individual sand grains, and with this the ability to ascertain if the extent of bleaching at the time of deposition was sufficient to completely zero the luminescence signal. Since the method is based on sunlight exposure, it is a burial dating technique that can be applied to colluvial, alluvial and aeolian deposits.

An additional method based on radiation dosimetry that is applied to archaeological deposits is electron spin resonance (ESR) dating. ESR is a signal that accumulates in crystalline substances as a result of radiation damage, and has a similar theoretical basis to luminescence dating (see Grün (2007) for detailed methodological descriptions of ESR). The target material is biological, crystalline material (teeth/shell) that at the time of formation is devoid of radiation damage.

Radiocarbon dating (Libby et al., 1949) has also been used to date alluvial, colluvial and aeolian sediment landforms in southern Africa. The application assumes that contemporary organic carbon is incorporated into the sediment at the time of deposition, and that it remains *in situ* without undergoing chemical alterations through time. This assumption is seldom met in soil formation, as microbial action mobilises and alters the chemical structure of soil organic carbon (Boström et al., 2007). Provided this dynamic is considered, it is possible to date stratigraphic horizons such as palaeosols, but caution is advised in the interpretation of radiocarbon dates on unstructured soil carbon. Sediment radiocarbon dates do not date the same event as luminescence dating on the same sediments. Instead of dating the deposition of the sediment, radiocarbon analysis is dating the deposition of the soil carbon, normally derived from surface plant production, which may substantially postdate sediment deposition on stable land surfaces.

In some instances, such as the accumulation of peat deposits, the assimilation of carbon into the sediment profile tracks the rate of surface aggradation and, provided rootlet penetration into the sedimentary profile is minimal, radiocarbon ages can provide a good approximation of the age of formation of the sediments. In other instances, carbon incorporated into sedimentary features has a biogeochemical history that makes it problematic for radiocarbon dating. Inorganic carbonates, such as calcretes, are often formed as subsurface features and therefore they postdate the sediment matrix (but they may form in response to particular climatic conditions that may be the desired dating event). At the time of formation, calcretes should retain the radiocarbon signature of the soil atmosphere, which is near-contemporary with the atmosphere. Provided the carbonate has a distinct formation event, the radiocarbon method should yield an appropriate age, but soil carbonates are typically open systems in which carbonate precipitates, dissolves, migrates and re-precipitates intermittently through time. Radiocarbon dates are thus seldom a true reflection of the age of target climate event behind calcrete formation (Callen et al., 1983).

In some locations carbonate precipitation takes place under ideal conditions for dating. The precipitation of speleothems and flowstones in karstic systems normally yields a highly crystalline matrix without problems of re-precipitation or other forms of contamination for radiocarbon dating. However, the source of the carbon in karstic carbonates includes a component from a geological source that is effectively devoid of radiocarbon. This 'dead carbon' component, and also the residence time of the water in the overburden, needs to be taken into account when interpreting karstic carbonate radiocarbon analysis. The limitation of radiocarbon dating in this application is the applicable age range. Many karstic carbonate deposits of interest in southern Africa are older than 50 kyr. The stability of the crystal structures that make them suitable for radiocarbon analysis also make them suitable for uranium series disequilibrium (U/Th) dating and, for older specimens, uranium-lead (U/Pb) dating. Uranium is highly soluble relative to Th and thus carbonate precipitates should contain no Th at the time of formation. The U/Th dating technique is based on the radioactive decay of ^{238}U and ^{234}U into ^{234}Th and ^{230}Th respectively (see Zhao et al. (2009) for a detailed description of the technique). The eventual decay product of U is stable lead (Pb) and the measurement of U/Pb systematics allows dating in over billions of years time scales (Wilde et al., 2001), but it can also be applied to younger materials (Pickering et al., 2010). Cave deposits in southern Africa contain globally important palaeoanthropological fossils and archaeological remains, and it is fortunate that most of the dating techniques described earlier can be applied to provide appropriate chronological controls.

7.3 Dating the evolution of the southern African landscape

The geological map of southern Africa portrays a landscape that resulted from a generally erosive regime, and has been in much the same state over the last 5 Ma. The evolution of this landscape is described by McCarthy and Rubidge (2005).

7.3.1 Denuding landscapes

The age of a landscape that is undergoing erosion is notional: a finite erosion rate remains finite however short the measurement time, and an eroding landscape is therefore always infinitesimally young. Instead of reporting an age, it is more accurate to calculate the rate of denudation, and cosmogenic dating has been applied to quantify this rate. Kounov et al. (2007) report denudation rates for the interior of South Africa of 1.5–3.0 m/Ma while Decker et al. (2011) also report rates of <4.0 m/Ma. Landscapes that denude at >5 m/Ma are typically subject to *in situ* weathering rather than fluvial erosion (Baldwin et al., 2003). This is supported by Decker et al. (2013) who compared denudation rates for Karoo dolerite surfaces of <4.0 m/Ma using a climate-dependent weathering model and concluded that high weathering rates limit landscape denudation. However, Decker et al. (2011) reviewed the available data for erosion rates and concluded that weathering is presently happening at a rate that is orders of magnitude slower than the rate of soil loss. These results indicate that chemical weathering and physical erosion of the southern African landscape must be closely matched on a very slowly denuding land surface. This is illustrated in Kruger National Park where the denudation rate across a rainfall gradient is more-or-less constant at 3–6 m/Ma, but that the degree of landscape dissection is differentiated by the relative contribution of chemical versus physical erosion (Chadwick et al., 2013).

The role of chemical versus physical erosion of a land surface is mediated by water, and where relief is steep, it is anticipated that physical erosion would dominate. This is verified in the high Drakensberg which is eroding at a rate of 6 m/Ma (Flemming et al., 1999) while the Gamsberg, which is subject to a much drier regime, is reducing at only 0.4 m/Ma (Cockburn et al., 2000). The retreat of the Gamsberg escarpment is taking place at about 10 m/Ma while the Drakensberg escarpment is retreating at 50–95 m/Ma. The hyper-arid Namib Desert below the escarpment in Namibia is denuding at 0.5–1.0 m/Ma whilst a wetter period commencing at ~2.8 Ma brought about the incision of prominent canyon features in this landscape (van der Wateren and Dunai, 2001).

High relief and high erosion rates are typically coupled (Portenga and Bierman, 2011) and accordingly ancient landscapes should self-level in the absence of faulting or tectonic uplift that alters the dynamics. The mountains of the southern Cape are enigmatic because they possess high relief and steep slope gradients and yet they are eroding at about 5.2 m/Ma, approximately 100 times slower than regions of the Alps and the Andes ranges with similar relief (Scharf et al., 2013). This is attributed to the chemically inert and physically robust rocks that make up the orogen, and accordingly implies that the denudation is not weathering-limited. Although rock mechanics explain the slow erosion rate, Scharf et al. (2013) also report low relief dependence of denudation, as both interfluves and slopes were eroding at the same rate. The implication is that the southern Cape mountain relief is not self-levelling, but that all aspects of the relief are eroding at the same rate, and accordingly that there has been tectonic and topographic stability through the

Cenozoic (Scharf et al., 2013). In contrast to this, Bierman et al. (2014) report that pediment surfaces in the same region are eroding at a rate that is approximately an order of magnitude lower than the 'landscape as a whole'. The maintenance of steep relief is attributed to a continental margin effect in which uplift is of isostatic rather than tectonic origin. As the landscape loses mass through erosion, it rises, and in the case of the southern Cape mountains this slow isostatic uplift is sufficient to maintain and even slowly enhance the characteristic relief of the region (Bierman et al., 2014). Comparing the long-term incision rate of the Sundays River over the last 4 Ma (16.1 m/Ma) with a 4.26 Ma elevated marine terrace near Durban, yields an isostatic uplift rate of 9.4 m/Ma (Erlanger et al., 2012), confirming that isostasy and not tectonic uplift is the dominant mechanism in southern Africa during the Neogene. A common theme in the study of landscape denudation is that denudation rates are highest in the wettest areas. This has also been demonstrated on short time scales (e.g. Tooth et al., 2002, 2004; Rodnight et al., 2005; Keen-Zebert et al., 2013), although these conclusions were reached on the basis of luminescence and radiocarbon dating of resulting sediments rather than on landscape erosional features themselves.

7.3.2 Aggrading deposits

The sedimentary by-product of landscape denudation is transported at different times by colluvial, fluvial and aeolian processes. All of these processes lend themselves to luminescence dating, as the transport process usually presents some degree of sunlight exposure that may be sufficient to re-set the luminescence clock to zero.

One of the earliest applications of luminescence dating in southern Africa was on the colluvial succession at St. Paul's Mission in KwaZulu-Natal (Wintle et al., 1993). This was done using the IRSL technique on the fine (<10 μm) sediment fraction, but subsequent comparisons with radiocarbon dates from the same site (Wintle et al., 1995a) demonstrated that colluvial transport was inadequate to fully zero the luminescence signal. A re-analysis of larger sediment grains using TL and IRSL dates, which have different bleaching characteristics, as well as radiocarbon dates, concluded that the IRSL method yielded more accurate ages. The results suggested that episodic deposition took place 100 kyr, 56–50 kyr, post-45 kyr, and within the last 2 kyr (Wintle et al., 1995b). The St. Paul's study illustrates the complexity that besets application of luminescence dating as a result of inadequate bleaching of the luminescence signal during sediment deposition, and the necessity to independently validate (here using radiocarbon dating) if results reflect the true age of the target event. Since this study, OSL methods that make use of a very rapidly bleaching luminescence signal have been developed, enabling colluvial processes to be more accurately dated. For example, Lyons et al. (2013) dated colluvial and donga (gully) formation processes between 22 and 1.6 kyr BP and concluded that the most recent period of donga formation (since 890 years BP) is likely to have been driven by climate change and not poor land management practices.

Although not strictly aggrading deposits, different river terraces in the Sundays River, Eastern Cape, have been dated to between 4.0 and 0.3 Ma using cosmogenic dating (Erlanger et al., 2012). The same technique suggests that the oldest alluvial deposits in the Vaal River (Rietputs Formation) are approximately 1.57 Ma (Gibbon et al., 2009). This deposit contains Early Stone Age (Acheulean) hand-axes that are presumed to have concentrated through alluvial and colluvial processes, and as a secondary deposit the age is presumed to be a minimum estimate. Further luminescence dating of alluvial deposits has focused on extreme flood event histories (Boshoff et al., 1993; Benito et al., 2011; Grodek et al., 2013), or the geomorphological controls on landscape evolution through alluvial processes (Tooth et al., 2002, 2004, 2013; Bourke et al., 2003; Eitel et al., 2005, 2006; Rodnight et al., 2005; Marren et al., 2006; Srivastava et al., 2006).

When rivers reach the sea, the fate of the sediment load is determined by ocean current and wave energetics. The feature of interest here is the almost ubiquitous terrestrial sand dunes that are found around the coast of southern Africa. The deposition of these dunes involves a period of aeolian transport that almost guarantees that the sand grains were exposed to sufficient sunlight at the time of deposition, and that the luminescence signal is completely zeroed. Accordingly, dunes are easy to date using luminescence techniques and their chronological patterns reflect climatic as well as sea-level changes through time. Luminescence dating of coastal dunes also established the ages of intra-formational human footprints preserved in the Nahoon dunes (124 kyr BP; Jacobs and Roberts, 2009) and Langebaan dunes (~120 kyr BP; Roberts, 2008), or the similar aged anatomically modern femur found in the dunes near Blind River (Wang et al., 2008). Most dating of coastal dunes is aimed at determining dune dynamics and formation history, and has been undertaken in Mozambique (Armitage et al., 2006), KwaZulu-Natal (Botha et al., 2003; Sudan et al., 2004; Cawthra et al., 2012), the Southern Cape (Carr et al., 2006, 2007, 2010) and Western Cape (Fuchs et al., 2008; Roberts et al., 2009).

Alluvial deposits in the Namib Desert have been dated at several locations to <130 kyr BP but most date within the last 50 kyr and are related to Pleistocene climate changes in precipitation (Stone and Thomas, 2013). The residence time of sand in the Namib Desert is estimated at 1 Ma based on cosmogenic dating (Vermeesch et al., 2010), but the dynamics of sand dune migration yields much younger luminescence dates for specific sand bodies. The age-frequency of sand dunes was thought to indicate ubiquitous climate changes through time (e.g. Stokes et al., 1997) but this was an artefact of inadequate sample densities, and age gaps have subsequently been filled (Stone and Thomas, 2013; Thomas and Burrough,

2013). Detailed age mapping within dunes, combined with structural evidence from ground penetrating radar, demonstrates that dune ages are controlled by the pace and direction of their migration (Stone and Thomas, 2013; Thomas and Burrough, 2013). The maximum age is determined by the reconstitution time (in which 100% of the sand body has moved as part of the dune migration). Luminescence dating is now questioning the underlying assumption that dune turnover reflects landscape-wide climate forcing. Examining lacustrine (Burrough and Thomas, 2008) and fluvial deposits within dune systems (Stone and Thomas, 2013), that are dated using the same approach as that used for dunes, may be a more direct way to identify the palaeo-moisture regime.

7.3.3 Palaeoanthropology and archaeological deposits

The southern African landscape preserves some of the richest hominid fossil-bearing deposits in the world, and in addition a record of human development is preserved in the Early Stone Age (ESA), Middle Stone Age (MSA), and Later Stone Age (LSA) archaeological records. Initial dating of hominid fossil sites relied on comparisons between the associated fauna, and the well-dated faunal assemblages from east Africa (e.g. Vrba, 1975). The first attempt to obtain a more direct date for the Sterkfontein fossils was the application of ESR dating that yielded a date of 2.1 Ma (Schwarcz et al., 1994). The prominence of the fossil StW 573 (Littlefoot) discovered at Sterkfontein Cave demanded better dating, and Partridge et al. (2003) generated an age of 4.02 Ma using cosmogenic dating constrained by evidence for palaeomagnetic reversals at the site. Using the same approach, the age of *Australopithecis sediba* from the Melapa site was assigned an age of 1.95–1.78 Ma (Dirks et al., 2010) and was later revised to 1.977 Ma (Pickering et al., 2011). Subsequently, the application of U/Pb dating yielded an age of 2.2 Ma for the flowstone associated with the Littlefoot fossil (Pickering and Kramers, 2010). Some reconciliation between the cosmogenic and U/Pb dates can be reached by using an updated half-life for ^{10}Be (Partridge, 2005), but the two dating approaches still yield different results.

Despite the abundance of Acheulian hand-axes in the ESA, there are very few locations at which they have been dated. The assemblage in the Vaal River gravels predates 1.57 Ma based on cosmogenic dating (Gibbon et al., 2009). The only site at which the ESA is directly dated is at Wonderwerk Cave where palaeomagnetic dating, luminescence and cosmogenic dating yielded an age range of 1.96–0.78 Ma (Chazan et al., 2008). The Fauresmith Industry, considered to be transitional between the ESA and MSA, is dated to 542–464 kyr BP using luminescence and combined U-series/ESR dating at Wonderwerk Cave (Porat et al., 2010). This appears to predate the late Acheulian assemblage from Duinefontein II, which has IRSL dates of 292 kyr and

265 kyr BP (Feathers, 2002). These apparently inverted dates challenge the perception that the ESA transition to the MSA was a temporally coherent phenomenon, while also raising questions regarding the technological classifications that have been applied in assigning stone assemblages to different industries.

Problems with early attempts to date MSA sites in southern Africa arose because the deposits were older than the 50 kyr limit of radiocarbon dating. One of the first MSA stone tool technological complexes identified was the Howiesons Poort (named after the type-site) which yielded a finite radiocarbon date of 29,400±675 ^{14}C BP at Diepkloof Cave (Parkington and Poggenpoel, 1987). This date was interpreted literally in the absence of any other dated assemblages from this time period (Parkington, 1990). In contrast, the equivalent assemblage from Border Cave was dated by extrapolation of deposition rates of the uppermost layers at the site that were radiocarbon dated. This yielded an age estimate of 95 kyr for the Howiesons Poort and 195 kyr for the base of the MSA deposits (Butzer et al., 1978). The date was treated with scepticism because of the implied age of 115–90 kyr BP for the human remains recovered at the site. However, the application of isoleucine epimerisation dating to the Howiesons Poort industry at Border Cave, Apollo 11 Cave, and Boomplaas Cave placed the industry in the age range 80–56 kyr BP with a most likely age of 66±5.0 kyr BP (Miller et al., 1999). Further application of U/Th dating at Boomplaas and Klasies River Mouth placed the Howiesons Poort at 70–60 kyr BP (Vogel, 2001).

The emerging antiquity of the Howiesons Poort and the MSA from southern Africa more generally has important global implications. In particular the recovery of anatomically modern human remains from Border Cave and Klasies River Mouth as well as subsequent discoveries of evidence for modern cognitive behaviour at Blombos Cave (Henshilwood et al., 2002) highlighted the significance of this geochronometric record. Subsequently, Howiesons Poort sediment at Rose Cottage Cave was dated to 70–60 kyr BP (Pienaar et al., 2008) and burned lithics from the same site to 60.4–56.3 kyr BP (Valladas et al., 2005). At Klasies River Mouth, Howiesons Poort sediments were dated to 60–50 kyr BP (Feathers, 2002) and 56±3.0 kyr BP from the same site using burned quartz samples (Tribolo et al., 2005). At Diepkloof Shelter, luminescence dating yields Howiesons Poort ages of 65–55 kyr BP (Tribolo et al., 2005). Sibudu Cave yielded luminescence dates of 61.1±1.5 kyr BP (Wadley and Jacobs, 2006) and Blombos Cave yielded minimum ages of 70.1±1.9 kyr BP (Jacobs et al., 2003a) and 68.8±3.0 kyr BP (Jacobs et al., 2003b).

While the luminescence dating of the Howiesons Poort appears to be converging, there remains a lack of precision that results in a chronological 'haze' (Jacobs et al., 2008). Part of this is attributed to the different assumptions and laboratory protocols implemented by different practitioners, and accordingly systematic re-dating of sites containing Howiesons Poort and the preceding Still Bay industries was conducted at a sub-continental scale. This indicates that the Howiesons Poort is 64.8–59.5 kyr BP and this is clearly distinguished from the Still Bay of 71.9–71.0 kyr BP (Jacobs et al., 2008). To a large degree, this result is corroborated by ESR and radiocarbon dates from Border Cave (Grün and Beaumont, 2001; Bird et al., 2003).

The transition from the MSA to the LSA in southern Africa is characterised by stone tool assemblages that are poorly defined and are assigned to the umbrella Middle Stone Age III (MSA III) and the Early Later Stone Age (ELSA). Although such deposits occur at many of the stratified MSA sites, spatial inter-comparisons are hampered by the lack of typologically diagnostic stone tool assemblages. The date brackets for this transition is ~55–37 kyr BP at Border Cave (Grün and Beaumont, 2001; Bird et al., 2003), and ~52–26 kyr BP at Sibudu Cave (Wadley and Jacobs, 2004).

The LSA in southern Africa has been recovered from many well-stratified sites (including many of the MSA sites mentioned above), and better preservation of appropriate organics in these deposits has allowed accurate and precise application of radiocarbon dating. A key element here has been the improvement in the calibration dataset. In 1998 the calibration of radiocarbon dates up to 10.3 kyr BP was based on a robust tree ring chronology, and alternative datasets extended the calibration to 24 kyr with increasing levels of uncertainty (Stuiver et al., 1998). Dates older than 24 kyr could not be calibrated. By 2004 the calibration limit was pushed back to 26 kyr BP (Reimer et al., 2004), but by 2013 reasonable calibrations became possible for the full 50 kyr range of the technique (Reimer et al., 2013). A southern hemisphere radiocarbon reservoir offset demanded a specific calibration dataset (Hogg et al., 2013), and this offset was factored in to the early Pretoria calibration (Vogel et al., 1993) that was employed by most radiocarbon users in southern Africa. Many of the dates from the earlier part of the LSA that were inadequately calibrated could later be placed into an absolute dating reference when the calibration datasets were extended. Together this led to high spatial and temporal resolution in the LSA for southern Africa, and the chronology for the last 30 kyr is relatively robust.

The Iron Age in southern Africa is the part of the archaeological heritage that is most relevant to the movement and economies of the peoples that eventually led to the current human geography of the region. It has powerful political ramifications in the identities, cultures and land rites in the region. Dating has mainly been accomplished using radiocarbon dating with exceptional precision. A summary of Iron Age dating is provided by Huffman (2007) where the association between traditions of pottery design and motifs is correlated with age. The use of pottery motifs as an age-indicator in the Iron Age eclipses radiocarbon dating because of its ease of application, but it is still necessary to acknowledge the radiocarbon dates on which the pottery chronology is based. This presents problems in the dating of certain sites such as furnaces where ceramics may not be recovered and where the problem of old wood may influence the use of radiocarbon dating. Luminescence dating may offer an alternative dating technique provided the dose rate for the site can be sufficiently constrained, and attempts to do this have been undertaken, but have not yet been published.

Within the Iron Age, the radiocarbon dates allow sufficient precision such that they can be analysed in their own right. Huffman (2010) considered the dating evidence for the burning of cattle byres (kraals) and, combined with ethnographic evidence, showed that such practices are associated with severe droughts, identifying 13 episodes of kraal burning across the subcontinent over the last 2000 years. These were correlated with El Niño events from southern Ecuador to suggest a common climatic origin. Although the Iron Age dates from southern Africa can be analysed in this way when aggregated, there are often problems with inadequate dating within individual sites. Examples of this are highlighted below.

7.3.4 *Palaeoclimate*

There are a number of depositional sequences in southern Africa that yield evidence of past climate. Some of these, such as speleothems, have excellent chronological control. The most prominent speleothem records are the 25 kyr record from Cold Air Cave (Holmgren et al., 1994, 2003), the 50 kyr record from Cango Cave (Talma and Vogel, 1992) and the 90–50 kyr Pinnacle Point record (Bar-Matthews et al., 2010). In each case, speleothems have been dated using U-series disequilibrium dating, and this technique does not commonly yield age inversions unlike many sedimentary sequences dated using radiocarbon. The 20 kyr Wonderkrater spring mound record yielded a number of radiocarbon age inversions (Scott et al., 2003), which is likely to be a function of the carbon systematics on the site. In other applications, such as the rock hyrax midden palaeoclimate records that extend back to 20 kyr BP (Chase et al., 2009, 2010, 2011), radiocarbon provides a coherent chronology. Dating of the Tswaing impact crater record has also been improved by a radiocarbon chronology that extends to 50 kyr BP, but this only covers the deposit to about 18 m depth in a total sequence of 80 m, and the basal date is fixed on a single fission-track date of 220 kyr BP (Kristen et al., 2007). The original age model used the basal date and radiocarbon dates to establish an accumulation rate, and relied on orbital tuning to refine the chronology (Partridge et al., 1997). There is clearly scope for improved dating resolution in the 50–220 kyr BP range.

7.4 Discussion

The chronology of both landscape and human evolution in southern Africa appears to be well-constrained and in places and at times probably very accurate. There are also examples of chronologies that are entirely fallacious that have led to incorrect interpretations of the past. These examples can guide future research by highlighting the most common shortcomings of different dating techniques, and a critique of past applications of dating can also present a checklist against which to assess new emerging evidence.

7.4.1 Interpretation of the dating context

Despite the maturity of some dating techniques, anomalous ages still emerge either within sequences that are dated by a single method, or between results for a single 'event' obtained using different methods. An example is a radiocarbon date of 28,880±170 ^{14}C years BP on a notched bone recovered at Sibudu Cave, but it is attributed to an age of 61.3±2.0 kyr BP based on the OSL method (Wadley and Jacobs, 2006). Another radiocarbon date from the same site of 34,300±2000 ^{14}C years BP on charcoal is rejected for the same reason, whilst another of 24,200±290 ^{14}C years BP is rejected because of an inverted age sequence (Wadley and Jacobs, 2004). These examples illustrate the problem of presuming which dates are valid and which are invalid. The Sibudu Cave authors presume that there is an error in the radiocarbon analysis, arguing that the standard pre-treatment of samples may not adequately remove contaminants that accumulate in charcoal (see Bird et al. (2014) for discussion of this protocol). They claim that 'Until multiple radiocarbon ages have been obtained for the same layer from samples pre-treated in an appropriate way, as was the case at Border Cave (Bird et al., 2003), meaningful comparisons cannot be made to other dating techniques, such as OSL dating' (Wadley and Jacobs, 2006, p5). The cited 'case at Border Cave' is directly comparable with the radiocarbon dates at Sibudu Cave. Radiocarbon dates produced for Border Cave by the same laboratory that provided the rejected Sibudu dates and using the same pre-treatment protocols, fell in a range 37.7±0.6 kyr to 39.8±0.6 ^{14}C kyr BP, while the specialised pre-treatment that is presumed to distinguish the Border Cave results yields a date of 38.5 ^{14}C kyr BP (Bird et al., 2003). Thus, pre-treatment of the samples may be rejected as a significant source of error. The rejected Sibudu date of 34,300 ^{14}C years BP calibrates to 34–39 kyr BP (rounded to the nearest thousand years, using the calibration of Hogg et al. (2013)). This is the age of the immediately overlying sediments (dated by OSL) at Sibudu Cave, and it is very likely that the charcoal that yielded the rejected date may be derived from the younger occupation of the site.

Whereas the Sibudu example centres on the question of rejecting dates, the provenance of dates that are accepted is also a critical part of the chronological framework. Recently, Chirikure et al. (2013) proposed a revision of the chronology for Great Zimbabwe using a Bayesian approach to the existing dataset of radiocarbon dates. They make use of the 2004 southern hemisphere calibration dataset

(McCormac et al., 2004) and presume that this is better than the calibration used by Huffman and Vogel (1991) which involved using the northern hemisphere calibration curve in a southern hemisphere application. Chirikure et al. (2013) argue from their re-analysis that the emergence of Great Zimbabwe as a regional capital took place around AD 1225, coeval with Mapungubwe. A key element here is identifying the correct stratigraphic position of stone wall structures that is indicative of the development of the site within the dating sequence. While Chikure et al. (2013) place this interface between the stratigraphic units called ‘Period II’ and ‘Period III’, Huffman (2015) places it between ‘Period IVa’ and ‘Period IVb’. The problem that leads to this discrepancy in the Iron Age chronology of southern Africa is not the result of ‘dating hygiene’ as suggested by Chirikure et al. (2013), but rather a problem of archaeological interpretation.

These examples suggest that greater confidence should be placed on the veracity of measured dates, provided that the context of the sample is fully understood. However, it is possible that contradictory dates will arise when the dating technique links to the target event via a set of assumptions that are not necessarily applicable in the given context.

7.4.2 Insight into the suitability of the dating technique to the target event

Early applications of radiocarbon dating to pedogenic calcretes were compared with the age results obtained with TL and U/Th dating (e.g. Geyh and Eitel, 1998). This analysis emphasised that each method was dating a different formational event, but there has commonly been a literal and uncritical interpretation of radiocarbon ages derived from calcrete deposits. Many of these ages have been overturned with the advent of luminescence techniques and this has led to after-the-fact accommodation of the problems associated with radiocarbon dating of pedogenic carbonates. In this instance it is not possible to ascertain if radiocarbon dates are accurate unless an independent ageing technique is applied, because under ideal circumstances the radiocarbon dates may be indeed correct (Geyh and Eitel, 1998). The consistency of radiocarbon dates within a sequence, and the consistency of associated data, is often used as a defence of the radiocarbon method on pedogenic carbonates. The dating of *heuweltjies* (round pedological features up to 5 m in diameter) in the western Cape has yielded systematic age/depth relationships, and the associated isotopic values of the carbonate are believed to retain a climatic signal, implying minimal post-depositional alteration of the carbonate matrix (Midgley et al., 2002; Potts et al., 2009). In such instances, both radiocarbon ages and isotopic values should be treated with caution until independent data confirm these. Radiocarbon dating of glacial and periglacial features in the Drakensberg to the last glacial period (Mills and Grab, 2005) has a similar criticism. These authors argue why the radiocarbon method is dating the target event, but the range of influences on soil carbon that might nullify the result cannot be disproven. In this instance, the results should be verified by alternative dating techniques such as cosmogenic 'burial dating'.

In any circumstance in which multiple dating techniques are applied, there exists a possibility that discrepancies will arise in the results. These may be because of systematic errors inherent in the technique, or that the techniques are dating different events. In the foregoing discussion, radiocarbon dating was demonstrated to be problematic in some applications, but the radiocarbon dating community has regular blind inter-laboratory performance comparisons using known-age samples, and there is a high level of consistency of results between laboratories (Scott et al., 2003). However, this confidence does not extend to all dating techniques (see Jacobs et al., 2008) and thus multiple dating techniques can be applied to situations in which radiocarbon is known to yield a good chronology. For example, at Rose Cottage Cave both radiocarbon and luminescence dating yielded very similar results (Pienaar et al., 2008). Whereas all the radiocarbon dates used in that study provided a coherent increase in age with depth, it was the luminescence dates that yielded age inversions. Examples like this raise queries about the assumptions that are made in different dating techniques, such as the effects of changing radiation dose rates used in dosimetry-based techniques like luminescence and ESR dating.

7.4.3 Reproducibility, consistency, and a body of evidence

Identifying erroneous dates usually only emerges from age inversions, or when different techniques produce different dates on the same target event. *Post hoc* accommodation is the most common means of rejecting erroneous dates, but it relies on the overwhelming evidence from other consistent dates. For example, the large number of luminescence dates for dunes in the Namib and Kalahari deserts (Stone and Thomas, 2013; Thomas and Burrough, 2013) led to a paradigm shift in the use of dunes as a palaeoclimate proxy. Similarly, a total of 71 ESR dates was used to provide the chronology of different MSA layers at Border Cave (Grün and Beaumont, 2001; Millard, 2006). This is in contrast to the dating of highly significant sites such as Mapungubwe and Great Zimbabwe. Mapungubwe was dated with only 24 radiocarbon dates with an additional 7 on gold artefacts from the site (Woodborne et al., 2009), and Great Zimbabwe has only 21 radiocarbon dates (Chirikure et al., 2013). Thus, the emerging chronology requires many more dates in order to reject inappropriate interpretations. While reproducibility and consistency in age determinations within a sequence, and between the same context in different locations, can provide grounds for the rejection of outliers, this approach is based on the precision of the body of evidence, and rejecting of dates infers knowledge about the accuracy (inaccuracy) of the rejected date. It is more appropriate to flag these dates for closer scrutiny in order to identify independent reasons why the date is deemed inaccurate.

7.5 Summary

The confidence with which the southern African landscape is dated is related to the concept of 'dating hygiene' (Spriggs, 1989) which refers to how and why dates may be incorrect, or should be rejected in order to avoid spurious chronologies (see Mulrooney (2013) for a good example of this). While it is necessary to believe that laboratories produce good results, a 'dating hygiene' criterion is to trust only laboratories that are able to verify the quality of their analyses. Interlaboratory benchmarking in radiocarbon dating is an example of this, but it is also needed in the field of luminescence dating (Jacobs et al., 2008). However, the major hygiene factor that affects the reliability of dates in southern Africa is the accuracy and precision of the context from which the samples have been recovered. The 'dating hygiene' concept is useful to avoid the perpetuation of dating myths, and it is important that researchers understand the range of dating techniques available, to comprehend the limitations of these techniques, and select those that are most appropriate for their circumstance. The range of different dating techniques applied to the southern African landscape reflects its diversity of evolutionary stages, environments, and the episodic impacts of human activities over long time periods.

References

Armitage, S. J., Botha, G. A., Duller, G. A. T., Wintle, A. G., Rebêlo, L. P. and Momade, F. J. (2006). The formation and evolution of the barrier islands of Inhaca and Bazaruto, Mozambique. *Geomorphology*, 82, 295–308.

Baldwin, J. A., Whipple, K. X. and Tucker, G. E. (2003). Implications of the shear stress river incision model for the timescale of postorogenic decay of topography. *Journal of Geophysical Research*, 108, 2158, doi:10.1029/2001JB000550.

Bar-Matthews, M., Marean, C. W., Jacobs, Z., Karkanas, P., Fisher, E. C., Herries, A. I. R., Brown, K., Williams, H. M., Bernatchez, J., Ayalon, A. and Nilssen, P. J. (2010). A high resolution and continuous isotopic speleothem record of paleoclimate and paleoenvironment from 90 to 53 ka from Pinnacle Point on the south coast of South Africa. *Quaternary Science Reviews*, 29, 2131–2145.

Benito, G., Thorndycraft, V. R., Rico, M. T., Sánchez-Moya, Y., Sopeña, A., Botero, B. A., Machado, M. J., Davis, M. and Pérez-González, A. (2011). Hydrological response of a dryland ephemeral river to southern African climatic variability during the last millennium. *Quaternary Research*, 75, 471–482.

Bierman, P. R., Coppersmith, R., Hanson, K., Neveling, J., Portenga, E. W. and Rood, D. H. (2014). A cosmogenic view of erosion, relief generation, and the age of faulting in southern Africa. *GSA Today*, 24 (9), 4–11.

Bird, M. I., Fifield, L. K., Santos, G. M., Beaumont, P. B., Zhou, Y., di Tada, M. L. and Hausladen, P. A. (2003). Radiocarbon dating from 40 to 60 ka BP at Border Cave, South Africa. *Quaternary Science Reviews*, 22, 943–947.

Bird, M. I., Levchenko, V., Ascough, P. L., Meredith, W., Wurster, C. M., Williams, A., Tilston, E. L, Snape, C. E. and Apperley, D. C. (2014). The efficiency of charcoal decontamination for radiocarbon dating by three pre-treatments – ABOX, ABA and hypy. *Quaternary Geochronology*, 22, 25–32.

Boshoff, P., Kovacs, Z., Van Bladeren, D. and Zawada, P. K. (1993). Potential benefits from palaeoflood investigations in South Africa: Technical note. *Journal of the South African Institution of Civil Engineers*, 35, 25–26.

Boström, B., Comstedt, D. and Ekblad, A. (2007). Isotope fractionation and ^{13}C enrichment in soil profiles during the decomposition of soil organic matter. *Oecologia*, 153, 89–98.

Botha, G. A., Bristow, C. S., Porat, N., Duller, G., Armitage, S. J., Roberts, H. M., Clarke, B. M., Kota, M. W. and Schoeman, P. (2003). Evidence for dune reactivation from GPR profiles on the Maputaland coastal plain, South Africa. In *Ground Penetrating Radar in Sediments*, ed. C. S. Bristow and H. M. Jol. London: Geological Society of London, Special Publications, 211, pp. 29–46.

Bourke, M. C., Child, A. and Stokes, S. (2003). Optical age estimates for hyper-arid fluvial deposits at Homeb, Namibia. *Quaternary Science Reviews*, 22, 1099–1103.

Burrough, S. L. and Thomas, D. S. G. (2008). Late Quaternary lake-level fluctuations in the Mababe Depression: Middle Kalahari palaeolakes and the role of Zambezi inflows. *Quaternary Research*, 69, 388–403.

Butzer, K. W., Beaumont, P. B. and Vogel, J. C. (1978). Lithostratigraphy of Border Cave, KwaZulu, South Africa: A Middle Stone Age sequence beginning *c.* 195,000 b.p. *Journal of Archaeological Science*, 5, 317–341.

Callen, R. A., Wasson, R. J. and Gillespie, R. (1983). Reliability of radiocarbon dating of pedogenic carbonate in the Australian arid zone. *Sedimentary Geology*, 35, 1–14.

Carr, A. S., Bateman, M. D. and Holmes, P. J. (2007). Developing a 150 ka luminescence chronology for the barrier dunes of the southern Cape, South Africa. *Quaternary Geochronology*, 2, 110–116.

Carr, A. S., Bateman, M. D., Roberts, D. L., Murray-Wallace, C. V., Jacobs, Z. and Holmes, P. J. (2010). The last interglacial sea-level high stand on the southern Cape coastline of South Africa. *Quaternary Research*, 73, 351–363.

Carr, A. S., Thomas, D. S. G., Bateman, M. D., Meadows, M. E. and Chase, B. (2006). Late Quaternary palaeoenvironments of the winter-rainfall zone of southern Africa: Palynological and sedimentological evidence from the Agulhas Plain. *Palaeogeography, Palaeoclimatology, Palaeoecology*, 239, 147–165.

Cawthra, H. C., Uken, R. and Ovechkina, M. N. (2012). New insights into the geological evolution of the Durban Bluff and adjacent Blood Reef, South Africa. *South African Journal of Geology*, 115, 291–308.

Chadwick, O. A., Roering, J. J., Heimsath, A. M., Levick, S. R., Asner, G. P. and Khomo, L. (2013). Shaping post-orogenic landscapes by climate and chemical weathering. *Geology*, 41, 1171–1174.

Chase, B. M., Meadows, M. E., Carr, A. S. and Reimer, P. J. (2010). Evidence for progressive Holocene aridification in southern Africa recorded in Namibian hyrax middens: Implications for African monsoon dynamics and the "African Humid Period". *Quaternary Research*, 74, 36–45.

Chase, B. M., Meadows, M. E., Scott, L., Thomas, D. S. G., Marais, E., Sealy, J. and Reimer, P. J. (2009). A record of rapid Holocene climate change preserved in hyrax middens from southwestern Africa. *Geology*, 37, 703–706.

Chase, B. M., Quick, L. J., Meadows, M. E., Scott, L., Thomas, D. S. G. and Reimer, P. J. (2011). Late glacial interhemispheric climate dynamics revealed in South African hyrax middens. *Geology*, 39, 19–22.

Chazan, M., Ron, H., Matmon, A., Porat, N., Goldberg, P., Yates, R., Avery, M., Sumner, A. and Horwitz, L. K. (2008). Radiometric dating of the Earlier Stone Age sequence in Excavation I at Wonderwerk Cave, South Africa: Preliminary results. *Journal of Human Evolution*, 55, 1–11.

Chirikure, S., Pollard, M., Manyanga, M. and Bandama, F. (2013). A Bayesian chronology for Great Zimbabwe: Re-threading the sequence of a vandalised monument. *Antiquity*, 87, 854–872.

Cockburn, H. A. P., Brown, R. W., Summerfield, M. A. and Seidl, M. A. (2000). Quantifying passive margin denudation and landscape development using a combined fission-track thermochronology and cosmogenic isotope analysis approach. *Earth and Planetary Science Letters*, 179, 429–435.

Decker, J. E., Niedermann, S. and de Wit, M. J. (2011). Soil erosion rates in South Africa compared with cosmogenic ^{3}He-based rates of soil production. *South African Journal of Geology*, 114, 475–488.

Decker, J. E., Niedermann, S. and de Wit, M. J. (2013). Climatically influenced denudation rates of the southern African plateau: Clues to solving a geomorphic paradox. *Geomorphology*, 190, 48–60.

Dirks, P. H. G. M., Kibii, J. M., Kuhn, B. F., Steininger, C., Churchill, S. E., Kramers, J. D., Pickering, R., Farber, D. L., Mériaux, A.-S., Herries, A. I. R., King, G. C. P. and Berger, L. R. (2010). Geological setting and age of *Australopithecus sediba* from southern Africa. *Science*, 328, 205–208.

Dunai, T. J. (2010). *Cosmogenic Nuclides: Principles, Concepts and Applications in the Earth Surface Sciences*. Cambridge: Cambridge University Press, 198pp.

Eitel, B., Kadereit, A., Blümel, W. D., Hüser, K. and Kromer, B. (2005). The Amspoort Silts, northern Namib desert (Namibia): Formation, age and palaeoclimatic evidence of river-end deposits. *Geomorphology*, 64, 299–314.

Eitel, B., Kadereit, A., Blümel, W.-D., Hüser, K., Lomax, J. and Hilgers, A. (2006). Environmental changes at the eastern Namib Desert margin before and after the Last Glacial Maximum: New evidence from fluvial deposits in the upper Hoanib River catchment, northwestern Namibia. *Palaeogeography, Palaeoclimatology, Palaeoecology*, 234, 201–222.

Erlanger, E. D., Granger, D. E. and Gibbon, R. J. (2012). Rock uplift rates in South Africa from isochron burial dating of fluvial and marine terraces. *Geology*, 40, 1019–1022.

Feathers, J. K. (2002). Luminescence dating in less than ideal conditions: Case studies from Klasies River main site and Duinefontein, South Africa. *Journal of Archaeological Science*, 29, 177–194.

Fleming, A., Summerfield, M. A., Stone, J. O., Fifield, L. K. and Cresswell, R. G. (1999). Denudation rates for the southern Drakensberg escarpment, SE Africa, derived from in-situ-produced cosmogenic ^{36}Cl: initial results. *Journal of the Geological Society*, 156, 209–212.

Fuchs, M., Kandel, A. W., Conard, N. J., Walker, S. J. and Felix-Henningsen, P. (2008). Geoarchaeological and chronostratigraphical investigations of open-air sites in the Geelbek Dunes, South Africa. *Geoarchaeology*, 23, 425–449.

Geyh, M. A. and Eitel, B. (1998). Radiometric dating of young and old calcrete. *Radiocarbon*, 40, 795–802.

Gibbon, R. J., Granger, D. E., Kuman, K. and Partridge, T. C. (2009). Early Acheulean technology in the Vaal River gravels, South Africa, dated with cosmogenic nuclides. *Journal of Human Evolution*, 56, 152–160.

Gosse, J. C. and Phillips, F. M. (2001). Terrestrial *in situ* cosmogenic nuclides: Theory and application. *Quaternary Science Reviews*, 20, 1475–1560.

Granger, D. E. and Muzikar, P. F. (2001). Dating sediment burial with *in situ*-produced cosmogenic nuclides: Theory, techniques, and limitations. *Earth and Planetary Science Letters*, 188, 269–281.

Grodek, T., Benito, G., Botero, B. A., Jacoby, Y., Porat, N., Haviv, I., Cloete, G. and Enzel, Y. (2013). The last millennium largest floods in the hyperarid Kuiseb River basin, Namib Desert. *Journal of Quaternary Science*, 28, 258–270.

Grün, R. (2007). Electron Spin Resonance Dating. In *Encyclopedia of Quaternary Science*, ed. S. A. Elias. Amsterdam: Elsevier, 2, pp. 1505–1516.

Grün, R. and Beaumont, P. (2001). Border Cave revisited: A revised ESR chronology. *Journal of Human Evolution*, 40, 467–482.

Henshilwood, C. S., d'Errico, F., Yates, R., Jacobs, Z., Tribolo, C., Duller, G. A. T., Mercier, N., Sealy, J. C., Valladas, H., Watts, I. and Wintle, A. G. (2002). Emergence of modern human behavior: Middle Stone Age engravings from South Africa. *Science*, 295, 1278–1280.

Hogg, A. G., Hua, Q., Blackwell, P. G., Niu, M., Buck, C. E., Guilderson, T. P., Heaton, T.J., Palmer, J. G., Reimer, P.J., Reimer, R. W., Turney, C. S. M. and Zimmerman, S. R. H. (2013). SHCal13 Southern Hemisphere calibration, 0–50,000 years cal BP. *Radiocarbon*, 55, 1889–1903.

Hogg, A. G., McCormac, F. G., Higham, T. F., Reimer, P. J., Baillie, M. G. and Palmer, J. G. (2002). High-precision radiocarbon measurements of contemporaneous tree-ring dated wood from the British Isles and New Zealand: AD 1850–950. *Radiocarbon*, 44, 633–640.

Holmgren, K., Lauritzen, S.-E. and Possnert, G. (1994). ^{230}Th ^{234}U and ^{14}C dating of a late Pleistocene stalagmite in Lobatse II Cave, Botswana. *Quaternary Science Reviews*, 13, 111–119.

Holmgren, K., Lee-Thorp, J. A., Cooper, G. R. J., Lundblad, K., Partridge, T. C., Scott, L., Sithaldeen, R., Talma, A. S. and Tyson, P. D. (2003). Persistent millennial-scale climatic variability over the past 25,000 years in Southern Africa. *Quaternary Science Reviews*, 22, 2311–2326.

Huffman, T. N. (2007). *Handbook to the Iron Age: The Archaeology of Pre-Colonial Farming Societies in Southern Africa.* Scottsville: University of KwaZulu-Natal, 518pp.

Huffman, T. N. (2010). Intensive El Niño and the Iron Age of South-eastern Africa. *Journal of Archaeological Science*, 37, 2572–2586.

Huffman, T. N. (2015). Social complexity in southern Africa. *African Archaeological Review*, 32, 71–91.

Huffman, T. N. and Vogel, J. C. (1991). The chronology of Great Zimbabwe. *South African Archaeological Bulletin*, 46, 61–70.

Ivy-Ochs, S. and Kober, F. (2008). Surface exposure dating with cosmogenic nuclides. *Eiszeitalter und Gegenwart*, 57, 179–209.

Jacobs, Z., Duller, G. A. T. and Wintle, A. G. (2003b). Optical dating of dune sand from Blombos Cave, South Africa: II – single grain data. *Journal of Human Evolution*, 44, 613–625.

Jacobs, Z. and Roberts, D. L. (2009). Last Interglacial Age for aeolian and marine deposits and the Nahoon fossil human footprints, Southeast Coast of South Africa. *Quaternary Geochronology*, 4, 160–169.

Jacobs, Z., Roberts, R. G., Galbraith, R. F., Deacon, H. J., Grün, R., Mackay, A., Mitchell, P., Vogelsang, R. and Wadley, L. (2008). Ages for the Middle Stone Age of southern Africa: Implications for human behavior and dispersal. *Science*, 322, 733–735.

Jacobs, Z., Wintle, A. G. and Duller, G. A. T. (2003a). Optical dating of dune sand from Blombos Cave, South Africa: I—multiple grain data. *Journal of Human Evolution*, 44, 599–612.

Keen-Zebert, A., Tooth, S., Rodnight, H., Duller, G. A. T., Roberts, H. M. and Grenfell, M. (2013). Late Quaternary floodplain reworking and the preservation of alluvial sedimentary archives in unconfined and confined river valleys in the eastern interior of South Africa. *Geomorphology*, 185, 54–66.

Kounov, A., Niedermann, S., de Wit, M. J., Viola, G., Andreoli, M. and Erzinger, J. (2007). Present denudation rates at selected sections of the South African escarpment and the elevated continental interior based on cosmogenic ^{3}He and ^{21}Ne. *South African Journal of Geology*, 110, 235–248.

Kristen, I., Fuhrmann, A., Thorpe, J., Röhl, U., Wilkes, H. and Oberhänsli, H. (2007). Hydrological changes in southern Africa over the last 200 Ka as recorded in lake sediments from the Tswaing impact crater. *South African Journal of Geology*, 110, 311–326.

Lal, D. (1991). Cosmic ray labeling of erosion surfaces: *In situ* nuclide production rates and erosion models. *Earth and Planetary Science Letters*, 104, 424–439.

Libby, W. F., Anderson, E. C. and Arnold, J. R. (1949). Age determination by radiocarbon content: world-wide assay of natural radiocarbon. *Science*, 109, 227–228.

Lyons, R., Tooth, S. and Duller, G. A. T. (2013). Chronology and controls of donga (gully) formation in the upper Blood River catchment, KwaZulu-Natal, South Africa: Evidence for a climatic driver of erosion. *The Holocene*, 23, 1875–1887.

Marren, P. M., McCarthy, T. S., Tooth, S., Brandt, D., Stacey, G. G., Leong, A. and Spottiswoode, B. (2006). A comparison of mud- and sand-dominated meanders in a downstream coarsening reach of the mixed bedrock-alluvial Klip River, eastern Free State, South Africa. *Sedimentary Geology*, 190, 213–226.

McCarthy, T. and Rubidge, B. (2005). *The Story of Earth & Life; A Southern African Perspective on a 4.6-billion-year Journey*. Cape Town: Struik Nature, 335pp.

McCormac, F. G., Hogg, A. G., Blackwell, P. G., Buck, C. E., Higham, T. F. G. and Reimer, P. J. (2004). ShCal04 Southern Hemisphere calibration, 0–11.0 cal kyr BP. *Radiocarbon*, 46, 1087–1092.

Midgley, J. J., Harris, C., Hesse, H. and Swift, A. (2002). Heuweltjie age and vegetation change based on δ^{13}C and ^{14}C analyses. *South African Journal of Science*, 98, 202–204.

Millard, A. R. (2006). Bayesian analysis of ESR dates, with application to Border Cave. *Quaternary Geochronology*, 1, 159–166.

Miller, G. H., Beaumont, P. B., Deacon, H. J., Brooks, A. S., Hare, P. E. and Jull, A. J. T. (1999). Earliest modern humans in southern Africa dated by isoleucine epimerization in ostrich eggshell. *Quaternary Science Reviews*, 18, 1537–1548.

Mills, S. C. and Grab, S. W. (2005). Debris ridges along the southern Drakensberg escarpment as evidence for Quaternary glaciation in southern Africa. *Quaternary International*, 129, 61–73.

Mulrooney, M. A. (2013). An island-wide assessment of the chronology of settlement and land use on Rapa Nui (Easter Island) based on radiocarbon data. *Journal of Archaeological Science*, 40, 4377–4399.

Parkington, J. E. (1990). A critique of the consensus view on the age of Howieson's Poort assemblages in South Africa. In *The Emergence of Modern Humans: An Archaeological Perspective*, ed. P. A. Mellars. Edinburgh: Edinburgh University Press, pp. 34–55.

Parkington, J. E. and Poggenpoel, C. (1987). Diepkloof rock shelter. In *Papers in the Prehistory of the Western Cape, South Africa*, J. Parkington and M. Hall (eds). Oxford: BAR International Series, 332, pp. 269–293.

Partridge, T. C. (2005). Dating of the Sterkfontein hominids: Progress and possibilities. *Transactions of the Royal Society of South Africa*, 60, 107–109.

Partridge, T. C., Demenocal, P. B., Lorentz, S. A., Paiker, M. J. and Vogel, J. C. (1997). Orbital forcing of climate over South Africa: A 200,000-year rainfall record from the Pretoria Saltpan. *Quaternary Science Reviews*, 16, 1125–1133.

Partridge, T. C., Granger, D. E., Caffee, M. W. and Clarke, R. J. (2003). Lower Pliocene hominid remains from Sterkfontein. *Science*, 300, 607–612.

Pickering, R., Dirks, P. H. G. M., Jinnah, Z., de Ruiter, D. J., Churchill, S. E., Herries, A. I. R., Woodhead, J. D., Hellstrom, J. C. and Berger, L. R. (2011). *Australopithecus sediba* at 1.977 Ma and implications for the origins of the genus *Homo*. *Science*, 333, 1421–1423.

Pickering, R. and Kramers, J. D. (2010). Re-appraisal of the stratigraphy and determination of new U-Pb dates for the Sterkfontein hominin site, South Africa. *Journal of Human Evolution*, 59, 70–86.

Pickering, R., Kramers, J. D., Partridge, T., Kodolanyi, J. and Pettke, T. (2010). U–Pb dating of calcite–aragonite layers in speleothems from hominin sites in South Africa by MC-ICP-MS. *Quaternary Geochronology*, 5, 544–558.

Pienaar, M., Woodborne, S. and Wadley, L. (2008). Optically stimulated luminescence dating at Rose Cottage Cave. *South African Journal of Science*, 104, 65–70.

Porat, N., Chazan, M., Grün, R., Aubert, M., Eisenmann, V. and Horwitz, L. K. (2010). New radiometric ages for the Fauresmith industry from Kathu Pan, southern Africa: Implications for the Earlier to Middle Stone Age transition. *Journal of Archaeological Science*, 37, 269–283.

Portenga, E. W. and Bierman, P. R. (2011). Understanding Earth's eroding surface with ^{10}Be. *GSA Today*, 21 (8), 4–10.

Potts, A. J., Midgley, J. J. and Harris, C. (2009). Stable isotope and ^{14}C study of biogenic calcrete in a termite mound, Western Cape, South Africa, and its palaeoenvironmental significance. *Quaternary Research*, 72, 258–264.

Reimer, P. J., Baillie, M. G. L., Bard, E., Bayliss, A., Beck, J. W., Bertrand, C. J. H., Blackwell, P.G., Buck, C. E., Burr, G. S., Cutler, K. B., Damon, P. E., Edwards, R. L., Fairbanks, R. G., Friedrich, M., Guilderson, T. P., Hogg, A. G., Hughen, K. A., Kromer, B., McCormac, G., Manning, S., Ramsey, C. B., Reimer, R. W., Remmele, S., Southon, J. R., Stuiver, M., Talamo, S., Taylor, F. W., van der Plicht, J. and Weyhenmeyer, C. E. (2004). IntCal04 terrestrial radiocarbon age calibration, 0–26 cal kyr BP. *Radiocarbon*, 46, 1029–1058.

Reimer, P. J., Bard, E., Bayliss, A., Beck, J. W., Blackwell, P. G., Ramsey, C. B., Buck, C. E., Cheng, H., Edwards, R. L., Friedrich, M., Grootes, P. M., Guilderson, T. P., Haflidason, H., Hajdas, I., Hatté, C., Heaton, T. J., Hoffmann, D. L., Hogg, A. G., Hughen, K. A., Kaiser, K. F., Kromer, B., Manning, S. W., Niu, M., Reimer, R. W., Richards, D. A., Scott, E. M., Southon, J. R., Staff, R. A., Turney, C. S. M. and van der Plicht, J. (2013). IntCal13 and Marine13 radiocarbon age calibration curves 0–50,000 years cal BP. *Radiocarbon*, 55, 1869–1887.

Roberts, D. L. (2008). Last interglacial hominid and associated vertebrate fossil trackways in coastal eolianites, South Africa. *Ichnos*, 15, 190–207.

Roberts, D. L., Bateman, M. D., Murray-Wallace, C. V., Carr, A. S. and Holmes, P. J. (2009). West coast dune plumes: Climate driven contrasts in dunefield morphogenesis along the western and southern South African coasts. *Palaeogeography, Palaeoclimatology, Palaeoecology*, 271, 24–38.

Rodnight, H., Duller, G. A. T., Tooth, S. and Wintle, A. G. (2005). Optical dating of a scroll-bar sequence on the Klip River, South Africa, to derive the lateral migration rate of a meander bend. *The Holocene*, 15, 802–811.

Scharf, T. E., Codilean, A. T., de Wit, M., Jansen, J. D. and Kubik, P. W. (2013). Strong rocks sustain ancient postorogenic topography in southern Africa. *Geology*, 41, 331–334.

Schwarcz, H. P., Grün, R. and Tobias, P. V. (1994). ESR dating studies of the australopithecine site of Sterkfontein, South Africa. *Journal of Human Evolution*, 26, 175–181.

Scott, E. M., Bryant, C., Cook, G. T. and Naysmith, P. (2003). Is there a fifth international radiocarbon intercomparison (VIRI)? *Radiocarbon*, 45, 493–495.

Scott, L., Holmgren, K., Talma, A. S., Woodborne, S. and Vogel, J. C. (2003). Age interpretation of the Wonderkrater spring sediments and vegetation change in the Savanna Biome, Limpopo province, South Africa. *South African Journal of Science*, 99, 484–488.

Spriggs, M. (1989). The dating of the Island Southeast Asian Neolithic: An attempt at chronometric hygiene and linguistic correlation. *Antiquity*, 63, 587–613.

Srivastava, P., Brook, G. A., Marais, E., Morthekai, P. and Singhvi, A. K. (2006). Depositional environment and OSL chronology of the Homeb silt deposits, Kuiseb River, Namibia. *Quaternary Research*, 65, 478–491.

Stokes, S., Thomas, D. S. G. and Washington, R. (1997). Multiple episodes of aridity in southern Africa since the last interglacial period. *Nature*, 388, 154–158.

Stone, A. E. C. and Thomas, D. S. G. (2013). Casting new light on late Quaternary environmental and palaeohydrological change in the Namib Desert: A review of the application of optically stimulated luminescence in the region. *Journal of Arid Environments*, 93, 40–58.

Stuiver, M., Reimer, P. J., Bard, E., Beck, J. W., Burr, G. S., Hughen, K. A., Kromer, B., McCormac, G., van der Plicht, J. and Spurk, M. (1998). INTCAL98 radiocarbon age calibration, 24,000–0 cal BP. *Radiocarbon*, 40, 1041–1083.

Sudan, P., Whitmore, G., Uken, R. and Woodbourn, S. (2004). Quaternary evolution of the coastal dunes between Lake Hlabane and Cape St Lucia, KwaZulu-Natal. *South African Journal of Geology*, 107, 355–376.

Talma, A. S. and Vogel, J. C. (1992). Late Quaternary paleotemperatures derived from a speleothem from Cango Caves, Cape Province, South Africa. *Quaternary Research*, 37, 203–213.

Thomas, D. S. and Burrough, S. L. (2013). Luminescence-based dune chronologies in southern Africa: Analysis and interpretation of dune database records across the subcontinent. *Quaternary International*, doi:10.1016/j.quaint.2013.09.008.

Tooth, S., Brandt, D., Hancox, P. J. and McCarthy, T. S. (2004). Geological controls on alluvial river behaviour: A comparative study of three rivers on the South African Highveld. *Journal of African Earth Sciences*, 38, 79–97.

Tooth, S., Hancox, P. J., Brandt, D., McCarthy, T. S., Jacobs, Z. and Woodborne, S. (2013). Controls on the genesis, sedimentary architecture, and preservation potential of dryland alluvial successions in stable continental interiors: Insights from the incising Modder River, South Africa. *Journal of Sedimentary Research*, 83, 541–561.

Tooth, S., McCarthy, T. S., Brandt, D., Hancox, P. J. and Morris, R. (2002). Geological controls on the formation of alluvial meanders and floodplain wetlands: The example of the Klip River, eastern Free State, South Africa. *Earth Surface Processes and Landforms*, 27, 797–815.

Tribolo, C., Mercier, N. and Valladas, H. (2005). Chronologie des technofaciès Howieson's Poort et Still Bay (Middle Stone Age, Afrique du Sud): bilan et nouvelles données de la luminescence. *Bulletin de la Société préhistorique française*, 102, 855–866.

Valladas, H., Wadley, L., Mercier, N., Froget, L., Tribolo, C., Reyss, J. L. and Joron, J. L. (2005). Thermoluminescence dating on burnt lithics from Middle Stone Age layers at Rose Cottage Cave. *South African Journal of Science*, 101, 169–174.

van der Wateren, F. M. and Dunai. T. J. (2001). Late Neogene passive margin denudation history – cosmogenic isotope measurements from the central Namib desert. *Global and Planetary Change*, 30, 271–307.

Vermeesch, P., Fenton, C. R., Kober, F., Wiggs, G. F. S., Bristow, C. S. and Xu, S. (2010). Sand residence times of one million years in the Namib Sand Sea from cosmogenic nuclides. *Nature Geoscience*, 3, 862–865.

Vogel, J. C. (2001). Radiometric dates for the Middle Stone Age in South Africa. In *Humanity from African Naissance to Coming Millennia*, ed. P. V. Tobias, M. A. Raath, J. Moggi-Cecchi and G. A. Doyle. Johannesburg: University of the Witwatersrand Press, pp. 261–268.

Vogel, J. C., Fuls, A., Visser, E. and Becker, B. (1993). Pretoria calibration curve for short-lived samples, 1930–3350 BC. *Radiocarbon*, 35, 73–85.

Vrba, E. S. (1975). Some evidence of chronology and palaeoecology of Sterkfontein, Swartkrans and Kromdraai from the fossil Bovidae. *Nature*, 254, 301–304.

Wadley, L. and Jacobs, Z. (2004). Sibudu Cave, KwaZulu-Natal: Background to the excavations of Middle Stone Age and Iron Age occupations. *South African Journal of Science*, 100, 145–151.

Wadley, L. and Jacobs, Z. (2006). Sibudu Cave: Background to the excavations, stratigraphy and dating. *Southern African Humanities*, 18, 1–26.

Wang, Q., Tobias, P. V., Roberts, D. L. and Jacobs, Z. (2008). A re-examination of a human femur found at the Blind River Site, East London, South Africa: Its age, morphology, and breakage pattern. *Anthropological Review*, 71, 43–61.

Wilde, S. A., Valley, J. W., Peck, W. H. and Graham, C. M. (2001). Evidence from detrital zircons for the existence of continental crust and oceans on the Earth 4.4 Gyr ago. *Nature*, 409, 175–178.

Wintle, A. G. (2010). Future directions of luminescence dating of quartz. *Geochronometria*, 37, 1–7.

Wintle, A. G., Botha, G. A., Li, S. H. and Vogel, J. C. (1995b). Chronological framework for colluviation during the last 110 KYR in Kwazulu-Natal. *South African Journal of Science*, 91, 134–139.

Wintle, A. G., Li, S. H. and Botha, G. A. (1993). Luminescence dating of colluvial deposts from Natal, South Africa. *South African Journal of Science*, 89, 77–82.

Wintle, A. G., Li, S. H., Botha, G. A. and Vogel, J. C. (1995a). Evaluation of luminescence-dating procedures applied to late-Holocene colluvium near St Paul's Mission, Natal, South Africa. *The Holocene*, 5, 97–102.

Woodborne, S., Pienaar, M. and Tiley-Nel, S. (2009). Dating the Mapungubwe Hill gold. *Journal of African Archaeology*, 7, 99–105.

Zhao, J.-X., Yu, K.-F. and Feng, Y.-X. (2009). High-precision ^{238}U–^{234}U–^{230}Th disequilibrium dating of the recent past: A review. *Quaternary Geochronology*, 4, 423–433.

8

Glacial and periglacial geomorphology

STEFAN W. GRAB AND JASPER KNIGHT

Abstract

The glacial and periglacial record of southern Africa during the Quaternary is limited to the highest-altitude areas of the Drakensberg and Cape Fold Belt, where late Pleistocene temperature depression in addition to uncertain changes in precipitation regime were sufficient in combination to develop small cirque glaciers and/or a range of periglacial features. This chapter reflects on past debates for and against Quaternary glaciations, and identifies research gaps in these debates. Geomorphological and sedimentary evidence for glacial and periglacial landforms is summarised in this chapter, and the climatic and environmental contexts in which they developed, where known, are explained. There remain significant gaps in our understanding of cold Quaternary events in high mountain areas of southern Africa, mainly due to an absence of reliable palaeoclimatic indicators, the sometimes inconclusive climatic signatures offered by periglacial landforms, and poor dating control. The Drakensberg and Cape Fold Belt still experience marginal periglacial climate conditions today, but are currently undergoing change due to both regional climate change and human-induced landscape alterations, thus future periglacial activity is likely to become further constrained in location and vigour.

8.1 Introduction

Many of the southern African proxy records discussed in this book are valuable in their ability to provide a chronological sequence of climatic and environmental events over periods spanning up to several thousands of years. To this end, Quaternary-age glacial and periglacial landforms are restricted, as favourable climatic conditions have only been present during narrow time-windows. Although there is considerable debate over past climatic conditions in upland areas of southern Africa where glacial and periglacial landforms would have developed (e.g. Partridge et al., 1999; Holmgren et al., 2003; Mills et al., 2009a, 2012), these landforms are difficult to date precisely and may have developed over multiple cool climate phases, such that their preserved morphologies cannot be taken uncritically as a measure of climate severity. Notwithstanding such difficulties, correct geomorphological interpretation and dating can allow for inferences of palaeoclimate and palaeoenvironmental conditions.

A wide range of relict Quaternary glacial and periglacial landforms exists in the Western and Eastern Cape mountains, and the Drakensberg sector of the Great Escarpment which is located in the border area between Lesotho and KwaZulu-Natal Province in South Africa (Fig. 8.1). However, there is disagreement as to the origin, climatic context and age of many of these landforms. This chapter reflects on (1) the historical development of understanding Quaternary-age glaciation in southern Africa, and hypothetical concerns of such a proposed glaciation; (2) the distribution and morphology of glacial and periglacial landforms; and (3) their palaeoclimatic and palaeoenvironmental implications. This chapter highlights the changing role of the cryosphere (*sensu lato*) in the development of southern Africa's high mountain landscapes, and that cold-climate processes are becoming less significant as a consequence of regional climate changes in the Anthropocene.

8.2 Setting the scene

The earliest glacial evidence in southern Africa dates to the Permo-Carboniferous when the African subcontinent was located farther south as part of Gondwanaland. This glacial episode is manifested in the Dwyka Group of the Karoo Sequence, consisting mainly of glaciomarine deglacial sequences which in places include boulder or conglomeratic tillites (Visser and Hall, 1985; von Brunn, 1996; Isbell et al., 2008). Surface-exposed glacially striated pavements with crescentic, circular and oval shaped markings and fractures have helped identify former ice flow patterns and calculate the original thickness of Karoo strata (Master, 2012). The Dwyka Group sediments extend over ~600,000 km^2 and attain a maximum thickness of ~800 m along the southern margins of the basin, but thin northwards to between 0 and 600 m thick (Isbell et al., 2008). This early glacial evidence, from across large areas of southern Africa, is important to understand late Paleozoic glaciations, climate dynamics, and continental drift (e.g. du Toit, 1937). Significant evidence exists, however, for the impacts of cold-climate (glacial and periglacial) processes in southern Africa during the Quaternary. In this chapter, evidence for late Pleistocene-age glaciation

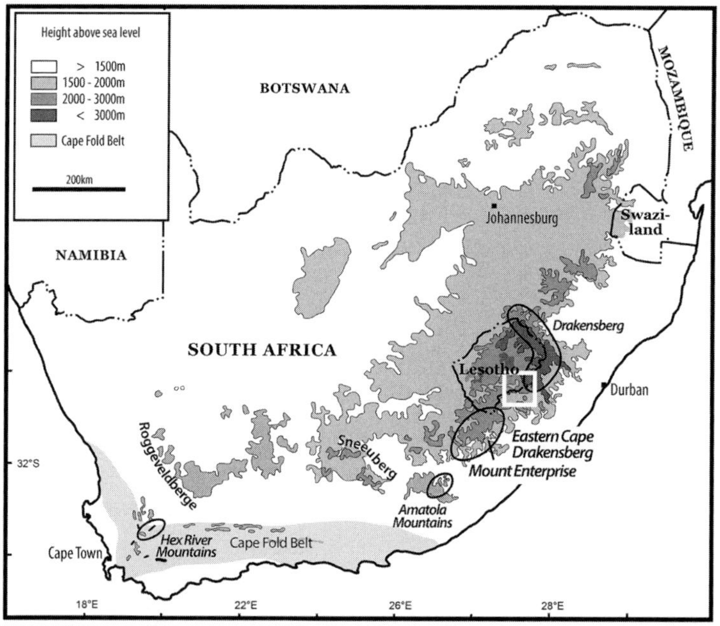

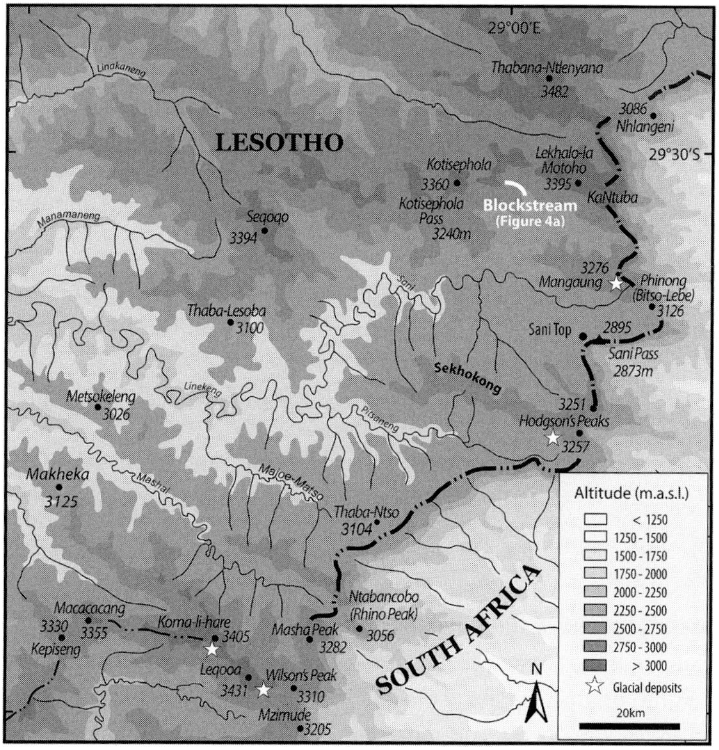

Fig. 8.1. (A) Location of key mountainous areas in southern Africa, and (B) close up of the study area (boxed in part A) showing the locations of places named in the text.

and periglaciation, within the Quaternary period, is considered from two regions in southern Africa (the Drakensberg and southwestern Cape).

In the Drakensberg sector of the Great Escarpment (between 28°30′ and 31°S), fluvial and aeolian deposition continued through the Triassic and Jurassic, laying down a succession of variably textured sediments of the Karoo Supergroup (Carboniferous to Jurassic), and culminating in the fine-grained Clarens Formation which reflects increased aridity and sand dune deposition (Holzförster, 2007). These geological processes were important to develop sufficient relief to induce cold climate conditions on mountain tops. The sediments were then overlain by the relatively short-lived Drakensberg Group volcanic event at ~182 Ma (Jourdan et al., 2007). Strike-slip faulting and rapid denudation of the continental margins during continental rifting of the early Cretaceous (~136 Ma) is manifested today as the Great Escarpment, located up to several hundred kilometres inland of the contemporary continental margin. In the Drakensberg sector, mountain peaks rise to over 3400 m a.s.l. – the highest in southern Africa – which made them the most climatically susceptible locations for Quaternary glaciation.

In the southwestern Cape, the Cape Supergroup (Ordovician to Devonian) comprises a 6–10 km thick sandstone sequence, deposited along a subsiding shelf. Subduction and accretion of the palaeo-Pacific plate beneath Gondwana subsequently caused regional folding of the sandstone strata around 330 Ma, forming the Cape Fold Belt (Booth, 2011). These mountains today extend east–west across the southern Eastern Cape and Western Cape provinces (Swarteberge, Outeniqua and Langeberg ranges; between ~25 and 19°E), and south–north in the Western Cape (Cederberg range, between ~34 and 31°S). The highest peaks reach up to 2325 m a.s.l. Here, the Great Escarpment is located 140–220 km farther inland, and has a lower maximum altitude (~1600 m a.s.l.) than the Cape Fold Mountains (Fig. 8.1). The highest peaks here are mainly underlain by resistant sandstones and quartzites.

8.3 Proponents for southern African glaciation during the late Pleistocene: a brief history

In the 1940s and 1950s, several workers considered the possibility of Quaternary glaciations in southern Africa (e.g. Kojot, 1948; du Toit, 1954), but it was generally thought that the climate would have been too dry to sustain any glaciers. Flint (1959) proposed that late Pleistocene glaciers did exist in southern Africa, but were confined to the highest mountains. Sparrow (1964, 1967) identified slope elements consistent with glaciation, such as U-shaped valleys, but these were localised and not compatible with any widespread glaciation. Subsequently, he reinterpreted such upland hollows and valleys as 'non-glacial cirques' formed by enhanced rock surface weathering under a range of climatic conditions (Sparrow, 1974). Marker and

Whittington (1971) examined the morphometry of 577 slope hollows located at high elevation positions (>2900 m a.s.l.) in the Drakensberg sector of the Great Escarpment, in Lesotho. They found hollows to have a consistent morphology, spatial variation and a preferred north-facing aspect (see Dyer and Marker, 1979). The work was later expanded to cover 12 different morphometric parameters tested upon 628 hollows covering 156,000 km^2 of the Lesotho Highlands (Marker, 1991). Cold Pleistocene climate conditions and associated glaciations were thought to have been the main driver for hollow development (Marker, 1991).

In the Cape Fold Belt, evidence for proposed plateau, cirque and valley glaciation during the Pleistocene was based on morphometric air photo analysis (cirques and moraines) and rock surface observations (glacial polishing and striae) (Borchert and Sänger, 1981; Sänger, 1987). Sänger (1988) proposed a snowline depression to ~1700 m a.s.l. during the Last Glacial Maximum (LGM) over the Hex River mountains, which are located in the direct path of moisture-laden mid-latitude cyclones approaching from the southwest. Several subsequent studies, based on a range of proxy indicators, have suggested a northerly displacement of pressure belts during the LGM (e.g. Gellert, 1991; Gasse et al., 2008), which would indeed have increased snowfall across southeastern African mountains.

Given the problems of misidentification, and the attribution of glaciation to erosional features alone, especially when based on small scale phenomena such as striae, more recent studies have focussed on identifying and interpreting possible depositional evidence for glaciation. Several studies have examined apparent terminal moraine ridges that may correspond to cirque glaciers in the Drakensberg sectors of the Eastern Cape (Lewis and Illgner, 2001) and KwaZulu-Natal/Lesotho Highlands (Mills and Grab, 2005; Mills et al., 2009a, b). The elevations of these moraine ridges have been used to calculate glacier equilibrium line altitudes (ELAs) and thus inform on LGM climate (Mills et al., 2012).

8.4 The opponent view to southern African glaciation during the late Pleistocene

The geomorphic evidence used to argue for past Quaternary glaciations in southern Africa has been challenged and debated at length over the years (see Boelhouwers and Meiklejohn, 2002; Hall, 2004; Osmaston and Harrison, 2005). Opponent views have challenged the scientific rigour and theoretical basis of Quaternary glaciations, although largely without alternative hypotheses based on field evidence (Knight, 2012). For example, the Western Cape ‘moraines’ have apparent bedrock remnants within them and thus are likely structural phenomena, while striated pebbles are argued to be associated with Silurian-age Pakhuis Tillites (Hall, 2004). There is also uncertainty on the climatic conditions of the LGM, with most discussion focusing on the precipitation (snow) required to sustain glaciers. There is a lack of independent evidence from other climate proxies to precisely evaluate precipitation conditions, but changes in synoptic climatological patterns (and thus moisture sources) are also uncertain (cf. Chase and Meadows, 2007). Arguments for and against the presence of Quaternary glaciers invoke high and low precipitation values, respectively, but both viewpoints are based on uncritical circular arguments in which, for example, the interpreted presence of a former glacier is assumed to also imply cold temperatures and wet conditions, even if there is no independent evidence for this (see Boelhouwers and Meiklejohn, 2002; Hall and Meiklejohn, 2011). Therefore, the climatology of the Pleistocene LGM must be considered independently of any inferences of climate at this time, based on geomorphological evidence. In the Drakensberg, depositional features interpreted as terminal moraines have been used to suggest the presence of late Pleistocene glaciers. This evidence and its implications are now discussed.

8.5 Glacial depositional landforms in the Drakensberg

Some depositional landforms (mainly debris ridges) have been identified in several Drakensberg locations, and interpreted as terminal moraines. Such ridges, however, have as yet only been reported from four sites: Mount Enterprise (~31°11′00″S, 28°00′00″E), Leqooa Valley (~29°44′35″S, 29°06′57″E), Sekhokong Range (29°36′51″S, 29°13′51″E) and Tsatsa-La-Mangaung Range (29°33′35″S, 29°17′39″E) (Lewis and Illgner, 2001; Mills and Grab, 2005; Mills et al., 2009a, b) (Fig. 8.1). The Mount Enterprise site is located in the southernmost end of the Drakensberg range, Eastern Cape Province. Here, the maximum summit altitude is 2565 m a.s.l., but escarpment cliffs generally terminate between 2200–2300 m a.s.l. The moraine ridges located at more northerly sites are all found in eastern Lesotho, on summit sites close to the Great Escarpment edge. The mountains west (inland) of the Great Escarpment also have summits exceeding 3300 m a.s.l., but ridges interpreted as glacigenic are conspicuously absent north of the Tsatsa-La-Mangaung Range.

All of the identified moraine ridges in eastern Lesotho, from different locations, occur within a narrow altitudinal range of 3005–3112 m a.s.l., which suggests a strong climatic control on their formation and thus does not favour alternative interpretations such as slope failure or colluviation (Fig. 8.2). However, the more southerly Mount Enterprise ridges are located between 2040–2059 m a.s.l. Notably, this region today receives considerably more frequent and heavier snowfalls than the higher altitude sites to the north. Maximum summit elevations (and calculated ELAs) above sea level are 3257 m (3095 m) at Sekhokong; 3431 m (3298 m) at Leqooa (Mills et al., 2012); and 2302 m

Fig. 8.2. Dissected glacigenic ridge forms, Leqooa valley, Lesotho (photo reproduced courtesy of Stephanie Mills). The ridges are buttressed (centre of photo) against bedrock slopes behind.

(2109 m) at Mount Enterprise (Lewis and Illgner, 2001) (Fig. 8.3). The large ELA differences between Lesotho and adjoining (within ~2° latitude) areas of the Eastern Cape beg questioning. This difference may imply possible misidentification of some landforms, or substantial palaeoclimatic gradients over southeastern Africa, which is not the case today, even though synoptic circulation patterns during the LGM were likely different to present (e.g. Cockcroft et al., 1987; Mills et al., 2012). At the sites in Lesotho, moraines are between 100–300 m in length, 15–70 m in width and 4–16 m in height. Thus, the glaciers would have all been small (0.05–0.31 km^2), located well below adjacent summits, with limited accumulation areas, and a mass balance strongly reliant on snowblow (Mills et al., 2012). Lack of sedimentological exposures through the ridges has limited analysis, but where trenched, the ridges generally consist of poorly sorted diamictons composed of clasts in a finer matrix. The covariance plot of C_{40} and RA indices fits within the morainic sediment envelopes according to Benn and Ballantyne (1994) (see Mills et al., 2009a, b). These macrosedimentological characteristics likely support both passive and active glacigenic

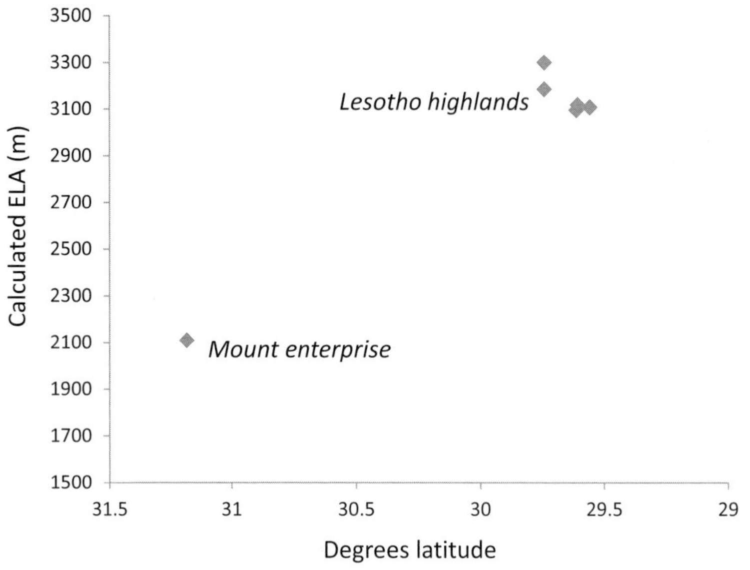

Fig. 8.3. Graph showing the relationship between ELA value and location of glacial sites in southern Africa. The Lesotho Highlands sites (two from the Sekhokong Range, two from Leqooa Valley, one from Tsata-La-Mangaung Range) show internal consistency.

transport modes (Mills et al., 2009a, b). Micromorphological analyses from the Lesotho sites display turbate structures, 'halo' structures and pressure shadows, which imply particle rotation during transport. The presence of *in situ* fractured quartz grains indicates high normal stress conditions, and subglacial stresses in particular (Mills et al., 2009a, b). Farther south, the Mount Enterprise ridges consist of clast- and matrix-supported clasts, some of which are striated (Lewis and Illgner, 2001). Although the sediment coincides with particle-size distribution envelopes associated with glacial lodgement and melt-out tills (see Sladen and Wrigley, 1983), no macro-sedimentological analyses have as yet been undertaken which may either support or refute a glacial origin.

8.6 Palaeoclimate implications of late Pleistocene glaciation in southeastern Africa

Radiocarbon-dated soil organic matter from within five moraines in eastern Lesotho provides some indication of minimum moraine ages. Calibrated ages range between 13,820–20,530 cal yr BP, and confirm ridge deposition during the last glacial period (Mills et al., 2009a). Although no reliable dates are available for the Mount Enterprise deposits, these are assumed to 'be of Last Glacial Maximum or younger age' (Lewis and Illgner, 2001, p370). A caveat, however, is that a temperature depression of 17°C below present values would be required in order to generate a LGM cirque glacier at Mount Enterprise (Lewis and Illgner, 2001), which does not correspond with other regional palaeotemperature proxies for this period.

To generate small glaciers, climate models are commonly tuned to different values (relative to present) of palaeotemperature and palaeoprecipitation. By these means, the most likely climatic conditions required to generate and sustain a small glacier can be established. Based on regional climate reconstructions from a variety of proxy records (Tyson and Partridge, 2000), Mills and Grab (2005) plotted likely LGM summer temperature (controlling ablation) and winter precipitation (controlling accumulation) for some of the eastern Lesotho glacier sites. In contrast to the Drakensberg sector of the Eastern Cape, Mills et al. (2012) calculated that with a temperature reduction of ~6°C (as indicated by proxy data), palaeoprecipitation would have been ~1100±200 mm/pa (~0–60% higher than at present). Further, the Lesotho sites are located in topographic positions which coincide with areas of longest-lasting winter snowcover (Grab et al., 2009), which supports their association with palaeo-snow accumulation sites (and thus possible glacial ice).

According to Holmgren et al. (2003), the LGM of southern Africa experienced an average and sustained temperature depression of ~6°C, with cooler pulses between 23–21, 19.5–17.5 and 15.0–13.5 kyr BP. Conditions during any of these periods would have favoured niche or cirque glaciers to develop in suitable topographic locations in the southern Drakensberg (Mills et al., 2009a, b), as is now supported by geomorphic evidence, outlined earlier. Previous concerns that the region was too dry during the LGM to support small glaciers (Hall, 2004) were largely based on an assumption presented by Partridge et al. (1999) that palaeoprecipitation along the Drakensberg was reduced by ~30% during the LGM, and that there was no evidence to support wetter conditions. However, spatial patterns of palaeoprecipitation values during the LGM are now known to be spatially variable across southern Africa, based on pollen and other biological records (e.g. Gasse et al., 2008; Scott et al., 2012). A key unknown factor, however, is the role of orography (e.g. the Great Escarpment) in modifying past precipitation patterns, including its seasonality. This means that palaeoclimate records from lowland sites might not be suitable analogues for precipitation at upland sites. Using the HadAM3h climate model and focusing on time slices at 21, 18 and 15 kyr BP, Mills et al. (2012) argued that the Lesotho Highland sites would have experienced greater precipitation seasonality than at present, with more spring–winter– autumn precipitation (much of it as snow), and less summer precipitation. This provides climatic support for the development of cirque glaciers during the LGM.

8.7 Periglacial phenomena

Mountains of the Western Cape, Eastern Cape and Lesotho/Drakensberg all host contemporary and relict periglacial phenomena, and thus can be considered as the 'core periglacial regions' of southern Africa. Contemporary (active) periglacial features include earth hummocks (thúfur), miniature varieties of sorted patterned ground, stone- and turf-banked lobes/sheets, terraces, and soil frost phenomena associated with needle ice. Soil frost and other subsurface processes can also operate synergistically with surface processes (e.g. fluvial, aeolian) to contribute to ongoing denudation of mountain slopes across southern Africa, which is a major contemporary management issue (Grab, 2010).

A variety of landforms interpreted as relict periglacial features has been reported from the core periglacial regions, including rock glaciers, pronival ('protalus') ramparts, ice wedge casts, nivation hollows, asymmetric valleys, openwork block deposits (blockslopes, blockstreams), large sorted patterned ground, thúfur, and turf-/stone-banked lobes (e.g. Lewis and Hanvey, 1993; Boelhouwers, 1994; Grab, 2000, 2002; Sumner, 2004) (Fig. 8.4). However, in many cases the correct identification of these features is not always straightforward. Some features are still under debate in terms of their origin and significance, and the environmental and climatic interpretation of such features remains controversial (see Grab et al., 2012, for a full review).

The most convincing and prominent periglacial landforms are those associated with block deposits. Openwork block deposits in the form of blockslopes and blockstreams are relatively common and the most conspicuous large-scale cryogenic landforms around higher summits of the core periglacial regions (Fig. 8.4A). Most blockstreams occur on colder, high altitude, south-facing slopes of the high Drakensberg. Blockstream dimensions in the order of several tens of metres to ~1.1 km in length, are up to 75 m wide and typically consist of subangular blocks 0.2–1.5 m long (see Boelhouwers, 1999; Grab, 1999; Boelhouwers et al., 2002). In the Matroosberg (Western Cape), blockstream clasts are platy and exhibit well-developed imbrication and downslope alignment, but in places with reorientation into a transverse fabric, dipping upslope at the blockstream frontal lobe (Boelhouwers, 1999). The most blocky clasts within the Drakensberg blockstreams tend to have a bi- or tri-modal clast orientation pattern, with primary alignment parallel to local slope gradient (Boelhouwers et al., 2002).

Large stone-banked lobes with tread lengths of up to 30 m and frontal banks 2–10 m wide and up to 3 m high have been reported from Drakensberg summits above 3400 m a.s.l. (Grab, 2000) (Fig. 8.4B). In places, secondary (active) solifluction features such as stone-banked sheets have developed on the stone-banked treads (Boelhouwers, 1994). The secondary sorting is to a depth of ~20 cm

Fig. 8.4. Periglacial phenomena from the Lesotho Highlands. (A) Views of the blockstream near Sani Pass (as described by Boelhouwers et al., 2002) showing its well defined margins and weathered and reoriented large clasts; (B) stone bank lobes (arrows); (C) sorted circle; (D) thúfur (photos A,D: Jasper Knight; B,C: Stefan Grab). A black and white version of this figure will appear in some formats. For the colour version, please refer to the plate section.

and is possibly a good indication of site-specific contemporary freeze depth. The primary lobes are sorted to depths of up to 60 cm and exhibit imbricated blocks dipping downslope.

Relict large sorted patterned ground has been reported from the Hex River Mountains, Western Cape (Boelhouwers, 1999), and KwaZulu-Natal/Lesotho Drakensberg (Boelhouwers, 1994; Grab, 2002; Sumner, 2004) (Fig. 8.4C). The widespread openwork block accumulations in the Hex River Mountains host large sorted stripes up to 150 m long and 2 m wide (Boelhouwers, 1999). In contrast, large sorted circles in the high Drakensberg contain gravel to cobble centres, surrounded by openwork borders (Grab, 2002). Larger sorted circles are ~1–5 m in diameter and sorted to over 1 m depth. This exceeds contemporary freeze depth, which is 40–50 cm during 'normal' winters on the high mountain interfluves where the patterns occur (Grab, 2004). This freezing regime also influences the formation and longevity of thúfur (earth hummocks) (Fig. 8.4D).

8.8 Discussion

In terms of reconstructing palaeoclimatic and palaeoenvironmental conditions in southern Africa during the Quaternary, the available geologic and isotopic records are highly variable in terms of length, resolution, location and environmental setting. As such, it is often difficult to compare one dataset to another or to extrapolate from one location to a wider region. This is particularly the case in mountainous regions of southern Africa where Quaternary climate records are uncommon and where climatic data from mountain-foot locations cannot be extrapolated to mountain tops. Consequently, the preserved record of glacial and periglacial phenomena often represents the only data source for Quaternary climates and environments in these mountain regions, limiting the extent to which the climatic boundary conditions of such analogues can be used.

The lack of constraining palaeoclimatic records and the absence of radiometric dating are significant limiting factors when using glacial and periglacial evidence from southern Africa. Radiocarbon dating is only possible where organic matter is present but this can provide only a minimum age of the underlying materials (see Woodborne, this volume). Cosmogenic nuclide dating of rock surfaces that have been exposed by glacier erosion does not suit the situation in southern Africa where the rocks are highly weathered and where the amount of glacier erosion is uncertain or may be minimal. The possible polygenic or multi-phase evolution of cryogenic landforms has not been fully explored, and even well-studied landforms such as blockstreams have yielded little conclusive evidence concerning their origins and palaeoenvironmental setting (e.g. Boelhouwers et al., 2002). It is also likely that high-altitude areas of the Drakensberg and Cape Fold Belt experienced significant differences in microclimate, controlled by aspect, snowcover duration and prevailing winds. These microscale controls are very difficult to resolve with sparse field data or by downscaled climate models, and are thus a significant challenge for future research.

8.9 Summary

There remains considerable uncertainty in the status of southern African glacial and periglacial records. This is largely because the evidence for cryogenic processes and associated landform development is often ambiguous, which is partly due to their marginal climatic setting, even during the coldest parts of the Quaternary. As such, there is disagreement on the identification and interpretation of many mountain landforms. Evidence of glacial moraines in the Lesotho sector of the Drakensberg range suggests that conditions were cold and wet enough during the LGM to sustain small glaciers, with a temperature depression of ~6°C and

ELAs down to ~3100 m a.s.l. Glacier reconstruction and mass balance modelling has enabled the reconstruction of regional LGM rainfall patterns, which at this time was enhanced during autumn-winter-spring. Evidence for enhanced periglacial activity at this time is uncertain, due to the absence of secure age-determinations, but is likely to have been widespread. Better identification and age determination of diagnostic periglacial landforms are important future research strands.

Acknowledgements

We thank Stephanie Mills, Rick Shakesby and Martin Kirkbride for their comments on a previous version of this chapter.

References

Benn, D. I. and Ballantyne, C. K. (1994). Reconstructing the transport history of glaciogenic sediments: A new approach based on the co-variance of clast shape indices. *Sedimentary Geology*, 91, 215–227.

Boelhouwers, J. C. (1994). Periglacial landforms at Giant's Castle, Natal Drakensberg, South Africa. *Permafrost and Periglacial Processes*, 5, 129–136.

Boelhouwers, J. C. (1999). Block deposits in southern Africa and their significance to periglacial autochthonous blockfield development. *Polar Geography*, 23, 12–22.

Boelhouwers, J. C. and Meiklejohn, I. (2002). Quaternary periglacial and glacial geomorphology of southern Africa: Review and synthesis. *South African Journal of Science*, 98, 47–55.

Boelhouwers, J. C., Holness, S., Meiklejohn, I. and Sumner, P. (2002). Observations on a blockstream in the vicinity of Sani Pass, Lesotho Highlands, southern Africa. *Permafrost and Periglacial Processes*, 13, 251–257.

Booth, P. W. K. (2011). Stratigraphic, structural and tectonic enigmas associated with the Cape Fold Belt: Challenges for future research. *South African Journal of Geology*, 114, 235–248

Borchert, G. and Sänger, H. (1981). Research findings of a Pleistocene glaciation of the Cape mountain-ridge in South Africa. *Zeitschrift für Geomorphologie*, 25, 222–224.

Chase, B. M. and Meadows M. E. (2007). Late Quaternary dynamics of southern Africa's winter rainfall zone. *Earth-Science Reviews*, 84, 103–138.

Cockcroft, M. J., Wilkinson, M. J. and Tyson, P. D. (1987). The application of a present-day climatic model to the late Quaternary in southern Africa. *Climatic Change*, 10, 161–181.

du Toit, A. L. (1937). *Our Wandering Continents: An Hypothesis of Continental Drifting*. London: Oliver & Boyd, 366pp.

du Toit, A. L. (1954). *The Geology of South Africa*. London: Oliver & Boyd, 611pp.

Dyer, T. G. J. and Marker, M. E. (1979). On some aspects of Lesotho hollows. *Zeitschrift für Geomorphologie*, 23, 256–270.

Flint, R. F. (1959). Pleistocene climates of eastern and southern Africa. *Bulletin of the Geological Society of America*, 70, 343–374.

Gasse, F., Chalié, F., Vincens, A., Williams, M. A. J. and Williamson, D. (2008). Climatic patterns in equatorial and southern Africa from 30,000 to 10,000 years ago reconstructed from terrestrial and near-shore proxy data. *Quaternary Science Reviews*, 27, 2316–2340.

Gellert, J. F. (1991). Pleistozän-kaltzeitliche Vergletscherungen im Hochland von Tibet und im südafrikanischen Kapgebirge. *Eiszeitalter und Gegenwart*, 41, 141–145.

Grab, S. W. (1999). Block and debris deposits in the High Drakensberg, Lesotho, Southern Africa: Implications for high altitude slope processes. *Geografiska Annaler*, 81A, 1–16.

Grab, S. W. (2000). Stone-banked lobes and environmental implications, high Drakensberg, southern Africa. *Permafrost and Periglacial Processes*, 11, 177–187.

Grab, S. W. (2002). Characteristics and palaeoenvironmental significance of relict sorted patterned ground, Drakensberg plateau, southern Africa. *Quaternary Science Reviews*, 21, 1729–1744.

Grab, S. W. (2004). Thermal regime through a sorted circle and stone-banked lobe, Drakensberg, Southern Africa. *Zeitschrift für Geomorphologie*, 48, 501–518.

Grab, S. W. (2010). Alpine turf exfoliation pans in Lesotho, southern Africa: Climate-process-morphological linkages. *Geomorphology*, 114, 261–275.

Grab, S. W., Mulder, N. and Mills, S. C. (2009). Spatial associations between longest-lasting winter snow cover and cold region landforms in the high Drakensberg, southern Africa. *Geografiska Annaler*, 91A, 83–97.

Grab, S. W., Mills, S. C. and Carr, S. J. (2012). Periglacial and Glacial Geomorphology. In *The Geomorphology of Southern Africa*, ed. P. Holmes and M. E. Meadows. Bloemfontein: Sun Press, pp. 233–265.

Hall, K. (2004). Glaciation in southern Africa. In *Quaternary Glaciations: Extent and Chronology. Part III: South America, Asia, Africa, Australia, Antarctica*, ed. J. Ehlers and P. L. Gibbard. Amsterdam: Elsevier, pp. 337–338.

Hall, K. and Meiklejohn, I. (2011). Glaciation in southern Africa and in the sub-Antarctic. In *Quaternary Glaciations: Extent and Chronology. Part III: South America, Asia, Africa, Australia, Antarctica*, ed. J. Ehlers and P. L. Gibbard. Amsterdam: Elsevier, pp. 1081–1085.

Holmgren, K., Lee-Thorp, J. A., Cooper, G. R. J., Lundblad, K., Partridge, T. C., Scott, L., Sithaldeen, R., Talma, A. S. and Tyson, P. D. (2003). Persistent millennial-scale climatic variability over the past 25 000 years in southern Africa. *Quaternary Science Reviews*, 22, 2311–2326.

Holzförster, F. (2007). Lithology and depositional environments of the Lower Jurassic Clarens Formation in the eastern Cape, South Africa. *South African Journal of Geology*, 110, 543–560.

Isbell, J. L., Cole, D. I. and Catuneanu, O. (2008). Carboniferous-Permian glaciation in the main Karoo Basin, South Africa: Stratigraphy, depositional controls, and glacial dynamics. In *Resolving the Late Paleozoic Ice Age in Time and Space*, ed. C. R. Fielding, T. D. Frank and J. L. Isbell. Geological Society of America Special Publication, 441, pp. 71–82.

Jourdan, F., Féraud, G., Bertrand, H., Watkeys, M. K. and Renne, P. R. (2007). Distinct brief major events in the Karoo large igneous province clarified by new $^{40}Ar/^{39}Ar$ ages on the Lesotho basalts. *Lithos*, 98, 195–209.

Knight, J. (2012). The shape of glacial valleys: Comment on Hall (2010). *South African Geographical Journal*, 94, 1–3.

Kojot, D. F. (1948). *An Investigation into the Evidence Bearing on Recent Climate Changes over South Africa*. Pretoria: South Africa Irrigation Department Memoir, Government Printer, 160pp.

Lewis, C. A. and Hanvey, P. M. (1993). The remains of rock glaciers in Bottelnek, East Cape Drakensberg, South Africa. *Transactions of the Royal Society of South Africa*, 48, 265–289.

Lewis, C. A. and Illgner, P. M. (2001). Late Quaternary glaciation in southern Africa: Moraine ridges and glacial deposits at Mount Enterprise in the Drakensberg of the Eastern Cape Province, South Africa. *Journal of Quaternary Science*, 16, 365–374.

Marker, M. E. (1991). The evidence for cirque glaciations in Lesotho. *Permafrost and Periglacial Processes*, 2, 21–30.

Marker, M. E. and Whittington, G. (1971). Observations on some valley forms and deposits in the Sani Pass area, Lesotho. *South African Geographical Journal*, 53, 96–99.

Master, S. (2012). Hertzian fractures in the sub-Dwyka Nooitgedacht striated pavement, and implications for the former thickness of Karoo strata near Kimberley, South Africa. *South African Journal of Geology*, 115, 561–576.

Mills, S. and Grab, S. W. (2005). Debris ridges along the southern Drakensberg escarpment as evidence for Quaternary glaciation in southern Africa. *Quaternary International*, 129, 61–73.

Mills, S. C., Grab, S. W. and Carr, S. J. (2009a). Recognition and palaeoclimatic implications of late Quaternary niche glaciation in eastern Lesotho. *Journal of Quaternary Science*, 24, 647–663.

Mills, S. C., Grab, S. W. and Carr, S. J. (2009b). Late Quaternary moraines along the Sekhokong range, eastern Lesotho: Contrasting the geomorphic history of north- and south-facing slopes. *Geografiska Annaler*, 91A, 121–140.

Mills, S. C., Grab, S. W., Rea, B. R., Carr, S. C. and Farrow, A. (2012). Shifting westerlies and precipitation patterns during the Late Pleistocene in southern Africa determined using glacier reconstruction and mass balance modelling. *Quaternary Science Reviews*, 55, 145–159.

Osmaston, H. A. and Harrison, S. P. (2005). The late Quaternary glaciations of Africa: A regional synthesis. *Quaternary International*, 138–139, 32–54.

Partridge, T. C., Scott, L. and Hamilton, J. E. (1999). Synthetic reconstructions of southern African environments during the Last Glacial Maximum (21–18 kyr) and the Holocene Altithermal (8–6 kyr). *Quaternary International*, 57/58, 207–214.

Sänger, H. (1987). Pleistocene glaciations in the Western Cape Folded Belt of South Africa. In *Current Research in African Earth Sciences*, ed. G. Matheis and H. Schandelmeier. Rotterdam: Balkema, pp. 447–449.

Sänger, H. (1988). Vergletscherung der Kap-Ketten im Pleistozän. *Berliner Geographische Studien*, 26, 182.

Scott, L., Neumann, F. H., Brook, G. A., Bousman, C. B., Norström, E. and Metwally, A. A. (2012). Terrestrial fossil-pollen evidence of climate change during the last 26 thousand years in Southern Africa. *Quaternary Science Reviews*, 32, 100–118.

Sladen, J. A. and Wrigley, W. (1983). Geotechnical properties of lodgement till: A review. In *Glacial Geology; An Introduction for Engineers and Earth Scientists*, ed. N. Eyles. London: Permagon, pp. 184–212.

Sparrow, G. W. A. (1964). Pleistocene periglacial landforms in the southern hemisphere. *South African Journal of Science*, 60, 143–147.

Sparrow, G. W. A. (1967). Southern African cirques and aretes. *Journal of Geography*, 11, 9–11.

Sparrow, G. W. A. (1974). Non-glacial cirque formation in southern Africa. *Boreas*, 3, 61–68.

Sumner, P. D. (2004). Relict sorted patterned ground in Lesotho. *Permafrost and Periglacial Processes*, 15, 89–93.

Tyson, P. D. and Partridge, T. C. (2000). Evolution of Cenozoic Climates. In *The Cenozoic of Southern Africa*, ed. T. C. Partridge and R. R. Maud. Oxford: Oxford University Press, pp. 371–387.

Visser, J. N. J. and Hall, K. J. (1985). Boulder beds in the glaciogenic Permo Carboniferous Dwyka Formation in South Africa. *Sedimentology*, 32, 281–294.

von Brunn, V. (1996). The Dwyka Group in the northern part of Kwazulu/Natal, South Africa: Sedimentation during late Palaeozoic deglaciation. *Palaeogeography, Palaeoclimatology, Palaeoecology*, 125, 141–163.

9

Colluvial deposits and slope instability

GREG A. BOTHA, ARNAUD J.A.M. TEMME AND REBEKAH G. SINGH

Abstract

Slope instability results from the interplay between tectonic uplift, climate-controlled weathering, and slope form (relief). Sediments accumulate as colluvium on footslopes when under the influence of gravity, fluvial processes and mass movements, and may be affected by pedogenesis, ecosystem change and human activity. This chapter shows that colluvial form and stratigraphy cannot be uncritically assigned to particular late Pleistocene and Holocene climatic periods, because of the feedbacks associated with weathering rate, vegetation responses to climate, and slope angle. Moreover, land surface instability and colluvial formation/incision is taking place due to contemporary environmental degradation, closely related to overgrazing.

9.1 Introduction

Large amounts of unconsolidated sediment can accumulate on shallow hillslopes from transport mechanisms as diverse as aeolian and fluvial processes and mass movements. Many authors refer to hillslope deposits as 'colluvium', which tends to be defined differently in various contexts around the world (Leopold and Völkel, 2007). The term generally refers to locally derived sediments, transported by running water or under the influence of gravity. This distinguishes alluvial sediments which are commonly channel-bound, from the unconfined movement of sheetwash or mass movement deposits (Fuchs and Lang, 2009). Stratified colluvial sediments in southern Africa are widely distributed below steep hillslope segments where mass movement processes dominate (Moon, 1998). Thick regolith, weathering products or soil profiles are commonly the sources of sediments that are redistributed downslope. Accretion on slopes where the gradient might exceed the angle of repose for sediments of various sizes, textures and shapes renders such deposits inherently unstable and susceptible to redistribution downslope. Landslides occur within regolith or weathered bedrock that is rendered unstable due to settlement, consolidation, high moisture content, seismic vibration or anthropogenic destabilisation. This unsorted debris may then be reworked and transported onto lower slopes by sheetwash and shallow ephemeral channel processes. On lower gradient slopes or in climatic zones where vegetation cover is able to reduce the influences of rainsplash or overland flow, fine-textured colluvial sediments accrete and are subject to pedogenic modification when slope instability declines.

Comprehensive studies of mass movement process and landslide susceptibility modelling have been published at a national scale in South Africa by Singh et al. (2011) and at a regional scale such as for KwaZulu-Natal Province (Singh et al., 2008; Singh, 2013) (Fig. 9.1). The widespread distribution of fine-textured, sandy hillslope deposits, typified by the gully (donga) sidewall exposures of colluvium across KwaZulu-Natal and part of Free State Province have also been described (Botha et al., 1992, 1994; Botha, 1996; Botha and Partridge, 2000; Clarke et al., 2003; Temme et al., 2008). Similar sediments in parts of Swaziland were described by Watson et al. (1984) and Dardis (1989, 1990). Thick wedges of colluvial deposits on lower hillslopes can represent valuable archives for interpreting slope evolution and the effects of cyclical climate change on slope stability (Botha et al., 1994; Botha, 1996; Temme et al., 2008).

This chapter assesses the diverse deposits associated with slope instability in southern Africa. These hillslope sediments are assessed from the perspective of landscape evolution at various timescales, as well as their significance with respect to the palaeoenvironmental conditions that trigger hillslope instability and colluviation, relative to factors that result in stabilisation of the deposits.

9.2 Slope processes in the long-term landscape evolution of southern Africa

Research into the terrain morphology, Cenozoic geology and geomorphological history has contextualised the long-term landscape evolution of the southern Africa region (e.g. Fleming et al., 1999; Partridge and Maud, 1987, 2000). Different slope elements are shown in Figure 9.2. Interpretations of landscape evolution in southern Africa have speculated on the existence of ancient land surface remnants (King, 1944) and highlighted the significance of Great Escarpment backwearing by fluvial and slope processes in response to continental uplift over a period of about 140 Ma since the rifting of Gondwana during the early Cretaceous (Partridge and Maud, 1987, 2000).

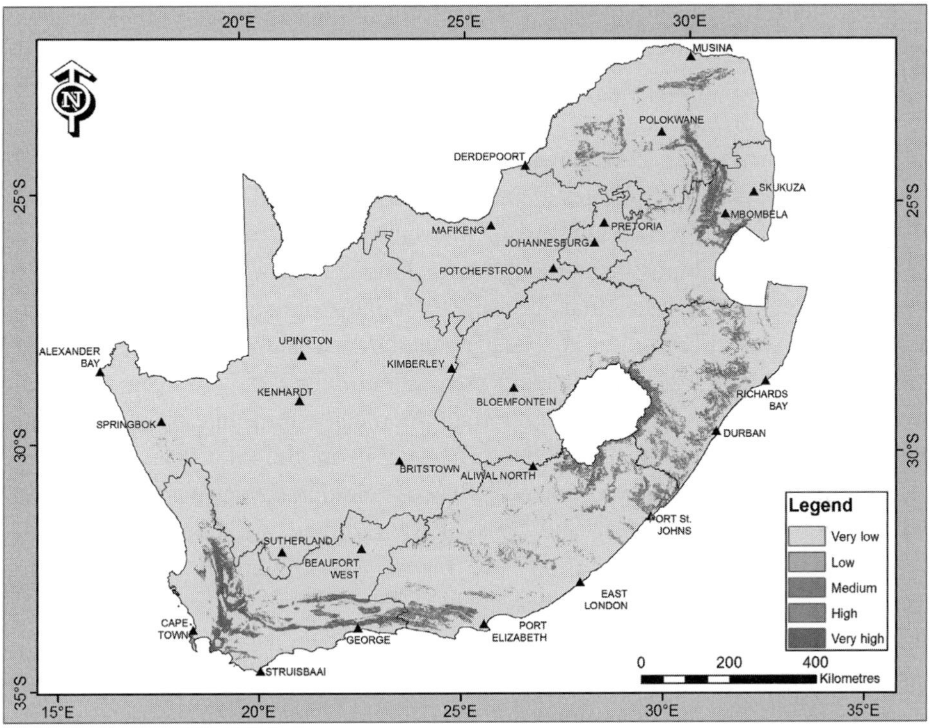

Fig. 9.1. Landslide susceptibility map for South Africa (adapted after Singh et al., 2011). A black and white version of this figure will appear in some formats. For the colour version, please refer to the plate section.

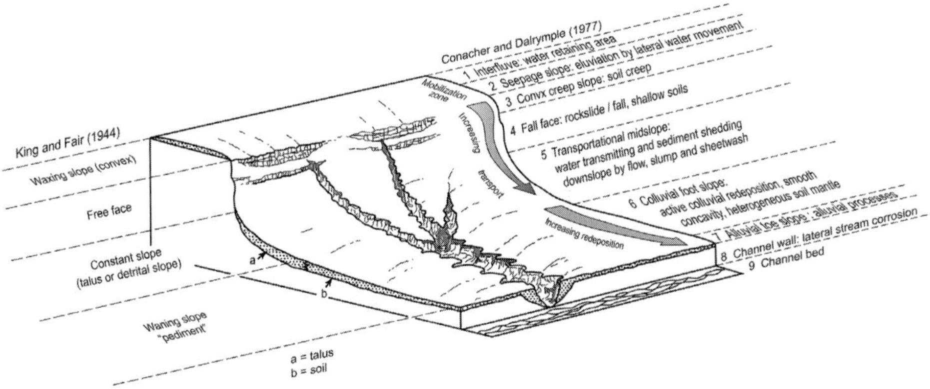

Fig. 9.2. Schematic longitudinal section through a hillslope showing the land surface and slope elements, as defined by King and Fair (1944) and Conacher and Dalrymple (1977) (adapted from Botha, 1996).

These morphotectonic models and patterns of regional epeirogenesis have been tested by quantitative assessments of bedrock denudation chronologies using cosmogenic ^{36}Cl methods on exposed rock surfaces along the Drakensberg sector of the Great Escarpment (Fleming et al., 1999), apatite fission-track analysis (Brown et al., 2000, Tinker et al., 2008), cosmogenic ^{26}Al and ^{10}Be isochron burial dating (Erlanger et al., 2012), and ^{26}Al and ^{10}Be exposure dating (Bierman et al., 2014). The analysis of resistant bedrock formations and alluvial or marine terrace deposits using these techniques has generated a range of averaged denudation rates that are able to contextualise geomorphic activity across the region. Fleming et al. (1999) calculated a scarp retreat rate of 50–95 m/Ma for sites along the Drakensberg escarpment over a 10^4–10^6 year timespan, which is an order of magnitude lower than the mean retreat rate of ~1.0–1.5 km/Ma implied by the Partridge and Maud (1987) model of Great Escarpment backwearing since the early Cretaceous. These calculated rates have implications for slope debris production and thus slope stability. Fleming et al. (1999) compare the rate of thin spalling (49–63 m/Ma) to that of block shedding (83–95 m/Ma) on the Drakensberg escarpment. The low calculated rate of summit lowering, only 6 m/Ma, points to the differential geomorphic impact of instability of steep slopes and areas of high relief, compared to low gradient slopes (*ibid.*). This is also supported by dating evidence elsewhere along the Great Escarpment (Chadwick et al., 2013).

By comparing terrestrial sediment supply to offshore deposition budgets since ~136 Ma, Tinker et al. (2008) showed that regional-scale denudation rates decreased by an order of magnitude during the Cenozoic, which may be due to a combination of climate change, tectonic quiescence and greater slope stability at this time. Using ^{10}Be in river sediments, Bierman et al. (2014) found rates of erosion in south-central South Africa of ~5 m/Ma, consistent with other tectonically quiescent areas. These long term landscape process rates refute the hypotheses of large scale, rapid uplift events (Partridge and Maud, 1987, 2000) and are consistent with slower rates of Cenozoic exhumation (e.g. Erlanger et al., 2012). These long term process rates also assume that bedrock responses are uniform spatially and temporally, although differences in rock mass strength, rock type and structure may also be important controls (Moon and Selby 1983; Moore and Blenkinsop, 2006). In addition, there are process–form feedbacks that mean that, as slope height, length and angle changes over time in response to headwall erosion, the rate of that erosion decreases (Moon and Munro-Perry, 1988). Although colluvial deposits related to early land surface remnants are rare, gravels cemented by siliceous and ferruginous duricrusts of Neogene age are preserved on pediment remnants in the Cape Fold Mountain foothills (Bierman et al., 2014). In most cases, however, even where the Pleistocene colluvial succession is well constrained by luminescence dating, derivation of short term rates of accretion is hampered by hiatuses.

9.3 Colluvial processes

In many parts of southern Africa, accumulation of slope deposits occurs below the transition from steep, relatively unstable upper hillslopes to the long, low-gradient pediment slopes. Here, thick colluvial wedges may develop below bedrock steps. Over the Pleistocene, global glacial–interglacial climatic changes have influenced the rate of regolith production on upper hillslopes through alternation of periods conductive to weathering, and subsequent failure of regolith through mass movements. The cyclical generation of erodible soil or landslide sediment that is transported to and deposited on mid slopes, and the hillslope environmental conditions that lead to accretion or stripping of pediment cover, is intrinsic to the understanding of spatial and temporal cycles of hillslope activity and stability across the region (Singh et al., 2008). Thick regolith can be destabilised and transported downslope by different hillslope processes.

9.3.1 Hillslope colluviation

Cyclical development of pediment surfaces during the Pleistocene has led to the accretion of sheetwash deposits, in particular in KwaZulu-Natal. The lack of widespread older colluvial mantles suggests that these sediments accrete episodically and their preservation is temporary over longer geological timescales, as long as uplift and drainage channel rejuvenation continue. These hillslope deposits in KwaZulu-Natal have been assigned as the Masotcheni Formation (Partridge et al., 2006) of broadly Cenozoic age, although the precise age and the spatial and temporal context of Masotcheni Formation-equivalent colluvia elsewhere in southern Africa remain uncertain.

In hilly terrain with marked relief and rock outcrops defining hill crests, the transition zone from convex to concave lower slopes exhibits a smooth landsurface form and subdued microrelief, commonly indicative of thick soil cover. Geomorphic models by King and Fair (1944) and Conacher and Dalrymple (1977) have characterised this transitional zone as the pediment or transportational midslope/colluvial footslope (Fig. 9.2). Across a broad range of climatic conditions, mainly on the eastern seaboard of southern Africa, erosional landscapes where drainage lines are incising through headcut erosion of rills and gullies (Fig. 9.3), infilled high-order tributary valleys and colluvial hollows have been exposed. These reveal profiles through poorly sorted, thinly bedded colluvial sediment and complex soil profiles that attest to the long-term degradation of this landscape (Dardis, 1989, 1990; Botha

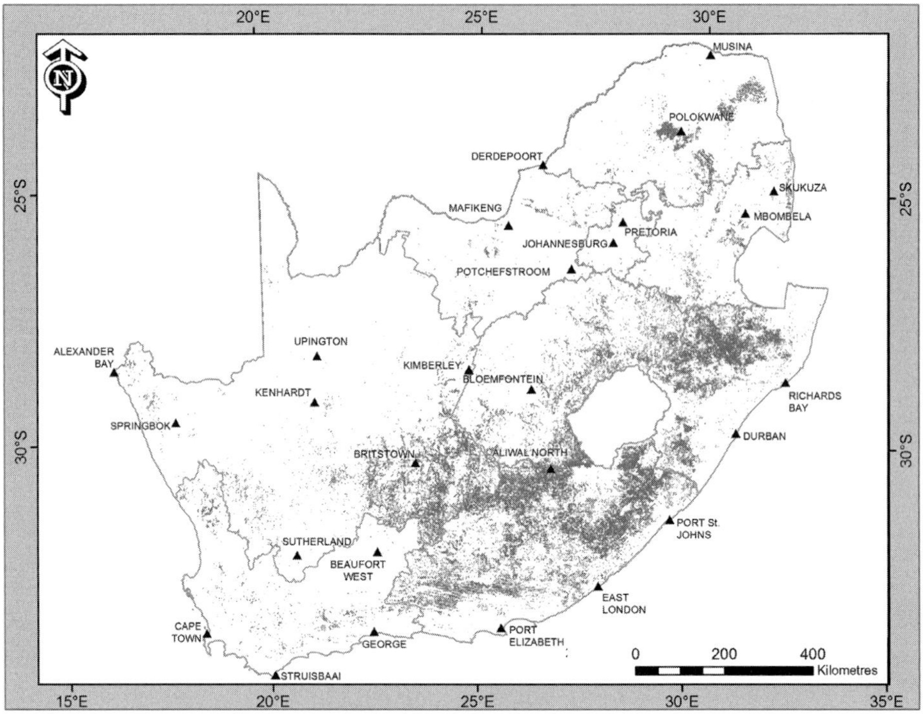

Fig. 9.3. The locations of gully systems in South Africa (adapted after le Roux, 2011). These are mainly cut into the Masotcheni Formation and equivalent colluvia. A black and white version of this figure will appear in some formats. For the colour version, please refer to the plate section.

et al., 1992). Such sediments are nonetheless widespread and in some locations may be stable over timescales of up to 150 kyr. Particularly in the central and western parts of KwaZulu-Natal, hillslopes are often mantled by Masotcheni Formation colluvial sediments that appear to have accreted in pulses rather than gradually (Botha, 1996). The main indicators for cyclical accretion and erosion include post-depositional pedogenesis and the existence of erosional hiatuses in the colluvial stratigraphy, represented by infilled palaeogullies and buried palaeosols (Fig. 9.4).

Research on colluvial sediments of the Masotcheni Formation has focussed on their stratigraphic subdivision by means of buried palaeosols and depositional layering, which allows for regional correlation (Botha et al., 1992; Botha and Fedoroff, 1995; Botha, 1996; Clarke et al., 2003; Temme et al., 2008, 2012). The palaeoenvironmental context for colluvial events can be informed by regional palaeoclimate records such as the 200-kyr rainfall record from the Pretoria Saltpan (Tswaing) (Partridge et al., 1997) and pollen/charcoal records from eastern South Africa (Scott, 2002). Although the Masotcheni Formation colluvia cover large areas of in particular eastern South Africa, the texture of the colluvial sediments is strongly linked to the weathered bedrock regolith. For example, sandstones and shales weather to medium to fine-grained,

Fig. 9.4. Photos of depositional and erosional elements of gully systems in KwaZulu-Natal. (a) sedimentary succession of slope sediments and buried palaeosols (photo courtesy of Mary Evans), (b) road-cut at Keate's Drift showing bedrock-cut gullies which have then been infilled by slope sediments (I) (photo: Greg Botha). A black and white version of this figure will appear in some formats. For the colour version, please refer to the plate section.

moderately sorted sands which may give rise to well-stratified slope deposits and possibly perched water tables. By contrast, sediments derived from weathered granite are coarser, have higher permeability and steeper angles of repose, but may also contain weathered core-stones which thus give rise to a bimodal sediment texture. Depositional laminations may be preserved locally when not obscured by post-depositional pedogenesis. Silt and clay derived from eroded regolith upslope is winnowed and generally transported downslope. At most sites the increase in silt and clay in colluvial sediments is closely related to the degree of soil formation. Sand deposition is influenced by the density of grass vegetation cover or surface roughness in the form of rills and shallow gullies. However, Telfer et al. (2014) interpreted stratified and massive fine sand deposits within hillslope deposits in the eastern Free State mountains as aeolian sediments interbedded with Masotcheni Formation colluvium. Pebble lenses within the sediment stratigraphy commonly represent shallow or ephemeral channel deposits. Colluvium may also contain clasts of reworked soil material or resistant materials like ferricrete pisoliths, which may accumulate as lags. Gravel lags sometimes contain stone tool artefacts, suggesting residual concentration of large clasts on the land surface. The relatively low energy sheetwash transport mechanism is illustrated by the intact burial of large 'knapping floor' middens from the Middle Stone Age (Botha, 1996).

9.3.2 Distribution of colluvium in southern Africa

Watson et al. (1984) regarded colluvium as mantling about 20% of the land surface of Africa south of the Zambezi, representing one of the most widespread geomorphic deposits on the subcontinent. Indeed, these slope deposits have been described from the Mpumalanga Lowveld through Swaziland and the drainage basins of central KwaZulu-Natal to the Eastern Cape. At higher elevations, hill-slope deposits occur in the Drakensberg foothills and extend through the eastern Free State and are widespread in the Maluti mountain foothills of Lesotho (Botha, 1996). These deposits have a more restricted distribution in the semi-arid parts of the northern provinces of South Africa, Zimbabwe and Botswana (Watson et al., 1984). The characteristics and distribution of colluvial deposits that are commonly exposed by preferential gully erosion have been reviewed by Botha (1996) and Botha and Partridge (2000). Coarse mass movement debris and crudely stratified boulder deposits are common on steep slopes in the Cape Fold Mountains (Partridge and Botha, 2000). The rolling hills of the Swartland region of the Western Cape Province also exhibit gullying of regolith (Boardman et al., 2012).

The lithological structural control on hillslope morphology defined by alternating sandstone/shale beds or the effect of resistant dolerite sills and shale beds or the influence of lithological steps in river gradients can influence the thickness and distribution of colluvial sediment mantles on adjacent hillslopes. Landscape-scale complexity, often modulated by rivers but affecting entire catchments, can influence the rate of colluviation and erosion independently of vegetation cover change or precipitation characteristics. Knickpoints on streams or slope breaks can reduce erosive forces or the headcutting retreat rate of gullies, thereby reducing colluviation and promoting pedogenesis on the land surface (Temme et al., 2008, 2012).

9.3.3 Dating of colluviation

In order to interpret the environmental and climatic controls on land surface stability, dating control on sediment layers or buried palaeosols is needed. Prior to the availability of radiocarbon dating, researchers interpreted hillslope erosion and colluviation events on the basis of the incorporated Early or Middle Stone Age artefacts, which lacked chronological bracketing (e.g. Davies, 1950), or with reference to the sequence of late Pleistocene global climatic cycles. Stone Age artefact assemblages and carbonate nodules were used to define the Bovu and Mphunga Colluvium units in the lower Middelveld of Swaziland, from which accumulation of the Mphunga sediment after ~40 kyr BP was inferred (Price Williams et al., 1982; Watson et al., 1984). Radiocarbon dating of colluvial sequences is often limited by the absence of deposited organic matter such as charcoal. Secondary soil carbonates or organics often represent post-depositional pedogenic modification, which means that the radiocarbon ages determined are a weighted mean of the longer period of formation (Botha, 1996). Subsequently, radiocarbon and luminescence dating can be deployed singly or in combination for age-control of palaeosols and colluvial deposits (e.g. Botha et al., 1992; Wintle et al., 1993, 1994; Botha, 1996; Telfer et al., 2012; Lyons et al., 2013).

A comprehensive chronological framework spanning over 135 kyr of Masotcheni Formation colluvial accretion on hillslopes in KwaZulu-Natal and the eastern Free State was based on radiocarbon and luminescence dates reported by Wintle et al. (1995) and Botha (1996). Multiple accretion phases of the Voordrag palaeosol succession over more than 100 kyr was defined by bracketing radiocarbon-dated palaeosols (Botha et al., 1992) with luminescence dates on the intervening colluvial sediments (Clarke et al., 2003). Colluvium and palaeosols in the Okhombe valley of the Drakensberg foothills were dated using the luminescence method by Temme et al. (2008), indicating eleven phases of colluvial accretion since ~42 kyr. In the higher elevation valleys of the Golden Gate area, Telfer et al. (2012) used the luminescence method to date slope sediment accretion during the period 47–16 kyr BP. Fuchs and Lang (2009) reviewed the use of luminescence dating of colluvium where inadequate duration, intensity and spectral composition of light exposure during transport can result in poorly bleached sediment, resulting in age over-estimation. Unfortunately, colluvial successions seldom contain materials that can be used to provide reliable independent chronological assessments against which the luminescence ages can be assessed.

9.4 Palaeoenvironmental interpretation of colluvial and slope deposits

Age estimates suggest that Pleistocene colluvium often accumulated during arid stages, while pedogenesis may have occurred under periods of relative stability influenced by greater humidity and vegetation cover. The conceptual model underpinning this suggestion was that of Knox (1972), that predicts a peak in redistribution (and hence colluviation) during transitions from dry to wet conditions, when precipitation has increased but whilst vegetation has not yet recolonised slopes. Temperature was not generally found to be a significant driving factor for colluviation (Clarke et al., 2003), although Temme et al. (2008) suggest that low temperatures may have been a factor controlling the occurrence of solifluction-derived colluvium at the base of steep slopes at higher elevations, close to the Drakensberg escarpment, where cold-air drainage played a role. During the same period (after ~43.1±0.5 kyr BP), Grab and Mills (2011) report pulses of colluvial sediment deposition associated with solifluction in the high Drakensberg plateau region. Correlation of phases of colluviation and pedogenesis between different locations in KwaZulu-Natal has been attempted by Botha (1996). This has to some extent been successful, and has led to the definition of allostratigraphic units that can be used to correlate colluvial strata between sites; and pedoderms, that can correlate palaeosols formed in the colluvium.

Luminescence dating of the deposits within Sibudu rock shelter in KwaZulu-Natal by Jacobs et al. (2008) identified three periods of occupation during the Middle Stone Age. These were associated with distinct cultural phases centred on 58.5, 47.7 and 38.6 kyr BP. Occupational hiatuses around 10.8±1.3 and 9.1±3.6 kyr BP, during Marine Isotope Stage (MIS) 3, were interpreted as representing dry periods when environmental conditions impacted human activities in the area. These periods coincide with colluvial deposition events on hillslopes in the interior of KwaZulu-Natal, bracketed by periods of hillslope stability associated with palaeosol development, that have been widely described from the Masotcheni Formation colluvial deposits (Botha, 1996, 2000; Botha and Partridge, 2000; Clarke et al., 2003). Periods of regional slope instability and soil erosion on hillslopes during MIS 4 and 3 were probably associated with altered rainfall patterns, environmental cooling and desiccation, and reduced vegetation cover. Evidence of significant change in vegetation structure and composition during this time has been revealed in pollen assemblages of wetland palaeosols exposed in the Voordrag donga within the Mkhuze River valley, northern KwaZulu-Natal (Botha et al., 1992; Clarke et al., 2003).

The relationship between colluvial slope deposits and floodplain alluvium indicates the changing balance between valley sides and fluvial channel activity in the interior of southeastern Africa. The steep-sided Mkhomazi River valley (KwaZulu-Natal) exposes thick colluvial deposits which buried two generations of alluvial terraces, represented by rapid/pool sedimentation and oxbow channel infill deposits. Accretion of sheetwash sediments over terrace alluvium at ~38 kyr BP, and gradual burial of floodplain palaeosols that formed prior to ~43 kyr BP, suggests that the gravelly colluvium derived from fissile shale exposed on the valley slopes accreted during the period preceding the coldest, driest phase of the last glacial period in this region.

Telfer et al. (2012) interpreted the sand accreted on slopes below Clarens Formation sandstone exposed in the Golden Gate area (Free State Province) as sand ramps comprising aeolian sand blown from the high altitude valleys during MIS 3/2 in the late Pleistocene. Studies over a broader area reported by Telfer et al. (2014) correlated the gully eroded hillslope deposits with the Masotcheni Formation of KwaZulu-Natal, and noted the inclusion of aeolian sand beds not recorded in occurrences elsewhere.

9.4.1 Present-day dynamics: land surface degradation

Since the early 20th Century, widespread erosion has been observed into colluvium across southern Africa (e.g. Botha, 1996). This problem has been linked to grazing pressure as well as cultivation of soils susceptible to erosion (Boardman et al., 2012; Mighall et al., 2012). This erosion causes losses in agricultural productivity and soil water retention capacity (Temme et al., 2012), more difficult access to fields (Sonneveld et al., 2005), siltation of downstream water courses, and damage to buildings and infrastructure. Based on pollen and charcoal preserved within recent sediments at Compassberg, Eastern Cape, Mighall et al. (2012) suggested that soil erosion was enhanced by agricultural practices since the 1930s. The onset of gullying within thick colluvial deposits at St Paul's (KwaZulu-Natal) is indicated by a surficial layer of laminated sand that buries a soil horizon dated at 1290±30 cal yr BP (Pta-4837) (Wintle et al., 1993), suggesting that gully incision has occurred within the past 1500 years. Gully sites at the Blood River vlei (KwaZulu-Natal) indicate final hillslope sedimentation at ~0.29 kyr BP for one site, with deposition of sandy sediment forming fans below incised gullies from ~0.89 kyr BP (Lyons et al., 2013). The timing of these indicators of hillslope instability was correlated with the Medieval Warm Period (~AD 900–1300) and Little Ice Age (~AD 1300–1800).

Past erosion phases are also evidenced in the colluvial stratigraphy (Botha et al., 1994), which raises the question as to what extent current erosion is a natural process influenced by climate changes (Lyons et al., 2013), or whether it is caused by human (land use) pressures. This debate fits within a wider discussion in southern Africa and elsewhere about the social context of erosion (e.g. Meadows, 2003; Boardman et al., 2012), yet remains largely unresolved. In the colluvium of the Okhombe valley (KwaZulu-Natal), Sonneveld et al. (2005) focussed on this issue, concluding that increasing population pressure, expressed as the number of homesteads within the catchment, did not coincide with an increasing amount of erosion features over the period AD 1945–2000, although abandonment of fields did seem to cause an increase in erosion. In addition, extensive colluvial mantles up to 13 m thick in the high Drakensberg are entirely a product of climate-driven geomorphic processes, without human influences (Grab and Mills, 2011). Supporting natural driving factors, Rienks et al. (2000) found that the erodibility of Masotcheni Formation sediments elsewhere in KwaZulu-Natal was correlated with soil electrical conductivity and sodium adsorption ratio values, and that landscape position and geomorphic threshold conditions also co-determine erosion. Whereas initial gully incision processes are often related to surficial soil chemistry, the catastrophic evacuation of bedrock depressions infilled with colluvium is related to the inherent erodibility of these sandy sediments, influenced partly by the chemical characteristics of buried palaeosols.

Many colluvial sequences in KwaZulu-Natal contain strata with varying textures, including layers with low hydraulic permeability. Consequently, in many cases there is considerable interflow which, combined with the high erodibility of sediments (Rienks et al., 2000), leads to a prevalence of subsurface erosion though tunnels or soil pipes (Beckedahl, 1998) (Fig. 9.5). Soil pipes often grow in size until their roofs

Fig. 9.5. A soil pipe and its collapse into a gully system (photo: Arnaud Temme).

collapse and a gully system is suddenly visible on the surface. It is currently unknown to what extent soil pipes have functioned as precursors of the extensive gully systems in colluvial deposits, and to what extent they continue to do so.

The absence of detailed regional mapping of colluvial deposits across the landscape hinders the interpretation of regional gully erosion maps such as those produced by le Roux et al. (2007, 2008). The majority of gullies are associated with thick colluvial/palaeosol deposits, but deep weathering profiles in high rainfall areas are also susceptible to gullying. The obvious disruption of dryland agriculture, sedimentation of river channels and increased turbidity of runoff water entering rivers, are widely acknowledged erosional aspects of colluvium.

9.5 Summary

Colluvium deposition in footslope locations is well marked across southern Africa. Studies of colluvial stratigraphy and dating of buried palaeosols have been undertaken in particular in KwaZulu-Natal where relatively steep slopes and weathering of sandstones and basalts of the Great Escarpment region have given rise to a complex interplay between fluvial action, slope processes and land-use change. Episodes of colluvium deposition generally correspond to cool Pleistocene climate phases, but incision and gullying of colluvial slopes is not so clearly linked to climate forcing, and may reflect a combination of climatic, geological and anthropogenic factors. These are as yet poorly understood, and colluvial deposits generally represent an under-exploited palaeoenvironmental archive in southern Africa.

References

Beckedahl, H. R. (1998). *Subsurface Soil Erosion Phenomena in South Africa*. Gotha: Perthes, 128pp.

Bierman, P. R., Coppersmith, R., Hanson, K., Neveling, J., Portenga, E. W. and Rood, D. H. (2014). A cosmogenic view of erosion, relief generation, and the age of faulting in southern Africa. *GSA Today*, 24(9), 4–11.

Boardman, J., Hoffman, M. T., Holmes, P. J. and Wiggs, G. F. S. (2012). Soil erosion and land degradation. In *Southern African Geomorphology: Recent Trends and New Directions*, ed. P. Holmes and M. Meadows. Bloemfontein: Sun Press, pp. 307–328.

Botha, G. A. (1996). *The geology and palaeopedology of late Quaternary colluvial sediments in northern KwaZulu-Natal, South Africa*. Pretoria: Council for Geoscience, Memoir of the Geological Survey of South Africa, 83, 165pp.

Botha, G. A. (2000). Paleosols and duricrusts. In *The Cenozoic of Southern Africa*, ed. T. C. Partridge and R. R. Maud. Oxford: Oxford University Press, pp. 131–144.

Botha, G. A. and Fedoroff, N. (1995). Palaeosols in late Quaternary colluvium, northern KwaZulu-Natal, South Africa. *Journal of African Earth Sciences, and the Middle East*, 21, 291–311.

Botha, G. A. and Partridge, T. C. (2000). Colluvial deposits. In *The Cenozoic of Southern Africa*, ed. T. C. Partridge and R. R. Maud. Oxford: Oxford University Press, pp. 88–99.

Botha, G. A., Scott, L., Vogel, J. C. and von Brunn, V. (1992). Palaeosols and palaeoenvironments during the Late Pleistocene Hypothermal in northern Natal. *South African Journal of Science*, 88, 508–512.

Botha, G. A., Wintle, A. G. and Vogel, J. C. (1994). Episodic Late Quaternary palaeogully erosion in northern Kwazulu/Natal, South Africa. *Catena*, 23, 327–340.

Brown, R. W., Gallagher, K., Gleadow, A. J. W. and Summerfield, M. A. (2000). Morphotectonic evolution of the South Atlantic margins of Africa and South America. In *Geomorphology and Global Tectonics*, ed. M. A. Summerfield. Chichester: John Wiley and Sons, pp. 255–281.

Chadwick, O. A., Roering, J. J., Heimsath, A. M., Levick, S. R., Asner, G. P. and Khomo, L. (2013). Shaping post-orogenic landscapes by climate and chemical weathering. *Geology*, 41, 1171–1174.

Clarke, M. L., Vogel, J. C., Botha, G. A. and Wintle, A. G. (2003). Late Quaternary hillslope evolution recorded in eastern South African colluvial badlands. *Palaeogeography, Palaeoclimatology, Palaeoecology*, 197,199–212.

Conacher, A. J. and Dalrymple, J. B. (1977). The nine-unit landsurface model: An approach to pedogeomorphic research. *Geoderma*, 18, 1–154.

Dardis, G. F. (1989). Quaternary erosion and sedimentation in badland areas of southern Africa. *Catena*, 14, 1–9.

Dardis, G. F. (1990). Late Holocene erosion and colluvium deposition in Swaziland. *Geology*, 18, 934–937.

Davies, O. (1950). Notes from Natal. *South African Archaeological Bulletin*, 5, 115–120.

Erlanger, E. D., Granger, D. E. and Gibbon, R. J. (2012). Rock uplift rates in South Africa from isochron burial dating of fluvial and marine terraces. *Geology*, 40, 1019–1022.

Fuchs, M. and Lang, A. (2009). Luminescence dating of hillslope deposits: A review. *Geomorphology*, 109, 17–26.

Fleming, A., Summerfield, M. A., Stone, J. O., Fifield, L. K. and Cresswell, R. G. (1999). Denudation rates for the southern Drakensberg escarpment, SE Africa, derived from in-situ-produced cosmogenic ^{36}Cl: Initial results. *Journal of the Geological Society*, 156, 209–212.

Grab, S. W. and Mills, S. C. (2011). Quaternary slope processes and morphologies in the upper Sehonghong Valley, eastern Lesotho. *Proceedings of the Geologists' Association*, 122, 179–186.

Jacobs, Z., Wintle, A. G., Duller, G. A. T., Roberts, R. G. and Wadley, L. (2008). New ages for the post-Howiesons Poort, late and final Middle Stone Age at Sibudu, South Africa. *Journal of Archaeological Science*, 35, 1790–1807.

King, L. C. (1944). Geomorphology of the Natal Drakensberg. *Transactions, Geological Society of South Africa*, 47, 255–282.

King, L. C. and Fair, T. J. D. (1944). Hill-slopes and dongas. *Transactions, Geological Society of South Africa*, 47, 1–4.

Knox, J. C. (1972). Valley alluviation in southwestern Wisconsin. *Annals of the Association of American Geographers*, 62, 401–410.

le Roux, J. J. (2011). *Monitoring Soil Erosion in South Africa at a Regional Scale*. Pretoria: Council for Geoscience, ARC-ISCW report no GW/A/2011/23, Project GW 59/004 task 50, 34 pp.

le Roux, J. J., Newby, T. S. and Sumner, P. D. (2007). Monitoring soil erosion in South Africa at a regional scale: Review and recommendations. *South African Journal of Science*, 103, 329–335.

le Roux, J. J., Morgenthal, T. L., Malherbe, J., Pretorius, D. J. and Sumner, P. D. (2008). Water erosion prediction at a national scale for South Africa. *Water SA*, 34, 305–314.

Leopold, M. and Völkel, J. (2007). Colluvium: Definition, differentiation, and possible suitability for reconstructing Holocene climate data. *Quaternary International*, 162–163, 133–140.

Lyons, R., Tooth, S. and Duller, G. A. T. (2013). Chronology of controls of donga (gully) formation in the upper Blood River catchment, KwaZulu-Natal, South Africa: Evidence for a climatic driver of erosion. *The Holocene*, 23, 1875–1887.

Meadows, M. E. (2003). Soil erosion in the Swartland, Western Cape Province, South Africa: implications of past and present policy and practice. *Environmental Science and Policy*, 6, 17–28.

Mighall, T. M., Foster, I. D. L., Rowntree, K.M. and Boardman, J. (2012). Reconstructing recent land degradation in the semi-arid Karoo of South Africa: A palaeoecological study at Compassberg, Eastern Cape. *Land Degradation and Development*, 23, 523–533.

Moon, B. P. (1998). Hillslope form and development. In *The Geomorphology of Southern Africa*, ed. B. P. Moon and G. F. Dardis. Johannesburg: Southern Book Publishers, pp. 57–77.

Moon, B. P. and Munro-Perry, P. M. (1988). Slope development on the Clarens Sandstone Formation in the northeastern Orange Free State. *South African Geographical Journal*, 70, 57–68.

Moon, B. P. and Selby, M. J. (1983). Rock mass strength and scarp forms in southern Africa. *Geografiska Annaler Series A-Physical Geography*, 65, 135–145.

Moore, A. and Blenkinsop, T. (2006). Scarp retreat versus pinned drainage divide in the formation of the Drakensberg escarpment, southern Africa. *South African Journal of Geology*, 109, 599–610.

Partridge, T. C., Botha, G. A. and Haddon, I. G. (2006). Cenozoic deposits of the interior. In *The Geology of South Africa*, ed. M. R. Johnson, C. R. Anhaeusser and R. J. Thomas. Johannesburg/Pretoria: GSSA/Council for Geoscience, pp. 585–604.

Partridge, T. C., deMenocal, P. B., Lorentz, S. A., Paiker, M. J. and Vogel, J. C. (1997). Orbital forcing of climate over South Africa: A 200,000-year rainfall record from the Pretoria Saltpan. *Quaternary Science Reviews*, 16, 1125–1133.

Partridge, T. C. and Maud, R. R. (1987). Geomorphic evolution of southern Africa since the Mesozoic. *South African Journal of Geology*, 90, 179–208.

Partridge, T. C. and Maud, R. R. (2000). Macro-scale geomorphic evolution of southern Africa. In *The Cenozoic of Southern Africa*, ed T. C. Partridge and R. R. Maud. Oxford: Oxford University Press, pp. 3–18.

Price Williams, D., Watson, A. and Goudie, A. S. (1982). Quaternary colluvial stratigraphy, archaeological sequences and palaeoenvironment in Swaziland, southern Africa. *Geographical Journal*, 148, 50–67.

Rienks, S. M., Botha, G. A. and Hughes, J. C. (2000). Sediment properties related to the erodibility of colluvial layers exposed in an erosion gully (donga) in northern KwaZulu-Natal, South Africa. *Catena*, 39, 11–31.

Scott, L. (2002). Microscopic charcoal in sediments: Quaternary fire history of the grassland and savanna regions in South Africa. *Journal of Quaternary Science*, 17, 77–86.

Singh, R. G. (2013). *Landslide Classification, Characterisation And Susceptibility Modelling in KwaZulu-Natal*. Pretoria: Council for Geoscience, Bulletin 144, 97 pp.

Singh, R. G., Botha, G. A., Richards, N. P. and McCarthy, T. C. (2008). Holocene landslides in KwaZulu-Natal, South Africa. *South African Journal of Geology*, 111, 39–52.

Singh, R. G., Forbes, C., Chiliza, G., Diop, S., Musekiwa, C. and Claassen, D. (2011). *Landslide Geohazards in South Africa: Landslide Susceptibility Mapping, Socio-Economic Impacts, Mitigation and Remediation Measures*. Pretoria: Council for Geoscience, Report No 2011–0016, 58pp.

Sonneveld, M. P. W., Everson, T. M. and Veldkamp, A. (2005). Multi-scale analysis of soil erosion dynamics in Kwazulu-Natal, South Africa. *Land Degradation and Development*, 16, 287–301.

Telfer, M. W., Mills, S. C. and Mather, A. E. (2014). Extensive Quaternary aeolian deposits in the Drakensberg foothills, Rooiberge, South Africa. *Geomorphology*, 219, 161–175.

Telfer, M. W., Thomas, Z. A. and Breman, E. (2012). Sand ramps in the Golden Gate Highlands national park, South Africa: Evidence of periglacial aeolian activity during the last glacial. *Palaeogeography, Palaeoclimatology, Palaeoecology*, 313/314, 59–69.

Temme, A. J. A. M., Baartman, J. E. M., Botha, G. A., Veldkamp, A., Jongmans, A. G. and Wallinga, J. (2008). Climate controls on late Pleistocene landscape evolution of the Okhombe valley, KwaZulu-Natal, South Africa. *Geomorphology*, 99, 280–295.

Temme, A. J. A. M., Schaap, J. D., Sonneveld, M. P. W. and Botha, G. A. (2012). Hydrological effects of buried palaeosols in eroding landscapes: A case study in South Africa. *Quaternary International*, 265, 32–42.

Tinker, J., de Wit, M. and Brown, R. (2008). Mesozoic exhumation of the southern Cape, South Africa, quantified using apatite fission track thermochronology. *Tectonophysics*, 455, 77–93.

Watson, A., Price Williams, D. and Goudie, A. S. (1984). The palaeoenviromental interpretation of colluvial sediments and palaeosols of the Late Pleistocene Hypothermal in southern Africa. *Palaeogeography, Palaeoclimatology, Palaeoecology*, 45, 225–249.

Wintle, A. G., Botha, G. A., Li, S. H. and Vogel, J. C. (1995). Chronological framework for colluviation during the last 110 ka in Natal. *South African Journal of Science*, 91, 134–139.

Wintle, A. G., Li, S. H. and Botha, G. A. (1993). Luminescence dating of colluvial deposits. *South African Journal of Science*, 89, 77–82.

Wintle, A. G., Li, S. H, Botha, G. A. and Vogel, J. C. (1994). Evaluation of luminescence dating applied to Late Holocene colluvium near St Paul's Mission, Natal, South Africa. *The Holocene*, 5, 97–102.

10

Desert dune environments

DAVID S.G. THOMAS

Abstract

The distribution and diverse nature of dunes in southern African drylands makes them a complex and enigmatic landscape component. While dunes are dominant landscape features in large areas of the Kalahari and Namib deserts, localised and topographically constrained dunes elsewhere are also an important part of the continental dune record. Over 600 published luminescence ages from dune sediments through the subcontinent should, in theory, make possible chronometrically robust analyses of the Quaternary palaeoclimate record. Compared to early aridity- and morphology-focussed interpretations of the dune record, in which three or four phases of dune development were proposed with a focus on the period of the Last Glacial Maximum, today's data demonstrate a complex and regionally variable record of dune development, with almost continuous dune accumulation through the last 30 kyr preserved in the southwest Kalahari and a more sporadic record in presently wetter eastern and northern areas. Dune geomorphological proxies require careful evaluation before Quaternary climate inferences can be usefully made.

10.1 Introduction

Dunes develop and change as a function of sediment supply together with environmental factors that make sediments, usually sand-sized (63–2000 μm in diameter), conducive to transport by wind (see Thomas (2011), Wiggs (2011) and others for explanations of the broad environmental and physical factors that control aeolian sediment transport and aeolian bedform development). Changes over time in the interactions between erodibility (surface protection, largely influenced by vegetation cover and therefore often linked to moisture availability or precipitation), erosivity (the ability to move sediment, linked to wind energy), and sediment supply, determine whether dunes develop, are active or inactive, or degrade over time, with external climate changes being a major forcing factor of long term shifts in dune behaviour. Establishing these changes in the state of dune systems and their timing has led to the use of dunes as Quaternary geomorphological proxies in the reconstruction of past climatic and environmental conditions (Thomas and Burrough, 2012). Making use of dune records in this way is not a simple process, but continental dunes in southern Africa, and in equivalent climatic contexts such as interior Australia (Fitzsimmons et al., 2013), often occur where other palaeoenvironmental proxies are limited. During the past decade or so, there has been considerable debate regarding the utility of desert dunes as climate proxies, improving their analysis and interpretation, and providing robust chronologies of dune sand accumulation. A large body of this work has taken place in southern Africa; from this, a complex picture of Quaternary dune development is emerging, as well as discussion of the best ways to use dune proxy records to infer Quaternary climates and environments.

Figure 10.1 shows the distribution of southern African continental dunes. The Kalahari and Namib deserts are major locations of dunes, with the former having been the primary nexus between contemporary and Quaternary dune research in southern Africa. As Figure 10.1 demonstrates, dunefields are widely distributed even within the Kalahari, and are dominated by linear dunes. In contrast, dunes in the Namib have been poorly utilised in palaeoenvironmental analyses despite the diversity of dune forms present (Livingstone et al., 2010). Other more isolated dunes, particularly lunettes on the margins of pans and playas, have been a significant focus of investigation in a range of locations, including within the southern Kalahari where pans are widespread within and beyond the linear dunefield; in Etosha, Namibia (Brook et al., 2007), and Barotseland, Zambia (O'Connor, 1997).

10.2 Southern African dunes as Quaternary palimpsests

10.2.1 Quaternary aridity

Passarge (1904) first mentioned that vegetated Kalahari dunes might be evidence of past drier climates in the region. It was over half a century later before scientific interest in the subcontinental interior dune palimpsests began in earnest. Flint and Bond (1968) noted that the vegetated and degraded linear ridges in western Zimbabwe were much lower and gentler-sloped than equivalently spaced dune ridges in the hyperarid Namib Desert, this difference being attributed to runoff erosion since the features ceased to be shaped by aeolian action. Grove (1969)

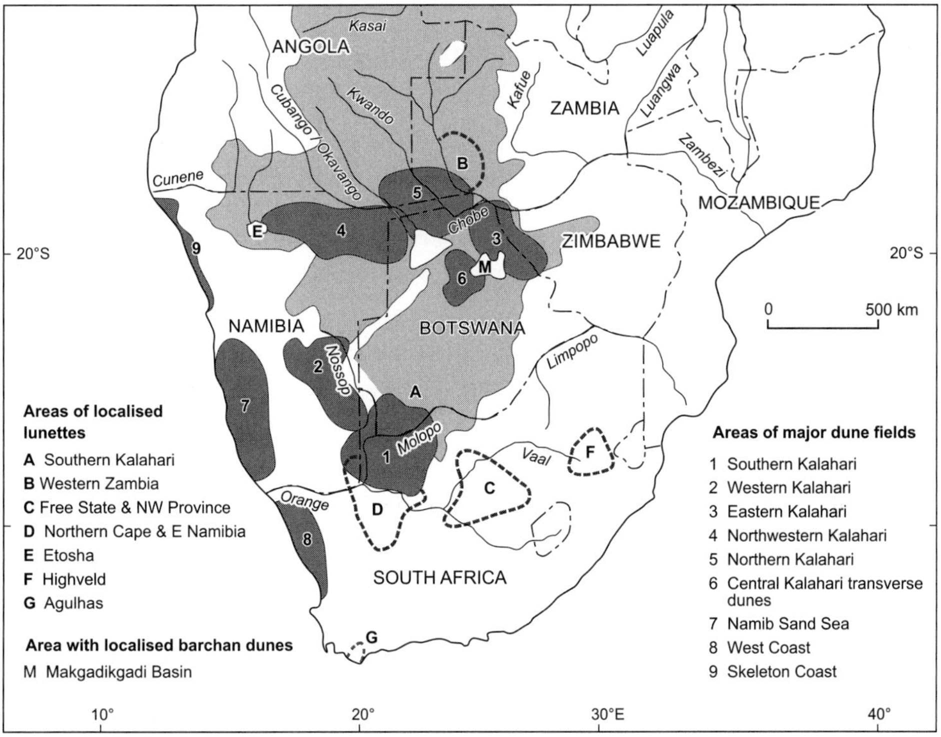

Fig. 10.1. Continental dunefields and localised dune occurrences in southern Africa. Dune areas 7 (Central Kalahari transverse dunes) and 9 (Skeleton Coast) have not yet been subject to Quaternary studies.

published an outstanding Kalahari-wide overview and analysis of landforms in this region, based on air photograph interpretation and field affirmation (see Thomas, 2011a, for an appraisal). Grove demonstrated the wide geographical distribution of dunes in the Kalahari, showing that the majority, but not all, were linear forms with morphologies differing markedly across the regional aridity gradient, which today ranges from mean annual precipitation values of ~120–150 mm in the southwest to over 600 mm in the northeast. This study also recorded lunette dunes on pan margins in the southern Kalahari, a field of barchan dunes on the floor of the Makgadikgadi depression, and large transverse ridges to the west of the large palaeoshoreline that marks the maximum extent of this basin (see Burrough et al. (2009) for a recent assessment of palaeolakes in the Kalahari). To date, all but these transverse ridges have been analysed in the context of Quaternary research in the region.

These early studies set out the palaeoenvironmental importance of degraded dunes in the region, and contributed significantly to the subsequent use of dunes as proxy evidence of Quaternary precipitation changes. Lancaster (1981) and Thomas

(1984) went on to make inferences about the degree of precipitation changes that had occurred since dune formation, with precipitation, via vegetation cover, seen as the dominant limit on dune activity. However, a much more complex situation has since emerged (e.g. Bailey and Thomas, 2014), whereby multiple controls on dune development are recognised, where the role of vegetation is complex and not consistent in inhibiting dune surface dynamics, and where simple climatic interpretations of dune accumulation records are now being challenged (Chase and Meadows, 2007). These issues with respect to the contributory factors of dune formation are now discussed.

10.2.2 Wind

While sufficient wind energy is a prerequisite for dune activity, initial studies focused on relationships between dune orientations and changes in atmospheric circulation since formation. Linear dunes generally develop parallel to resultant sand transport regimes, often under the interaction of two intersecting seasonal winds (Fryberger, 1979; see Andreotti et al. (2009), Thomas (2013a) and du Pont et al. (2014) for more recent discussion of controls on dune orientation). Differences in orientation of Kalahari linear dunes compared to today's dominant wind direction led to significant inferences regarding shifts in late Quaternary circulation over the subcontinent (Lancaster, 1981). Distinct phases of linear dune formation, resulting in the emplacement of the more northerly dune systems but affecting the southern dunes on multiple occasions, were inferred and were reinforced by distinctions in pan-margin lunette dune orientations in the southern Kalahari (Lancaster, 1978).

10.2.3 Dune origins

Sediment supply and availability have never been a significant factor debated in Kalahari dune studies, given the ubiquity of sand in the substantial post-Gondwanaland Kalahari basin, and much of the sediment owes its origin to basement weathering (McFarlane et al., 2007) and fluvial inflow (Thomas and Shaw, 1991). The antiquity of the Namib Desert (Van Zinderen Bakker, 1975) also means sediment supply, from marine and inland escarpment contexts, has not been a limiting factor in dune development (Besler, 1980; Walden and White, 1997; Vermeesch et al., 2010). The West Coast dunefield, south of the Namib, is more supply-limited, with postglacial sea-level rise contributing to onshore sediment supply that was then transported by strong westerly winds (Chase and Thomas, 2006) to form a dune system that is neither truly continental nor coastal. At more local scales, isolated dunes or small dune systems may be closely allied to sediment availability. Thus, the formation of pan-margin lunettes (e.g. in the southern Kalahari and on the Agulhas Peninsula) is associated with either drying of small water bodies (Lancaster, 1978) or the supply of sediment to downwind margins by wave action that may be linked to climate seasonality (e.g. Carr et al., 2006). Similarly, in linear dunefields, localised sediment supply may come from the reactivation and recycling of existing dune sands, which in turn is impacted by erodibility and erosivity controls on dune surfaces.

10.2.4 A date with dunes

In the absence of suitable dating techniques, initial studies could not establish the timing of dune formation directly. Dunefields were regarded as products of global Quaternary cold stages, when arid and windy conditions were thought to have occurred in subtropical-tropical contexts. Rare occurrences of material within or below dunes suitable for radiocarbon dating were exceptions (e.g. Lancaster, 1979; Rust, 1984); otherwise relative dating dominated, based on the degree of dune degradation, differences in dune orientation, and degree of dune pedogenic reddening (Grove, 1969; Lancaster, 1981; Thomas, 1984). In 1992 the first thermoluminescence (TL) ages were published from lunette dune sediments on the margin of Etosha Pan, Namibia (Buch et al., 1992; Heine, 1992). Stokes et al. (1997) provided the first application of the more robust optically stimulated luminescence (OSL) technique, with a Kalahari-wide analysis comprising almost 50 ages from southern, eastern and northern Kalahari dunefields. Multiple periods of late Quaternary dune accumulation were determined (ages from 1.1±0.2 kyr to 164±32 kyr BP). Significantly, it was shown that none of the major dunefields had formed in a single time period; i.e. dune proxies preserve records of multiple accumulation phases. Since that study, OSL dating has become a standard component of Quaternary dune investigations in southern Africa, with over 620 ages from 31 studies published by October 2013 (Thomas and Burrough, 2013). Table 10.1 shows the source of the geochronometric data that now contribute to the dune record, in terms of the regions covered and dune types sampled.

10.3 Dune sensitivity and complex development

With improved direct geochronometric control and concomitant advances in aeolian geomorphology and dune process studies during the past twenty years (Thomas and Wiggs, 2008; Thomas, 2013b), the potential for utilising dune data in southern African Quaternary studies is now greatly enhanced. However, the interpretation of these multiple data types is now more complex, and has been more controversial, than when only limited data and simple environmental interpretations

Table 10.1 *Luminescence age records for southern African dune systems, by dunefield and dune type (data from Thomas and Burrough, 2013). Numbers are the total number of luminescence age records in each dune category.*

	Kalahari (all dunefields)	Namib Sand Sea	West Coast dunefield	Karoo	Agulhas Peninsula	Free State	Other minor locations	Total
Linear dunes	254	38						292
Barchan dunes	23				5			28
Lunette dunes				10	17	29	40	47
Network dunes	2		56					58
Climbing dunes				6				6
Sand ramps						6		6
Parabolic dunes	2							2
Sand sheets	17		30					
Other forms	3							3
Total	387	38	86	16	22	35	40	624

were published. Two primary issues affecting multiple data interpretations are geographic and temporal variability in dune formation controls (dune sensitivity) and the preservation of accumulated sediments. A third issue, the generation and quality of ages, is discussed elsewhere (Thomas and Burrough, 2013; Woodborne, this volume).

While degraded linear dune forms may indicate inactivity, the presence of vegetation on dune surfaces does not necessarily do so, especially in semiarid to arid contexts. This is borne out by a range of process studies that show that linear dunes which, due to their predominantly extending nature (Telfer et al., 2011) are less mobile than other dune types, can be important sites for vegetation growth, even in arid locations. In situations where vegetation is sparse and variable across dune bodies, and liable to seasonal and longer-term variability in erosivity and erodibility controls (e.g. Wiggs et al., 1994; Thomas and Leason, 2005), surficial dune activity can mobilise dune surfaces and affect components of the preserved record.

Thus, linear dunes in such situations, regardless of wind energy variations, are not strictly inactive and can exist in states that are more than simply either stabilised or active (Livingstone and Thomas, 1993). Extracting a robust palaeoenvironmental signal is therefore questionable (Chase and Meadows, 2007), with repeated dune dating programmes from the southern and western Kalahari showing, when data are combined, that dune accumulation has been a relatively constant process over the late Quaternary, especially in the Holocene (Stone and Thomas, 2008; Thomas and Burrough, 2012). In contrast, linear dunes in wetter contexts today, which may be morphologically degraded and covered in dense vegetation including woodland, such as the dunes first described by Flint and Bond (1968), are features with a strong palaeoclimatic signal. Therefore, in the context of Quaternary studies utilising dunes as climate proxies, all linear dunes are not necessarily the same. For example, those in the southern Kalahari might be regarded as too sensitive to surface mobility driven by decadal-century climatic variability to be significant proxies for past hydrological changes, whereas those to the north provide much more robust evidence of substantial and significant past rainfall changes, with reactivation today requiring major changes in climatic parameters (and therefore in climate systems) that impact on erodibility and erosivity (Thomas et al., 2005). This does not mean that the dated dune records from the southern Kalahari dunes are not useful in a Quaternary context. Rather, the utility of the dated records is not simply as a hydrological proxy, but requires careful interpretation in localised contexts where detailed analyses can provide significant data on Quaternary landscape development. This is revealed, for example, in integrated sedimentary and geochronometric analyses from lunette, linear dune and pan-floor contexts at Witpan, Northern Cape Province of South

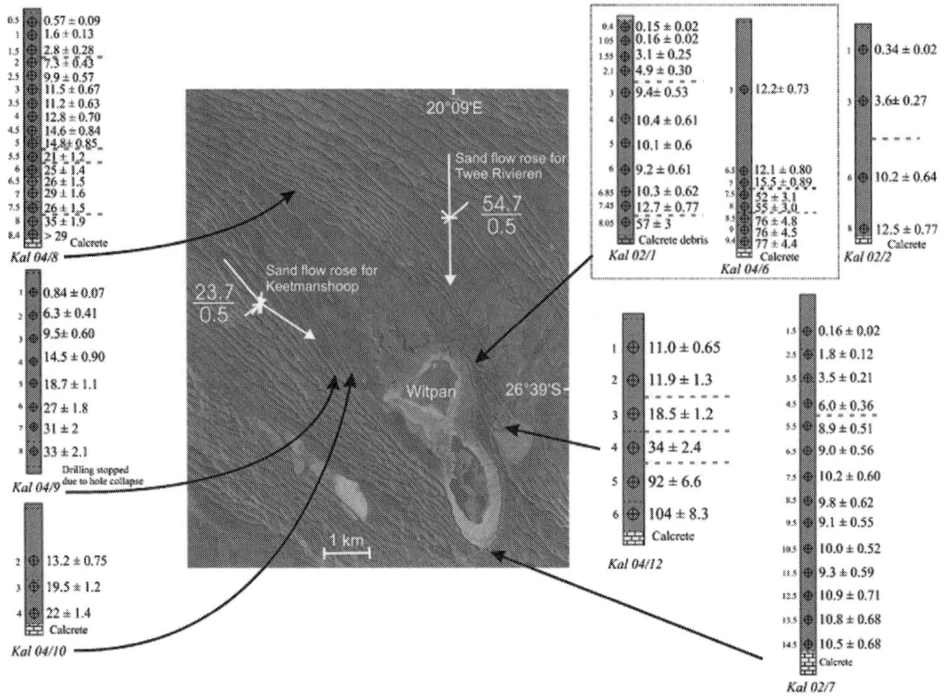

Fig. 10.2. Linear dune ages (kyr BP) from sites close to Witpan, Northern Cape (from Telfer and Thomas, 2007; reproduced with permission from Elsevier). Dashed lines highlight significant hiatuses in the preserved record. Potential sand flow roses are derived from the nearest meteorological stations and are calculated using the method of Fryberger (1979). Keetmanshoop is located 200 km west of Witpan, Twee Rivieren is 60 km northeast. Resultant vector directions (arrows) show the variability in modern potential sand flow, where dune trends show notable curvature through the southern Kalahari dunefield.

Africa (Telfer and Thomas, 2006, 2007; Telfer et al., 2009) (Fig. 10.2). Different morphologies of dune types, seen in the field, are shown in Figure 10.3.

Furthermore, dunes are not landforms that simply accumulate; they are also subject to erosion which can impact on the record that is preserved. This is evidenced by degraded forms, such as in the low linear ridges of northwest Botswana (McFarlane et al., 2005) and the gullied lunettes of the Northern Cape (Hürkamp et al., 2011). However, some dune types such as linear dunes can experience spatially variable erosion of their profiles during times of wind activity, as part of the normal processes of dune dynamics (Munyikwa, 2005). This means that a single sampled location within a dune or dunefield is extremely unlikely to capture the full sedimentary accumulation history of that system (Telfer et al., 2010). Even with sampling of multiple locations in a dunefield for dating, the frequency of ages in a particular period (e.g. in a millennial 'bin')

Fig. 10.3. Different dune morphologies seen in the field. (a) Linear dunes in the SW Kalahari, which may support significant grass cover, especially in wet years; (b) SW Kalahari dunes exhibiting surface instability due to grazing pressure and drought; (c) sand ramp near Sossus Vlei, Namibia; (d) pan-fringing dunes at Witpan, Northern Cape (photos: David Thomas). A black and white version of this figure will appear in some formats. For the colour version, please refer to the plate section.

should not be used as an indication of the significance of an accumulation episode (Thomas and Burrough, 2012), as sampling may have played a role in the ages that are generated (Stone and Thomas, 2008), or the accumulation units that are missed in sampling. Consequently, full consideration needs to be given to sampling strategies, and potentially to the employment of techniques such as ground-penetrating radar, if the maximum utility of dune records in Quaternary research is to be achieved (see Bailey and Thomas (2014) for full discussion and exemplification of key issues).

10.4 Regional records

Given the points raised earlier, the production of simple palaeoclimatic interpretations from dune records, or the allocation of dune formation in some regions of southern Africa to distinct climatic phases, needs care. Early models of three or four phases of dune development, and the importance of the Last Glacial Maximum as an arid period of dune development, need reappraisal (e.g. Chase and Brewer, 2009; Thomas and Burrough, 2012). These issues are borne in mind in the review of regional dune accumulation records and their climatic interpretation that follows.

10.4.1 Kalahari dunes

Stokes et al. (1997) inferred four dune accumulation periods from Kalahari linear dunefield data, at 95–115, 41–46, 20–26 and 9–16 kyr BP, as determined by precipitation changes across the interior SW–NE rainfall gradient, modulated by Atlantic sea surface temperatures. Subsequent increases in the number of dated dune records (Table 10.2) have complicated this picture, particularly for the southern and western Kalahari where linear dune accumulation is recorded through most of the past 20 kyr, with more punctuated records, when mean ages are placed in millennial bins, back to 100 kyr BP. If one sigma errors are added to linear dune ages, then there is overlap in all records from these locations back to 110 kyr BP (Fig. 10.4). Thus, this does not mean continuous dune activity and accumulation, but that ages are currently unable to distinguish with certainly discreet accumulation phases. Dated lunette dune records also show significant accumulation in the southern Kalahari during the last 30 kyr, with a break in dated preserved records (when both mean ages and ages with one sigma errors are considered) from 18–21 kyr BP (Thomas and Burrough, 2013).

The situation is marginally better resolved for more northern and eastern Kalahari locations, but here many older published age records have greater uncertainties attached to them. However, other dated records, including from lake shorelines, provide a context in which humidity changes can be placed (Burrough et al., 2009; Thomas and Burrough, 2012; Burrough and Thomas, 2013). An additional new record comes from dated barchan dunes from the floor of palaeo-lake Makgadikgadi (Burrough and Thomas, 2013), where barchan dune stabilisation is recorded within the periods 4.9–4.4 and 3.4–1.7 kyr BP. McFarlane and Long (2015) have recently suggested that these features may be spring mounds, but recent new field observations, including of formative features in the adjacent Sua Pan, suggest that an aeolian origin is more appropriate.

10.4.2 Namib and West Coast

The Namib sand sea offers considerable potential as an archive of dune accumulation episodes, given the size of many dunes and the diversity of dune forms within the system. To date, only 38 OSL ages have been published, from major linear ridges in the northern (Bristow et al., 2007) and southern (Bubenzer et al., 2007) extremes of the main dunefield. While ages from upper dune sections fall variously

Table 10.2 *The spatial and morphological variations in dated dune records from dunefields and other locations in the Kalahari (data from Thomas and Burrough, 2013). Numbers are the total number of luminescence age records in each dune category.*

	Southern	Western	Eastern	Northeastern	Northern	Makgadikgadi	Etosha	Zambia
Linear dunes	136	47	33	19	19			
Barchan dunes						23		
Network dunes	2							
Lunette dunes	67						8	11
Parabolic dunes								
Sandsheet	7		10					
Other/indeterminate	3							
Total	217	47	43	19	19	23	8	11

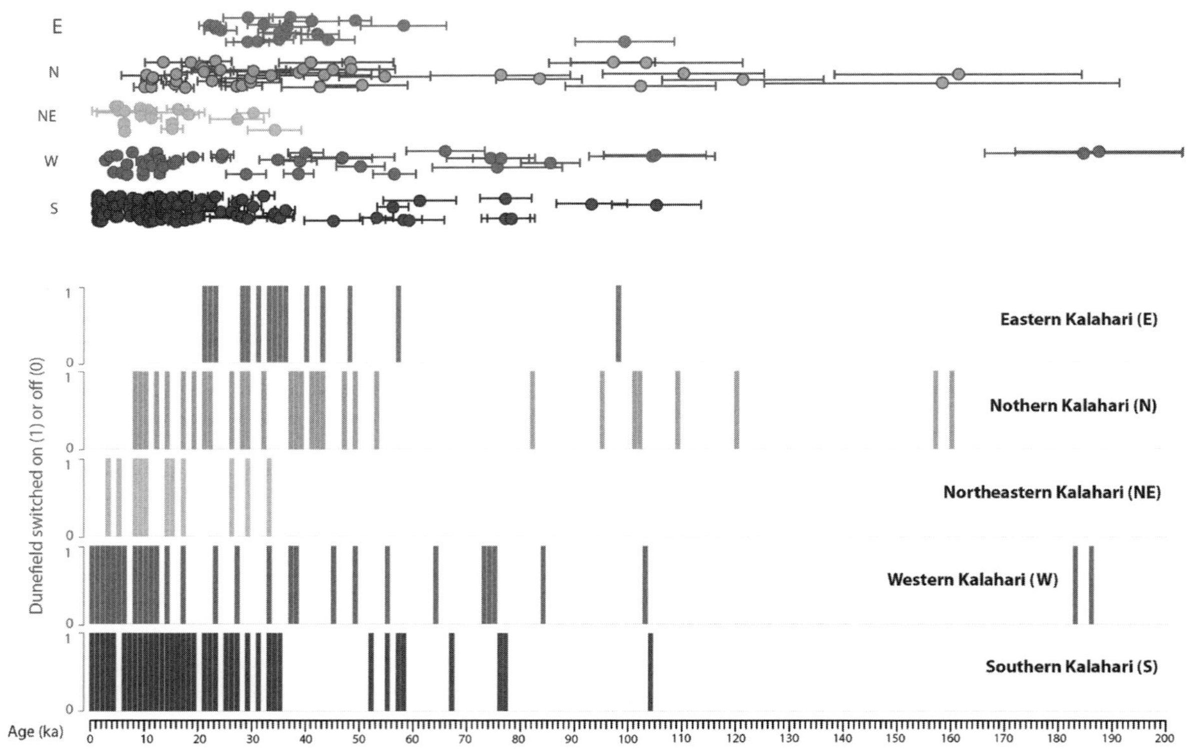

Fig. 10.4. Dated open field (linear dune and sand sheet) records from the Kalahari (adapted from Thomas and Burrough, 2014), see Table 10.2 and Figure 10.1 for locations. The data are presented by dunefield in two formats: (top) mean ages are each plotted with their respective one sigma errors; (bottom) mean ages only are plotted in a binary format in 1000-year bins. Empty bins mean no dune ages are recorded for that particular millennium, full bins indicate one or more ages are recorded for that millennium. Both approaches show near continuous dune accumulation through the last 30 kyr in presently drier southern and western areas, and a more nuanced accumulation record elsewhere, though high age errors (from older analyses) in northern dune samples precludes a clear record of development emerging. A black and white version of this figure will appear in some formats. For the colour version, please refer to the plate section.

within the last 20 kyr, other ages from deeper-drilled locations fall within the 65–120 kyr BP range. With so few samples it is, however, difficult to make inferences regarding dune system accumulation.

To the south, the West Coast dunefield has been more systematically analysed with a detailed record of almost 90 OSL ages from sandsheet and reticulate forms in the shallow aeolian system (Chase and Thomas, 2006, 2007). With individual dune age outliers of over 120 kyr, most accumulation is recorded within the last 80 kyr, with a propensity of ages in the last 20 kyr. Given the context of the sand sea, Chase and Thomas (2007) have primarily interpreted the development of this system as a response to sediment supply issues related to sea-level rise and periods of increased windiness, with interpretation assisted by comparison with ecological and marine core records.

10.4.3 Other dunes

In southern Africa, other dunes, outside these major aeolian systems, also contribute to the Quaternary record. While pan-margin lunettes are prevalent within the southern Kalahari, they are also notable in parts of South Africa's Free State (Holmes et al., 2008), Karoo (Thomas et al., 2002) and Agulhas Peninsula (Carr et al., 2006). Sand ramps also provide promising lines of investigation (Telfer et al., 2011). In all these locations, more humid than the main sand seas, other Quaternary proxy records may be found in close proximity, providing additional data than can contextualise dune records (e.g. Carr et al., 2006).

10.5 Summary

Studies of palaeo-dune records are most likely to advance Quaternary research when they both embrace detailed geochronometric analyses and a robust understanding of how aeolian processes and dune dynamics can impact on preserved records (Thomas and Wiggs, 2008). The need to conduct these elements in tandem is demonstrated by the significant advances now being achieved through studies using empirically grounded modelling to identify how accumulated records are impacted by intrinsic dune dynamics and external forcing processes (e.g. Bailey and Thomas, 2014). In the absence of other suitable proxies, large tracts of the southern African interior require the robust use of geomorphological proxy records, dominated by dune archives, to reconstruct Quaternary palaeoenvironments, and to avoid the inappropriate extrapolation of records from peripheral locations as representative of these variable and complex dryland regions (Thomas and Burrough, 2012; Burrough and Thomas, 2013). Southern Africa has one of the most geochronometrically detailed Quaternary dune records (Thomas and Burrough, 2013), but within this is a requirement to reinterpret records utilising some of the advances now being achieved through new modes of analysis being developed in other dune system contexts (e.g. Leighton et al., 2013), and through modelling.

References

Andreotti, B., Fourrière, A., Ould-Kaddour, F., Murray, B. and Claudin, P. (2009). Giant aeolian dune size determined by the average depth of the atmospheric boundary layer. *Nature*, 457, 1120–1123.

Bailey, R. M. and Thomas, D. S. G. (2014). A quantitative approach to understanding dated dune stratigraphies. *Earth Surface Processes and Landforms*, 39, 614–631.

Besler, H. (1980). *Die Dünen-Namib: Entstehung und Dynamik eines Ergs*. Stuttgarter Universität Institut für Geographie, 96, 241pp.

Bristow, C. S., Duller, G. A. T. and Lancaster, N. (2007). Age and dynamics of linear dunes in the Namib Desert. *Geology*, 35, 555–558.

Brook, G. A., Marais, E., Srivastava, P. and Jordan, T. (2007). Timing of lake-level changes in Etosha Pan, Namibia, since the middle Holocene from OSL ages of relict shorelines in the Okondeka region. *Quaternary International*, 175, 29–40.

Bubenzer, O., Bödeker, O. and Besler, H. (2007). A transcontinental comparison between the southern Namib Erg (Namibia) and the southern Great Sand Sea (Egypt). *Zentralblatt für Geologie und Paläontologie, Teil 1 Heft*, 1/4, 7–23.

Buch, M. W., Rose, D. and Zöller, L. (1992). A TL-calibrated pedostratigraphy of the western lunette dunes of Etosha Pan/northern Namibia: Palaeoenvironmental implications for the last 140 ka. *Palaeoecology of Africa*, 23, 129–147.

Burrough, S. L. and Thomas, D. S. G. (2013). Central southern Africa at the time of the African Humid Period: A new analysis of Holocene palaeoenvironmental and palaeoclimate data. *Quaternary Science Reviews*, 80, 29–46.

Burrough, S. L., Thomas, D. S. G. and Singerayer, J. S. (2009). Late Quaternary hydrological dynamics in the Middle Kalahari: Forcing and feedbacks. *Earth-Science Reviews*, 96, 313–326.

Carr, A. S., Thomas, D. S. G. and Bateman, M. D. (2006). Climatic and sea level controls on Late Quaternary eolian activity on the Agulhas Plain, South Africa. *Quaternary Research*, 65, 252–263.

Chase, B. M. and Brewer, S. (2009). Last Glacial Maximum dune activity in the Kalahari Desert of southern Africa: Observations and simulations. *Quaternary Science Reviews*, 28, 301–307.

Chase, B. M. and Meadows, M. E. (2007). Late Quaternary dynamics of southern Africa's winter rainfall zone. *Earth-Science Reviews*, 84, 103–138.

Chase, B. M. and Thomas, D. S. G. (2006). Late Quaternary dune accumulation along the western margin of South Africa: Distinguishing forcing mechanisms through the analysis of migratory dune forms. *Earth and Planetary Science Letters*, 251, 318–333.

Chase, B. M. and Thomas, D. S. G. (2007). Multiphase late Quaternary aeolian sediment accumulation in western South Africa: Timing and relationship to palaeoclimatic changes inferred from the marine record. *Quaternary International*, 166, 29–41.

du Pont, S. C., Narteau, C. and Gao, X. (2014). Two modes for dune orientation. *Geology*, 42, 743–746.

Fitzsimmons, K. E., Cohen, T. J., Hesse, P. P., Jansen, J., Nanson, G. C., May, J.-H., Barrows, T. T., Haberlah, D., Hilgers, A., Kelly, T., Larsen, J., Lomax, J. and Treble,

P. (2013). Late Quaternary palaeoenvironmental change in the Australian drylands. *Quaternary Science Reviews*, 74, 78–96.
Flint, R. F. and Bond, G. (1968). Pleistocene sand ridges and pans in western Rhodesia. *Bulletin of the Geological Society of America*, 79, 299–314.
Fryberger, S. G. (1979). Dune form and wind regime. *USGS Professional Paper*, 1052, 137–169.
Grove, A. T. (1969). Landforms and climate change in the Kalahari and Ngamiland. *Geographical Journal*, 135, 191–212.
Heine, K. (1992). On the ages of humid Late Quaternary phases in southern African arid areas (Namibia, Botswana). *Palaeoecology of Africa*, 23, 149–164.
Holmes, P. J., Bateman, M. D., Thomas, D. S. G., Telfer, M. W., Barker, C. H. and Lawson, M. P. (2008). A Holocene-late Pleistocene aeolian record from lunette dunes of the western Free State panfield, South Africa. *The Holocene*, 18, 1193–1205.
Hürkamp, K., Völkel, J., Heine, K., Bens, O., Leopold, M. and Winkelbauer, J. (2011). Late Quaternary environmental changes from aeolian and fluvial geoarchives in the southwestern Kalahari, South Africa: Implications for past African climate dynamics. *South African Journal of Geology*, 114, 459–474.
Lancaster, N. (1978). Composition and formation of southern Kalahari pan margin dunes. *Zeitschrift für Geomorphologie, N.F.*, 22, 148–169.
Lancaster, I. N. (1979). Evidence for a widespread late Pleistocene humid period in the Kalahari. *Nature*, 279, 145–146.
Lancaster, N. (1981). Palaeoenvironmental implications of fixed dune systems in Southern Africa. *Palaeogeography, Palaeoclimatology, Palaeoecology*, 33, 327–346.
Leighton, C. L., Bailey, R. M. and Thomas, D. S. G. (2013). The utility of desert sand dunes as Quaternary chronostratigraphic archives: Evidence from the northeastern Rub' al Khali. *Quaternary Science Reviews*, 78, 303–318.
Livingstone, I., Bristow, C., Bryant, R. G., Bullard, J., White, K., Wiggs, G. F. S., Baas, A. C. W., Bateman, M. D. and Thomas, D. S. G. (2010). The Namib Sand Sea digital database of aeolian dunes and key forcing variables. *Aeolian Research*, 2, 93–104.
Livingstone, I. and Thomas, D. S. G. (1993). Modes of linear dune activity and their palaeoenvironmental significance: An evaluation with reference to southern African examples. In *The Dynamics and Environmental Context of Aeolian Sedimentary Systems*, ed. K. Pye. Bath: Geological Society Special Publication, 72, pp. 91–101.
McFarlane, M. J., Coetzee, S. H., Kuhn, J. R., Vanderpost, C. H. M. and Eckardt, F. D. (2007). In situ rounding of quartz grains within an African surface weathering profile in North West Ngamiland, Botswana. *Zeitschrift für Geomorphologie, N.F.*, 51, 269–286.
McFarlane, M. J., Eckardt, F. D., Ringrose, S., Coetzee, S. H. and Kuhn, J. R. (2005). Degradation of linear dunes in Northwest Ngamiland, Botswana and the implications for luminescence dating of periods of aridity. *Quaternary International*, 135, 83–90.
McFarlane, M. J. and Long, C. W. (2015). Pan floor 'barchan' mounds, Ntwetwe Pan, Makgadikgadi, Botswana: Their origin and palaeoclimatic implications. *Quaternary International*, 372, 108–119.
Munyikwa, K. (2005). The role of dune morphogenetic history in the interpretation of linear dune luminescence chronologies: A review of linear dune dynamics. *Progress in Physical Geography*, 29, 317–336.
O'Connor, P. W. (1997). *Aeolian Activity and Environmental Change in the Central Mega Kalahari: Implications for the Timing, Nature and Causes of Late Quaternary Aridity*. Unpublished PhD thesis, University of Sheffield.
Passarge, S. (1904). *Die Kalahari*. Berlin: Riemer, 823pp.

Rust, U. (1984). Geomorphic evidence of Quaternary environmental change in Etosha, South West Africa/Namibia. In *Late Cainozoic Palaeoclimates of the Southern Hemisphere*, ed. J. Vogel. Rotterdam: Balkema, pp. 279–286.

Stokes, S., Thomas D. S. G. and Washington, R. (1997). Multiple episodes of aridity in southern Africa since the last interglacial period. *Nature*, 388, 154–158.

Stone, A. E. C. and Thomas, D. S. G. (2008). Linear dune accumulation chronologies from the southwest Kalahari, Namibia: Challenges of reconstructing late Quaternary palaeoenvironments from aeolian landforms. *Quaternary Science Reviews*, 27, 1667–1681.

Telfer, M. W., Bailey, R. M., Burrough, S. L., Stone, A. E. S., Thomas, D. S. G. and Wiggs, G. S. F. (2010). Understanding linear dune chronologies: Insights from a simple accumulation model. *Geomorphology*, 120, 195–208.

Telfer, M. W. and Thomas, D. S. G. (2006). Complex Holocene lunette dune development, South Africa: Implications for paleoclimate and models of pan development in arid regions. *Geology*, 34, 853–856.

Telfer, M. W. and Thomas, D. S. G. (2007). Late Quaternary linear dune accumulation and chronostratigraphy of the southwestern Kalahari: Implications for aeolian palaeoclimatic reconstructions and predictions of future dynamics. *Quaternary Science Reviews*, 26, 2617–2630.

Telfer, M. W., Thomas, D. S. G., Parker, A. G., Walkington, H. and Finch, A. A. (2009). Optically Stimulated Luminescence (OSL) dating and palaeoenvironmental studies of pan (playa) sediment from Witpan, South Africa. *Palaeogeography, Palaeoclimatology, Palaeoecology*, 273, 50–60.

Telfer, M. W., Thomas, Z. A. and Breman, E. (2011). Sand ramps in the Golden Gate Highlands National Park, South Africa: Evidence of periglacial aeolian activity during the last glacial. *Palaeogeography, Palaeoclimatology, Palaeoecology*, 313–314, 59–69.

Thomas, D. S. G. (1984). Ancient ergs of the former arid zones of Zimbabwe, Zambia and Angola. *Transactions, Institute of British Geographers, NS*, 9, 75–88.

Thomas, D. S. G. (2011). Aeolian landscapes and bedforms. In *Arid Zone Geomorphology: Process, form and Change in Drylands*, 3rd edition, ed. D. S. G. Thomas. Chichester: Wiley, pp. 429–453.

Thomas, D. S. G. (2013a). Aeolian Palaeoenvironments of desert landscapes. In *Treatise on Geomorphology, Vol. 11, Aeolian Geomorphology*, ed. J. F. Schroder. Oxford: Elsevier, pp. 356–374.

Thomas, D. S. G. (2013b). Reconstructing paleoenvironments and palaeoclimates in drylands: what can landform analysis contribute? *Earth Surface Processes and Landforms*, 38, 3–16.

Thomas, D. S. G. and Burrough, S. L. (2012). Interpreting geoproxies of late Quaternary climate change in African drylands: Implications for understanding environmental and early human behaviour. *Quaternary International*, 253, 5–17.

Thomas, D. S. G. and Burrough, S. L. (2013). Luminescence-based dune chronologies in southern Africa: analysis and interpretation of dune database records across the subcontinent. *Quaternary International,* doi:10.1016/j.quaint.2013.09.008.

Thomas D. S. G., Holmes, P. J., Bateman, M. D. and Marker, M. E. (2002). Geomorphic evidence for late Quaternary environmental change from the eastern Great Karoo margin, South Africa. *Quaternary International*, 89, 151–164.

Thomas, D. S. G., Knight, M. and Wiggs, G. F. S. (2005). Remobilization of southern African desert dune systems by twenty-first century global warming. *Nature*, 435, 1218–1221.

Thomas, D. S. G. and Leason, H. C. (2005). Dunefield activity response to climate variability in the southwest Kalahari. *Geomorphology*, 64, 117–132.

Thomas, D. S. G. and Shaw, P. A. (1991). *The Kalahari Environment.* Cambridge: Cambridge University Press, 284pp..

Thomas, D. S. G. and Wiggs, G. F. S. (2008). Aeolian system responses to global change: Challenges of scale, process and temporal integration. *Earth Surface Processes and Landforms*, 33, 1396–1418.

Van Zinderen Bakker, E. M. (1975). The origin and palaeoenvironment of the Namib Desert biome. *Journal of Biogeography*, 2, 65–73.

Vermeesch, P., Fenton, C. R., Kober, F., Wiggs, G. F. S., Bristow, C. S. and Xu, S. (2010). Sand residence times of one million years in the Namib Sand Sea from cosmogenic nuclides. *Nature Geoscience*, 3, 862–865.

Walden, J. and White, K. (1997). Investigation of the controls on dune colour in the Namib Sand Sea using mineral magnetic analyses. *Earth and Planetary Science Letters*, 152, 187–201.

Wiggs, G. F. S. (2011). Sediment mobilisation by the wind. In *Arid Zone Geomorphology: Process, form and Change in Drylands*, 3rd edition, ed. D. S. G. Thomas. Chichester: Wiley, pp. 455–486.

Wiggs, G. F. S., Thomas, D. S. G., Bullard, J. E. and Livingstone, I. (1994). Dune mobility and vegetation cover in the southwest Kalahari Desert. *Earth Surface Processes and Landforms*, 20, 515–529.

11

Changes in fluvial systems during the Quaternary

STEPHEN TOOTH

Abstract

Despite their potential importance, southern Africa's fluvial systems have yet to be fully exploited as archives of Quaternary palaeoenvironmental changes. This chapter outlines the regional environmental setting, and reviews the nature and drivers of southern African fluvial system responses, focusing mainly on examples where a secure geochronology exists. A key message is the need for recognition of diversity in the nature and drivers of Quaternary fluvial system change. Across the subcontinent, physiographic, tectonic, geological, climatic, vegetative – and more recently, anthropogenic – factors have combined in various ways to produce a variety of fluvial system responses that may incorporate aspects of channel lateral migration, avulsion, aggradation and/or incision. Depending on factors such as catchment scale and geomorphic setting, different fluvial systems exhibit differing sensitivities to natural and anthropogenic drivers, in turn influencing the production and preservation of fluvial sedimentary archives.

11.1 Introduction

It has long been recognised that southern Africa's fluvial systems (Fig. 11.1) are important archives for establishing palaeoenvironmental changes that have occurred prior to and during the Quaternary, and so can play a key role in the reconstruction of long-term landscape development, palaeoclimatic change, faunal (including hominin) evolution, and human cultural developments. This recognition is evident in early debates over the palaeoenvironmental significance of alluvial terraces along the Vaal and Orange rivers (for a summary, see Dollar, 1998), many

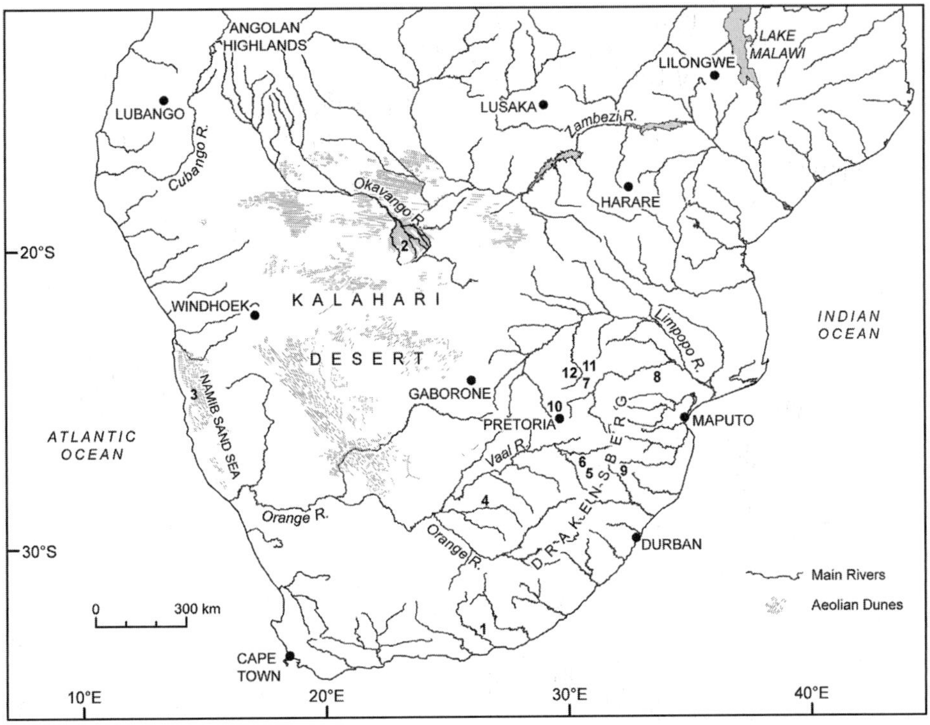

Fig. 11.1. Map of southern Africa showing key fluvial systems and other physical features (adapted from Thomas and Burrough, 2012); numbers refer to sites named in the text or figures: (1) lower Sundays River; (2) Okavango Delta; (3) Tsauchab River; (4) middle Modder River; (5) Klip River; (6) Schoonspruit; (7) Nyl/Mogalakwena River; (8) Olifants River; (9) Blood River; (10) Tswaing Crater; (11) Makapansgat valley; (12) Wonderkrater spring mound.

of which host important fossil fauna and archaeological artefacts. Despite this recognition and the rapid developments in Quaternary science, however, these fluvial archives arguably have yet to be fully exploited. The majority of palaeoenvironmental investigations in southern Africa have tended to focus on colluvial, aeolian and lacustrine features (see Botha et al., Ellery et al., this volume) or on more highly resolved isotope or organic records, such as those derived from speleothems or spring mounds (see Holmgren and Shaw, this volume). There are two potential reasons for the relative neglect of southern African fluvial archives. First, geochronologies for fluvial deposits have tended to be limited. Traditionally, establishment of fluvial chronologies relied mainly on radiocarbon (^{14}C) dating, but this technique has a maximum age limit of ~35–40 kyr BP, and can be problematic in some fluvial contexts owing to flood reworking of older carbon into younger deposits, or root penetration and contamination of older deposits by younger carbon. Second, fluvial landforms and sediments pose interpretative challenges, as they rarely provide unambiguous, unequivocal records of palaeoenvironmental change.

Over recent decades, however, developments in geochronology and more detailed geomorphological and sedimentological investigations of southern Africa's fluvial systems mean that some restrictions no longer apply. Following preliminary applications of luminescence techniques to southern African colluvial and fluvial deposits (e.g. Wintle et al., 1995; Bateman and Holmes, 1999; Scheepers and Rust, 1999), pioneering work by Rodnight et al. (2005, 2006) in South African floodplain wetlands contributed to improved sampling and analytical protocols for the application of optically stimulated luminescence (OSL) techniques to poorly-bleached fluvial sediments. Using OSL dating, late Quaternary fluvial chronologies now have been established for fluvial systems across the subcontinent (e.g. Eitel et al., 2005, 2006; Srivastava et al., 2005, 2006; Brook et al., 2006; Tooth et al., 2007, 2009, 2014; Keen-Zebert et al., 2013; Lyons et al., 2014), complementing and extending Holocene fluvial chronologies based on ^{14}C dating (e.g. Damm and Hagedorn, 2010). Other luminescence approaches (e.g. thermally transferred-OSL) and geochronological techniques (e.g. cosmogenic burial dating, palaeomagnetism) offer the prospect of establishing fluvial chronologies further back in time (Gibbon et al., 2009; Brink et al., 2012; Erlanger et al., 2012; Duller et al., 2015).

Although some of this research is still ongoing, these developments enable a reassessment of the potential of southern Africa's fluvial systems as archives of Quaternary environmental change. As yet, however, there are not sufficient numbers of securely dated fluvial chronologies from enough locations to enable palaeoenvironmental assessment on a regional basis or by time slices. Instead, this chapter discusses selected river and floodplain chronologies in order to illustrate broader issues in the use of fluvial archives. The chapter provides an overview of the regional environmental setting and then considers the nature and drivers of pre-Quaternary and Quaternary fluvial system responses. This provides the context for considering fluvial system sensitivity to palaeoenvironmental change, and the relative importance of natural and anthropogenic drivers of fluvial systems in the late Holocene.

11.2 Regional environmental setting

Fluvial systems are the integrated product of different environmental factors that include tectonics, geology, climate, vegetation and, increasingly, anthropogenic inputs. Together, these influence gross physiography, runoff and sediment supply, thereby influencing fluvial system characteristics. Several generalised types of fluvial systems in southern Africa can be identified, with examples numbered in Figure 11.1. These types include: (1) moderate-size, moderate-gradient rivers that drain seaward from the Drakensberg or other coast-parallel ranges to the ocean, such as the Blood and Sundays rivers; (2) larger, moderate-gradient rivers associated with the extensive Orange, Limpopo and Zambezi river systems; and (3) moderate-size, moderate- to low-gradient rivers of the internally draining Kalahari basin. Across the extensive interior, intermittent and ephemeral rivers are common, but there are also large perennial rivers that arise in more humid zones (e.g. Orange and Okavango Rivers). Many rivers have become closely adjusted to the variable geological structures and lithologies that they traverse, leading to a variety of bedrock-confined and bedrock-influenced fluvial styles (Fig. 11.2), the latter including ingrown meanders, anabranching rivers, and meandering rivers flanked

Fig. 11.2. Photos of typical bedrock and alluvial features in southern Africa fluvial systems. (a) Ingrown meander of upper Orange River, incised into basalt, highlands of Lesotho. Flow direction from upper left to middle right; (b) terrace of the Modder River (site 4) with OSL ages (kyr BP) indicated. From bottom to top, the succession comprises: lower grey palaeosol (LGP), red palaeosol (RP), upper grey palaeosol (UGP), brown palaeosol (BP), and weathered silty sand (SC); (c) incised Schoonspruit (site 6). Flow direction from middle right to lower left (photo by T.S. McCarthy). OSL ages (kyr BP) are indicated; (d) Olifants River (site 8) showing the effects of extreme flooding in January 2012. Flow direction from left to right (photos a,b,d: Stephen Tooth). A black and white version of this figure will appear in some formats. For the colour version, please refer to the plate section.

by seasonally inundated floodplain wetlands (Heritage et al., 1999; McCarthy and Tooth, 2004; Tooth and McCarthy, 2004).

11.3 Nature and drivers of fluvial system responses

All of the environmental factors influencing fluvial systems are subject to various degrees of change on different spatial and temporal scales. For instance, tectonic activity involving land surface uplift or depression can wax and wane, different lithologies can be uncovered during landscape denudation, climate can vary through time, and vegetation cover can fluctuate in density, composition or health, all of which can impact directly and indirectly on fluvial system processes, form and response. Consequently, fluvial system characteristics – for example, channel morphology, or the nature of floodplain or terrace sedimentary deposits – are in part a reflection of their environmental histories, but also of ongoing environmental changes. The relationship between environmental and fluvial system change, however, is not straightforward, with catchment scale and geomorphic setting being important determinants of how different systems will respond to changing external conditions (e.g. Nanson and Tooth, 1999; Tooth and Nanson, 2011).

Fluvial system responses focus on one or more of the following processes, some of which can operate simultaneously: (1) lateral migration; (2) avulsion; (3) vertical aggradation; and/or (4) vertical incision. Some processes may occur essentially independent of coeval environmental change (e.g. lateral migration, avulsion), as they can take place even under conditions of steady discharge and sediment supply. Commonly, however, the rate and/or frequency of such processes may be driven by environmental change and expressed through an altered discharge-sediment supply balance. For instance, decreased discharge and/or increased sediment supply can lead to aggradation and/or avulsion, while increased discharge and/or decreased sediment supply can lead to incision. All the above processes may result in changes to channel-floodplain morphology and associated sediments. Establishing a geochronology of such sediments through the use of radiocarbon or luminescence dating enables comparison with other, independently dated palaeoenvironmental proxies, and can provide a robust basis for interpretation of possible driving factors of environmental change.

11.3.1 Pre-Quaternary and Quaternary fluvial system responses in southern Africa: an overview

Various studies illustrate the range of fluvial system responses to Quaternary palaeoenvironmental change in southern Africa. The uplift history of the subcontinent (Knight and Grab, this volume) means that during the Cenozoic most fluvial

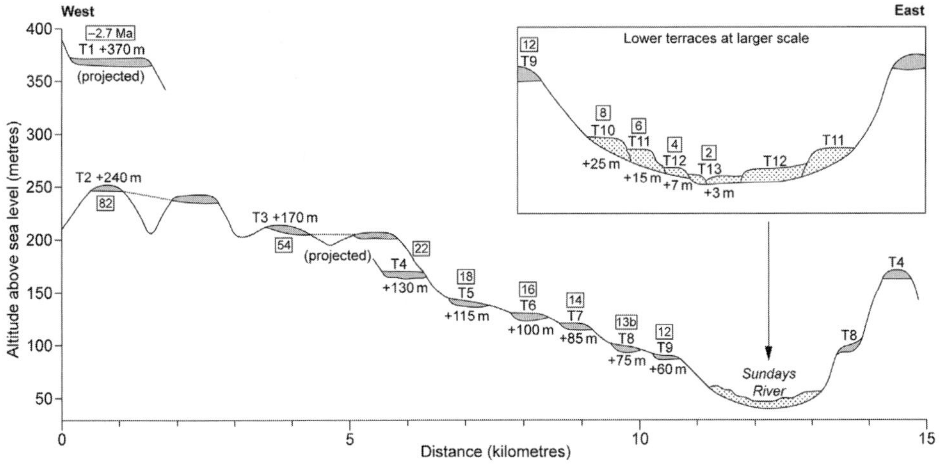

Fig. 11.3. Schematic cross section of alluvial terraces (T1–T13) in the lower Sundays River (site 1) (after Hattingh and Rust, 1999; Bridgland and Westaway, 2008). Numbers in boxes indicate Marine Isotope Stages, based on Bridgland and Westaway's incision/uplift modelling.

systems outside of the Kalahari basin have undergone long-term incision into bedrock, albeit with shorter term, commonly climate-driven, phases of lateral migration, aggradation, and/or avulsion superimposed upon this response.

Around the margins of the subcontinent, well-preserved flights of alluvial terraces can be found along many fluvial systems, as spectacularly demonstrated in the lower Sundays River valley (Fig. 11.3). Geochronology is limited, however, and formation of these terraces has been subject to competing interpretations. Hattingh and Rust (1999) interpreted the terraces as a response to late Miocene to Holocene tectonics, climatic changes, and sea-level fluctuations. Bridgland and Westaway (2008) proposed that terrace formation occurred in response to uplift and climate changes during the late Pliocene to Pleistocene. Erlanger et al. (2012) argued for progressive river incision over the past 4 Ma and refute hypotheses of recent, rapid uplift. Pre-Quaternary and Quaternary sea-level fluctuations have impacted on the lower reaches of many rivers on the western margin of the subcontinent (Scheepers and Rust, 1999; de Wit et al., 2000; Ward et al., 2002; Krapf et al., 2003). In this region, the long-term Cenozoic trend towards drier and more variable climates also led to widespread contraction of formerly more extensive drainage networks, many of which are now marked by ephemeral pans and playas (e.g. Goudie and Wells, 1995; Tooth and McCarthy, 2007). In the Namib Sand Sea, western Namibia, some former ocean-bound river systems now terminate on salt pans amongst obstructing aeolian dunefields many tens of kilometres from the Atlantic coastline, and thus have been isolated from sea-level changes (e.g. Jacobson et al., 1995; Brook et al., 2006).

In the interior of southern Africa, along reaches of the Vaal and Orange rivers, long-term incision into erodible lithologies has occurred in association with lateral migration and/or periodic aggradation (e.g. Fig. 11.2a), also generating flights of alluvial terraces (Helgren, 1977; de Wit et al., 2000; Ward et al., 2002; McCarthy and Tooth, 2004; Tooth et al., 2013). Terrace formation probably extends back to pre-Quaternary times but this cannot be confirmed by existing geochronologies. Cosmogenic nuclide burial dating of the lower Vaal River terraces (Gibbon et al., 2009) indicates that gravel and coarse sand aggradation occurred between ~1.80 and 1.15 Ma, which provides a minimum age for associated early Acheulean stone tools. By contrast, over similar timescales, the Kalahari basin has been infilling with colluvial, lacustrine, aeolian and fluvial sediments, as exemplified by the extensive (~40,000 km^2) Okavango Delta (site 2), which has formed within a depression related to the southwesterly extension of the East African Rift System (McCarthy, 2013). Along the characteristic vegetation-lined, sand-bed channels in the delta, a combination of Quaternary river flow regime changes, subtle tectonic activity (minor faulting and tilting) and autogenic changes (channel sedimentation and vegetation encroachment) has resulted in a complex but poorly constrained history of lateral migration, aggradation and avulsion, leading to the generation of a thick sedimentary succession. Historically, repeated large-scale avulsive shifts in the distribution of water and sediment across the delta have occurred on 100–200 year timescales (McCarthy, 2013) but it is unknown as to whether such avulsions occurred with lesser or greater frequency earlier in the Quaternary.

11.3.2 Late Quaternary fluvial system responses in southern Africa: climatic versus non-climatic drivers

For some fluvial systems, geochronology is more extensive for the late Quaternary. This provides greater insight into the processes, timings and drivers of fluvial system response. Along the lower reaches of the Tsauchab River (site 3) in the Namib Sand Sea for instance, fluvial system responses have been characterised by aeolian and fluvial interactions. Episodes of increased flooding and river activity at ~25 kyr, 9–7 kyr and 0.9–0.3 kyr BP appear to have been controlled by periods of warmer southeast Atlantic sea surface temperatures, which led to higher rainfall along the coast and inland (Brook et al., 2006).

Farther inland, climatic and non-climatic (especially geological) drivers have interacted to influence fluvial system response. Along the partially bedrock-confined, <500 m-wide middle Modder River valley (site 4), climatically driven shifts in discharge and sediment supply, mediated through vegetation cover changes, have led to a late Quaternary history of channel lateral migration, aggradation and incision. These responses have generated complex fluvial successions hosting fossil fauna and archaeological artefacts (Churchill et al., 2000; Tooth et al., 2013). At the Erfkroon site, Lyons et al. (2014) used OSL dating, mineral magnetism and diffuse reflectance spectroscopy to demonstrate that broadly continuous, fine-grained, overbank aggradation and concomitant palaeosol formation (Fig. 11.2b) occurred from at least ~44 kyr BP, under changing climates that were at times warmer/wetter and cooler/drier than the present-day semiarid conditions. These climatic changes correspond with other regional and southern hemisphere climatic proxies (Fig. 11.4), and in large part reflect the dynamic interplay of westerly (Atlantic Ocean) and easterly (Indian Ocean) moisture sources.

At Erfkroon, after ~0.83 kyr BP, two phases of deep channel incision and gully (donga) formation, and an intervening short phase of fine-grained aggradation, appear to have been initiated by rapid climatic and associated vegetation fluctuations, probably during the latter stages of the Mediaeval Warm Period (MWP: ~900–1300 AD/~1.10–0.81 kyr BP) or the Little Ice Age (LIA: ~1300–1800 AD/ ~0.81–0.21 kyr BP). Climate changes during these periods may have impacted more widely on fluvial landscapes across southern Africa, as evidenced by increased flooding, channel incision, donga formation, and/or fine-grained aggradation at this time (e.g. Vogel and Rust, 1987; Smith, 1991; Zawada, 2000; Heine, 2004; Benito et al., 2011; Heine and Völkel, 2011; McCarthy et al., 2011; Lyons et al., 2013). Critically, however, while recent incision of the Modder River in the vicinity of Erfkroon appears to have been initiated by climatic fluctuations, the full depth of incision (~20 m through alluvium and into bedrock) has been controlled by the partial breaching of a resistant dolerite barrier (i.e. a sill or dyke cropping out in the river bed) farther downstream (Tooth et al., 2013). Ongoing incision marks the end of net sediment accumulation – a key feature of the late Quaternary record – and instead probably signals the onset of major sediment evacuation and bedrock valley deepening that will continue until the channel encounters a more resistant part of the same barrier, or another resistant barrier lower in the rock profile.

By contrast with fluvial responses along the middle Modder valley, the <1500 m-wide Klip River valley (site 5) has experienced channel lateral migration and periodic avulsion without any pronounced vertical aggradation or incision (Tooth et al., 2007, 2009). At least five avulsion events have occurred in the last ~30 kyr, but climatic drivers appear to have been muted. Palaeochannel morphologies show no discernable differences from the present-day channel, and avulsion timings do not show any consistent correlations with regional or hemispheric palaeoenvironmental proxies, having occurred both during cooler/drier and warmer/wetter periods (Fig. 11.5). Although avulsion is commonly triggered by an individual flood or series of floods, the available evidence suggests that avulsions have not been climatically driven, but

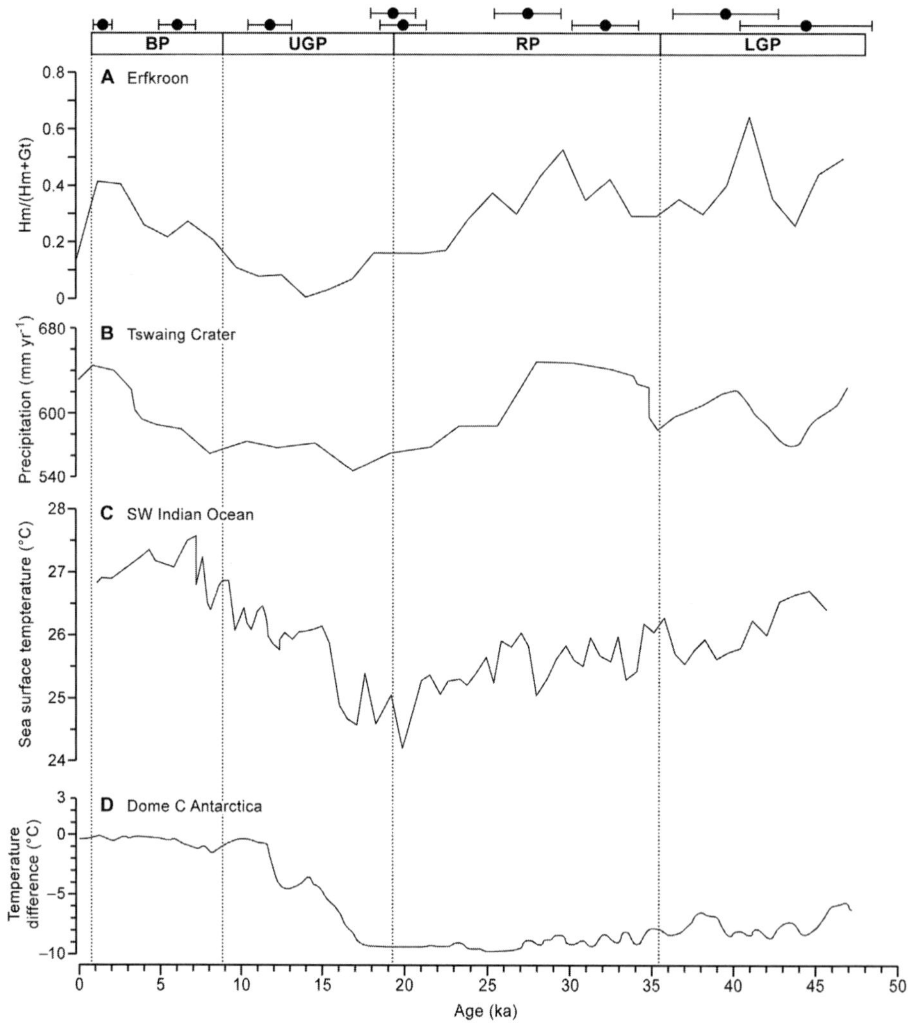

Fig. 11.4. Comparison of the palaeosol succession at Erfkroon (Modder River, site 4) with other palaeoenvironmental records (after Lyons et al., 2014; original references cited therein). Palaeosol designations and OSL ages (with 1 σ error) are shown at the top of the diagram: lower grey palaeosol (LGP), red palaeosol (RP), upper grey palaeosol (UGP), brown palaeosol (BP). Vertical dotted lines indicate the approximate boundaries between palaeosols. (a) Hm/(Hm+Gt) ratios, indicating the relative proportion of hematite (Hm) to goethite (Gt), where hematite formation is favoured in seasonally dry (<~600 mm/yr) and warm environments, and goethite formation in wetter (>~600 mm/yr) and cooler environments; (b) Tswaing Crater palaeorainfall record; (c) reconstructed southwest Indian Ocean sea surface temperatures; (d) temperature variations at Dome C, Antarctica, smoothed using a 700-year binomial filter.

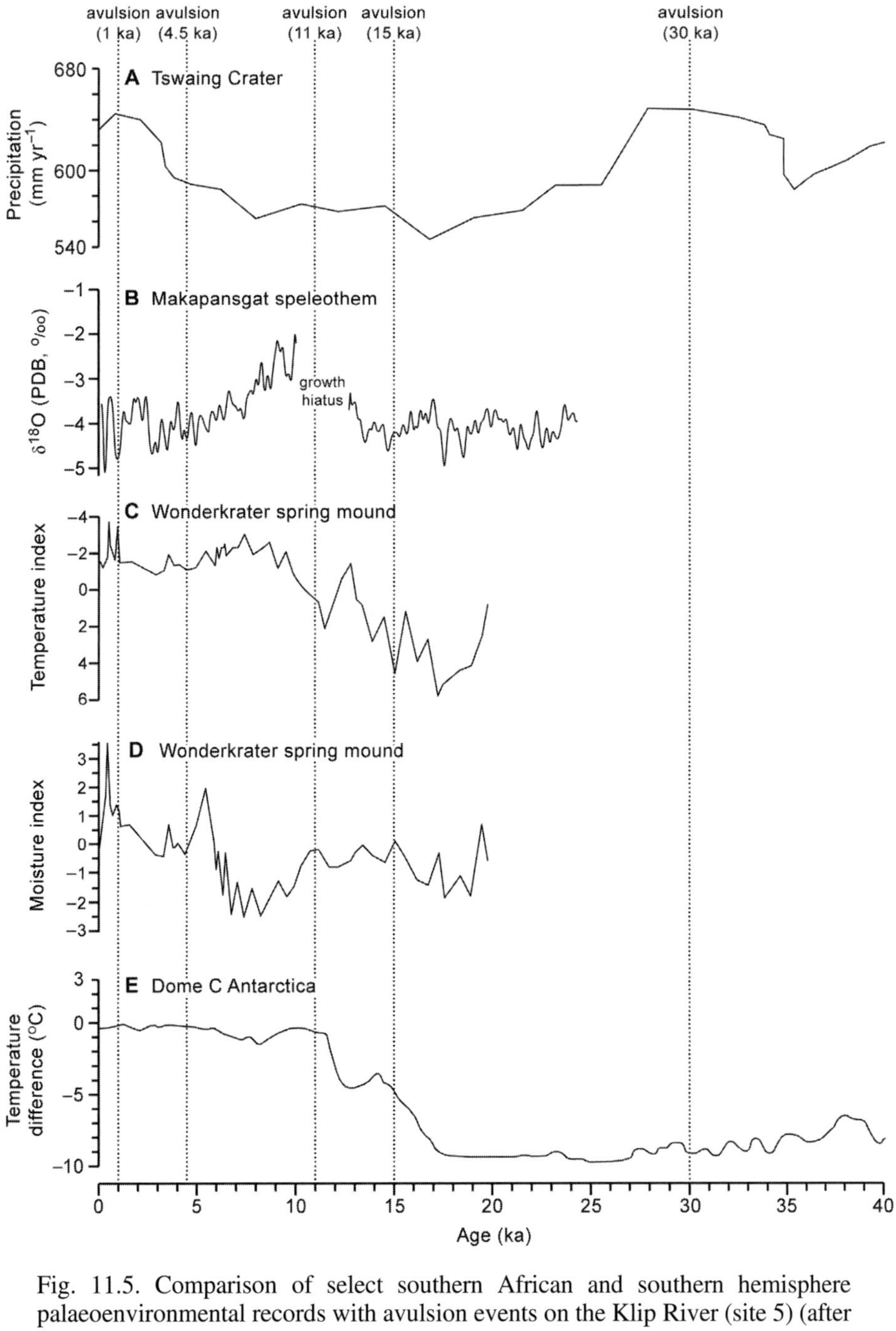

Fig. 11.5. Comparison of select southern African and southern hemisphere palaeoenvironmental records with avulsion events on the Klip River (site 5) (after Tooth et al., 2007; original references cited therein). (a) Tswaing Crater palaeorainfall record; (b) 50-year interval $\delta^{18}O$ record from a stalagmite in the Makapansgat valley (smoothed with a 9-term binomial filter); (c) temperature index derived from Principal Component Analysis of pollen in Wonderkrater spring mound sediments; (d) moisture index derived from Principal Components Analysis of pollen in Wonderkrater spring mound sediments; (e) temperature variations at Dome C, Antarctica, smoothed using a 700-year binomial filter.

rather have occurred autogenically during meander-belt development (Tooth et al., 2007). Tooth et al. (2009, 2013) attribute long-term vertical stability on rivers like the Klip to their typically wide valleys and geologic controls within the study reaches. In such settings, extensive meander belts and infrequent avulsion events distribute sediment laterally rather than vertically, while the channel bed remains anchored to relatively stable bedrock floors, thereby limiting the potential for incision.

As the Modder River example demonstrates, however, stable barriers in dolerite and other resistant lithologies cannot persist indefinitely, for fluvial erosion eventually leads to partial or full breaching. Breaching generates bedrock knickpoints that retreat headward, leading to incision in upstream reaches that may thus occur independently of any palaeoclimatic changes. For instance, on the Schoonspruit (site 6), located 85 km northwest of the Klip River study site, dolerite barrier breaching has resulted in upstream river incision, essentially abandoning a floodplain that was active as recently as 1.28 kyr BP (Tooth et al., 2004; Keen-Zebert et al., 2013). Extensive dongas are now forming on the desiccated former floodplain and in adjacent sediments (Fig. 11.2c), providing local exposures of fossiliferous sediments (including an early *Homo* upper molar) that have been dated by palaeomagnetism to ~1 Ma (Brink et al., 2012).

11.3.3 Holocene fluvial system responses in southern Africa: large and extreme flood events

In hydroclimatically variable settings like southern Africa, many fluvial systems are subject to periodic extreme floods generated by intense and/or prolonged rainfall. These may be singular, 'random' events that occur under stationary climates, or be part of longer term, climatically driven changes in flood frequency and/or magnitude. Holocene ^{14}C- and luminescence-dated palaeoflood records have been reconstructed from deposits (including slackwater sediment sequences) along a number of southern Africa's fluvial systems (e.g. Zawada, 1994, 1997, 2000; Heine, 2004; Eitel et al., 2005; Benito et al., 2011), and accounts have been provided of the erosional and depositional impacts of historical flood events (e.g. Stear, 1985; Zawada and Smith, 1991; Smith, 1992; Rountree et al., 2000). Extreme floods can have significant effects on the production and preservation of fluvial sedimentary archives. For instance, McCarthy et al. (2011) demonstrated how periodic floods have led to tributary fan progradation across the <1 km-wide Nyl/Mogalakwena River valley (site 7), thereby impeding down-valley water and sediment transfer. Sediment logs and limited ^{14}C and OSL dating (McCarthy et al., 2011; Colarossi, 2013) suggest that up to 35 m of Quaternary alluvium has accumulated as a consequence of these trunk-tributary interactions, thereby contributing to formation of floodplain wetlands upstream. In settings where trunk-tributary coupling is less pronounced, extreme

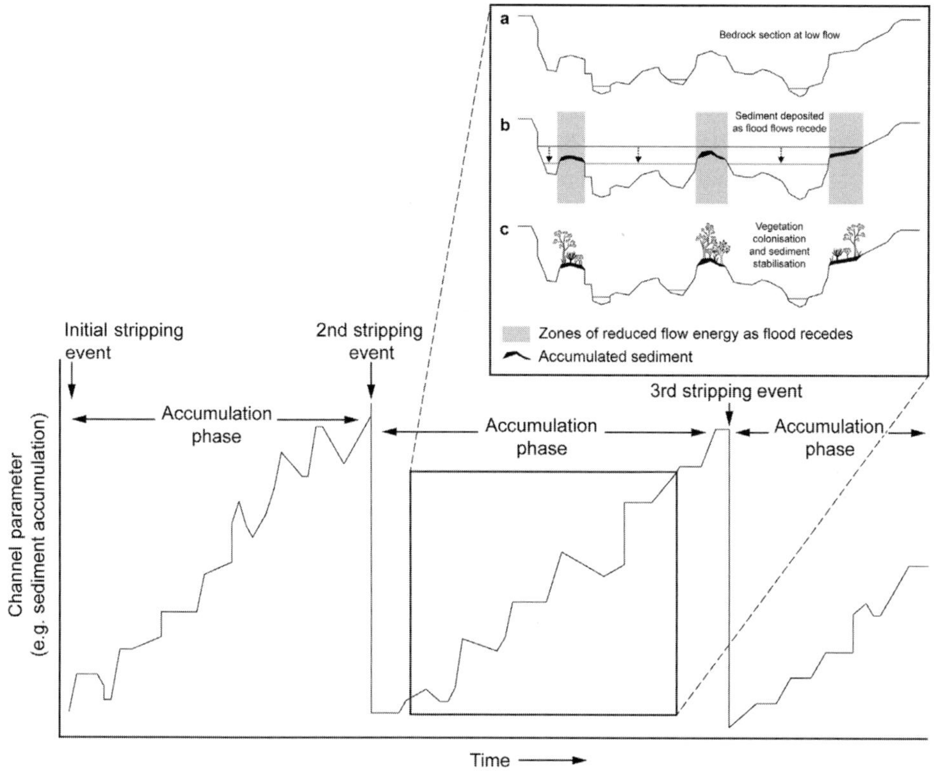

Fig. 11.6. Conceptual model of the relationships between alluvial deposition/ erosion cycles and river behaviour (adapted after van Niekerk et al., 1999; Rountree et al., 2000).

floods tend to restrict long-term fluvial sedimentation. For example, on the Olifants River (site 8), periodic cyclone-driven floods can result in catastrophic 'stripping' of alluvium and widespread riparian vegetation removal (Fig. 11.2d). In the intervals between such floods, net sediment accumulation occurs, typically in association with vegetation regrowth (van Niekerk et al., 1999; Rountree et al., 2000). Preliminary OSL ages from the basal parts of floodplains and islands (Heritage et al., 2014) indicate that alluvium rarely survives for longer than 500–600 years, suggesting that cycles of stripping and accumulation occur on decadal to centennial timescales (Fig. 11.6).

11.4 Discussion

The foregoing evidence for the range of Quaternary fluvial system forms and responses in southern Africa contributes to broader debates in the research literature, including: (1) sensitivity to palaeonvironmental change; and (2) the relative importance of natural and anthropogenic drivers in the dynamics of fluvial systems.

11.4.1 Fluvial system sensitivity to palaeoenvironmental change

Fluvial system 'sensitivity' is commonly defined in terms of the propensity for change or the limited ability to recover from change. In the case of dryland rivers, some authors have argued that even subtle climatic changes can bring about significant, sustained adjustments (e.g. Hereford, 1984, 1986), whereas others have suggested that they are relatively insensitive to all but substantial climatic changes (e.g. Reid, 1994, 2009). Reviews by Nanson and Tooth (1999) and Nanson et al. (2002) concluded that dryland rivers are difficult to typify in terms of climate change sensitivity, since catchment scale, geomorphic setting, and boundary resistance are also important controls. Relatively large, low gradient, lowland rivers tend to respond to sustained and long term (10^3–10^4 yr) climatic changes rather than minor, short term changes. Smaller, higher gradient, upland or piedmont rivers, however, can be sensitive even to relatively minor, short term climatic changes, unless confined by resistant bedrock, inset within indurated or clay-rich alluvium, or stabilised by riparian vegetation. The examples from southern Africa's fluvial systems, discussed above, support these conclusions.

11.4.2 Relative importance of natural and anthropogenic drivers in the late Holocene

The sensitivity concept can be expanded to consider the degree to which fluvial systems have been prone to anthropogenic changes. With the onset of European colonisation, human impact on southern Africa's fluvial systems has greatly intensified, particularly over the last 100–150 years. A few relatively pristine fluvial systems remain (e.g. the Okavango), but in comparison to the variable sensitivity to late Quaternary climatically driven changes, many fluvial systems have been very sensitive to anthropogenic impacts, with dramatic and possibly irreversible changes in process rates, channel-floodplain forms, and sedimentary products (McCarthy et al., 2007; Tooth et al., 2009).

Natural and anthropogenic drivers of fluvial system change, however, are difficult to disentangle (Fig. 11.7). Anthropogenic changes (e.g. sediment impoundment in reservoirs, riparian vegetation degradation) may prime the system for accelerated response to natural environmental drivers such as extreme weather events, or the reverse may apply such that natural and anthropogenic drivers act in tandem. This may pose additional challenges for interpreting fluvial system adjustments to environmental change.

11.5 Summary

Similar to Thomas and Burrough's (2012) general assessment of terrestrial palaeoenvironmental records in interior southern Africa, this chapter has

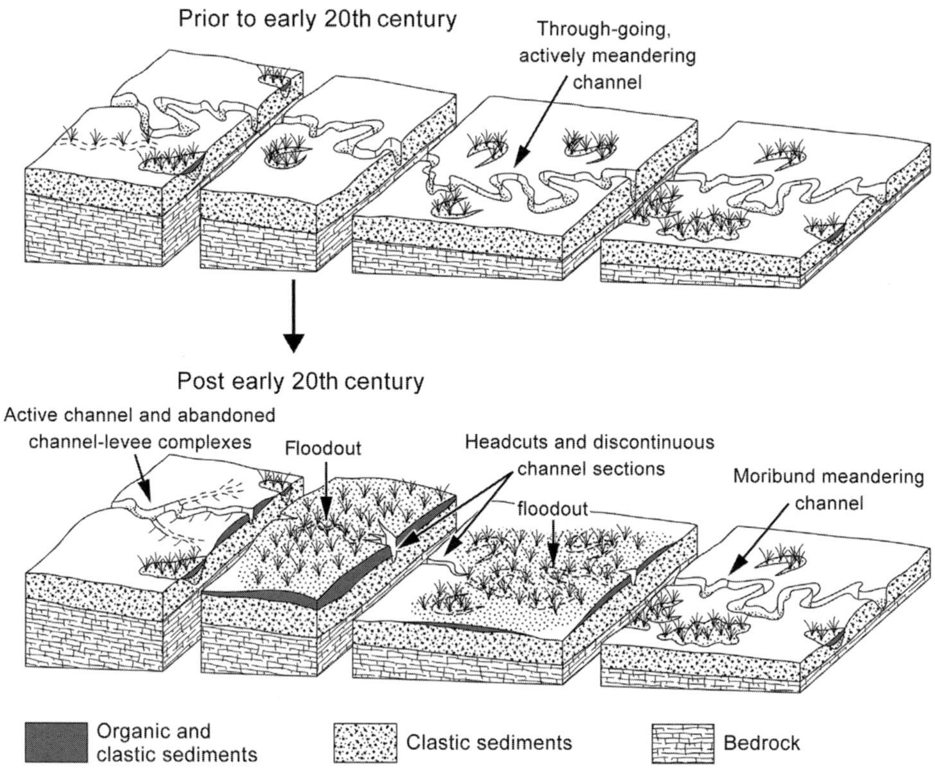

Fig. 11.7. Schematic illustration of changes to the channel and floodplain morphology of the Blood River (site 9) (adapted after Tooth et al., 2014). The relative roles of natural and anthropogenic drivers of these changes remain unclear.

demonstrated that there are not necessarily 'good' or 'bad' fluvial archives. Rather, different archives contain different suites of information about palaeoenvironments at various spatial and temporal scales, and this can complement and extend the information derived from other sources (cf. Nanson and Tooth, 1999). Continued development and application of geochronology and other analytical techniques can contribute to our understanding of fluvial landforms and sediments. For example, a combination of robust luminescence chronologies and geochemical analyses for alluvial-palaeosol successions can complement data from lacustrine or aeolian settings (Lyons et al., 2014). With greater regional coverage and improved geochronology for more Quaternary time slices, fluvial archives are likely to make an even greater contribution to a range of interdisciplinary scientific endeavours, including an improved understanding of the tempo and patterns of landscape development, palaeoclimatic change, faunal evolution, and human cultural developments.

Acknowledgements

The author's investigations into southern Africa's fluvial systems and Quaternary palaeoenvironmental changes have been conducted in collaboration with a number of colleagues worldwide and supported by funding from various sources including the Skye Foundation, Aberystwyth University, University of the Witwatersrand, British Society for Geomorphology, The Royal Society and The Natural Environment Research Council.

References

Bateman, M. D. and Holmes, P. J. (1999). Orange River alluvial terraces: Is luminescence dating a useful tool? *South African Journal of Science*, 95, 57–58.

Benito, G., Thorndycraft, V. R., Rico, M. T., Sánchez-Moya, Y., Sopeña, A., Botero, B. A., Machado, M. J., Davis, M. and Pérez-González, A. (2011). Hydrological response of a dryland ephemeral river to southern African climatic variability during the last millennium. *Quaternary Research*, 75, 471–482.

Bridgland, D. and Westaway, R. (2008). Climatically controlled river terrace staircases: A worldwide Quaternary phenomenon. *Geomorphology*, 98, 285–315.

Brink, J. S., Herries, A. I. R., Moggi-Cecchi, J., Gowlett, J. A. J., Bousman, C. B., Hancox, J. P., Grün, R., Eisenmann, V., Adams, J. W. and Rossouw, L. (2012). First hominin remains from a ~1.0 million year old bone bed at Cornelia-Uitzoek, Free State Province, South Africa. *Journal of Human Evolution*, 63, 527–535.

Brook, G. A., Srivastava, P. and Marais, E. (2006). Characteristics and OSL minimum ages of relict fluvial deposits near Sossus Vlei, Tsauchab River, Namibia, and a regional climate record for the last 30 ka. *Journal of Quaternary Science*, 21, 347–362.

Churchill, S. E., Brink, J. S., Berger, L. R., Hutchison, R. A., Rossouw, L., Stynder, D., Hancox, P. J., Brandt, D., Woodborne, S., Loock, J. C., Scott, L. and Ungar, P. (2000). Erfkroon: A new Florisian fossil locality from fluvial contexts in the western Free State, South Africa. *South African Journal of Science*, 96, 161–163.

Colarossi, P. (2013). *Geology and geochronology of the Nyl River floodplain sediments, Limpopo Province, South Africa*. Unpublished MSc thesis, University of the Witwatersrand, 178pp.

Damm, B. and Hagedorn, J. (2010). Holocene floodplain formation in the southern Cape region, South Africa. *Geomorphology*, 122, 213–222.

de Wit, M. C. J., Marshall, T. R. and Partridge, T. C. (2000). Fluvial deposits and drainage evolution. In *The Cenozoic of Southern Africa*, eds T. C. Partridge and R. R. Maud. Oxford: Oxford University Press, 40, pp. 55–72.

Dollar, E. S. J. (1998). Palaeofluvial geomorphology in southern Africa: A review. *Progress in Physical Geography*, 22, 325–349.

Duller, G. A. T., Tooth, S., Barham, L. and Tsukamoto, S. (2015). New investigations at Kalambo Falls, Zambia: luminescence chronology, site formation, and archaeological significance. *Journal of Human Evolution*, 85, 111–125.

Eitel, B., Kadereit, A., Blümel, W.-D., Hüser, K. and Kromer, B. (2005). The Amspoort Silts, northern Namib Desert (Namibia): Formation, age and palaeoclimatic evidence of river-end deposits. *Geomorphology*, 64, 299–314.

Eitel, B., Kadereit, A., Blümel, W.-D., Hüser, K., Lomax, J. and Hilgers, A. (2006). Environmental changes at the eastern Namib Desert margin before and after the Last Glacial Maximum: New evidence from fluvial deposits in the upper Hoanib River catchment, northwestern Namibia. *Palaeogeography, Palaeoclimatology, Palaeoecology*, 234, 201–222.

Erlanger, E. D., Granger, D. E. and Gibbon, R. J. (2012). Rock uplift rates in South Africa from isochron burial dating of fluvial and marine terraces. *Geology*, 40, 1019–1022.

Gibbon, R. J., Granger, D. E., Kuman, K. and Partridge, T. C. (2009). Early Acheulean technology in the Rietputs Formation, South Africa, dated with cosmogenic nuclides. *Journal of Human Evolution*, 56, 152–160.

Goudie, A. S. and Wells, G. L. (1995). The nature, distribution and formation of pans in arid zones. *Earth-Science Reviews*, 38, 1–69.

Hattingh, J. and Rust, I. C. (1999). Drainage evolution and morphological development of the late Cenozoic Sundays River, South Africa. In *Varieties of Fluvial Form*, eds A. J. Miller and A. Gupta. Chichester: John Wiley and Sons, pp.145–166.

Heine, K. (2004). Little Ice Age climatic fluctuations in the Namib Desert, Namibia, and adjacent areas: Evidence of exceptionally large floods from slack water deposits and desert soil sequences. In *Palaeoecology of Quaternary Drylands*, eds W. Smykatz-Kloss and P. Felix-Henningsen. Berlin: Springer, Lecture Notes in Earth Sciences, 102, pp. 137–165.

Heine, K. and Völkel, J. (2011). Extreme floods around AD 1700 in the northern Namib Desert, Namibia, and in the Orange River catchment, South Africa – were they forced by a decrease of solar irradiance during the Little Ice Age? *Geographia Polonica*, 84, 61–80.

Helgren, D. M. (1977). Geological context of the Vaal River faunas. *South African Journal of Science*, 73, 303–307.

Hereford, R. (1984). Climate and ephemeral-stream processes: Twentieth-century geomorphology and alluvial stratigraphy of the Little Colorado River, Arizona. *Geological Society of America Bulletin*, 95, 654–668.

Hereford, R. (1986). Modern alluvial history of the Paria River drainage basin, southern Utah. *Quaternary Research*, 25, 293–311.

Heritage, G. L., van Niekerk, A. W. and Moon, B. P. (1999). Geomorphology of the Sabie River, South Africa: An incised bedrock-influenced channel. In *Varieties of Fluvial Form*, eds A. J. Miller and A. Gupta. Chichester: John Wiley and Sons, pp. 53–79.

Heritage, G., Entwistle, N., Tooth, S. and Milan, D. J. (2014). Long-term flood controls on semi-arid river form: Evidence from the Sabie and Olifants rivers, eastern South Africa. In *Sediment Dynamics from the Summit to the Sea*, Proceedings of a Symposium held in New Orleans, Louisiana, December 2014. IAHS, 367, pp. 141–146.

Jacobson, P. J., Jacobson, K. M., Seeley, M. K. (1995). *Ephemeral Rivers and Their Catchments: Sustaining People and Development in Western Namibia.* Windhoek: Desert Research Foundation of Namibia, 160pp.

Keen-Zebert, A., Tooth, S., Rodnight, H., Duller, G. A. T., Roberts, H. M. and Grenfell, M. (2013). Late Quaternary floodplain reworking and the preservation of alluvial sedimentary archives in unconfined and confined river valleys in the eastern interior of South Africa. *Geomorphology*, 185, 54–66.

Krapf, C. B. E., Stollhofen, H. and Stanistreet, I. G. (2003). Contrasting styles of ephemeral river systems and their interactions with dunes of the Skeleton Coast erg (Namibia). *Quaternary International*, 104, 41–52.

Lyons, R., Tooth, S. and Duller, G. A. T. (2013). Chronology and controls of donga (gully) formation in the upper Blood River catchment, KwaZulu-Natal, South Africa: Evidence for a climatic driver of erosion. *The Holocene*, 23, 1875–1887.

Lyons, R., Tooth, S. and Duller, G. A. T. (2014). Late Quaternary climatic changes revealed by luminescence dating, mineral magnetism and diffuse reflectance spectroscopy of river terrace palaeosols: A new form of geoproxy data for the southern Africa interior. *Quaternary Science Reviews*, 95, 43–59.

McCarthy, T. S. (2013). The Okavango Delta and its place in the geomorphological evolution of southern Africa. *South African Journal of Geology*, 116, 1–54.

McCarthy, T. S. and Tooth, S. (2004). Incised meanders along the mixed bedrock-alluvial Orange River, Northern Cape Province, South Africa. *Zeitschrift für Geomorphologie*, 48, 273–292.

McCarthy, T. S., Tooth, S., Jacobs, Z., Rowberry, M. D., Thompson, M., Brandt, D., Hancox, P. J., Marren, P. M., Woodborne, S. and Ellery W. N. (2011). The origin and development of the Nyl River floodplain wetland: Trunk-tributary river interactions in a dryland setting. *South African Geographical Journal*, 93, 172–190.

McCarthy, T. S, Venter, J., Arnold, V. and Ellery, W. N. (2007). The collapse of Johannesburg's Klip River wetland. *South African Journal of Science*, 103, 391–397.

Nanson, G. C. and Tooth, S. (1999). Arid-zone rivers as indicators of climate change, In *Paleoenvironmental Reconstruction in Arid Lands*, eds A. K. Singhvi and E. Derbyshire. Rotterdam: A.A. Balkema, pp. 175–216.

Nanson, G. C., Tooth, S. and Knighton, A. D. (2002). A global perspective on dryland rivers: perceptions, misconceptions and distinctions. In *Dryland Rivers: Hydrology and Geomorphology of Semi-Arid Channels*, eds L. J. Bull and M. J. Kirkby. Chichester: John Wiley and Sons, pp. 17–54.

Reid, I. (1994). River landforms and sediments: Evidence of climatic change. In *Geomorphology of Desert Environments*, eds A. D. Abrahams and A. J. Parsons. London: Chapman and Hall, pp. 571–592.

Reid, I. (2009). River landforms and sediments: Evidence of climatic change. In: *Geomorphology of Desert Environments*, eds A.J. Parsons and A.D. Abrahams, 2nd edn. London: Springer, pp. 695–721.

Rodnight, H., Duller, G. A. T., Tooth, S. and Wintle, A. G. (2005). Optical dating of a scroll-bar sequence on the Klip River, South Africa, to derive the lateral migration rate of a meander bend. *The Holocene*, 15, 802–811.

Rodnight, H., Duller, G. A. T., Wintle, A. G. and Tooth, S. (2006). Assessing the reproducibility and accuracy of optical dating of fluvial deposits. *Quaternary Geochronology*, 1, 109–120.

Rountree, M. W., Rogers, K. H. and Heritage, G. L. (2000). Landscape state change in the semi-arid Sabie River, Kruger National Park, in response to flood and drought. *South African Geographical Journal*, 82, 173–181.

Scheepers, A. C. T. and Rust, I. C. (1999). The Uniab River fan: An unusual alluvial fan on the hyper-arid Skeleton Coast, Namibia. In *Varieties of Fluvial Form*, eds A. J. Miller and A. Gupta. Chichester: John Wiley and Sons, pp. 273–294.

Smith, A. M. (1991). Extreme palaeofloods: Their climatic significance and the chances of floods of similar magnitude recurring. *South African Journal of Science*, 87, 219–220.

Smith, A. M. (1992). Palaeoflood hydrology of the lower Umgeni River from a reach south of the Inanda Dam, Natal. *South African Geographical Journal*, 74, 63–68.

Srivastava, P., Brook, G. A. and Marais, E. (2005). Depositional environment and luminescence chronology of the Hoarusib River Clay Castles sediments, northern Namib Desert, Namibia. *Catena*, 59, 187–204.

Srivastava, P., Brook, G. A., Marais, E., Morthekai, P. and Singhvi, A. K. (2006). Depositional environment and OSL chronology of the Homeb silt deposits, Kuiseb River, Namibia. *Quaternary Research*, 65, 478–491.

Stear, W. M. (1985). Comparison of the bedform distribution and dynamics of modern and ancient sandy ephemeral flood deposits in the southwestern Karoo region, South Africa. *Sedimentary Geology*, 45, 209–230.

Thomas, D. S. G. and Burrough, S. L. (2012). Interpreting geoproxies of late Quaternary climate change in African drylands: Implications for understanding environmental change and early human behaviour. *Quaternary International*, 253, 5–17.

Tooth, S. and McCarthy, T. S. (2004). Anabranching in mixed bedrock-alluvial rivers: The example of the Orange River above Augrabies Falls, Northern Cape Province, South Africa. *Geomorphology*, 57, 235–262.

Tooth, S. and McCarthy, T. S. (2007). Wetlands in drylands: Key geomorphological and sedimentological characteristics, with emphasis on examples from southern Africa. *Progress in Physical Geography*, 31, 3–41.

Tooth, S. and Nanson, G. C. (2011). Distinctiveness and diversity of arid zone rivers. In *Arid Zone Geomorphology: Process, Form and Change in Drylands*, ed. D. S. G. Thomas. Chichester: Wiley, pp. 269–300.

Tooth, S., Brandt, D., Hancox, P. J. and McCarthy, T. S. (2004). Geological controls on alluvial river behaviour: A comparative study of three rivers on the South African Highveld. *Journal of African Earth Sciences*, 38, 79–97.

Tooth, S., Rodnight, H., McCarthy, T. S., Duller, G. A. T. and Grundling, A. (2009). Late Quaternary dynamics of a South African floodplain wetland and the implications for assessing recent human impacts. *Geomorphology*, 106, 278–291.

Tooth, S., Hancox, P. J., Brandt, D., McCarthy, T. S., Jacobs, Z. and Woodborne, S. M. (2013). Controls on the genesis, sedimentary architecture, and preservation potential of dryland alluvial successions in stable continental interiors: Insights from the incising Modder River, South Africa. *Journal of Sedimentary Research*, 83, 541–561.

Tooth, S., McCarthy, T. S., Rodnight, H., Keen-Zebert, A., Rowberry, M. D. and Brandt, D. (2014). Late Holocene development of a major fluvial discontinuity in floodplain wetlands of the Blood River, eastern South Africa. *Geomorphology*, 205, 128–141.

Tooth, S., Rodnight, H., Duller, G. A. T., McCarthy, T. S., Marren, P. M. and Brandt, D. (2007). Chronology and controls of avulsion along a mixed bedrock-alluvial river. *Geological Society of America Bulletin*, 119, 452–461.

van Niekerk, A. W., Heritage, G. L., Broadhurst, L. W. and Moon, B. P. (1999). Bedrock anastomosing channel systems: Morphology and dynamics of the Sabie River, Mpumulanga Province, South Africa. In *Varieties of Fluvial Form*, eds A. J. Miller and A. Gupta. Chichester: John Wiley and Sons, pp. 33–51.

Vogel, J. C. and Rust, U. (1987). Environmental changes in the Kaokoland Namib Desert during the present millennium. *Madoqua*, 15, 5–16.

Ward, J. D., Jacob, R. J., De Wit, M. J. C., Spaggiari, R. I. and Bluck, B. J. (2002). *Post-Gondwana Evolution of the Vaal–Orange Drainage System: Economic Implications*. Unpublished excursion guide, 16th International Sedimentological Conference, Rand Afrikaans University, South Africa, 75pp.

Wintle, A. G., Li, S.-H., Botha, G. A. and Vogel, J. C. (1995). A chronological framework for colluviation during the last 110 kyr in KwaZulu/Natal. *South African Journal of Science*, 91, 134–139.

Zawada, P. K. (1994). Palaeoflood hydrology of the Buffels River. Laingsburg, South Africa: was the 1981 flood the largest? *South African Journal of Geology*, 97, 21–32.

Zawada, P. K. (1997). Palaeoflood hydrology: Method and application in flood-prone southern Africa. *South African Journal of Science*, 93, 111–132.

Zawada, P. K. (2000). *Palaeoflood Hydrological Analysis of Selected South African Rivers*. Pretoria: Council for Geoscience, 173pp.

Zawada, P. K. and Smith, A. M. (1991). The 1988 Orange River flood, Upington region, Northwestern Cape Province, RSA. *Terra Nova*, 3, 317–324.

12

Wetlands in southern Africa

WILLIAM N. ELLERY, SUZANNE E. GRENFELL, MICHAEL C. GRENFELL, REBECCA POWELL, DONOVAN C. KOTZE, PHILIP M. MARREN AND JASPER KNIGHT

Abstract

In southern Africa, wetlands of different types are an integral part of the drainage network, yet evolve and are sensitive to different combinations of geologic, climatic, geomorphic, edaphic and hydrologic controls. Understanding of these controls can help in the interpretation of environmental and climatic records from different wetland types, given that wetland sensitivity to environmental and climatic changes may vary throughout their 'life cycle'. The chapter discusses inland wetland records from dated sites in South Africa in order to consider their significance for reconstructing late glacial and Holocene climates; and the relationship of wetlands to preservation of the Pleistocene archaeological record. Wetlands are sensitive to degradation under contemporary environmental and climatic changes, which may impact on their hydrological and ecological function as well as the integrity of associated archaeological sites.

12.1 Introduction

Wetlands are defined in a South African context as 'land which is transitional between terrestrial and aquatic systems where the water table is usually at or near the surface, or the land is periodically covered with shallow water, and which land in normal circumstances supports or would support vegetation typically adapted to life in saturated soil' (Republic of South Africa, National Water Act, 1998). This definition, therefore, shows that hydrology and ecosystems are key controls on the functioning and classification of wetlands, similar to that defined by the Ramsar Convention. In southern Africa, a range of wetland types exists. These include permanent (lakes, riparian edges), seasonal (pans, slope-foot locations) and ephemerally-flooded areas (McCarthy and Hancox, 2000; Rowberry et al., 2011). An alternative hierarchical wetland classification by Ollis et al. (2013) distinguishes marine, estuarine and inland wetlands, the latter of which are then disaggregated based on their topographical position (valley floor, slope, plain and bench), and by constituent hydrogeomorphic units (e.g. floodplain, valley bottom, depression, seep and flat). This classification scheme, therefore, focuses on geomorphic properties of individual wetlands, rather than their hydrological, ecological and biogeochemical properties. Such an integrated classification, which includes both the attributes of wetlands and their varied origins and evolution, may be more appropriate.

Across southern Africa, wetlands of different types are found in different regions (Fig. 12.1). For example, depressions of deflational origin – which may be intermittently flooded – are widespread in the semi-arid northwest of the region, including northwest South Africa, Botswana and southwest Zimbabwe (Goudie and Thomas, 1985; Allan et al., 1995) and along palaeo-drainage lines of today's major rivers (Holmes and Barker, 2006). Floodplain and valley-bottom wetlands are concentrated in the moister eastern and southern parts of southern Africa, such as Maputaland (Rogers, 1997; McCarthy and Hancox, 2000; Ellery et al., 2012). Valley-bottom wetlands typically occur near low-order streams where the valley is narrow or impounded (M.C. Grenfell et al., 2008, 2009), while floodplain wetlands can only develop when discharge is sufficient to cause streams to laterally migrate, which occurs on higher-order streams and on shallower slopes. Wetlands can also develop where the groundwater table is perched or as a result of high recharge (Dippenaar, 2014; Weitz and Demlie, 2014). Tooth and McCarthy (2007) have discussed the occurrence and diversity of wetlands in southern Africa, contrasting those in dryland areas with those in humid regions.

Wetlands are significant palaeoenvironmental archives by virtue of the fact that they can act as both short- and long-term sediment sinks, and may contain both minerogenic and organic components (McCarthy et al., 2010). Mineral components, contributed mainly by fluvial action but also by slope and aeolian processes, can inform on land surface aridity/humidity and source region. Organic components (including micro- and macrofossils) can inform on species and ecosystem type, and may in some cases be used for radiocarbon dating. Wetlands are thus an important palaeoenvironmental archive, not least because they often have fairly good preservation potential, and are found in locations where other evidence is scarce. This chapter (1) reviews the different types of wetlands, their properties and distributions in southern Africa; (2) discusses some environmental records contained within wetlands, with a particular focus on valley-bottom wetlands; and (3) considers some of the main threats to exploiting such archives in the future, including management and climate change issues.

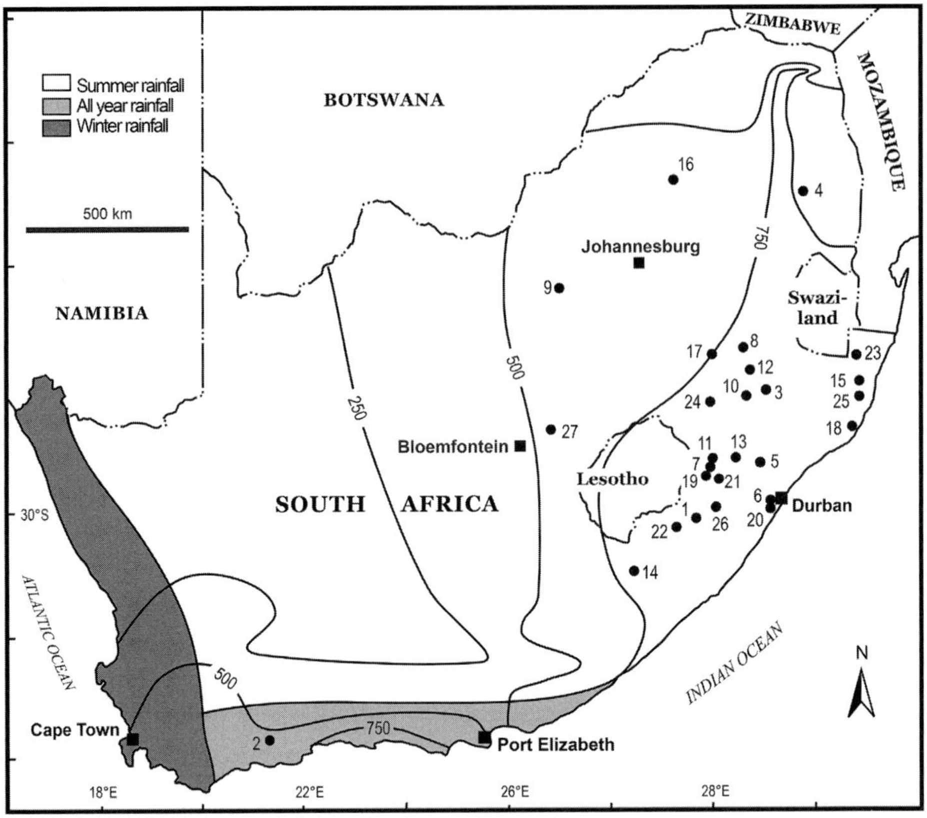

Fig. 12.1. Mean annual rainfall isohyets (mm) and rainfall seasonality across South Africa (after Schulze, 1997). The locations of valley-bottom and floodplain wetlands used in Fig. 13.2 are: (1) Ntsikeni, (2) Goukou, (3) Blood River Vlei, (4) Craigieburn, (5) Mvoti Vlei, (6) Fredville, (7) Hlatikulu, (8) Wakkerstroom, (9) Schoonspruit, (10) Gladstone, (11) Hlatikhulu-Nsonge floodplain, (12) Boschoffsvlei, (13) Kruisfontein, (14) Weatherley, (15) Mkuze, (16) Nylsvlei/Wonderkrater, (17) Klip River, (18) Mfolozi, (19) Stillerust, (20) Chelmsford, (21) Mgeni Vlei, (22) Mzimvubu, (23) Pongola. Other wetlands discussed in this chapter are: (24) Braamhoek, (25) Mfabeni Mire, (26) Mahwaqa Mountain, (27) Florisbad.

12.2 Controls on wetland types and their dynamics

Macroscale controls on wetlands include tectonics and sea-level change, climate/hydrology, catchment properties such as channel type, sediment supply and, more recently, anthropogenic factors (McCarthy and Hancox, 2000). At the scale of individual freshwater wetland systems, hydrological and thus hydrochemical regimes strongly control wetland dynamics and properties. For example, changes in groundwater contribution to valley foot positions by recharge can dramatically impact on wetland chemistry and ecosystem structure and function (Humphries et al., 2011a; Riddell et al., 2013; Bird and Day, 2014; Weitz and Demile, 2014). High rates of evapotranspiration and a negative water balance from wetlands can lead to salinisation of wetland waters, with impacts on the chemical and ecological status of wetlands (Humphries et al., 2011b). This is a significant factor in the dry and hot northwest of the region, including the Okavango Delta (McCarthy and Hancox, 2000); and in eastern South Africa, such as the Mkuze River wetland system which drains into Lake St Lucia and the wider iSimangaliso World Heritage Site (Humphries et al., 2011b). Variations in sediment input into wetland systems, although these are linked to rainfall and weathering/erosion within the catchment, show a complex relationship to wetland dynamics, because high sediment accumulation may reduce the area available for water and wetland ecosystems. However, shifting channels or in-channel bars can block or restrict water outflow and thus increase the area flooded (Grenfell et al., 2008) and favour wetland development (Ellery et al., 2012). Variations in catchment connectivity, drainage density and land surface properties can trigger sediment and biogeochemical cascades within river basins that can result in footslope water impounding and wetland development (M.C. Grenfell et al., 2009). Such geomorphic developments have potential to influence wetland type, structure and function over timescales of decades to millennia (McCarthy and Ellery, 1995; Ashley et al., 2004; S. E. Grenfell et al., 2009, 2010; Humphries et al., 2010; McCarthy et al., 2011a; Ellery et al., 2012). An exception is the Okavango Delta wetland which has developed in a subsiding half-graben which has provided long-term accommodation space and thus ensured the persistence of this system (McCarthy, 2013).

The examination of wetland longitudinal slope offers a useful framework for investigating wetland evolution (Fig. 12.2). Longitudinal slope (*S*) is governed by the relationship between discharge (*Q*), sediment supply (*Qs*) and particle size (*D*):

$$S \alpha \frac{Qs}{Q} D$$

Typically, long-profile concavity develops over time, because *Q* typically increases downstream as drainage basin area increases; *Qs* increases, but at a declining rate due to sediment storage; and *D* tends to become finer due to abrasion and changes in sediment source (Lane, 1955). However, transmission losses that result in *Q* decreasing with distance downstream, and erosion-resistant substrates, can lead to localised steeping of the long profile. This has potential to increase stream power and make wetlands vulnerable to incision.

Once an equilibrium longitudinal profile is attained on a river and wetland reach, available energy is then used to erode laterally by meander migration (e.g. Tooth et al., 2002), thereby expanding wetland size. Therefore, although resistant lithology may act as a control on wetland gradient in the short–medium term (10^2–10^4

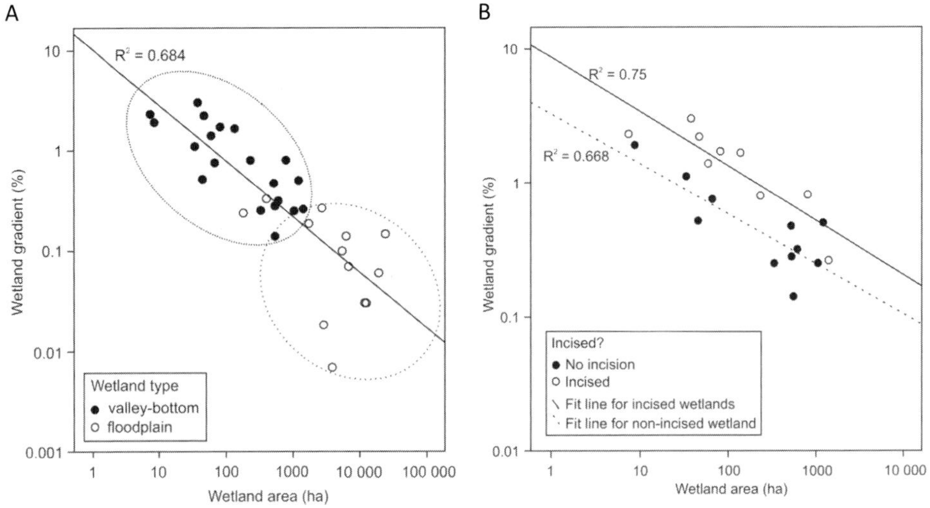

Fig. 12.2. The relationship between (A) wetland longitudinal gradient and area of valley-bottom and floodplain wetlands; and (B) valley-bottom wetland gradient as a function of wetland area.

years), these wetlands are degradational in the long-term ($>10^4$ years), with the rate of degradation determined by the rate of substrate weathering and erosion. The size of the wetland thus produced is dependent upon stream discharge, as larger rivers will erode longer and broader valleys even as the downstream local base level adjusts slowly through erosion. In cases where local base level increases, sediment may be deposited along the blocked valley due to a loss of transport capacity. Therefore, such wetlands are aggradational, and wetland area increases over time. In floodplain wetlands, localised steepening may lead to meander cut-offs or channel avulsion (Slingerland and Smith, 1998), as is the case of the Mfolozi River wetland (Grenfell and Ellery, 2009). On the Klip River floodplain, however, climate changes have had no impact on processes such as meander migration and avulsion rate (Rodnight et al., 2005; Tooth et al., 2007). Valley-bottom wetlands, however, are characterised by smaller, discontinuous or diffuse flows and have limited capacity to rework sediment laterally.

The potential vulnerability of a valley-bottom wetland to incision is shown in Figure 12.3. As wetlands develop, which results in long profile steepening, they become more vulnerable to incision (deposition moves the wetland along path A–B over time). Conversely, natural climate cycles or human activities such as hardening of surfaces in the catchment may increase runoff, moving the wetland along path A–C. Wetlands can thus be viewed as components of a larger process-response system in a state of dynamic equilibrium, whereby wetlands are not the endpoints of landscape development, but rather a temporal step in a recurring cycle

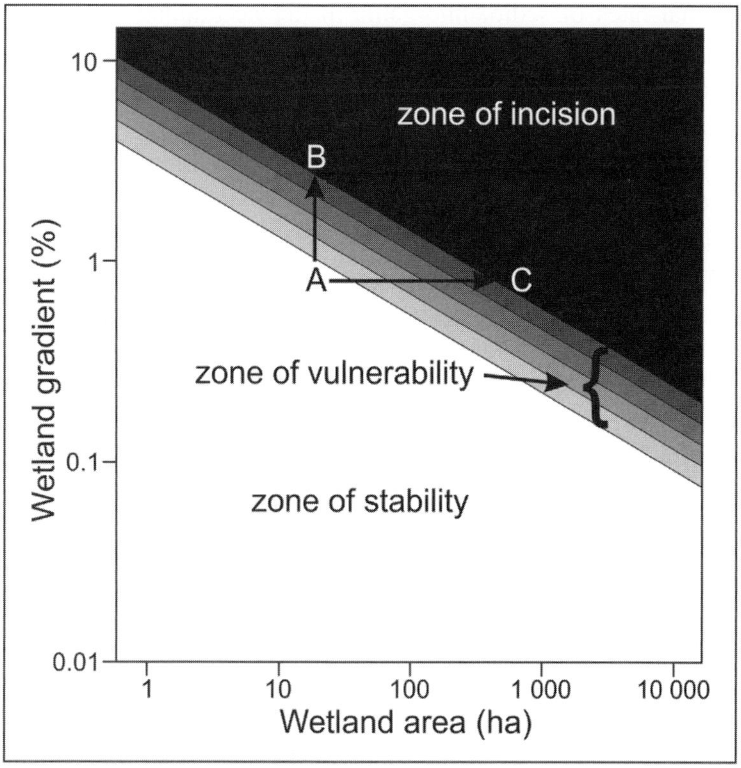

Fig. 12.3. Zones of stability, vulnerability and incision for southern African valley-bottom wetlands.

of erosion and deposition. This conceptual model has implications for the formation, persistence and preservation potential of palaeoenvironmental records within wetlands.

12.2.1 Valley-bottom and floodplain wetlands

Valley-bottom and floodplain wetlands have been extensively studied in southern Africa, and are thus focused on in this chapter (Fig. 12.1). These are significant because valley-bottom and footslope locations have potential to be depositional sinks and thus act as palaeoenvironmental archives. There is also a clear relationship between slope gradient and wetland area (Fig. 12.2A), suggesting that, over time, as sediments accumulate in valley bottoms, footslope locations or floodplains, wetlands get larger. An implication of wetland growth is that the macroscale wetland structure breaks down over time due to continued sediment accumulation, which causes stream blockage and water diversion. The integrated wetland system then becomes compartmentalised with wetter and dryer patches, and differential rates of sediment/organic matter accumulation and preservation (Tooth et al., 2009), as is the case in the Okavango. Incision of wetland sediments can be initiated by local base level fall, changes in catchment land use or climate (Boardman, 2014). This demonstrates a clear relationship to wetland gradient, where steeper wetlands are more vulnerable to incision (Fig. 12.2B). Where wetlands have high organic matter accumulation, they may develop a domed surface (e.g. Wonderkrater; McCarthy et al., 2010) which therefore may become steeper over time.

Inland wetland types show different physical properties. Floodplain wetlands are generally larger (>1800 ha) than valley-bottom ones (<200 ha), and have shallower slopes (Fig. 12.2A). Both wetland types may occupy areas in the range 200–1800 ha. Particularly small floodplain wetlands include Hlatikhulu Vlei (403 ha; Nsonge River) and Stillerust Vlei (186 ha; Mooi River). In contrast, the valley-bottom wetlands of Lynspruit (1380 ha), Schoonspruit (1200 ha) and Wakkerstroom (1051 ha) are unusually large. The relationship between wetland slope and size (R^2=0.684, $p<0.05$) shows the link between climate and geomorphology. In terms of climatic controls, mean annual catchment runoff is the most frequently described variable (e.g. Vandaele et al., 1996; Morgan and Mngomezulu, 2003; Zucca et al., 2006), although Schumm (1979) used catchment size as a proxy for catchment runoff. In southern Africa, wetland size appears to be a stronger indicator of valley-bottom wetland gradient than catchment area. Wetland size is also useful from an applied perspective because it is easier to measure than either catchment area or runoff. The long term viability of wetlands in inland locations, independent of those connected to river mouths at the coast, is closely related to climate variability and wetlands as hydrochemical and biogeochemical buffers. Several studies have shown that, over certain spatial and temporal scales, inland wetlands are able to withstand climate variability (e.g. Mohamed and Savenije, 2014), which is key to understanding the ways in which wetlands may have potential to preserve Holocene palaeoenvironmental and palaeoclimatic records.

12.3 Environmental and climatic records from wetlands

Variations in wetland composition (organic vs inorganic), macro- and microfossil content, ash/charcoal content, and growth rate (based on radiocarbon dating) can all provide environmental and climatic information. Reconstructed growth rates of different wetlands across South Africa, based on age–depth relationships (Fig. 12.4), show relatively high variability, which is likely related to moisture availability, with coastal wetlands (e.g. Mfabeni Mire; Grundling et al., 2013) experiencing higher accumulation rates than inland and mountain footslope

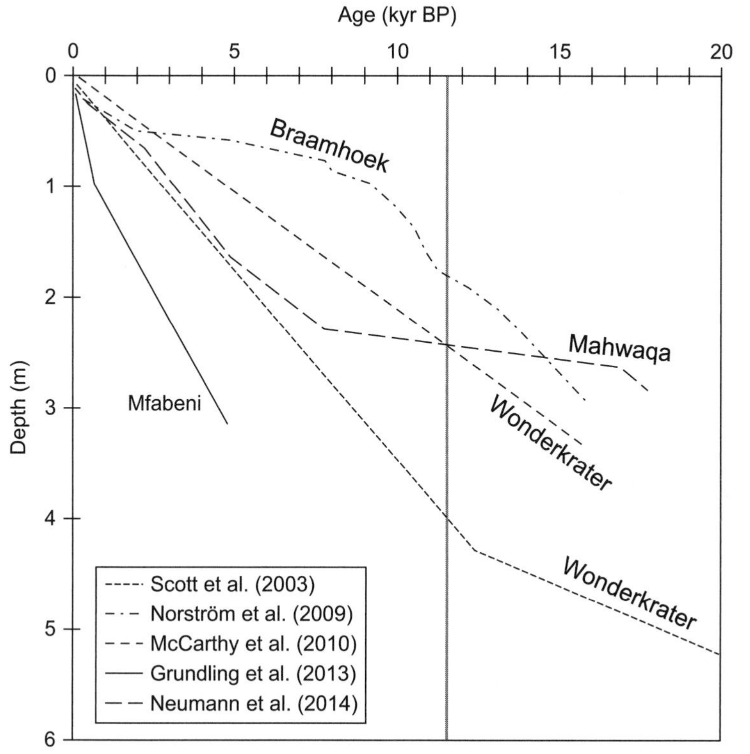

Fig. 12.4. Plot of the relationships between radiocarbon age and depth within selected South African wetlands. Sites are Wonderkrater (Scott et al., 2003; McCarthy et al., 2010), Braamhoek (Norström et al., 2009), Mfabeni Mire, St. Lucia (Grundling et al., 2013), and Mahwaqa Mountain, KwaZulu-Natal (Neumann et al., 2014). The solid grey vertical line represents the lateglacial–Holocene boundary at 11.6 kyr BP.

wetlands (e.g. Braamboek; Norström et al., 2009; Mahwaqa Mountain; Neumann et al., 2014). Although average accumulation rates are argued to have increased from the lateglacial into the Holocene (Grundling et al., 2013), such calculations are based on relatively sparse radiocarbon data points and there is likely high spatial and temporal variability not captured by such calculations.

There have been relatively few detailed or multidisciplinary accounts of palaeoenvironmental changes based on wetland records from southern Africa. Studies by Norström et al. (2009) and Finné et al. (2010) from Braamhoek; Scott et al. (2003) and McCarthy et al. (2010) from Wonderkrater; and Neumann et al. (2014) from Mahwaqa Mountain are notable exceptions (Fig. 12.1). These records provide insight into regional climatic and environmental changes from the lateglacial onwards and, in some instances, extend back to around 20 kyr BP. Records from some locations that were or may have been lakes/wetlands at periods in their depositional history, such as Florisbad and Tswaing, are also available (e.g. Kristen et al., 2007; Rabumbula and Holmes, 2012), and these may extend back to >200 kyr BP, but lie outside of the scope of discussion in this chapter. During the lateglacial–Holocene transition, wetland records are highly variable. It is notable that, despite being in relatively close proximity, wetland ecosystem responses to temperature at Mahwaqa and Braamhoek are different during the lateglacial and become more similar during the Holocene, whereas the opposite is the case with respect to moisture (Fig. 12.5). At a finer scale, the frequency of charcoal (thus burning) is greater at Braamhoek in the lateglacial, suggesting more arid conditions, which is supported by lower tree pollen and higher Asteraceae during this time and in the mid-Holocene (Fig. 12.5B). Part of the difficulty with interpreting wetland records lies in identifying local depositional system controls *versus* regional climatic controls. This may be a significant issue in developing regional correlations between wetland records, although the use of Principal Component Analysis (PCA) can help in this regard (e.g. Fig. 12.5A). More complex analytical methods such as Fourier or Bayesian analysis are limited by the number of data points available.

12.4 Sensitivity of wetlands to climate and environmental changes

The environmental records contained within wetlands reflect the sensitivity of wetland systems to be affected by climate and environmental changes and processes. For example, as wetlands aggrade over time, resulting in perched water tables, the wetlands become isolated from groundwater systems and thus more ombrotrophic, reflecting a changed relationship between wetland processes and climate, which can be denominated as their 'life cycle'. If wetlands start forming at different times or form at different rates, and thus reflect different stages of their 'life cycle', the preserved climate records of different wetlands will suggest different climate regimes even if climate was, in fact, uniform across the region. Geomorphic relationships as outlined in Figure 12.2 may therefore conceal complexity with respect to climate.

This changing sensitivity of wetlands to climate forcing can also help in the understanding of patterns of human activity around key wetland sites such as Florisbad. Sedimentary evidence from this site shows significant changes in processes and environments over time, including periods that are dominated by spring-fed pan, aeolian dune, and fluvial environments, both during the Holocene and earlier (e.g. Scott and Nyakale, 2002; Douglas, 2006; Rabumbulu and Holmes, 2012). Detailed analysis of sediments and geochemistry reveals the interactions between spring waters and surface (lake/river) waters in the environmental chemistry of the site (Douglas et al., 2010). Late Holocene sediments comprise organic

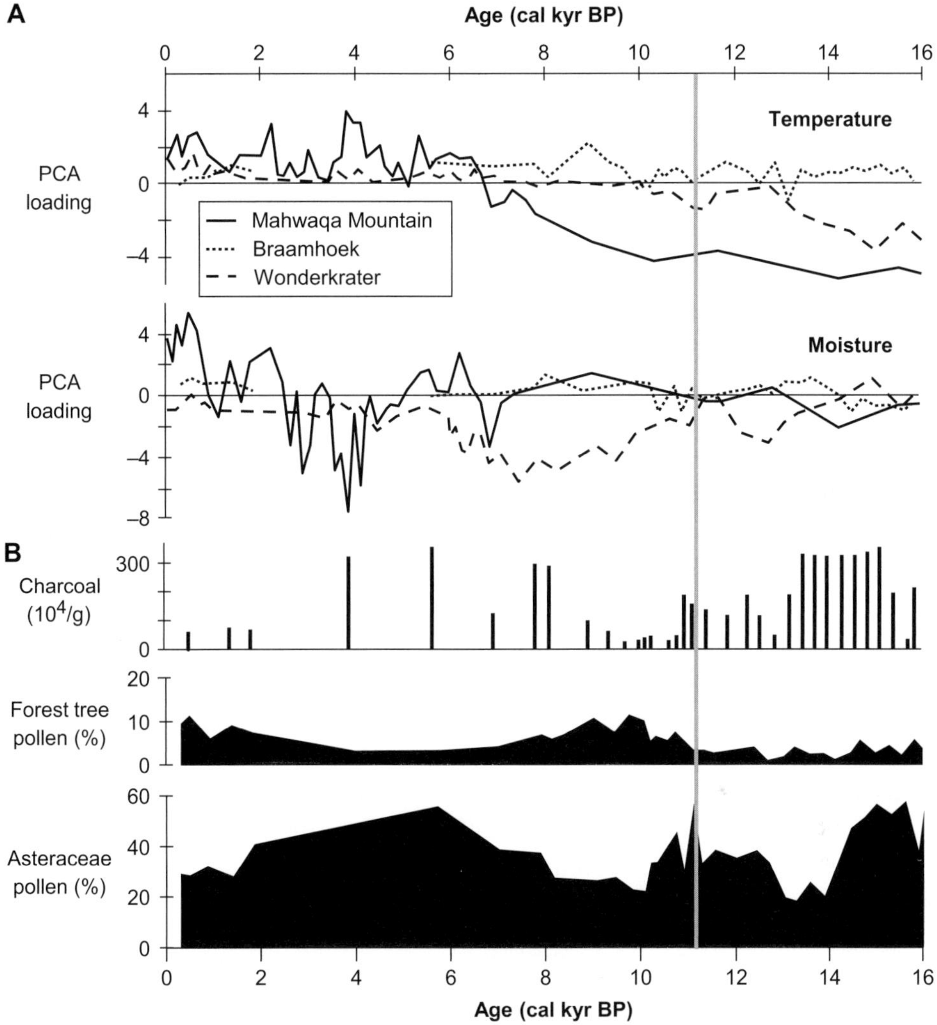

Fig. 12.5. Plots of certain palaeoenvironmental variables from selected wetland case studies in South Africa for the period 16 cal kyr BP to present. (A) Principal component analysis (PCA) loadings for temperature and moisture from three South African wetlands (redrawn after Neumann et al., 2014); (B) abundance of charcoal, forest tree pollen and Asteraceae pollen from Braamhoek wetland (redrawn after Norström et al., 2009). The solid grey vertical line represents the lateglacial–Holocene boundary at 11.6 kyr BP.

sands containing abundant Poaceae, Cyperaceae and Asteraceae, indicative of a lake/pan environment with variations in water area and depth over time (Scott and Nyakale, 2002). The archaeological record at Florisbad has a particular focus on the Middle Stone Age (MSA), dated at this site to 279±47 kyr BP (Kuman et al.,

1999). The evidence suggests hunting and butchery activities adjacent to the pan, where artefacts were slowly buried as the pan expanded into a larger wetland, presumably in response to late Pleistocene climate change (Kuman and Clarke, 1996; Kuman et al., 1999). Reoccupation of the site around 157±21 and 121±6 kyr BP (Kuman et al., 1999), including a MSA hearth, may reflect dryer climatic periods or where lunette dunes may have transgressed across the site (e.g. Douglas, 2006). Geochemical results support covariability between chemical processes within and adjacent to the wetland, water table perching, and fossilisation of artefacts and faunal remains, although the precise role of climate in driving these processes is uncertain (Douglas et al., 2010). Butzer (1984) summarised the linked environmental and archaeological records of many southern African pan and wetland sites, including Florisbad, its adjacent site of Vlakkraal, some of the smaller western Free State pans, and the much larger Kathu and Etosha pans in the southern Kalahari basin. Butzer (1984) argued that the environmental and archaeological records of these sites show no clear spatial or temporal patterns, because of regional climatic gradients, but that these records are controlled by moisture availability. This clearly indicates the climatic and hydrological context for human activities around these sites.

The sensitivity of wetlands to contemporary climate and environmental changes, as well as those of the past, presents a number of issues for wetland management. These arise due to groundwater fluctuations and contamination, river water pollution and water management upstream, aquatic biodiversity loss, and urbanisation within the catchment (Tooth et al., 2009; Mitchell, 2013). Key issues include degradation through incision, drying out of the wetland surface, maintaining wetland biodiversity, invasive species, inappropriate planning and land-use, and river system management. These issues may negatively impact upon wetland hydrological and ecological functioning, as well as the integrity of any associated archaeological sites.

12.5 Summary

Wetlands are sensitive to environmental and climatic changes over decadal to millennial timescales because they develop as a result of sediments, water and ecosystems in combination. Southern African wetlands are highly diverse in terms of their size, geomorphic settings and controls, and this diversity results in variable levels of sensitivity to external forcing, which has implications for the accumulation and interpretation of the sedimentary, biotic and chemical records preserved within wetland sequences. Understanding the complex relationships wetlands have with respect to climate is fundamental to the use of such sequences in Quaternary palaeoenvironmental reconstruction.

References

Allan, D. E. G., Seaman, M. T. and Kaletja, B. (1995). The endorheic pans of South Africa. In *Wetlands of South Africa*, ed. G. Cowan. Pretoria: Department of Environmental Affairs and Tourism, pp. 75–101.

Ashley, G. M., Mworia, J. M., Muasya, A. M., Owen, R. B., Driese, S. G., Hover, V. C., Renaut, R. W., Goman, M. F., Mathai, S. and Blatt, S. H. (2004). Sedimentation and recent history of a freshwater wetland in a semi-arid environment: Loboi Swamp, Kenya, East Africa. *Sedimentology*, 51, 1–21.

Bird, M. S. and Day, J. A. (2014). Wetlands in changed landscapes: The influence of habitat transformation on the physico-chemistry of temporary depression wetlands. *PLoS ONE*, 9(2): e88935, doi:10.1371/journal.pone.0088935.

Boardman, J. (2014). How old are the gullies (dongas) of the Sneeuberg uplands, Eastern Karoo, South Africa? *Catena*, 113, 79–85.

Butzer, K. W. (1984). Archeogeology and Quaternary environment in the interior of southern Africa. In *Southern African Prehistory and Paleoenvironments*, ed. R. G. Klein. Rotterdam: Balkema, pp. 1–64.

Dippenaar, M. A. (2014). Towards hydrological and geochemical understanding of an ephemeral palustrine perched water table "wetland" (Lanseria Gneiss, Midrand, South Africa). *Environmental Earth Sciences*, 72, 2447–2456.

Douglas, R. M. (2006). Formation of the Florisbad spring and fossil site – an alternative hypothesis. *Journal of Archaeological Science*, 33, 696–706.

Douglas, R. M., Holmes, P. J. and Tredoux, M. (2010). New perspectives on the fossilization of faunal remains and the formation of the Florisbad archaeological site, South Africa. *Quaternary Science Reviews*, 29, 3275–3285.

Ellery, W. N., Grenfell, S. E., Grenfell, M. C., Humphries, M. S., Barnes, K., Dahlberg, A. and Kindness, A. (2012). Peat formation in the context of the development of the Mkuze floodplain on the coastal plain of Maputaland, South Africa. *Geomorphology*, 141–142, 11–20.

Finné, M., Norström, E., Risberg, J. and Scott, L. (2010). Siliceous microfossils as late-Quaternary paleo-environmental indicators at Braamhoek wetland, South Africa. *The Holocene*, 20, 747–760.

Goudie, A. S. and Thomas, D. S. G. (1985). Pans in southern Africa with particular reference to South Africa and Zimbabwe. *Zeitschrift für Geomorphologie, N.F.*, 29, 1–19.

Grenfell, M. C., Ellery, W. and Grenfell, S. E. (2008). Tributary valley impoundment by trunk river floodplain development: A case study from the KwaZulu-Natal Drakensberg foothills, eastern South Africa. *Earth Surface Processes and Landforms*, 33, 2029–2044.

Grenfell, M. C., Ellery, W. N. and Grenfell, S. E. (2009). Valley morphology and sediment cascades within a wetland system in the KwaZulu-Natal Drakensberg Foothills, Eastern South Africa. *Catena*, 78, 20–35.

Grenfell, S. E. and Ellery, W. N. (2009). Hydrology, sediment transport dynamics and geomorphology of a variable flow river: The Mfolozi River, South Africa. *Water SA*, 35, 271–282.

Grenfell, S. E., Ellery, W. N. and Grenfell, M. C. (2009). Geomorphology and dynamics of the Mfolozi River floodplain, KwaZulu-Natal, South Africa. *Geomorphology*, 107, 226–240.

Grenfell, S. E., Ellery, W. N., Grenfell, M. C., Ramsay, L. F. and Flügel, T. J. (2010). Sedimentary facies and geomorphic evolution of a blocked-valley lake: Lake Futululu, northern KwaZulu-Natal, South Africa. *Sedimentology*, 57, 1159–1174.

Grundling, P., Grootjans, A. P., Price, J. S. and Ellery, W. N. (2013). Development and persistence of an African mire: How the oldest South African fen has survived in a marginal climate. *Catena*, 110, 176–183.

Holmes, P. and Barker, C. H. (2006). Geological and geomorphological controls on the physical landscape of the Free State. *South African Geographical Journal*, 88, 3–10.

Humphries, M. S., Kindness, A., Ellery, W. N., Hughes, J. and Benitez-Nelson, C. R. (2011a). ^{137}Cs and ^{210}Pb derived sediment accumulation rates and their role in the long-term development of the Mkuze River floodplain, South Africa. *Geomorphology*, 119, 88–96.

Humphries, M. S., Kindness, A., Ellery, W. N. and Hughes, J. C. (2011b). Sediment geochemistry, mineral precipitation and clay neoformation on the Mkuze River floodplain, South Africa. *Geoderma*, 157, 15–26.

Kristen, I., Fuhrmann, A., Thorpe, J., Röhl, U., Wilkes, H. and Oberhänsli, H. (2007). Hydrological changes in southern Africa over the last 200 Ka as recorded in lake sediments from the Tswaing impact crater. *South African Journal of Geology*, 110, 311–326.

Kuman, K. and Clarke, R. J. (1996). Florisbad – New investigations at a Middle Stone Age Hominid site in South Africa. *Geoarchaeology*, 1, 103–125.

Kuman, K., Inbar, M. and Clarke, R. J. (1999). Palaeoenvironments and cultural sequence of the Florisbad Middle Stone Age Hominid site, South Africa. *Journal of Archaeological Science*, 26, 1409–1425.

Lane, E. W. (1955). *The importance of fluvial morphology in hydraulic engineering.* Proceedings, American Society of Civil Engineers, 81, Paper 745, 17pp.

McCarthy, T. S. (2013). The Okavango Delta and its place in the geomorphological evolution of southern Africa. *South African Journal of Geology*, 116, 1–54.

McCarthy, T. S. and Ellery, W. N. (1995). Sedimentation on the distal reaches of the Okavango Fan, Botswana, and its bearing on calcrete and silcrete (ganister) formation. *Journal of Sedimentary Research*, A65, 77–90.

McCarthy, T. S., Ellery, W. N., Backwell, L., Marren, P., de Klerk, B., Tooth, S., Brandt, D. and Woodborne, S. (2010). The character, origin and palaeoenvironmental significance of the Wonderkrater spring mound, South Africa. *Journal of African Earth Sciences*, 58, 115–126.

McCarthy, T. S. and Hancox, P. J. (2000). Wetlands. In *The Cenozoic of Southern Africa*, ed. T. C. Partridge and R. R. Maud. Oxford: Oxford University Press, pp. 218–235.

McCarthy, T. S., Tooth, S., Jacobs, Z., Rowberry, M. D., Thompson, M., Brandt, D., Hancox, P. J., Marren, P. M., Woodborne, S. and Ellery, W. N. (2011). The origin and development of the Nyl River floodplain wetland, Limpopo Province, South Africa: Trunk-tributary river interactions in a dryland setting. *South African Geographical Journal*, 93, 172–190.

McCarthy, T. S., Tooth, S., Kotze, D. C., Collins, N. B., Wandrag, G. and Pike, T. (2010). The role of geomorphology in evaluating remediation options for floodplain wetlands: The case of Ramsar-listed Seekoeivlei, eastern South Africa. *Wetlands Ecology and Management*, 18, 119–134.

Mitchell, S. A. (2013). The status of wetlands, threats and the predicted effect of global climate change: The situation in Sub-Saharan Africa. *Aquatic Science*, 75, 95–112.

Mohamed, Y. and Savenije, H. H. G. (2014). Impact of climate variability on the hydrology of the Sudd wetland: Signals derived from long term (1900–2000) water balance computations. *Wetlands and Ecological Management*, 22, 191–198.

Morgan, R. P. C. and Mngomezulu, D. (2003). Threshold conditions for initiation of valley-side gullies in the Middle Veld of Swaziland. *Catena*, 50, 401–414.

Neumann, F. H., Botha, G. A. and Scott, L. (2014). 18,000 years of grassland evolution in the summer rainfall region of South Africa: Evidence from Mahwaqa Mountain, KwaZulu-Natal. *Vegetation History and Archaeobotany*, 23, 665–681.

Norström, E., Scott, L., Partridge, T. C., Risberg, J. and Holmgren, K. (2009). Reconstruction of environmental and climate changes at Braamhoek wetland, eastern escarpment South Africa, during the last 16,000 years with emphasis on the Pleistocene–Holocene transition. *Palaeogeography, Palaeoclimatology, Palaeoecology*, 271, 240–258.

Ollis, D., Snaddon, K., Job, N. and Mbona, N. (2013). *Classification system for wetlands and other aquatic ecosystems in South Africa*. Pretoria: SANBI Biodiversity Series 22, 110pp.

Rabumbulu, M. and Holmes, P. J. (2012). Depositional environments of the Florisbad Spring site and surrounds: A revised synthesis. *South African Geographical Journal*, 94, 191–207.

Republic of South Africa (1998). *The National Water Act. Act No. 36 of 1998*. Pretoria: Government Gazette.

Riddell, E. S., Everson, C., Clulow, A. and Mengistu, M. (2013). The hydrological characterisation and water budget of a South African rehabilitated headwater wetland system. *Water SA*, 39, 57–66.

Rodnight, H., Duller, G. A. T., Tooth, S. and Wintle, A. G. (2005). Optical dating of a scroll-bar sequence on the Klip River, South Africa, to derive the lateral migration rate of a meander bend. *The Holocene*, 15, 802–811.

Rogers, K. H. (1997). Freshwater wetlands. In *Vegetation of Southern Africa*, ed. R. M. Cowling, D. M. Richardson and S. M. Pierce. Cambridge: Cambridge University Press, pp. 322–347.

Rowberry, M. D., McCarthy, T. S., Thompson, M., Nomnganga, A. and Moyo, L. (2011). The spatial and temporal characterisation of flooding within the floodplain wetland of the Nyl River, Limpopo Province, South Africa. *Water SA*, 37, 445–451.

Schumm, S. A. (1979). Geomorphic thresholds: The concept and its applications. *Transactions of the Institute of British Geographers, NS*, 4, 485–515.

Scott, L., Holmgren, K., Talma, A. S., Woodborne, S. and Vogel, J. C. (2003). Age interpretation of the Wonderkrater spring sediments and vegetation change in the Savanna Biome, Limpopo Province, South Africa. *South African Journal of Science*, 99, 484–488.

Scott, L. and Nyakale, M. (2002). Pollen indications of Holocene palaeoenvironments at Florisbad spring in the central Free State, South Africa. *The Holocene*, 12, 497–503.

Slingerland, R. and Smith, N. D. (1998). Necessary conditions for a meandering-river avulsion. *Geology*, 26, 435–438.

Tooth, S., Brandt, D., Hancox, P. J. and McCarthy, T. S. (2004). Geological controls on alluvial river behaviour: A comparative study of three rivers on the South African Highveld. *Journal of African Earth Sciences*, 38, 79–97.

Tooth, S. and McCarthy, T. S. (2007). Wetlands in drylands: Geomorphological and sedimentological characteristics, with emphasis on examples from southern Africa. *Progress in Physical Geography*, 31, 3–41.

Tooth, S., McCarthy, T. S., Brandt, D., Hancox, P. J. and Morris, R. (2002). Geological controls on the formation of alluvial meanders and floodplain wetlands: The example of the Klip River, eastern Free State, South Africa. *Earth Surface Processes and Landforms*, 27, 797–815.

Tooth, S., Rodnight, H., Duller, G. A. T., McCarthy, T. S., Marren, P. M. and Brandt, D. (2007). Chronology and controls of avulsion along a mixed bedrock-alluvial river. *Geological Society of America Bulletin*, 119, 452–461.

Tooth, S., Rodnight, H., McCarthy, T. S., Duller, G. A. T. and Grundling, A. T. (2009). Late Quaternary dynamics of a South African floodplain wetland and the implications for assessing recent human impacts. *Geomorphology*, 106, 278–291.

Vandaele, K., Poesen, J., Govers, G. and van Wesemael, B. (1996). Geomorphic threshold conditions for ephemeral gully incision. *Geomorphology*, 16, 161–173.

Weitz, J. and Demlie, M. (2014). Conceptual modelling of groundwater–surface water interactions in the Lake Sibayi Catchment, Eastern South Africa. *Journal of African Earth Sciences*, 99, 613–624.

Zucca, C., Canu, A. and Della Peruta, R. (2006). Effects of land use and landscape on spatial distribution and morphological features of gullies in an agropastoral area in Sardinia (Italy). *Catena*, 68, 87–95.

13

Sandy coasts

HAYLEY C. CAWTHRA AND MARK D. BATEMAN

Abstract

The South African coastline is variable in geomorphic character, but at least two thirds of the coastline is dominated by sandy and part-lithified sandy intertidal shorelines. Part-lithified sandy intertidal shorelines consist of a mosaic of rocky outcrops on dominantly sandy beaches. The rocks on these mixed shores are generally composed of cemented Quaternary deposits. Carbonate cemented Quaternary aeolianite/beachrock complexes are relatively common along the South African coastline and on the continental shelf, where they form prominent reefs. The Quaternary deposits along the coastline are characterised mainly by transgressive dune cordons. These systems record sensitive palaeoenvironmental fluctuations, as archived in their structure, diagenetic history, and relationship to past sea-levels and ancient shorelines.

13.1 Introduction

The South African coastline is >2700 km in length and covers a wide range of climate zones, wave/tidal regimes and physiographic settings (Fig. 13.1). The Great Escarpment forms the backdrop to the coastal plain on the west coast which is ~60 km wide around Cape Town and ~190 km wide near the Orange River mouth. Off the south coast lies the extensive Agulhas Bank where the relatively flat continental shelf extends up to 130 km in a water depth of <200 m. On this coastline, the coastal plain is generally narrower at <40 km wide (Flemming, 1981) but broadens to the west due to the Agulhas Bank. It is also narrow in KwaZulu-Natal but widens northwards reaching 440 km width in Mozambique (Armitage et al., 2006).

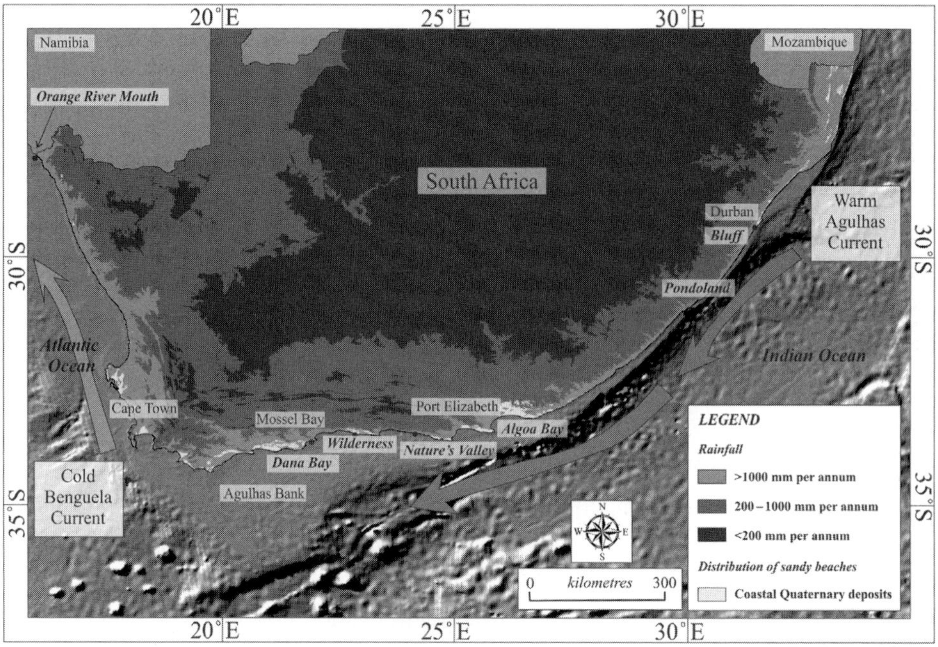

Fig. 13.1. Map of the South African coastline, the paths of the Agulhas and Benguela Currents and the extensive coastal Quaternary deposits. Note also the extensive shallow offshore Agulhas Bank, parts of which would have been exposed during Quaternary sea-level lowstands.

The South African shoreline is composed of approximately 29% rocky coast, 39% sandy coast, 32% mixed coast (Harris et al., 2011), defined here as 'part-lithified sandy intertidal shorelines'. The distribution of coastline types is strongly governed by the highly variable geological substrate (Musekiwa et al., 2015). In places, resistant lithologies (such as the quartzites of the Palaeozoic Cape Supergroup on the west and south coasts and the Natal Group and Msikaba Formation sandstones on the east coast) extend to the coast and form the basis for relatively steep sea-cliffs or resistant rocky shorelines (Roberts et al., 2014) (Fig. 13.2).

This chapter concentrates on sandy coasts, including part-lithified sandy intertidal shorelines composed of cemented Quaternary deposits. Volumetrically, calcareous aeolian sediments dominate the coastal successions, with an age range of Palaeogene to Recent. Where lithologies are resistant, coastal topography is steep and thus limits sediment accommodation space for beaches. As beaches constitute the major sediment source of coastal dunes (Tinley, 1985; Porat and Botha, 2008; Roberts et al., 2014), the paucity of beaches and the steep topography limit the size and create obstructions for inland transgression of dunes. Broad coastal plains tend to develop in areas of softer lithological substrates (typically shale-dominated)

Fig. 13.2. Examples of coastline types found on the southern Cape coast. (A) Cliffed coastlines in Table Mountain Group sandstone at Pondoland, (B) beachrock dominated foreshore with active back beach dunes at Reebok, (C) barrier dunes and backbarrier lagoons at Wilderness (photos: Hayley Cawthra). A black and white version of this figure will appear in some formats. For the colour version, please refer to the plate section.

through planation by subsequent Cenozoic sea-level highstands. In these regions, palaeoshorelines may ingress up to tens of kilometres inland (Pether et al., 2000; Roberts et al., 2006, 2014). The aeolian deposits along the west coast are dominated by dune plumes, whereas their counterparts along the south and east coasts comprise coast-parallel trending dune barriers or cordons (Roberts et al., 2014) (Fig. 13.1).

13.2 Regional setting

In terms of coastal climate, winds along the southern coastline between Cape Town and Port Elizabeth are mainly coast-parallel (Kruger, 2004), turning more north-northeasterly (summer) and southwesterly (winter) east of Port Elizabeth (Tinley, 1985; Ramsay, 1996). Sea surface temperatures (SST) are largely controlled by the cold Benguela Current and the warm Agulhas Current (Fig. 13.1), with nearshore upwelling of the former giving rise to a high marine productivity. Waves along the South African coastline have a long fetch and are of high energy. On the southern coastline, the waves are southwesterly (Heydorn and Tinley, 1980) and on average 2.1 m high and 11 s apart (Whitfield, 1983; Cooper, 2001). On the east coast, waves are dominantly from the south (Smith et al., 2010) with average heights of 2.07 m and spacing of 9 s (Cooper, 2001). The tidal range along most of the South African coast is microtidal and semidiurnal, with spring tidal ranges typically 1.8–2.0 m and neap tidal ranges falling within 0.6–0.8 m (Cooper, 2001). Longshore drift of sediment is northwards on the west coast (1.4 km^3/a^{-1}; de Decker, 1988) and eastward/northeastward (0.8 km^3/a^{-1}; Schoonees, 2000) on the southern and eastern coastlines (Davies, 1980).

Currently, the nearshore area on the west coast/western part of the south coast receives little terrestrial sediment (Dingle et al., 1987). In contrast, 78% of the sediment on the continental shelf on the eastern coast is of fluvial origin (Dingle et al., 1987). On the west coast only the Orange, Olifants and Berg rivers contribute sediment offshore, which is largely mud (Meadows et al., 2002). Littoral sediment varies slightly around the coastline with coarser sediments found in KwaZulu-Natal and the west coast, compared to the south coast (Cooper, 2001).

The west coast is heterogeneous, with contrasting rocky cliffs and extensive sandy beaches, sheltered bays and exposed straight coasts (Harris et al., 2011) (Fig. 13.2) and distinctive dunes associated with river mouths (Roberts et al., 2009). West coast beaches are generally steep and narrow with high berms (Cooper, 2001; Harris et al., 2011). In the southwestern Cape, a dominantly rocky coast is interspersed with smaller sandy beaches (Harris et al., 2011; Musekiwa et al., 2015). This coastline, due to a combination of fine sediments and high wave energy, is dominantly composed of flat and wide dissipative beaches with multiple offshore bars and a wide surf zone (Harris et al., 2011). The south coast comprises a cliffed or rocky coastline interspersed with a sequence of log-spiral bays that are largely structurally controlled by a series of half-grabens which trend SE–NW and extend onto the continental shelf (Partridge and Maud, 2000). In the bays of Wilderness and Nature's Valley, beaches are mostly dissipative, although less well developed than on the west coast (Harris et al., 2011). In many bays of the Eastern Cape, sand has commonly been blown over the rocky headland through mobile dunes, in the form of headland-bypass dune systems (Illenberger and Burkinshaw, 2008; Harris et al., 2011). The 50 km-long, 2 km-wide Alexandria dunefield in Algoa Bay (Fig. 13.1) is one of the largest active coastal dune fields in the world (McLachlan et al., 1982). The Wild Coast, which lies between the south- and east coasts, is predominantly rock or cliff, interspersed with pocket and embayed beaches (Harris et al., 2011). In the southern part of the 550 km-long KwaZulu-Natal coast, between Port Edward and Durban, beaches are predominantly part-lithified sandy, as well as rocky, intertidal shorelines with numerous small temporary open and closed estuaries. In northern KwaZulu-Natal, the coastal plain is of lower relief and wider, and sandy beaches become more dominant. These beaches tend to be reflective, being narrow and steeply sloping with a limited surf zone. The wave-dominated marine environment and relatively arid hinterland along much of the southern and Eastern Cape coasts has also produced a number of 'wave dominated' estuarine systems, often characterised by temporary connections to the Indian Ocean and the formation of perched estuaries (Cooper, 2001; de Lecea et al., this volume).

13.3 Lithified Quaternary sediments

Carbonate-cemented Quaternary deposits are relatively common along the South African coastline where they create part-lithified sandy intertidal shorelines, and on the continental shelf, where they form prominent reefs (Martin and Flemming, 1987; Cooper and Flores, 1991; Ramsay and Cooper, 2002; Roberts et al., 2006; Carr et al., 2010; Bosman, 2012; Cawthra et al., 2012). These deposits volumetrically dominate the part-lithified sandy intertidal shorelines. The oldest reported Quaternary cemented coastal sand deposits in South Africa (~390 kyr BP) crop out near Mossel Bay (Jacobs et al., 2011; Roberts et al., 2012) and Richards Bay (Porat and Botha, 2008), and the youngest (<100 yr BP) on the Durban Bluff (Cawthra and Uken, 2012).

As carbonate-cemented Quaternary deposits studied along the South African coast lack organic structures (Cooper and Flores, 1991; Cawthra et al., 2012; Roberts et al., 2012) (Fig. 13.3), the mechanism behind the cementation is inorganic. The most common modes of cementation are through the evaporation of interstitial seawater in response to climatic conditions, or dissolution of shell fragments by meteoric water. Beachrocks are consolidated sedimentary formations, consisting of beach material bonded together by *in situ* precipitated carbonate cements (calcite and/or aragonite) (Bathurst, 1975; Vousdoukas et al., 2007). Lithification occurs in the intertidal and/or supratidal zones either on the beach surface, or beneath a thin veneer of unconsolidated sediment (Neumeier, 1999; Vousdoukas et al., 2009). Constituent particles include clastic, biogenic and authigenic sands and gravels, as well as human artefacts (Milliman, 1974; Cawthra and Uken, 2012). Aeolianite (palaeodune) cementation is dominated by calcite spar, generally equant in form, infilling voids between constituent grains (Flügel, 2004). This process occurs in the vadose diagenetic zone and is facilitated by the percolation of rainwater through unconsolidated sand dunes.

Carbonate diagenesis has been used to infer relative ages of aeolianite on palaeo-coastlines, as carbonate overprinting, replacement and re-cementation potentially provide insight into diagenetic zones. As an example, cementation of

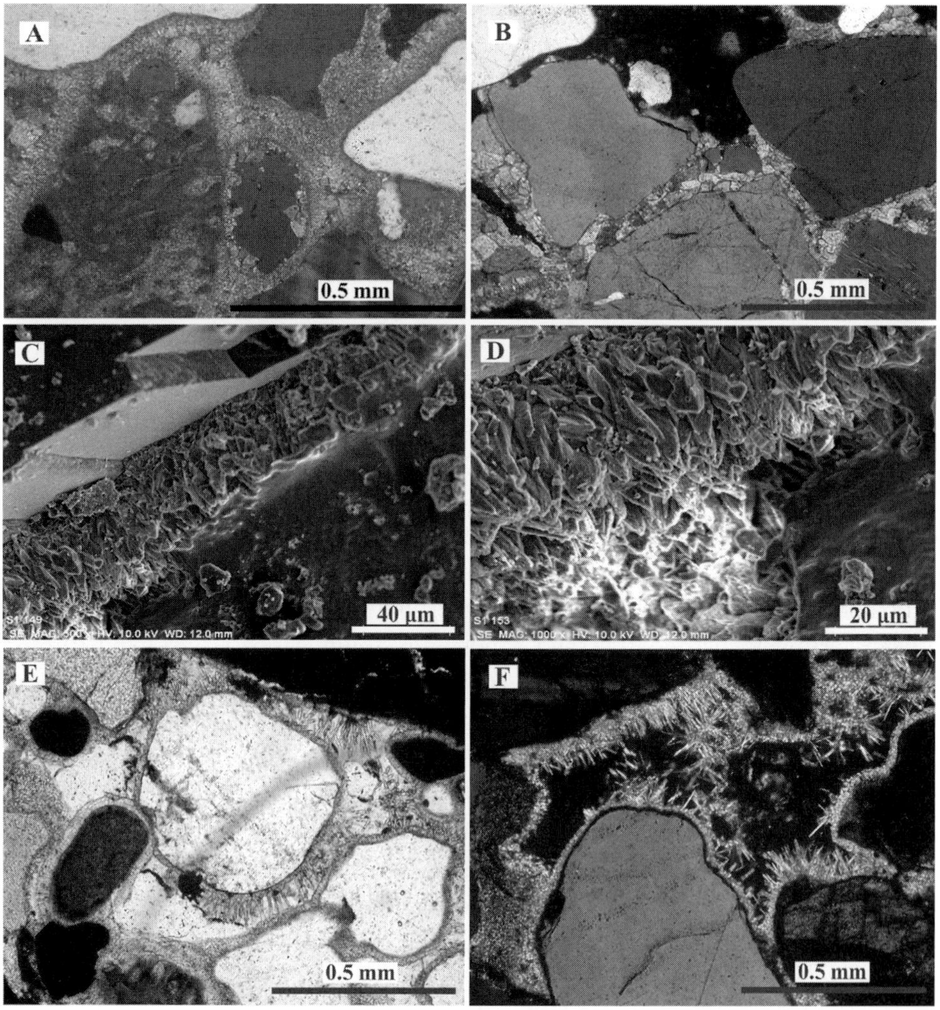

Fig. 13.3. (A) 'Dog tooth' calcite spar infilling a void, Blood Reef (15 m below sea level), transmitted light microscope image, (B) blocky equant calcite spar cement, Durban Bluff, transmitted light microscope image, (C) isopachous rim cement (micrite) and equant spar, Blood Reef (23 m below sea level), scanning electron microscope image, (D) micritic isopachous rim cement, Blood Reef (19 m below sea level), scanning electron microscope image, (E) minor isopachous micrite rims surrounding clasts, prominent aragonite radiating into voids, Durban Bluff, transmitted light microscope image, (F) magnified micrite rims and aragonite crystals, Durban Bluff, transmitted light microscope image (photos: Hayley Cawthra).

the wind-blown sands off the Durban Bluff and adjacent continental shelf (Ramsay and Cooper, 2002; Cawthra et al., 2012) initially occurred in the interstadial of Marine Isotope Stage (MIS) 7a. In response to a rise in sea level, the water table gradually migrated towards the landward margin of the basin. The raised deposits are characterised by a predominance of equant or drusy calcite spar, but a minor influence of micrite precipitation attests to the last interglacial (MIS 5). Offshore, MIS 4 deposits (forming Blood Reef), in thin section, reveal multiple cementation episodes, argued to be remnant of MIS 3 and the Holocene transgression of MIS 1.

13.3.1 Erosional features

The reliability of erosional coastal features of former sea level depends on identification and understanding of the processes that were active in developing the features (Guilcher, 1988). The most reliable proxies of former sea levels include sedimentary (e.g. beachrock/aeolianite) and erosional features (e.g. notches and erosional terraces) (e.g. Laborel et al., 1994; Elias, 2007; Lambeck et al., 2010). In the case of sandy carbonate-cemented coastal deposits, most erosion occurs on the seaward margins of deposits (Miller and Mason, 1994) due to the interaction of freshwater with sea water dissolving carbonate cements and mechanical erosion of outcrops by wave action.

In a South African context, erosional features of the intertidal environment provide relatively reliable sea-level indicators where associated with cemented Quaternary deposits. Ramsay (1996), Bosman (2012) and Cawthra et al. (2012) have recognised that the dominance of remnant erosional features on seafloor outcrops of the east coast relate to the Holocene transgression. Above present sea level, a well-developed palaeo-cliff-line occurs on the seaward side of the dune cordon on the north side of the Wilderness embayment (Illenberger, 1996). Erosion of pre-existing beachrock/aeolianite has also resulted in the incision of intertidal potholes, solution basins, undercuts and the development of joints/fractures (Marker and Holmes, 2010) (Fig. 13.4) as well as impressive aeolianite cliffs, e.g. 180+ m-high barrier dunes near Wilderness (Fig. 13.1).

13.3.2 Depositional features

Coastal dune systems form by the deflation of beach sediment, and therefore the composition of the dune sediments reflects marine and terrestrial influences, as well as bedrock lithology. Fragments of ancient sandy shorelines cemented into aeolianite and beachrock provide insight into the evolution of littoral zones and coastlines through time. Despite the distinctive preservation of the former beaches, cementation of beachrocks appears possible at a range of vertical elevations within and above the tidal range (Fig. 13.4). While hampering its use as a precise sea-level indicator (e.g. Kelletat, 2006; Vött et al., 2010), when used in conjunction with an accurate sedimentological context, it can be used as such (Cawthra and Uken,

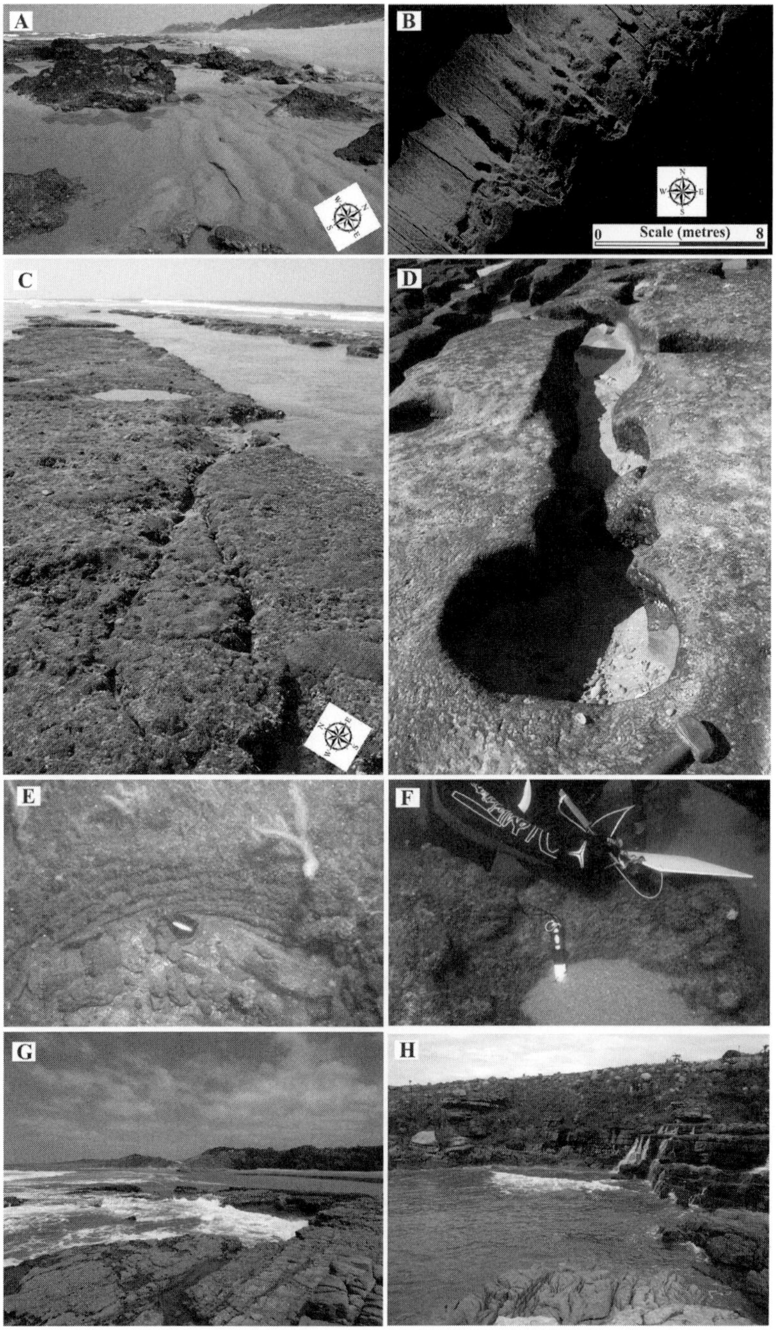

Fig. 13.4. (A) Erosional gullies on a rocky shoreline, trending in a coast-perpendicular orientation, Durban Bluff, (B) multibeam echosounder data from Blood Reef showing submerged erosional gullies 17 m below sea level, of comparable dimensions to (A), (C) fractured joint sets on an outcrop of beachrock, Durban Bluff, (D) intertidal zone potholes, Reebok (Mossel Bay), (E, F) erosional potholes, now submerged 18 and 24 m below sea level respectively, Blood Reef, (G) wave-cut platform in the Msikaba Formation, Pondoland, (H) coastal sea-cliffs and incised overhands, Pondoland (photos: Hayley Cawthra). A black and white version of this figure will appear in some formats. For the colour version, please refer to the plate section.

2012). In addition to indicating changes in coastline within aeolianites, the preserved rhizoliths indicate the presence of past vegetation during dune construction, and ichnofossils within the dune sequences may include the preserved footprints of now-extant or extinct animals. Examples of this include the Nahoon hominid footprint site, indicating that humans were in the coastal zone during MIS 5 (Jacobs and Roberts, 2009) and elephant trackways preserved in aeolian sediment at Still Bay (Roberts et al., 2008). Use of amino acid racemisation of cement within aeolianites has shown that eroded material from older barrier dunes has been recycled and incorporated into new barrier dunes (Roberts et al., 2009; Dunajko and Bateman, 2010).

The South African coastline has extensive Neogene and Quaternary sandy deposits of beach and aeolian origin. These range from extensive active coastal dune fields (e.g. Alexandria dunefield: Illenberger and Verhagen, 1990; False Bay and Duinefontyn dune plumes: Roberts et al., 2009), vegetated barrier or cordon dunes (e.g. Wilderness dunes: Illenberger, 1996; Bateman et al., 2011; the unconsolidated Sixteen Mile Dunes: Compton and Franceschini, 2005) and the aeolianites of the Waenhuiskrans Formation (Malan, 1987), to discrete aeolian units found within coastal cave stratigraphies (e.g. Blombos cave: Henshilwood et al., 2002; Pinnacle Point: Marean et al., 2007). In KwaZulu-Natal, seven different types of coastal dunes have been identified, from beach ridge dunes, barrier dunes to parabolic and climbing dunes (Jackson et al., 2014). The Isipingo, Kosi Bay and Port Durnford Formations are largely a product of preserved calcified dunes dating from the mid- to late Quaternary (Porat and Botha, 2008). In terms of preserved beach deposits, relict raised estuarine sediments (e.g. Cape Agulhas: Carr et al., 2010) and raised cemented beachrock (e.g. East London: Jacobs and Roberts, 2009; Mossel Bay: Jacobs et al., 2011) have been reported on the south coast. Raised beaches along the KwaZulu-Natal coast (including the Durban Bluff beaches) have been described in detail by Cooper and Flores (1991). These include the Dawson's Rocks and Reunion Members as outlined by Porat and Botha (2008) who also noted a number of presently submerged beach deposits.

On the south coast, preservation of Quaternary deposits was aided by log-spiral bays (e.g. Still Bay, Vlees Bay, Mossel Bay) with headlands sheltering deposits from waves and storm surges. The relative scarcity of older Quaternary deposits (MIS 11, 9, 7) may be a function of dune burial by younger material, or a result of their prior removal by erosion (Roberts et al., 2014). On the east coast, where climatic and environmental conditions are more favourable for carbonate cementation (Cooper and Flores, 1991; Cawthra and Uken, 2012), these part-lithified sandy intertidal shorelines have been preserved. Past sandy shorelines are also recorded on the continental shelf where the cementation of beach and aeolian sediments allowed the preservation of submerged beaches and dunes. Off the south coast, the broad continental shelf provides extensive accommodation space for deposits to form during sea-level lowstands (Cawthra et al., 2014). In the Durban area where accommodation space is limited by a steeply inclining coastal shelf, these deposits tend to be stacked (Botha and Porat, 2007; Green et al., 2013).

13.4 Quaternary coastal and sea-level change

Raised beach deposits, aeolianites and erosive features have provided evidence for at least two sea-level highstands during the middle to late Quaternary. On the west coast these two highstands form part of a series of marine terraces with transgressive maxima incorporating the Langebaan and Springfontyn Formations, and span back to the late Pliocene (Roberts et al., 2011). The earlier highstand is thought to have occurred during MIS 11 around 400 kyr BP. Similar highstands have been reported from Mossel Bay where a wave-cut platform extends 1 km inland from the present coast (Roberts et al., 2012) and around the Alexander Bay area (Gresse, 1988). Along the east coast, mid-Quaternary highstands are represented by coastal lake muds (Port Durnford Formation) and aeolian sands (Kosi Bay Formation; Botha and Porat, 2007). The latter on the Tshongwe–Sihangwane ridge underlie late Quaternary dunes and are an example of the buried former coastal aeolian sand landscape (Cooper and Kensley, 1991).

Evidence for the 6–8 m highstand attributed to MIS 5e is more widespread, such as relict estuarine sediments on the south coast (Carr et al., 2010). Preserved raised beach deposits along the KwaZulu-Natal coast have been dated using U-series to MIS 5 (Ramsay and Cooper, 2002). At the Nahoon hominid footprint site, shelly beach deposits at 6 m elevation date to MIS 5 (Jacobs and Roberts, 2009). Within the MIS 5e highstand there is some stratigraphic evidence that more than one transgressive event took place (Carr et al., 2010), which fits with recent findings from a variety of far-field locations (Hearty et al., 2007).

Dune formation was not limited to highstands of MIS 11 and 5e. Bateman et al. (2011), working in the Wilderness dunes on the south coast, suggested barrier dune formation started in MIS 9 with directly dated dunes (using the luminescence method) spanning back to MIS 7. Both highstands and lowstands have significantly altered the configuration and sediment dynamics of all parts of the coastline in the past (Fig. 13.5). This is particularly evident on gently sloping sections of the continental shelf. Modelling has demonstrated how these changes would have transformed the coastal plain over the last 420 kyr, and influenced how and where early humans would have lived (Fisher et al., 2010). Given the tectonic stability of the coastline and its far-field position, relative to Quaternary ice sheets, the association of dunes to sea levels reflects a eustatic control on shoreline position. Evidence for drowned dunes relating to sea-level lowstands during MIS 4–2 have

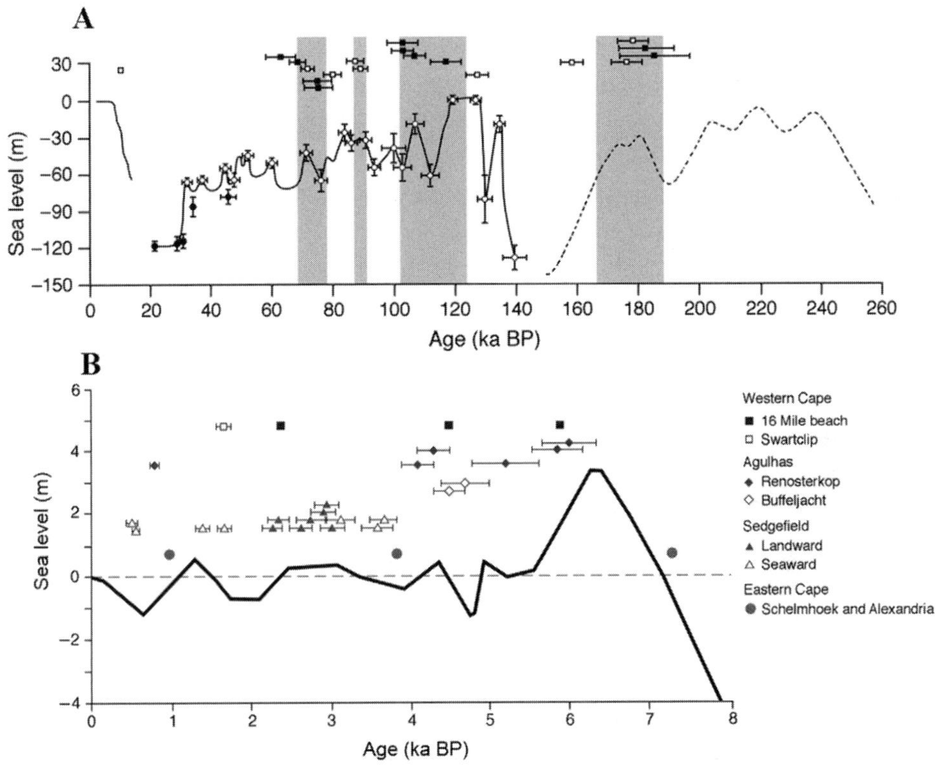

Fig. 13.5. Phases of dune building from western, southern and eastern South African coastlines. (A) Late Quaternary dunes (shaded bars) and their association with eustatic sea level (adapted from Bateman et al., 2004), (B) summarised regional patterns of Holocene coastal dunes (data from Illenberger and Verhagen, 1990; Compton and Franchescini, 2005; Carr et al., 2006) and based on Compton's (2001) South African Holocene sea-level curve.

been reported from the Wilderness area (Cawthra et al., 2014). Two distinct cemented sandy shorelines at -100 m and -60 m have been mapped on the narrow but steep KwaZulu-Natal shelf (Salzmann et al., 2013) where they have been associated with a preserved barrier dune complex which was overstepped during rapid sea-level transgressions around 11.5 kyr BP (Green et al., 2013).

Holocene coastal dunes appear to have developed after the mid-Holocene sea-level highstand around 6700 years BP (Compton, 2001) (Fig. 13.5). Most of the Holocene dune fields are not lithified and are a series of mobile parabolic dunes forming plumes moving inland (Roberts et al., 2009), or are stacked parabolic dunes remaining semi-permanently fixed behind beaches or atop older aeolianite dunes (e.g. Dana Bay and Wilderness).

13.5 Summary

The coastline of South Africa is dominated by sand, both in terms of the present-day beaches and coastal dune fields, but also through cementation with preserved beachrocks and aeolianites. The latter two contain a range of Quaternary sedimentary and geomorphic features that record changes in sea level and coastal configuration. These show that Quaternary sea levels around South Africa have varied from +13 m to -120 m relative to present in response to global climate change and regional eustatic adjustments. As a result of these changes, coastlines shifted and land was exposed or inundated. The extent of this movement was dependent on the gradient of onshore and nearshore coastal platforms.

References

Armitage, S. J., Botha, G. A., Duller, G. A. T., Wintle, A. G., Rebêlo, L. P. and Momade, F. J. (2006). The formation and evolution of the barrier islands of Inhaca and Bazaruto, Mozambique. *Geomorphology*, 82, 295–308.

Bateman, M. D., Carr, A. S., Dunajko, A. C., Holmes, P. J., Roberts, D. L., McLaren, S. J., Bryant, R. G., Marker, M. E. and Murray-Wallace, C. V. (2011). The evolution of coastal barrier systems: A case study of the Middle-Late Pleistocene Wilderness barriers, South Africa. *Quaternary Science Reviews*, 30, 63–81.

Bateman, M. D., Holmes, P. J., Carr, A. S., Horton, B. P. and Jaiswal, M. K. (2004). Aeolianite and barrier dune construction spanning the last two glacial-interglacial cycles from the southern Cape coast, South Africa. *Quaternary Science Reviews*, 23, 1681–1698.

Bathurst, R. G. C. (1975). *Carbonate Sediments and their Diagenesis*, 2nd Ed. Developments in Sedimentology, 12. New York: Elsevier, 660pp.

Botha, G. A. and Porat, N. (2007). Soil chronosequence development in dunes on the southeast African coastal plain, Maputaland, South Africa. *Quaternary International*, 162–163, 111–132.

Bosman, C. (2012). *The marine geology of the Aliwal Shoal, Scottburgh, South Africa.* Unpublished PhD thesis, University of KwaZulu-Natal, South Africa, 581pp.

Carr, A. S., Bateman, M. D., Roberts, D. L., Murray-Wallace, C. V., Jacobs, Z. and Holmes, P. J. (2010). The last interglacial sea-level high stand on the southern Cape coastline of South Africa. *Quaternary Research*, 73, 351–363.

Carr, A. S., Thomas, D. S. G. and Bateman, M. D. (2006). Climatic and sea level controls on Late Quaternary eolian activity on the Agulhas Plain, South Africa. *Quaternary Research*, 65, 252–263.

Cawthra, H. C., Bateman, M. D., Carr, A. S., Compton, J. S. and Holmes, P. J. (2014). Understanding Late Quaternary change at the land–ocean interface: A synthesis of the evolution of the Wilderness coastline, South Africa. *Quaternary Science Reviews*, 99, 210–223.

Cawthra, H. C. and Uken, R. (2012). Modern beachrock formation in Durban, KwaZulu-Natal. *South African Journal of Science*, 108, Art. #935, 5pp, doi:/10.4102/sajs.v108i7/8.935.

Cawthra, H. C., Uken, R. and Ovechkina, M. N. (2012). New insights into the geological evolution of the Durban Bluff and adjacent Blood Reef, South Africa. *South African Journal of Geology*, 115, 291–308.

Compton, J. S. (2001). Holocene sea-level fluctuations inferred from the evolution of depositional environments of the southern Langabaan Lagoon salt marsh, South Africa. *The Holocene*, 11, 395–405.

Compton, J. S. and Franceschini, G. (2005). Holocene geoarchaeology of the Sixteen Mile Beach barrier dunes in the Western Cape, South Africa. *Quaternary Research*, 63, 99–107.

Cooper, J. A. G. (2001). Geomorphological variability among microtidal estuaries from the wave-dominated South African coast. *Geomorphology*, 40, 99–122.

Cooper, J. A. G. and Flores, R. M. (1991). Shoreline deposits and diagenesis resulting from two Late Pleistocene highstands near +5 and +6 metres, Durban, South Africa. *Marine Geology*, 97, 325–343.

Cooper, M. R. and Kensley, B. F. (1991). An early Pleistocene decapod crustacean fauna from Zululand. *South African Journal of Science*, 87, 601–604.

Davies, J. L. (1980). *Geographical Variation in Coastal Development*. New York: Longman, 212pp.

de Decker, R. H. (1988). The wave regime on the inner shelf south of the Orange River and its implications for sediment transport. *South African Journal of Geology*, 91, 358–372.

Dingle, R. V., Birch, G. F., Bremner, J. M., de Decker, R. H., Du Plessis, A., Engelbrecht, J. C., Fincham, M. J., Fitton, B. W., Flemming, B. W., Gentle, R. I., Goodland, S. W., Martin, A. K., Mills, E. G., Moir, G. J., Parker, R. J., Robson, S. H., Rogers, J., Salmon, D. A., Siesser, W. G., Simpson, S. W., Summerhayes, C. P., Westall, F., Winter, A. and Woodborne, M. W. (1987). Deep-sea sedimentary environments around southern Africa (South-East Atlantic and South-West Indian Oceans). *Annals of the South African Museum*, 98, 1–27.

Dunajko, A. C. and Bateman, M. D. (2010). Sediment provenance of the Wilderness barrier dunes, southern Cape coast, South Africa. *Terra Nova*, 22, 417–423.

Elias, S. A. (2007). Sea level studies. In *Encyclopedia of Quaternary Science*, ed. S. A. Elias. Amsterdam: Elsevier, pp. 2967–3095.

Fisher, E. C., Bar-Matthews, M., Jerardino, A. and Marean, C. W. (2010). Middle and Late Pleistocene paleoscape modeling along the southern coast of South Africa. *Quaternary Science Reviews*, 29, 1382–1398.

Flemming, B. W. (1981). Factors controlling shelf sediment dispersal along the southeast African continental margin. *Marine Geology*, 42, 259–277.

Flügel, E. (2004). *Microfacies of Carbonate Rocks: Analysis, Interpretation and Application*. Berlin: Springer-Verlag, 984pp.

Green, A. N., Cooper, J. A. G., Leuci, R. and Thackeray, Z. (2013). Formation and preservation of an overstepped segmented lagoon complex on a high-energy continental shelf. *Sedimentology*, 60, 1755–1768.

Gresse, P. G. (1988). Washover boulder fans and reworked phosphorite in the Alexander Bay Formation. *South African Journal of Geology*, 91, 391–398.

Guilcher, A. (1988). *Coral Reef Geomorphology*. New York: John Wiley & Sons, 228pp.

Harris, L., Nel, R. and Schoeman, D. (2011). Mapping beach morphodynamics remotely: A novel application tested on South African sandy shores. *Estuarine Coastal and Shelf Science*, 92, 78–89.

Hearty, P. J., Hollin, J. T., Neumann, A. C., O'Leary, M. J. and McCulloch, M. (2007). Global sea-level fluctuations during the Last Interglaciation (MIS 5e). *Quaternary Science Reviews*, 26, 2090–2112.

Henshilwood, C. S., D'Errico, F., Yates, R., Jacobs, Z., Tribolo, C., Duller, G. A. T., Mercier, N., Sealy, J. C., Valladas, H., Watts, I. and Wintle, A. G. (2002). Emergence

of modern human behavior: Middle Stone Age engravings from South Africa. *Science*, 295, 1278–1280.
Heydorn, A. E. F. and Tinley, K. L. (1980). *Estuaries of the Cape Part 1: Synopsis of the Cape Coast, Natural Features, Dynamics and Utilisation*. Pretoria: CSIR Research Report, 380, 96pp.
Illenberger, W. K. (1996). The geomorphic evolution of the Wilderness dune cordons, South Africa. *Quaternary International*, 33, 11–20.
Illenberger, W. K. and Burkinshaw, J. R. (2008). Coastal dunes and dunefields. In *Geomorphology of the Eastern Cape*, ed. C. A. Lewis. Grahamstown: NICS, pp. 85–106.
Illenberger, W. K. and Verhagen, B. T. (1990). Environmental history and dating of coastal dune fields. *South African Journal of Science*, 86, 311–314.
Jackson, D. W. T., Cooper, J. A. G. and Green A. N. (2014). A preliminary classification of coastal sand dunes of KwaZulu-Natal. *Journal of Coastal Research,* Special Issue, 70, 718–722.
Jacobs, Z. and Roberts, D. L. (2009). Last Interglacial age for the Nahoon Fossil Human Footprints, Southeast Coast of South Africa. *Quaternary Geochronology*, 4, 160–169.
Jacobs, Z., Roberts, R. G., Lachlan, T. J., Karkanas, P., Marean, C. W. and Roberts, D. L. (2011). Development of the SAR TT-OSL procedure for dating Middle Pleistocene dune and shallow marine deposits along the southern Cape coast of South Africa. *Quaternary Geochronology*, 6, 491–513.
Kelletat, D. (2006). Beachrock as sea-level indicator? Remarks from a geomorphological point of view. *Journal of Coastal Research*, 22, 1558–1564.
Kruger, A. C. (2004). *Climate of South Africa: Climate Regions. WS 45*. Pretoria: South African Weather Service, 19pp.
Laborel, J., Morhange, C., Lafont, R., Le Campion, J., Laborel-Deguen, F. and Sartoretto, S. (1994). Biological evidence of sea-level rise during the last 4500 years on the rocky coasts of continental southwestern France and Corsica. *Marine Geology*, 120, 203–223.
Lambeck, K., Woodroffe, C. D., Antonioli, F., Anzidei, M., Gehrels, W. R., Laborel, J. and Wright, A. J. (2010). Palaeoenvironmental records, geophysical modelling and reconstruction of sea-level trends and variability on centennial and longer time scales. In *Understanding Sea-Level Rise and Variability*, ed. J. A. Church, P. L. Woodworth, T. Aarup and W. S. Wilson. Oxford: Wiley-Blackwell, pp. 61–121.
Malan, J. A. (1987). The Bredasdorp Group in the area between Gans Bay and Mossel Bay. *South African Journal of Science*, 83, 506–507.
Marean, C. W., Bar-Matthews, M., Bernatchez, J., Fisher, E., Goldberg, P., Herries, A. I. R., Jacobs, Z., Jerardino, A., Karkanas, P., Minichillo, T., Nilssen, P. J., Thompson, E., Watts, I. and Williams, H. M. (2007). Early human use of marine resources and pigment in South Africa during the Middle Pleistocene. *Nature*, 449, 905–908.
Marker, M. E. and Holmes P. J. (2010). The geomorphology of the Coastal Platform in the southern Cape. *South African Geographical Journal*, 92, 105–116.
Martin, A. K. and Flemming, B. W. (1987). Aeolianites of the South African coastal zone and continental shelf as sea-level indicators. *South African Journal of Science*, 83, 507–508.
McLachlan, A., Siebe, P. R. and Ascaray, C. (1982). *Survey of a Major Coastal Dunefield in the Eastern Cape*. University of Port Elizabeth Zoology Report Series, no. 10. Port Elizabeth: University of Port Elizabeth, 48pp.
Meadows, M. E., Rogers, J., Lee-Thorp, J. A., Bateman, M. D. and Dingle, R. V. (2002). Holocene geochronology of a continental-shelf mudbelt off southwestern Africa. *The Holocene*, 12, 59–67.

Miller, W. R. and Mason, T. R. (1994). Erosional features of coastal beachrock and aeolianite outcrops in Natal and Zululand, South Africa. *Journal of Coastal Research*, 10, 374–394.

Milliman, J. D. (1974). *Marine Carbonates: Recent Sedimentary Carbonates Part 1*. New York, Heidelberg and Berlin: Springer-Verlag, 375pp.

Musekiwa, C., Cawthra, H. C., Unterner, M. and van Zyl, F. W. (2015). An assessment of coastal vulnerability for the South African coast. *South African Journal of Geomatics*, 4, 123–137.

Neumeier, U. (1999). Experimental modelling of beachrock cementation under microbial influence. *Sedimentary Geology*, 126, 35–46.

Partridge, T. C. and Maud, R. R. (2000). *The Cenozoic of Southern Africa*. Oxford: Oxford University Press, 406pp.

Pether, J., Roberts, D. L. and Ward, J. (2000). Deposits of the West Coast. In *The Cenozoic of Southern Africa*, ed. T. C. Partridge and R. R. Maud. Oxford: Oxford University Press, pp. 33–54.

Porat, N. and Botha, G. (2008). The luminescence chronology of dune development on the Maputaland coastal plain, southeast Africa. *Quaternary Science Reviews*, 27, 1024–1046.

Ramsay, P. J. (1996). *Quaternary Marine Geology of the Sodwana Bay Continental Shelf, Northern KwaZulu-Natal*. Pretoria: Council for Geoscience Bulletin, 117, 86pp.

Ramsay, P. J. and Cooper, J. A. G. (2002). Late Quaternary sea-level change in South Africa. *Quaternary Research*, 57, 82–90.

Roberts, D. L., Bateman, M. D., Murray-Wallace, C. V., Carr, A. S. and Holmes, P. J. (2008). Fossil elephant trackways, sedimentation and diagenesis in OSL/AAR-dated Late Quaternary coastal aeolianites: Still Bay, South Africa. *Palaeogeography, Palaeoclimatology, Palaeoecology*, 257, 261–279.

Roberts, D. L., Bateman, M. D., Murray-Wallace, C. V., Carr, A. S. and Holmes, P. J. (2009). West coast dune plumes: Climate driven contrasts in dunefield morphogenesis along the western and southern South African coasts. *Palaeogeography, Palaeoclimatology, Palaeoecology*, 271, 24–38.

Roberts, D. L., Botha, G. A., Maud, R. R. and Pether, J. (2006). Coastal Cenozoic deposits. In *The Geology of South Africa*, ed. M. R. Johnson, C. R. Anhausser and R. J. Thomas. Pretoria: Geological Society of South Africa/Council for Geoscience, pp. 605–628.

Roberts, D. L., Cawthra, H. C. and Musekiwa, C. (2014). Dynamics of late Cenozoic aeolian deposition along the South African coast: A record of evolving climate and ecosystems. In *Sedimentary Coastal Zones from High to Low Latitudes: Similarities and Differences*, ed. I. P. Martini and H. R. Wanless. London: Geological Society of London Special Publication 388, pp. 353–387.

Roberts, D. L., Karkanas, P., Jacobs, Z., Marean, C. W. and Roberts, R. G. (2012). Melting ice sheets 400,000 yr ago raised sea level by 13 m: Past analogue for future trends. *Earth and Planetary Science Letters*, 357–358, 226–237.

Roberts, D. L., Matthews, T., Herries, A. I. R., Boulter, C., Scott, L., Dondo, C., Mthembi, P., Browning, C., Smith, R. M. H., Haarhoff, P. and Bateman, M. D. (2011). Regional and global context of the Late Cenozoic Langebaanweg (LBW) palaeontological site: West Coast of South Africa. *Earth-Science Reviews*, 106, 191–214.

Salzmann, L., Green, A. and Cooper, J. A. G. (2013). Submerged barrier shoreline sequences on a high energy, steep and narrow shelf. *Marine Geology*, 346, 366–374.

Schoonees, J. S. (2000). Annual variation in the net longshore sediment transport rate. *Coastal Engineering*, 40, 141–160.

Smith, A. M., Mather, A. A., Bundy, S. C., Cooper, J. A. G., Guastella, L. A., Ramsay, P. J. and Theron, A. (2010). Contrasting styles of swell-driven coastal erosion: Examples from KwaZulu-Natal, South Africa. *Geological Magazine*, 147, 940–953.

Tinley, K. L. (1985). *Coastal dunes of South Africa;* South African National Scientific Programmes report 109. Pretoria: CSIR, 300pp.

Vousdoukas, M. I., Velegrakis, A. F. and Karambas, T. V. (2009). Morphology and sedimentology of a microtidal beach with beachrocks: Vatera, Lesbos, NE Mediterranean. *Continental Shelf Research*, 29, 1937–1947.

Vousdoukas, M. I., Velegrakis, A. F. and Plomaritis, T. A. (2007). Beachrock occurrence, characteristics, formation mechanisms and impacts. *Earth-Science Reviews*, 85, 23–46.

Vött, A., Bareth, G., Brückner, H., Curdt, C., Fountoulis, I., Grapmayer, R., Hadlerm H., Hoffmeister, D., Klasen, N., Lang, F., Masberg, P., May, S. M., Ntageretzis, K., Sakellariou, D. and Willershäuser, T. (2010). Beachrock-type calcarenitic tsunamites along the shores of the eastern Ionian Sea (western Greece) case studies from Akarnania, the Ionian Islands and the western Peloponnese. *Zeitschrift für Geomorphologie, Supplementband*, 54, 1–50.

Whitfield, A. K. (1983). Effect of prolonged aquatic macrophyte senescence on the biology of the dominant fish species at Swartvlei. *South African Journal of Science*, 79, 153–157.

14

Environmental change during the Pleistocene and Holocene: Estuaries and lagoons of southern Africa

ANDER M. DE LECEA, ANDREW N. GREEN AND J. ANDREW G. COOPER

Abstract

The South African coast contains abundant estuaries and lagoons, most of which originated as river valleys incised during Quaternary sea-level fall and subsequently drowned and/or infilled during rising interglacial sea levels. This chapter discusses these changes and highlights the geomorphological and sedimentological evolution of several southern African estuaries during the Pleistocene to present. The development, infilling and positioning of incised valley systems is mainly controlled by sea-level variation as well as fluvial and marine sediment supply. Most contemporary estuaries in southern Africa show dramatic responses to shorter term sea-level fluctuations and sediment supply during the Holocene. Barring anthropogenic interference, the dynamic behaviour of estuaries, dictated by their transient position and geomorphic character within an incised valley, is considered to fluctuate on centennial to millennial timescales.

14.1 Introduction

The southern African coast contains abundant estuaries and lagoons formed at the junction of fluvial and marine environments (Fig. 14.1). Estuaries, by definition, receive sediment from both fluvial and marine sources and contain facies influenced by waves, tides and fluvial processes (Dalrymple et al., 1992; Dalrymple, 2006). From a geological, stratigraphic perspective, estuaries are considered as transgressive sediment packages that have infilled available accommodation space provided by underlying valleys incised into older stratigraphy formed during base level fall (Zaitlin et al., 1994). Alternatively, they may also form in topographic

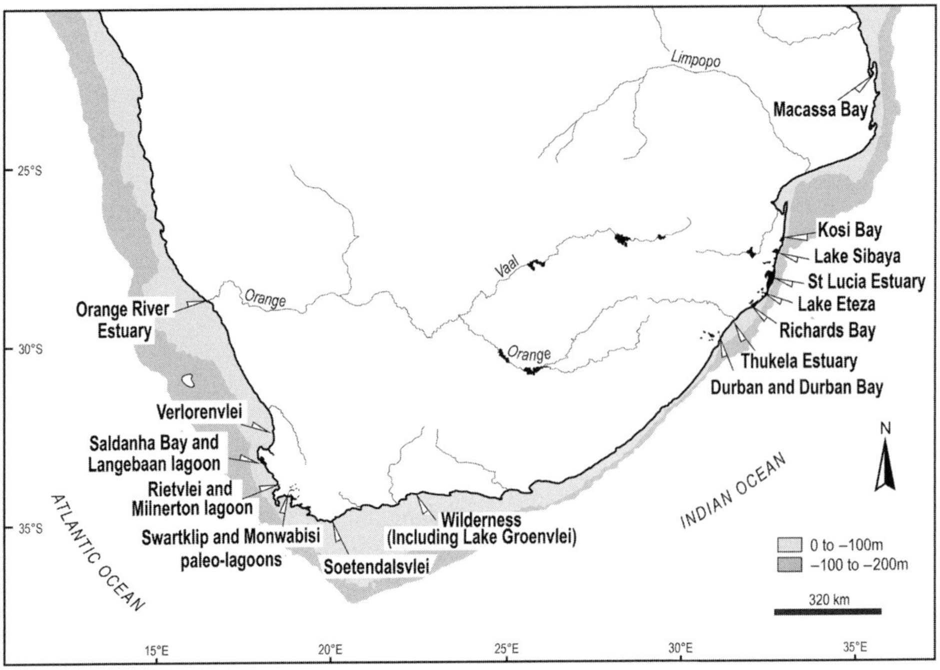

Fig. 14.1. The distribution of the South African estuaries discussed in this chapter. Inland lakes, wetlands and lagoons are also shown.

lows produced by compactional subsidence of the coastal plain in abandoned deltas (Dalrymple, 2006). These systems reflect cycles of sea-level fall and subsequent rise, and are thus closely related to changes in climate and glacioeustasy. Estuaries are important sedimentary archives of sea-level variability, particularly as they are one of only a few coastal elements that tend to escape erosion by transgressive ravinement (Nordfjord et al., 2005). Contemporary estuaries in southern Africa thus reflect the most recent episode of infilling of accommodation space during rising Holocene sea levels. During previous lowstands, estuaries were generally located offshore of their present positions and close to the modern shelf break (e.g. Green et al., 2013a). During previous interglacial highstands, many estuaries were in a similar position as today but occupied larger areas (e.g. Cooper et al., 1989). The geological record of these changes along the southern African coast is the focus of this chapter.

The ~3000 km long South African coastline contains approximately 289 estuaries, of which only 37 (12.8%) are permanently open to the sea (Reddering and Rust, 1990). The majority of estuaries share common characteristics in that (1) they have small tidal prisms (spring tidal ranges of 1.8–2.0 m); (2) they are mostly hosted by bedrock-incised river valleys (although many extend alongshore at the

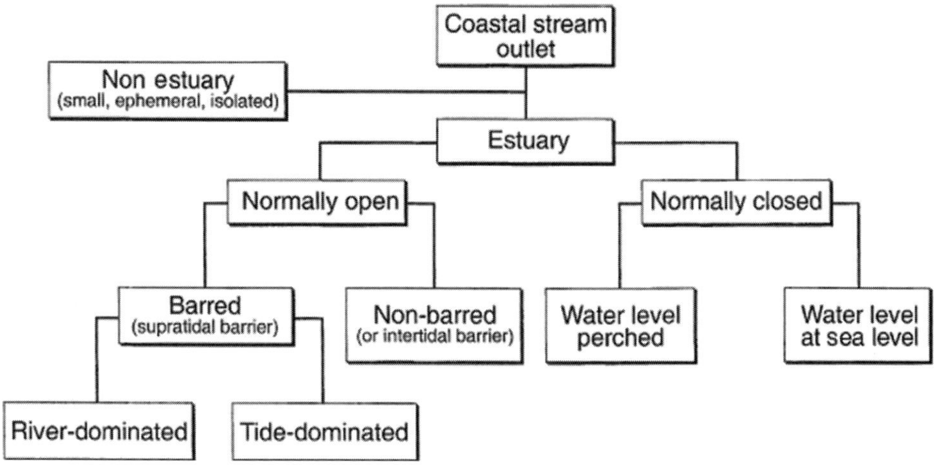

Fig. 14.2. Geomorphological classification of South African estuaries (after Cooper, 2001).

contemporary shoreline); (3) most lack a surface connection to the ocean for periods of time due to a barred entrance; and consequently (4) most lack ebb-tidal deltas, which is also due in part to the South African coast's high wave energy (Cooper, 2001).

Contemporary South African estuaries can be classified as permanently or temporarily open estuaries (Fig. 14.2). Permanently open estuaries may be either barred or non-barred. Barred open estuaries are further subdivided into river-dominated (e.g. Mvoti estuary) and tide-dominated systems (e.g. Mtentu estuary). Temporarily open estuaries are subdivided into perched (e.g. Mdloti estuary) and non-perched types (e.g. Seekoei estuary) (Cooper, 2001) according to the elevation of the enclosed water body in relation to the tide. In addition, there are several large lagoons formed in low-lying topography on coastal plains (e.g. Wilderness lakes and Northern KwaZulu-Natal lagoons) or low-lying rock coasts (e.g. Langebaan lagoon).

Estuary morphology is influenced by a combination of human and natural influences. Human-induced changes occur over short timescales (tens of years to centuries) but can have profound impacts on estuary form and function. Changes in land cover and land use can affect rates of soil loss and increased catchment sediment yield. Other human-induced changes can include:

- Artificial breaching of estuary barriers (e.g. the Bot River; Branch et al., 1985). Artificial splitting of the Mfolozi River from Lake St Lucia (Perissinotto et al., 2013) produced a sustained closure of the ocean-estuary connection without further artificial breaching;

- Alteration of sediment supply by river impoundments (Garland and Moleko, 2000) or changing catchment land use practice;
- Alteration of estuary dynamics by construction within the estuary (e.g. bridges, embankments, reclamation; Begg, 1978) which produces local flow constrictions and velocity accelerations.

14.2 Origins of South African estuaries

Most of southern Africa's estuaries and coastal lagoons originated as river valleys which were subsequently drowned and/or infilled during sea-level transgression (Cooper et al., 1999). As such, estuarine sediments are commonly a mixture of both terrestrial and marine materials, determined by the rate of sea-level rise, the shape of the estuarine valley and sediment availability (Cooper, 2001). Here, we consider changes in estuary morphology and the record of Quaternary environmental change within estuarine sediments.

The majority of estuaries in southern Africa developed from bedrock valleys incised during Quaternary sea-level lowstands, many of which originated in pre-Quaternary times. Dingle et al. (1983) and Green and Garlick (2011) presented evidence of Cretaceous and Pliocene incised valleys off the west and east coasts, respectively, that were repeatedly filled and re-incised throughout the Quaternary. Many of these systems were underfilled with sediments, remaining as topographic lows that were subsequently exploited by river rejuvenation. A consequence of this is the confinement of incised valleys; sediments of previous sea-level cycles are rarely preserved due to exhumation following repeated Pleistocene sea-level lowering past the shelf break (cf. Compton, 2011). Exceptions occur in wider floodplain systems such as Durban Bay (Mkhize, 2014), St Lucia (Van Heerden, 1987) and Maputo Bay (Green et al., 2015).

14.3 Quaternary estuarine evolution: Sea-level change and sedimentation

Over longer periods of time (thousands of years), estuaries can be profoundly affected by sea-level change (Fig. 14.3). The repeated excavation of incised valleys means that few of the preserved estuarine sequences pre-date the last major interglacial period (~80–100 kyr BP); those that do can provide windows into past sea-level states and climate. The best preserved records typically span the most recent sea-level cycle of the late Pleistocene to present (e.g. Nordfjord et al., 2005; Chaumillon et al., 2010).

In southern Africa, as sea level began to fall after the last interglacial highstand (Marine Isotope Stage (MIS) 5a), it caused the excavation of accumulated sediments within bedrock valleys as a consequence of base level lowering and river rejuvenation (Wright et al., 2000). Sea levels remained low for an extended period, reaching a minimum of -120 m at ~22 kyr BP (Compton, 2011). At this time, excavated sediment was transported across the shelf edge to form lowstand deltas, such as off the KwaZulu-Natal coast (Green, 2011; Green et al., 2013b). These comprise aggrading/prograding clinoforms located at the continental shelf break. The subsequent late Pleistocene/Holocene sea-level rise flooded the incised valleys and caused the shoreline to migrate landwards, filling and/or drowning the estuarine valleys. Evidence of this can be found offshore between Lake Sibaya (Miller, 1998) and St Lucia where the offshore portions of adjacent river systems have been completely infilled by estuarine and marine sediments (Green, 2009). After 7000 yr BP, regional sea levels were higher than present, reaching a maximum of +3.5 m at 4480 yr BP (Ramsay, 1995) (Fig. 14.3). Sea level then lowered to -2 m at ~3000 yr BP before rising to +1.5 m at 1610 yr BP, reaching its current position ~900 yr BP (Ramsay, 1995). Estuarine sedimentary records during each of these periods are now considered briefly.

14.4 Estuaries of previous interglacial highstands

Estuarine sediments from previous interglacial highstands occur at various locations around the southern African coast, usually fringing contemporary estuaries. During previous interglacials, sea levels were close to or exceeded present values (Ramsay and Cooper, 2002) (Fig. 14.3), and a number of elevated estuarine deposits are found along the coast.

At Swartklip in False Bay, Barwis and Tankard (1983) reported a 50 m thick succession of poorly cemented sandstone containing marine molluscs dated at 135 kyr BP. This was subsequently overlain by an estuarine or microtidal barrier lagoon facies that formed around 120 kyr BP (MIS 5e) at +5.1 m above present sea level. Sedimentary evidence of salinity gradient changes point to estuarine conditions. The abundant presence of the high salinity-tolerant gastropod *Tomichia ventricosa*, as well as gastropods less tolerant to salinity, indicate seasonal rainfall patterns similar to present.

Near Swartklip, the Monwabisi palaeo-lagoon preserves similar raised palaeo-estuarine and lagoonal sediments (Franceschini et al., 2005). Based on foraminifera assemblages, the depositional environments of the area include an intertidal flat from the penultimate interglacial (MIS 7–8), establishing a raised estuarine shoreline at that time. Most of the deposit relates to a lagoon, established near present sea level, which developed during MIS 5. The sediments contain lagoonal foraminifera and abundant *Assiminea globulus*, a diagnostic gastropod of modern estuaries (Pillay et al., 2009). These sediments are overlain by a thick calcrete

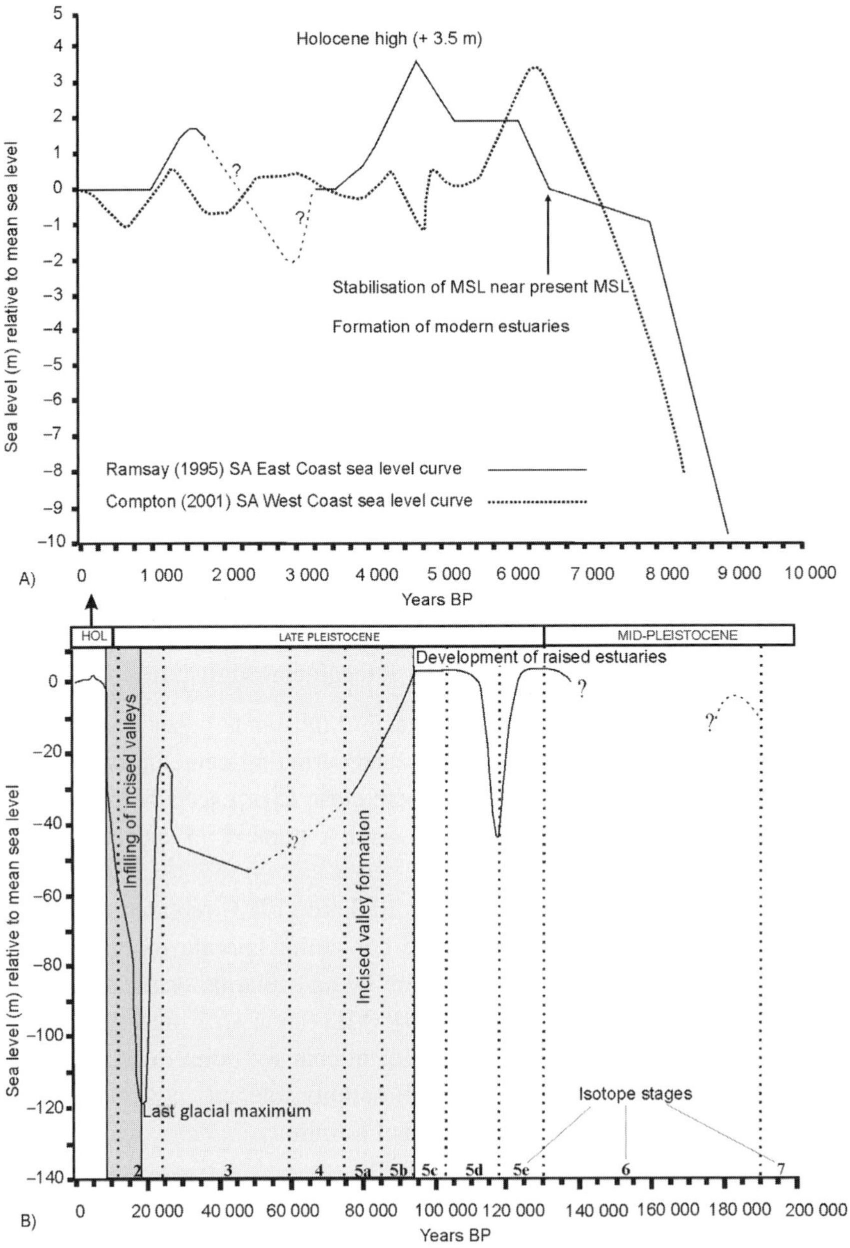

Fig. 14.3. (A) Holocene sea-level changes from the west (Compton, 2001) and east coast (Ramsay, 1995) of South Africa. Sea levels stabilised at their current position approximately 6–7 kyr BP, the time at which most modern estuaries in South Africa began to form. (B) Late Quaternary sea-level variability for the southern African coastline (after Ramsay and Cooper, 2002). The last interglacial period (MIS 5) is associated with several raised estuary deposits. However, most estuaries began to form in the period from MIS 5b to the last glacial maximum (LGM; MIS 2). Holocene infilling of these estuaries occurred later during the subsequent transgression, punctuated by sudden rises in sea level not depicted in this sea-level curve.

horizon of >100 kyr BP (Gile and Hawley, 1968), indicating lowered sea level and subsequent subaerial exposure, stranding the palaeo-lagoon since that time.

The Wilderness coastal lakes formed during the peak of the last interglacial highstand (MIS 5e). These estuaries were established by rising sea levels that drowned topographic depressions within a coastal dune landscape (Illenberger, 1996; Bateman et al., 2011). Beach/shallow marine and estuarine deposits at the base of the most seaward cordon suggest that the lakes were enclosed by a barrier that formed during MIS 5e when sea levels were at +5 m (Illenberger, 1996).

The Port Durnford Formation of northern KwaZulu-Natal (Hobday and Jackson, 1979) is considered to be a last interglacial analogue of a transgressive barrier/lagoonal complex. It includes overwash deposits overlain by lagoonal sediments and barrier sands attributed to sea-level transgression up to +8 m (Hobday and Jackson, 1979). Sediments at this elevation mark the development of coast-parallel palaeo-lagoons that were separated from the ocean by sandy barriers but connected by several estuarine inlets, similar to the present coastal configuration of northern KwaZulu-Natal (Wright et al., 2000). Raised estuarine deposits can also be found at Kosi Bay, where lagoonal molluscan assemblages occur on a raised shore terrace at +6 m (Cooper et al., 1989). These fauna indicate a palaeo-estuary system with large surface water area and deep lagoonal basins, with two connections to the ocean (Wright et al., 2000; Wright, 2002). This preservation of estuarine sediments from several sea-level highstands is due to their location along the higher margins of the incised valley, which escaped erosion during intervening lowstands.

14.5 Lowstand deltas, estuaries and incised valleys

With the exception of the Agulhas Bank and Orange River shelf, the South African shoreline was located close to the modern day continental shelf break during the last glacial maximum (LGM) around 22–18 kyr BP. In many instances these LGM-age fluvial valleys can be traced on the shelf where they form complex, preserved palaeo-drainage patterns (Green, 2009; Green et al., 2013a, b) (Fig. 14.4). In KwaZulu-Natal, the fluvial styles of many shelf rivers were different to those of today, as they were not restricted by a steep hinterland characterised by narrow valleys (Dladla, 2014). The wider coastal plain afforded by the exposed shelf promoted river meandering and the development of more complex drainage patterns.

Valleys offshore the modern Orange River and on the KwaZulu-Natal shelf, based on seismic stratigraphic and coring evidence, show cycles of river incision, estuarine development and burial since at least the late Cretaceous (Stevenson and McMillan, 2004; Green et al., 2013a). Valley incision during the Pliocene fed several large (100 m thick) lowstand deltas on the shelf edge (Green, 2011; Green

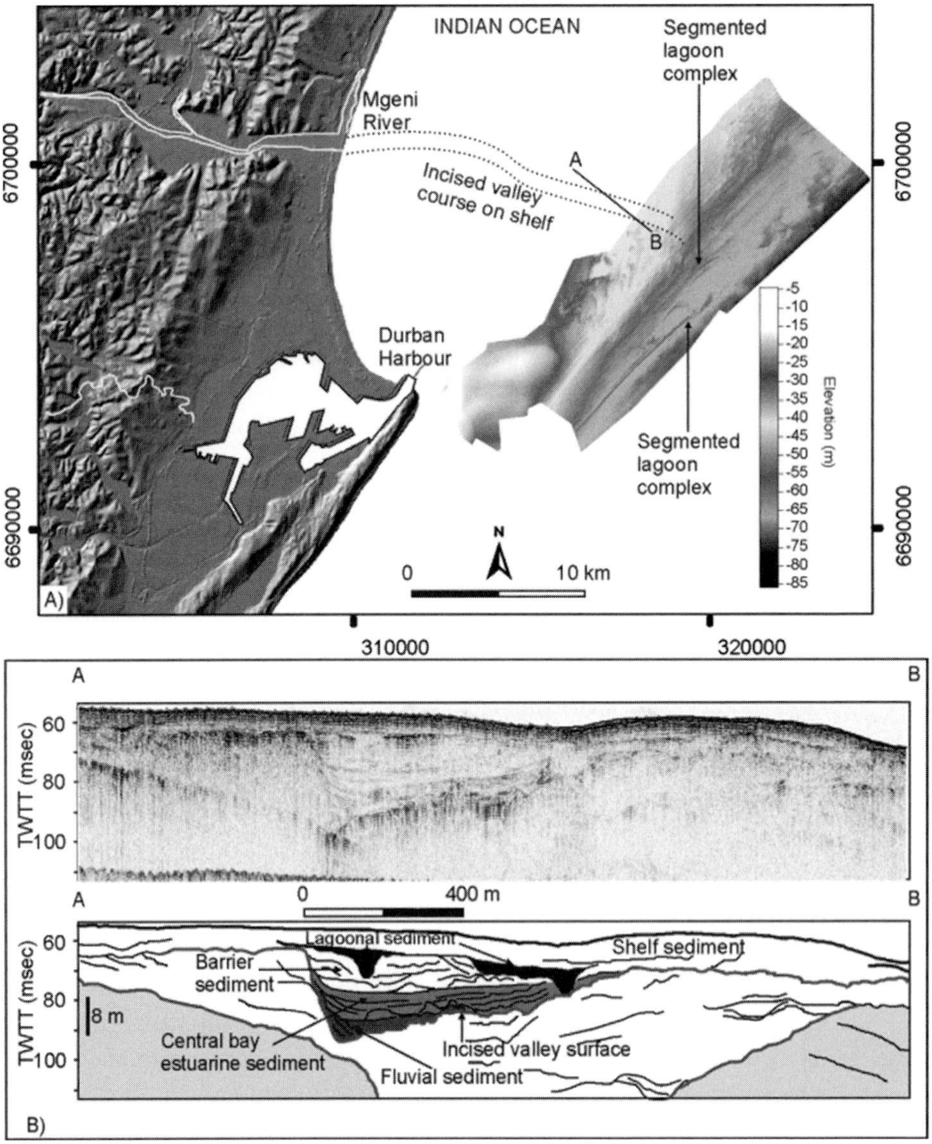

Fig. 14.4. (A): Sun-shaded digital terrain model detailing a segmented lagoon complex on the seafloor, offshore the Durban area. Note the course of the palaeo-Mgeni River's incised valley (data supplied courtesy of the eThekwini Municipality and Environmental Mapping and Surveying Pty Ltd). (B): Seismic profile and interpretative line drawing of the cross section A–B through the palaeo-Mgeni River's incised valley. This valley was formed during the LGM and was infilled during subsequent transgression by a succession of fluvial, estuarine, barrier and lagoonal sediments. This typical succession is capped by modern shoreface sands in this example.

et al., 2013a). Since then, regular cycles of Pleistocene sea-level fall (cf. Compton, 2011) have excavated most of these older drainage networks. In Durban Bay and the adjacent shelf, these are sporadically preserved as parts of several compound valley systems <100 m wide and 5 m deep (Green and Garlick, 2011; Mkhize, 2014).

14.6 Holocene estuarine deposits on the continental shelf

Contemporary South African estuaries originated in incisions associated with the LGM (Cooper, 2001). These LGM incised valleys are the best developed and can be observed throughout southern Africa (Stevenson and McMillan, 2004; Green, 2009; Cawthra et al., 2012; Green et al., 2013a, b). The valleys are <80 m deep and 2 km in width and extend across the shelf (Green, 2009; Green et al., 2013b).

Following the LGM, relative sea level began to rise rapidly but this took place by alternating periods of slow and rapid rise, the latter being ascribed to periods of proglacial lake drainage (Camoin et al., 2004; Liu and Milliman, 2004). These meltwater pulses are recorded in the development and preservation of several estuarine and lagoonal complexes on the South African shelf (Green et al., 2014). Meltwater pulse 1B (at 11.5–11.2 kyr BP) was preceded by a long period of slowly rising sea level during the Younger Dryas stadial (12.7–11.6 kyr BP) (Camoin et al., 2004). Meltwater pulse 1B caused rapid sea-level rise from -58 to -45 m (Liu and Milliman, 2004) and was responsible for the overstepping and preservation of several KwaZulu-Natal barrier shorelines, some associated with back-barrier estuaries (Salzmann et al., 2013). Offshore of Durban, a major shoreline complex preserved as beachrock and aeolianite records the development of a segmented lagoon during the Younger Dryas sea-level stillstand (Green et al., 2013a) (Fig. 14.4a).

Incised valleys on the continental shelf (Green, 2009; Green et al., 2013b) contain sediment fills that reflect the landward translation of estuaries during rising sea level. The infills show a basal fluvial unit overlain by sediments typical of river-dominated estuaries, including a muddy central estuarine basin overlain by sandy barrier and flood tide deltaic deposits (Fig. 14.4b). The similarity between these fills and those of contemporary estuaries indicates little change in the sedimentary conditions of South African estuaries from the LGM to present.

14.7 Late Holocene estuary evolution

With the stabilisation of sea level close to present around 6000–7000 yr BP, estuaries began to form in their modern locations. Subtle variations in sea level and sediment supply to these systems during the Holocene have, however, caused significant geomorphological change and promoted the variability observed in many modern estuary forms.

14.7.1 East coast

Diatom evidence from Macassa Bay, Mozambique, shows that this estuary alternated between marine and freshwater dominance during the last 6600 years (Norström et al., 2012). Ongoing work in Maputo Bay using stable isotopes ($\delta^{13}C$) and C:N ratios on sediment cores indicates a similar system change from fresh to seawater on at least two occasions (de Lecea et al., unpublished). In Macassa Bay, marine and brackish diatom species indicate that sea level was ~3 m above present about 6600 yr BP and in the period 4700–1000 yr BP. At 6300 yr BP, diatom records indicate a change to a more freshwater lagoon system that was likely caused by lowered sea level (Norström et al., 2012).

The Kosi Bay system comprises four interconnected lakes (Mkhawuleni, Mpungwini, Nhlange and Amanzimnyama) that occur behind a 130 m-high vegetated dune, and is almost permanently connected to the Indian Ocean by a narrow estuary mouth. Low fluvial and marine sedimentation rates mean that the bathymetry largely preserves the basal incised river valley (Cooper et al., 2012). The Holocene evolution of this system involved margin reworking and segmentation of a once-larger lagoon during the mid-Holocene highstand (+3.5 m at 4500 BP). A former inlet at Bhanga Nek was closed during subsequent sea-level regression and marked a shift towards contemporary estuarine lake conditions (Cooper, 1994; Ramsay, 1995).

At Lake Sibaya the basal incised valley is also well preserved beneath the modern system (Miller, 1998). Here, lagoonal conditions terminated by inlet closure and the formation of a 2 km wide dune system that isolated the lake from the sea (at ~5000 yr BP) and impounded it, such that its surface water level is now ~20 m above sea level (Wright, 2002). Most freshwater comes from the surrounding high groundwater table, aided by highly permeable sands. River input plays a small role in its hydrology (Wright et al., 1999).

In the St Lucia estuary-lake, the basal incised valley is covered with <40 m of lagoonal sediment from inflowing rivers. A former inlet close to Leven Point was sealed by aeolian deposits. The water levels inside the new lake rose and merged with the adjacent Mfolozi system, resulting in both estuaries flowing to the sea through a combined estuary where today's river mouth is located (Wright et al., 2000). Sediments in Lake Teza, a remnant of a former lagoon of the Mfolozi River, record the late Holocene highstand when an estuary occupied the area (Scott and Steenkamp, 1996).

14.7.2 West coast

Good examples of coastal responses to late Holocene sea-level changes are recorded in the estuaries, lakes and wetlands of the southern and western Cape. Lowering of sea level by about 3 m following the late Holocene highstand caused

Lake Groenvlei and other Wilderness lakes to lose (or modify) their connection to the ocean. The evolution of two wetlands, Rietvlei and Milnerton Lagoon in Cape Town, shows a complex response to sea-level fall (Schalke, 1973; Rogers, 1980; Grindley et al., 1989). The mid-Holocene Rietvlei is believed to have been an open estuarine embayment, sheltered by an outcrop of calcrete-capped Pleistocene deposits. As sea level fell, the developing modern barrier complex of Table Bay allowed for the formation of Milnerton Lagoon. This in turn linked Rietvlei to its present mouth farther south (Grindley et al., 1989).

Langebaan Lagoon (80 km NNE of Cape Town) comprises a ~15 km long and ~4 km wide lagoon that trends parallel to the Atlantic coast. The lagoon is separated from the sea by Pleistocene and Holocene (including present-day) coastal dunes, older granite headlands and aeolianites, with a single connection to the sea via a narrow tidal inlet at Saldanha Bay (Compton, 2001; Compton and Franceschini, 2005). Fossil records indicate that Langebaan Lagoon first flooded at 10–9 kyr BP, marking its transition from a river to an estuary (Compton, 2001). Fossil remains of *Ostrea atherstone* at -2 m in sediment cores indicate that at 6700 yr BP, sea level had reached 0 to 3 m above present, producing lagoonal conditions (Compton, 2001). The current lagoon morphology was established at ~700 yr BP with the onset of contemporary sea level and some segmentation of the coastal waterbody.

Verlorenvlei (north of Langebaan Lagoon) was a semi-sheltered open estuary by 8500 yr BP (Baxter and Meadows, 1999). A sea-level rise to +2 m saw the estuary becoming fully open prior to returning to a more dynamic estuarine-riverine environment by 7200 yr BP (Meadows et al., 1996; Baxter and Meadows, 1999). This was followed by a 3 m sea-level fall when subaerial exposure of much of the system led to the formation of coastal dunes (Baxter and Meadows, 1999). From 5000 yr BP the dominant sedimentation was of terrestrial origin until 4300 yr BP, when marine incursion occurred and estuarine conditions were re-established (Meadows et al., 1996). As present sea level stabilised ~750 yr BP, Verlorenvlei became a marine embayment with a well-developed saltmarsh (Baxter and Meadows, 1999).

14.8 Sediment supply, sea-level rise and valley geometry

The South African coast is an ideal natural laboratory to examine the influence of sediment supply, sea-level rise and valley geometry on estuary evolution. Although sea-level position influences the form and function of estuaries, sediment supply can be equally important in infilling patterns and final estuarine morphology. Broadly, estuaries can be classified into 'catch-up', 'keep-up' and 'give-up' types according to how well sedimentation keeps pace with sea level (Cooper et al., 2012). Examples of all three modes are preserved around the South African coast. The incised valley morphology in the Kosi and Sibaya lake systems reflects low sedimentation rates and transgressive drowning of the valley, resulting in a give-up estuary morphology marked by preservation of the basal incised valley reflected in the modern bathymetry. In contrast, high rates of Holocene sedimentation meant that many estuaries kept pace with sea level for much of their evolution. These systems, usually described as river-dominated (Cooper, 1993), include the Mgeni and Orange rivers. Catch-up estuaries are those envisaged by most geomorphic models (e.g. Zaitlin et al., 1994) and involve a period of initial flooding when estuaries are deep, but sedimentation 'catches up' during slowing sea-level rise or stillstand, and the estuary fill aggrades to sea level. The infill of such estuaries involves initial fluvial sedimentation, followed by deep water lagoonal and finally fluvial deposition. This is to some extent reflected in incised valleys buried on the shelf.

This classification is suited to a Holocene timescale of evolution, but higher frequency sea-level oscillations during the late Holocene have also been important. Minor reductions in sea level have caused isolation of lagoons and estuaries from marine influence, whereas minor increases tend to maintain or restore marine influences (such as in the Western Cape and Kosi Bay). In contrast, major fluctuations in sea level have left many estuarine systems stranded on the shelf or subaerially exposed, often without a full record of the estuarine infill. This demonstrates the importance of first-order controls such as sea level and sediment supply on estuarine development. The roles played by tides and waves in valley infilling cannot be ignored (e.g. Hayes, 1979; Boyd et al., 1992; Chaumillon et al., 2010), but this is secondary along most of the microtidal South African coastline.

14.9 Summary

Estuaries mark a continuum between fluvial and marine inputs, and throughout the Quaternary, South African estuaries have undergone cycles of incision and seaward migration of the estuary form, together with phases of infilling and transgression up the continental shelf. The nature of incision and infilling has largely been controlled by macroscale changes in sea level and sediment supply. Dramatic changes in estuarine form and function can be linked to small scale fluctuations of sea level and sediment supply during the Holocene, in addition to contemporary human activity. It is important to note the geologically ephemeral nature of estuaries, shown in their rapid response to environmental changes over decadal to millennial timescales.

References

Barwis, J. H. and Tankard, A. J. (1983). Pleistocene shoreline deposition and sea-level history at Swartklip, South Africa. *Journal of Sedimentary Petrology*, 53, 1281–1294.

Bateman, M. D., Carr, A. S., Dunajko, A. C., Holmes, P. J., Roberts, D. L., McLaren, S. J., Bryant, R. G., Marker, M. E. and Murray-Wallace, C. V. (2011). The evolution of coastal barrier systems: a case study of the Middle-Late Pleistocene Wilderness barriers, South Africa. *Quaternary Science Reviews*, 30, 63–81.

Baxter, A. J. and Meadows, M. E. (1999). Evidence for Holocene sea-level change at Verlorenvlei, Western Cape, South Africa. *Quaternary International*, 56, 65–79.

Begg, G. (1978). *The Estuaries of Natal. Report 41*. Pietermaritzburg: Natal Town and Regional Planning Commission, 657pp.

Boyd, R., Dalrymple, R. W. and Zaitlin, B. Z. (1992). Classification of clastic coastal depositional environments. *Sedimentary Geology*, 80, 139–150.

Branch, G. M., Bally, R., Bennett, B. A., de Decker, H. P., Fromme, G. A. W., Heÿl, C. W. and Willis, J. P. (1985). Synopsis of the impact of artificially opening the mouth of the Bot River Estuary: Implications for management. *Transactions of the Royal Society of South Africa*, 45, 465–483.

Camoin, G. F., Montaggioni, L. F. and Braithwaite, C. J. R. (2004). Late glacial to post glacial sea-levels in the Western Indian Ocean. *Marine Geology*, 206, 119–146.

Cawthra, H. C., Neumann, F. H., Uken, R., Smith, A. M., Guastella, L. A. and Yates, A. (2012). Sedimentation on the narrow (8 km wide), oceanic current-influenced continental shelf off Durban, Kwazulu-Natal, South Africa. *Marine Geology*, 323–325, 107–122.

Chaumillon, E., Tessier, B. and Reynaud, J.-Y. (2010). Stratigraphic records and variability of incised valleys and estuaries along French coasts. *Bulletin de la Société géologique de France*, 181, 75–85.

Compton, J. S. (2001). Holocene sea-level fluctuations inferred from the evolution of depositional environments of the southern Langebaan Lagoon salt marsh, South Africa. *The Holocene*, 11, 395–405.

Compton, J. S. (2011). Pleistocene sea-level fluctuations and human evolution on the southern coastal plain of South Africa. *Quaternary Science Reviews*, 30, 506–527.

Compton, J. S. and Franceschini, G. (2005). Holocene geoarchaeology of the sixteen mile beach barrier dunes in the Western Cape, South Africa. *Quaternary Research*, 63, 99–107.

Cooper, J. A. G. (1993). Sedimentation in a river dominated estuary. *Sedimentology*, 40, 979–1017.

Cooper, J.A.G. (1994). Lagoons and microtidal coasts. In *Coastal Evolution: Late Quaternary Shoreline Morphodynamics*, ed. R. W. G. Carter and C. D. Woodroffe. Cambridge: Cambridge University Press, pp. 219–265.

Cooper, J. A. G. (2001). Geomorphological variability among microtidal estuaries from the wave-dominated South African coast. *Geomorphology*, 40, 99–122.

Cooper, J. A. G., Green, A. N. and Wright, C. I. (2012). Evolution of an incised valley coastal plain estuary under low sediment supply: a "give-up" estuary. *Sedimentology*, 59, 899–916.

Cooper, J. A. G., Kilburn, R. N. and Kyle, R. (1989). A late Pleistocene molluscan assemblage from Lake Nhlange, Zululand, and its palaeoenvironmental implications. *South African Journal of Geology*, 92, 73–83.

Cooper, J. A. G., Wright, C. I. and Mason, T. (1999). Geomorphology and sedimentology. In *Estuaries of South Africa*, ed. B. R. Allanson and D. Baird. Cambridge: Cambridge University Press, pp. 5–25.

Dalrymple, R. W. (2006). Incised valleys in time and space: An introduction to the volume and an examination of the controls on valley formation and filling. In *Incised Valleys in Time and Space*, ed. R. W. Dalrymple, D. A. Leckie and R. W. Tillman. Tulsa, OK: SEPM Special Publication 85, pp. 5–10.

Dalrymple, R. W., Zaitlin, R. A. and Boyd, R. (1992). A conceptual model of estuarine sedimentation. *Journal of Sedimentary Petrology*, 62, 1130–1146.

Dingle, R. V., Siesser, W. G. and Newton, A. R. (1983). *Mesozoic and Tertiary Geology of Southern Africa*. Rotterdam: Balkema, 375pp.

Dladla, N. N. (2014). *Seismic stratigraphic models for Late Pleistocene/Holocene incised valley systems on the Durban continental shelf*. MSc Thesis, University of KwaZulu-Natal, Westville, 110pp.

Franceschini, G., McMillan, I. and Compton, J. (2005). Foraminifera of Langebaan Lagoon salt marsh and their application to the interpretation of late Pleistocene depositional environments at Monwabisi, False Bay coast, South Africa. *South African Journal of Geology*, 108, 285–296.

Garland, G. G. and Moleko, L. (2000). Geomorphological impacts of Inanda Dam on the Mgeni Estuary, north of Durban, South Africa. *Bulletin of Engineering, Geology and Environment*, 59, 119–126.

Gile, L. H. and Hawley, J. W. (1968). Age and comparative development of desert soils at the Gardner Spring radiocarbon site, New Mexico. *Soil Science Society of America Journal*, 32, 709–716.

Green, A. N. (2009). Palaeo-drainage, incised valley fills and transgressive systems tract sedimentation of the northern KwaZulu-Natal continental shelf, South Africa, SW Indian Ocean. *Marine Geology*, 263, 46–63.

Green, A. (2011). The late Cretaceous to Holocene sequence stratigraphy of a sheared passive upper continental margin, northern KwaZulu-Natal, South Africa. *Marine Geology*, 289, 17–28.

Green, A. N., Cooper, J. A. G., Leuci, R. and Thackeray, Z. (2013a). Formation and preservation of an overstepped segmented lagoon complex on a high-energy continental shelf. *Sedimentology*, 60, 1755–1768.

Green, A. N., Cooper, J. A. G. and Salzmann, L. (2014). Geomorphic and stratigraphic signals of postglacial meltwater pulses on continental shelves. *Geology*, 42, 151–154.

Green, A. N., Cooper, J., Wiles, E. and de Lecea, A. M. (2015). Seismic architecture, stratigraphy and evolution of a sub-tropical marine embayment: Maputo Bay, Mozambique. *Marine Geology*. 369, 300–309.

Green, A. N., Dladla, N. N. and Garlick, G. L. (2013b). Spatial and temporal variations in incised valley systems from the Durban continental shelf, KwaZulu-Natal, South Africa. *Marine Geology*, 335, 148–161.

Green, A. N. and Garlick, G. L. (2011). A sequence stratigraphic framework for a narrow, current-swept continental shelf: The Durban Bight, central KwaZulu-Natal, South Africa. *Journal of African Earth Sciences*, 60, 303–314.

Grindley, J. R., Rogers, J., Pether, J. and Woodborne, M. W. (1989). Holocene evolution of Rietvlei near Cape Town, deduced from the palaeoecology of some mid-Holocene estuarine Mollusca. *Palaeoecology of Africa*, 19, 347–353.

Hayes, M. O. (1979). Barrier island morphology as a function of tidal and wave regime. In *Barrier Islands*, ed. S. P. Leatherman. New York: Academic Press, pp. 1–28.

Hobday, D. and Jackson, M. (1979). Transgressive shore zone sedimentation and syndepositional deformation in the Pleistocene of Zululand, South Africa. *Journal of Sedimentary Research*, 49, 145–158.

Illenberger, W. (1996). The geomorphologic evolution of the Wilderness dune cordons, South Africa. *Quaternary International*, 33, 11–20.

Liu, J. P. and Milliman, J. D. (2004). Reconsidering melt-water pulses 1A and 1B: Global impacts of rapid sea-level rise. *Journal of Ocean University of China*, 3, 183–190.

Meadows, M., Baxter, A. and Parkington, J. (1996). Late Holocene environments at Verlorenvlei, Western Cape Province, South Africa. *Quaternary International*, 33, 81–95.

Miller, W. R. (1998). *The Bathymetry, Sedimentology and Seismic Stratigraphy of Lake Sibaya-Northern KwaZulu-Natal.* MSc Thesis, University of Natal, 94pp.

Mkhize, N. E. (2014). *Sequence and Lithostratigraphy of Incised Valley Fills: Evolution of the Durban Bay system.* MSc Thesis, University of KwaZulu-Natal, Westville, 126pp.

Nordfjord, S., Goff, J. A., Austin Jr, J. A. and Sommerfield, C. K. (2005). Seismic geomorphology of buried channel systems on the New Jersey outer shelf: Assessing past environmental conditions. *Marine Geology*, 214, 339–364.

Norström, E., Risberg, J., Gröndahl, H., Holmgren, K., Snowball, I., Mugabe, J. A. and Sitoe, S. R. (2012). Coastal paleo-environment and sea-level change at Macassa Bay, southern Mozambique, since c 6600 cal BP. *Quaternary International*, 260, 153–163.

Perissinotto, R., Stretch, D. D. and Taylor, R. H. (2013). *Ecology and Conservation of Estuarine Ecosystems: Lake St. Lucia as a Global Model.* Cambridge: Cambridge University Press, 511pp.

Pillay, D., Branch, G. M. and Steyn, A. (2009). Complex effects of the gastropod Assiminea globulus on benthic community structure in a marine-dominated lagoon. *Journal of Experimental Marine Biology and Ecology*, 380, 47–52.

Ramsay, P. J. (1995). 9000 years of sea-level change along the southern African coastline. *Quaternary International*, 31, 71–75.

Ramsay, P. J. and Cooper, J. A. G. (2002). Late Quaternary sea-level change in South Africa. *Quaternary Research*, 57, 82–90.

Reddering, J. S. V. and Rust, I. C. (1990). Historical changes and sedimentary characteristics of southern African estuaries. *South African Journal of Science*, 86, 425–428.

Rogers, J. (1980). *First report on the Cenozoic sediments between Cape Town and Eland's Bay.* Pretoria: Geological Survey of South Africa Open File Report 0136, 64pp.

Salzmann, L., Green, A. N. and Cooper, J. A. G. (2013). Submerged barrier shoreline sequences on a high energy, steep and narrow shelf. *Marine Geology*, 346, 366–374.

Schalke, H. (1973). The upper Quaternary of the Cape Flats area (Cape Province, South Africa). *Scripta Geologica*, 15, 1–57.

Scott, L. and Steenkamp, M. (1996). Environmental history and recent human disturbance at coastal Lake Teza, Kwazulu/Natal. *South African Journal of Science*, 92, 348–350.

Stevenson, I. R. and McMillan, I. K. (2004). Incised valley fill stratigraphy of the Upper Cretaceous succession, proximal Orange Basin, Atlantic margin of southern Africa. *Journal of the Geological Society*, 161, 184–208.

Van Heerden, I. L. (1987). Sedimentation in the greater St Lucia complex as related to palaeo-sea-levels. *6th National Oceanographic Symposium, Stellenbosch*, Paper 157, B-106.

Wright, C. I. (2002). *Aspects of the Cenozoic Evolution of the Northern KwaZulu-Natal Coastal Plain, South Africa.* Pretoria: Bulletin of the Council for Geoscience 132, 120pp.

Wright, C., Cooper, J. and Kilburn, R. (1999). Mid Holocene palaeoenvironments from Lake Nhlange, northern Kwazulu-Natal, South Africa. *Journal of Coastal Research*, 15, 991–1001.

Wright, C. I., Miller, W. R. and Cooper, J. A. G. (2000). The late Cenozoic evolution of coastal water bodies in Northern Kwazulu-Natal, South Africa. *Marine Geology*, 167, 207–229.

Zaitlin, B. A., Dalrymple, R. W. and Boyd, R. (1994). The stratigraphic organization of incised-valley systems with relative sea-level change. In *Incised Valley Systems: Origin and Sedimentary Sequences*, ed. R. W. Dalrymple, R. Boyd and B. A. Zaitlin. Tulsa, OK: SEPM Special Publication 51, pp. 45–60.

15

Soils and duricrusts

JÜRGEN RUNGE

Abstract

Weathering is a climate-dependent mechanism whereby different physical, chemical and biological processes lead to regolith and soil formation. Pleistocene and Holocene climatic changes caused significant modification of formerly weathered soils, and hardened layers of substrata (duricrusts) which are now closely associated with scarps and mesas. This chapter outlines the main pedogenic processes, the role of time, and regional soil patterns in southern Africa. Major types of duricrusts (calcretes, silcretes, ferricretes) are briefly described, and their significance for long-term landscape evolution is discussed. Duricrusts are important landscape markers for understanding past, present and possibly future weathering, soil and erosion dynamics in sensitive, climate-controlled landscapes.

15.1 Introduction

Changing environmental and climatic conditions in southern Africa during the Pleistocene and Holocene played a considerable role in the spatial and temporal dynamics of weathering and soil forming processes and ecosystems (Meadows, 2001). These edaphic changes, in combination with late Neogene tectonic uplift, exerted significant geomorphological and ecological effects on southern African landscapes in their totality (Coltorti et al., 2007), and which in turn exerted influence on evolutionary events, particularly on the emergence of early hominids in Africa (Partridge et al., 1995). Because southern Africa was not subject to direct Quaternary glaciations, except in isolated areas of the Lesotho highlands and the Cape Fold Mountains (cf. Lewis, 1994, 1996), older regoliths and soils have not been eroded and removed as was the case in periglaciated and glaciated regions of the northern hemisphere. Similarly, southern Africa was not largely covered by thick Pleistocene aeolian sediments such as the loess that formed the initial and fertile substrates for Holocene pedogenesis in North America, Europe and Asia. Due to long-lasting terrestrial denudation, weathering mantles in southern Africa are often thin and moderately fertile. However, in some areas (e.g. northern KwaZulu-Natal; Botha, 1996) the parent rock is now covered by thick and partially eroded colluvial and hillslope sediments. Such strata contain palaeosols, stone lines and slope deposits (hillwash) that provide evidence for Quaternary landscape history. It is advantageous that the study of soils and landscapes considers a palaeoenvironmental perspective, since soils can in some instances represent an important palaeoenvironmental archive. This chapter explores the relationship between climate, weathering, terrestrial denudation, soils and duricrusts with the aim of developing a better understanding of Pleistocene and Holocene land surface evolution in southern Africa.

15.2 Weathering and soils

Soil formation (pedogenesis) at any one place requires an understanding of present and past climate patterns that give rise to surface and subsurface weathering processes and patterns. Weathering comprises physical (mechanical), physico-chemical and biological (biochemical/biomechanical) processes that transform the composition, state and properties of near-surface rocks into fine grained textures, offering water and nutrients in pores and space for roots of plants (Hall, 1988; Ehlen, 2005). As such, climate is the driving force for the majority of weathering processes, mediated by geologic, land surface and edaphic properties (Phillips, 2005). Temperature and precipitation regimes are important controls on weathering and have dominance in different parts of southern Africa. For example, insolation and high diurnal temperature variations cause changes in volume, especially in rocks that are mineralogically variable, and can cause rock disintegration and lead to an increase in total weathered surface area. At high altitudes where seasonal frost is pronounced, mechanical frost-induced weathering can take place where sufficient moisture is present (Grab, 2007). Chemical weathering triggered by high temperatures and the presence of water is pervasive throughout many areas of southern Africa and is commonly more effective than physical processes alone. Soluble salts such as sodium carbonate (Na_2CO_3), sodium chloride ($NaCl$) and sulphate (SO_4) can be washed out or leached from weathered regolith. This process is reversible if the solvent (dew, rainfall, groundwater) evaporates or if the solution concentration becomes too high (saturation). Even more effective is hydrolysis which weakens and irreversibly destroys rock structures, whereas dissociated percolating water (i.e. through low pH) breaks up primary minerals into their corresponding bases and acids. Silicates such as feldspars will be fully disintegrated, and the recrystallisation of constituent molecules results in the formation of less complex clays such as kaolinite or gibbsite. With increasing depth of weathering – in some cases up to several hundreds of metres deep – an enrichment of quartz, sesquioxides (Fe_2O_3, Al_2O_3) and manganese can be observed. Often this is described as 'tropical weathering' under year-round high temperatures and constantly high rainfall (Thomas, 1994; Wirthmann, 2000; Gupta, 2011). This climatic and weathering context facilitates the formation of duricrusts.

Southern Africa has a significant west–east precipitation gradient, but experiences much less spatial variability of temperature than precipitation. Precipitation and moisture availability are the most likely controls on weathering processes and rates, and also strongly influence vegetation patterns. Today, rainfall in southern Africa is linked to the seasonal migration of the Intertropical Convergence Zone (ITCZ) and incoming cyclones from the Indian and Atlantic oceans. Pleistocene and earlier-Holocene precipitation (and temperature) patterns, however, are likely to have been significantly different, and thus result in different weathering processes and patterns. Soils and duricrusts may potentially provide a record of such past climates (Thomas, 1994; Meadows, 2001; Gupta, 2011).

15.2.1 Factors significant in pedogenesis

The effectiveness of soil formation is a function of the complex interplay between parent material, climate, topography, organisms and time. Past climate fluctuations can modify the intensity of weathering processes and thus regolith formation. Soil formation was generally more efficient and faster under humid and warm climates (e.g. after the breakup of Gondwana and during the Cenozoic, cf. Partridge and Maud, 1987), and slowed down when climate was colder and dryer. Conventional pedology considers soil formation as an *in situ* (autochthonous) transformative process; it works from top to bottom forming loose materials leading to a regolith that rests on bedrock. As soil formation takes place over time in a landscape context with climate-driven surface processes (water, wind, vegetation), loose particles may be transported downslope from higher elevation to lower-lying areas such as valley floors and plains. Consequently, the *in situ* soil profile can be incised and pedogenesis may continue elsewhere in the system as soil material is moved and redeposited as mass movement/solifluction deposits and colluvium. The surface vegetation cover and climate regime (rainfall, temperature, wind) in a region strongly control the effectiveness of these surface processes.

Complex multilayered soil horizons with sediment strata varying in space and time are typical features of weathering mantles in many tropical and subtropical landscapes (Phillips, 2005). To distinguish between *in situ*-formed soils and redeposited (allochthonous) 'soil' sediments, the terms 'pedisediment', or more recently 'pedosediment' were introduced (Fölster, 1964). This concept allows the integration of spatio-temporal patterns of weathered materials and/or soils into a dynamic morpho-pedological landscape framework (Phillips, 2001) (Fig. 15.1).

15.2.2 Soils and geomorphology

Prolonged surface and subsurface weathering can lead to the formation of saprolites which can yield great depths of weathering (>100 m; Sobanski and Valeton, 1994). They are widespread on Gondwana basement rocks and important in understanding the evolution of specific relief features of tropical geomorphology (i.e. rolling peneplains, palaeoplains, inselbergs and duricrusts) (Ehlen, 2005; Dixon et al., 2009). Environmental conditions that favour the formation of saprolite include moderate to flat relief, which minimises erosion and allows leaching of chemical weathering products (desilicification and enrichment of sesquioxides). Longer periods of tectonic stability also support their formation, whereas tectonic activity (uplift) and climate change can cause erosion (Thomas, 1994; Wirthmann, 2000).

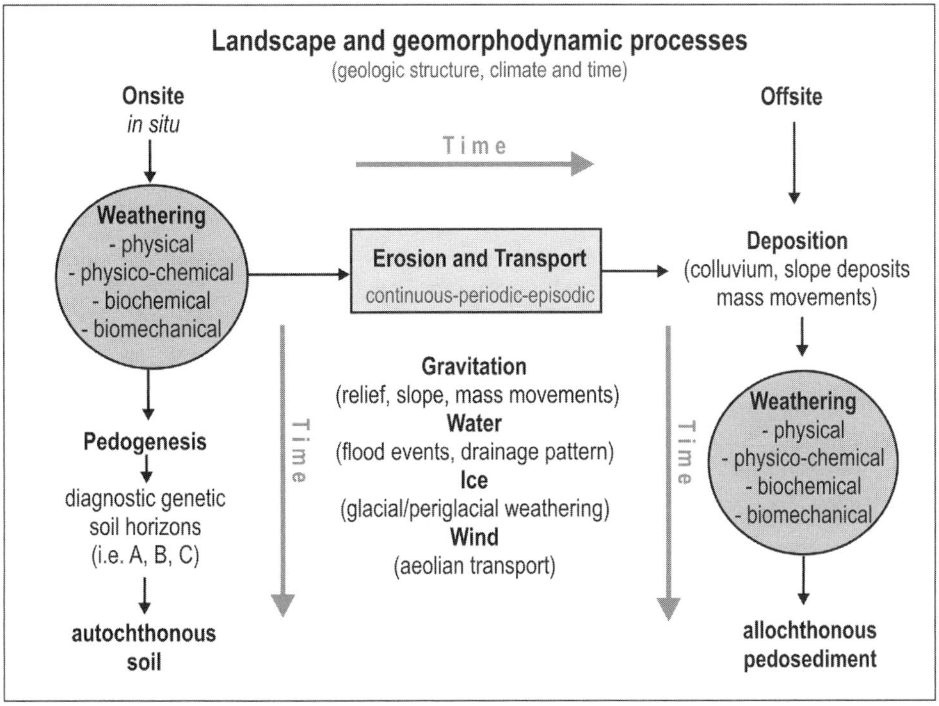

Fig. 15.1. A dynamic geomorpho-pedological concept for soils and pedosediments in tropical and subtropical environments.

The regional soil patterns across southern Africa depend on geological and geomorphological setting, present-day climate and Quaternary climate history (Fig. 15.2). A catena or a pedological transect (also known as a toposequence) typically shows complex successions of soils and pedosediments from hill summit to footslope. Shallower soils would be located on steeper slope sections because of enhanced erosion at these sites (e.g. Lithosols with rock fragments, and A–C soil profiles as Leptosols). In downslope and valley locations, thicker colluvial and alluvial materials predominate (e.g. Gleysols, Luvisols and Planosols). Catenas can also help reconstruct the Pleistocene and Holocene erosional history of the southern African landscape (Thomas, 1994). However, on deeply weathered saprolitic soils, many upslope positions, even near catchment divides, show deeply weathered kaolinitic clays. This is due to previous deep chemical weathering, etchplanation, and long-term land surface denudation. Such observations, in light of older Gondwana landscapes with specific geomorphological properties, often form a sharp contrast to the conventional extratropical interpretation of a catena and are in contrast to the classical shallow soils on upper slopes and deeply weathered, dark and humus-rich soils on footslopes (cf. Phillips, 2001; Rossetti, 2004).

15.3 Soils and landscape units in South Africa

The Soil Science Society of South Africa (SSSSA) and its sub-committee the Soil Classification Working Group (SCWG) are presently (from 2012) updating the existing soil classification for South Africa. In comparison with other international soil classifications, the South African system is comparatively simple because it was developed for the agricultural, environmental science and planning sectors (Turner, 2013; van Huyssteen et al., 2013). It is based mainly on recognition of morphological and physical properties which can easily be distinguished in the field (Fey, 2010; SCWG, 2012). These soils are classified using a hierarchical system containing two main levels; soil form and soil family. There are currently 73 soil forms and 400 families defined by the nature of the topsoil (organic, humic, vertic, melanic or orthic), and numerous diagnostic subsoil horizons (SCWG, 2012).

In recent years, a series of user-oriented 'land-type' maps at 1:250,000-scale, with explanatory notes, has been published for most of South Africa by the Institute for Soil, Climate and Water (ISCW). Climate–geomorphology–soil relationships are well-explained in these documents. The last generalised soil map with a report covering the entire South African territory dates back to van der Merwe in 1962 (cited in Laker, 2000). Here, a new outline map for southern Africa's soils, combining data from Laker (2000) with the soil type names from the FAO-UNESCO (1977, 2012) Soil Map of the World, is presented (Fig. 15.2). More detailed maps are presented for different soil groups by Fey (2010).

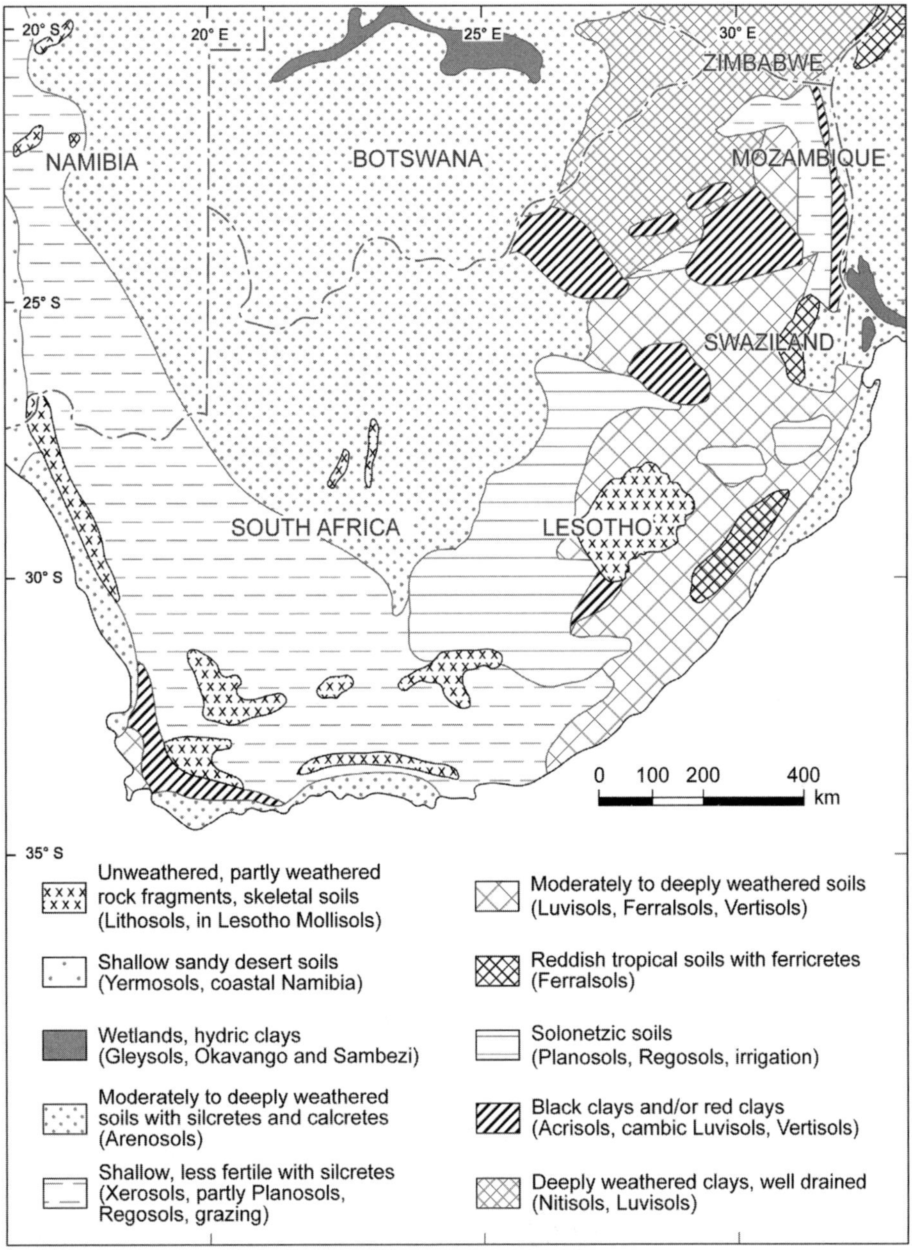

Fig. 15.2. Generalised soil map of southern Africa with FAO soil terminology and indication of dominating duricrusts (adapted from FAO-UNESCO, 1977, 2012; Laker, 2000). Also see Soil Map of the World, Soil Map of Africa, available from http://eusoils.jrc.ec.europa.eu/library/maps/. The term Mollisol originates from the USDA Soil Taxonomy.

Southern Africa exhibits large topographic barriers such as the Great Escarpment and isolated basaltic to sandstone plateaus of the Drakensberg (KwaZulu-Natal, Lesotho) that influence climate and vegetation. Cold (Benguela Current) and warm (Agulhas Current) ocean currents influence seasonal and annual rainfall distributions across southern Africa. Late Pliocene tectonic uplifting (Partridge and Maud, 1987; Partridge et al., 1995) followed by erosion and deposition during the Quaternary have created complex landforms with embedded multifold soils (Botha, 1996); these are influenced by underlying parent rock and climate-limited weathering. The Cape Fold Belt has shallow Litho- and Leptosol (A–C), sandy Areno- and deeper developed Luvisols (Fig. 15.2). Coastal plains which extend around the coast from Namibia (desertic Yermosols) to Mozambique have low to high rainfall and represent a narrow plain with partly fertile Areno-, Luvi-, Ferral- and Vertisols. The steepest slopes of the Great Escarpment, with isolated inselbergs and mountain ridges in north-central Namibia, eastern Botwana and Lesotho (>3000 m asl), are dominated by skeletal to shallow soils (Lithosols). In the humid highlands of the Drakensberg, with extended grass-covered Lithosols over basalt and andesite, humus-rich Mollisols can develop (Fig. 15.2). As geomorphological barriers hinder moisture reaching the interior, vegetation in the central Highveld is sparse and landscapes are mostly savanna to steppe-type grasslands with Xero-, Plano- and Regosols. Soils derived from basaltic or doleritic rocks (dykes) in the Highveld grasslands can be nutrient rich.

The Kalahari region including Namibia and Botswana represents an extensive livestock producing area with many calcrete outcrops (Nash et al., 1994). The region is covered by 50–>300 m thick sand layers (Thomas and Shaw, 1991). Corresponding soils are sandy in texture and moderately to deeply weathered (Arenosols) with calcareous and/or siliceous hardpan layers in the profile or at the surface (Figs. 15.2, 15.3). Deep boreholes and wells have enabled commercial grazers to become permanently established in this semi-arid region. Irrigation on Solonetzic soils such as Plano- and Regosols, however, can cause severe ecological problems like salinisation and subsequent soil degradation (FAO, 2009).

The savannas of the Lowveld are associated with gabbros and granites of the Bushveld igneous complex, forming planation surfaces with low relief. Acid basement rocks give rise to sandy soils with moderate nutrient levels, whereas the gabbros of the Mispah rock complex form potentially fertile soils (Suttie et al., 2005). Further, deeply weathered soils and colluvium dominate the area north of Lesotho into Swaziland and Zimbabwe with moderate to deep, semi-fertile Luvi- and Ferralsols. In low-lying poorly drained valleys and depressions (pans), Vertisols over nodular calcretes are common. Oxisols or Ferralsols form under abundant precipitation and high temperatures in undulated, low relief landscapes

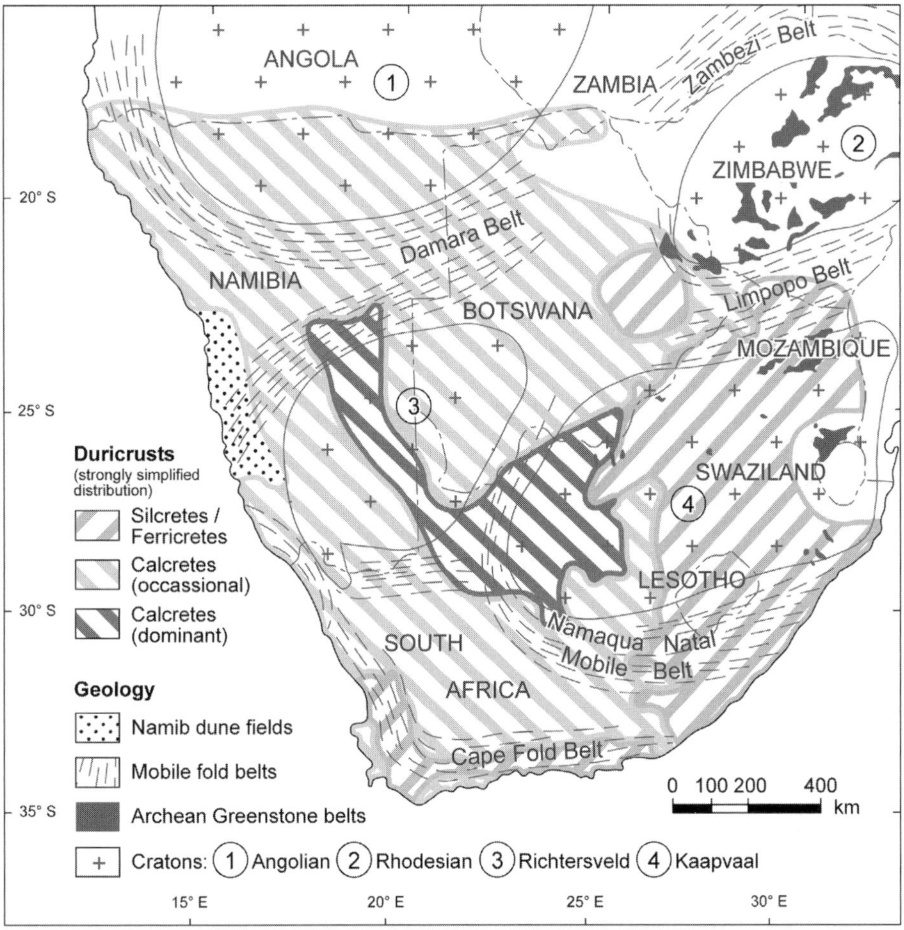

Fig. 15.3. Geological background and generalised duricrust occurrence in southern Africa (modified after Goudie (1973) and Twidale (1988)).

with good drainage conditions. Iron (III) oxide formation leads to characteristic red laterites. When the primary iron/alumina content in the parent rock and saprolite is high, ferricretes often occur as a result of climate aridification. Wetlands of the Okavango basin and Zambezi valley (Fig. 15.2) are regularly flooded and therefore show periodic reduction of clay minerals that leads to blue/grey, mottled and organic rich hydric clays (Gleysols).

All soils in most locations broadly reflect recent and past environmental conditions, strongly influenced by their geomorphological position, climate, rooted vegetation and soil fauna. In connection with slope drainage patterns, soil vegetation and fauna play a key role in soil organic content (Thomas, 1994).

15.4 Duricrusts

A result of persistent and intensive weathering, pedogenesis, climate change and terrestrial denudation is the formation of duricrusts. These are comprised of a thick hard crust sometimes several metres thick at or just below the ground surface. According to Nash and McLaren (2007), such 'geochemical sediments' and their residual deposits are extremely important for the terrestrial environment. Duricrusts can contribute to the understanding of the Neogene geomorphic landscape evolution by providing isotopic and sedimentary archives of environmental change (sometimes with incorporated artefacts), and in some cases they can act as relative or absolute dating tools (Nash and McLaren, 2007). Due to their petrological hardness and resistance to erosion, they often form residual reliefs (inversions) such as flat-topped mesas, and have an influence on the structure and form of slopes (McFarlane, 1976; Runge, 1993; Thomas, 1994). They can also represent economic value as 'orecretes' or metalliferous duricrusts (Dill et al., 2013). Three basic types of duricrusts are normally distinguished: Calcretes, Silcretes and Ferricretes (Goudie, 1973). The distribution of duricrusts across southern Africa is shown in Figure 15.3. Major features of duricrusts and environmental conditions for their evolution are summarised in Nash and McLaren (2007).

Calcretes are prevalent in the semi-arid to arid central parts of southern Africa, northern Namibia, central Botswana (Kalahari) and the upper Karoo and can reach >30 m thick. Silcretes are common in the Namib Desert (partly transitional forms between calcretes and silcretes also occur), semi-arid regions of the Cape Fold Belt, and scattered areas in eastern South Africa (Highveld and KwaZulu-Natal). Ferricretes with deeply weathered tropical soils are widespread in eastern southern Africa (Swaziland, Mozambique and Zimbabwe) (Fig. 15.3).

15.4.1 Calcretes

Pedogenic and groundwater calcretes are both considered to be relict products of Pleistocene climate changes (Blümel, 1982; Thomas, 1994; Wright, 2007). As the primary content of calcium in many parent rocks is too low to permit large *in situ* crust formation, chalky windblown dust may have contributed to their development. Calcretes and associated landforms may also develop in relation to former drainage networks (Nash et al., 1994; Moore et al., 2012), and may be linked to episodically flooded depressions (pans) (Netterberg, 1978; Burrough and Thomas, 2008). While precipitation of ascending calcite was formerly considered an explanation, lateral groundwater flows of calcium-rich solutions are now considered to be more likely. Some of the material suspected to be calcrete also consists of dolomite, which suggests strong evaporative (arid) conditions during crust formation (Atlhopheng and Ekosse, 2007). The dissolved carbonate (repeatedly recrystallised during periods of wetting and drying of regolith) may be derived from topographic highs, precipitated out at footslopes, and subsequently covered by slopewash deposits. Over time, indurated duricrusts can prove resistant to erosion, possibly ending up as a calcareous cap rock (mesa, Fig. 15.4A), causing an inversion of relief (Pain and Ollier, 1995). Besides the pedogenic origin of calcretes, Wright (2007, p11) emphasises that in 'some semi-arid to arid regions extensive precipitation occurs in the shallow phreatic zone, producing large bodies of authigenic carbonate with many characteristics in common with pedogenic calcrete'.

15.4.2 Silcretes

Silcrete is a hard, grey rock made by silicification of pre-existing deposits, especially quartz-rich alluvium (Pain and Ollier, 1995; Nash and Ullyott, 2007). Silcretes tend to form through a variety of weathering and environmental mechanisms, hence no single model is universally accepted (Watson and Nash, 1997). Silcrete formation is the result of silica precipitating due to a drop in pH (<3.75) that also makes titanium soluble and mobile, which can be enriched in these crusts by up to 2% (Thomas, 1994). Formation of silcrete requires strong evaporation and cooling, organic processes, and an annual rainfall of 250–1000 mm (Bond, 1964; van de Graaff, 1983). Silcretes may also form under alkaline oxidising conditions by crystobalitic-tridymitic silica cement (Atlhopheng and Ekosse, 2007). At the micro-scale, different types of silcretes can be identified: grain supported fabric (quartzitic and conglomeratic) silcretes, fabrics with more than 40% silica cement, and either quartzose, chalcedonic or opaline silcretes (Atlhopheng and Ekosse, 2007). In the Kalahari region, silcrete formation also occurs in non-pedogenic contexts under high pH. Environmental conditions for silcrete formation may include fluvial or lacustrine settings, and *in situ*, capillary, pedogenic or detrital origins (Goudie, 1973; Nash and Ullyott, 2007; Ringrose et al., 2009). In terms of landscape evolution, silcretes can form weathering-resistant layers that cause relief inversion and morphological accentuation during fluvial dissection of planation surfaces (Nash and McLaren, 2007) (Fig. 15.4). Provenancing of silcrete artefacts from the Middle Stone Age (MSA) in the Tsodilo Hills, northwest Botswana, has shown that specific silcrete sources were preferred by MSA peoples for tool manufacture, even transporting silcrete over distances exceeding 220 km (Nash et al., 2013).

15.4.3 Ferricretes

Ferricrete is a reddish-brown, compact and hard, nodular to pisolitic, surficial or near-surface material rich in iron – and partly aluminium – oxides and

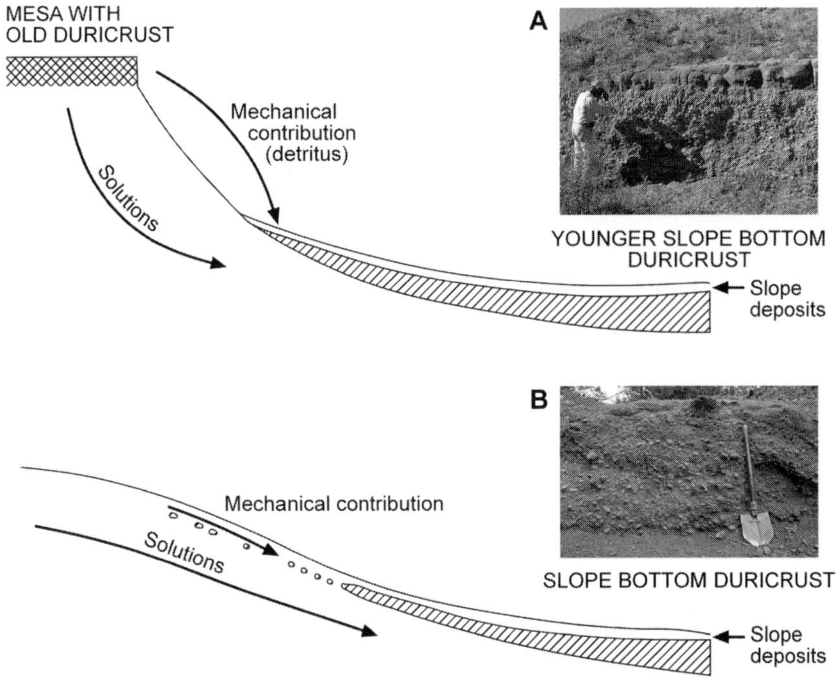

Fig. 15.4. Step-like and undulating topography formed by duricrust cappings. (A) Mechanical and chemical degradation of primary upper crust causes accumulation of calcite (nodular crusts) downslope, covered by clayey dark hillwash and vertisol (photo shows road cut, north of Gaborone, Botswana). (B) Low relief planation surface with lateral transport of ferruginous quartz pebbles and soluble iron from upper to lower slopes, forming an inclined and cemented duricrust with sandy to clayey slope deposits (photo shows a roadcut, southwest of Bangui, near Mbaiki/Ngotto, Central African Republic). Line drawings modified from McFarlane (1976), photos: J. Runge.

hydroxides (Thomas, 1994; Alexandre, 2002). It is mainly dominated by goethite and haematite minerals as well as remaining quartz from basement rocks. Below the ferricrete crust, kaolinitic clays within a mottled zone with deeply weathered saprolite are typical. The climate for ferricrete formation is usually tropical wet-dry to wet (>1200 mm annual rainfall) and high temperatures (>22°C). Former groundwater lowering and temperature variations under semi-arid to arid climates may have favoured ferricrete formation (Stevenson and McMillan, 2004). As with other duricrusts, fossil drainage channels and valleys could have been suitable locations for iron enrichment and later hardening of the residual weathering mantle. Following landscape lowering by tectonic uplift or climate-driven denudation, incision may have led to precipitation of iron as the water table lowered (or climate became drier). Sesquioxide movement in soils may be due to clay migration and surface erosion, reduced or redissolved by complexation with organic compounds (Atlhopheng and Ekosse, 2007). Sesquioxides remobilised on upper slopes or from 'older' primary duricrusts can migrate downslope and precipitate on lower slopes (forming 'younger' or secondary crusts) (Fölster, 1964; McFarlane, 1976). This forms a distinct lower duricrust level that contributes to a characteristic step-like relief, preserving older surfaces against erosion by resistant, chemically different morphopedological crusts. Such landscape features were formed during the Jurassic to Quaternary denudation history of southern Africa. Downslope, crusts are frequently covered by shallow slopewash deposits, giving evidence of climate-controlled Pleistocene to Holocene sediment dynamics (Fig. 15.4).

15.5 Economic importance of duricrusts

Apart from revealing climate history, landscape development and possible impact on human evolution, duricrusts may have economic importance. Calcretes and ferricretes and their corresponding weathering products have long been used for road construction in tropical and subtropical areas. High concentrations of gibbsite in tropical deeply weathered soils can give rise to aluminium ore (bauxite). Other special types are 'orecretes' with high concentrations of heavy metals accumulated near the surface. Dill et al. (2013) divide orecretes into oxicretes, carbocretes, silicacretes, halcretes (halogenides), sulcretes (sulphate plus aluminium-phosphate-sulphate minerals), phoscretes (phosphates) and others. Orecretes contain significantly higher concentrations of minerals such as Lead (Pb), Copper (Cu), Zinc (Zn), Indium (In), Iron (Fe), Manganese (Mn), Nickel (Ni), Cobalt (Co), Tungsten (W), rare earth elements (REE) and Silver (Ag). These deposits are commonly exploited through widespread artisanal to large-scale industrial mining in Africa (Runge and Shikwati, 2011). The specific chemical composition of all duricrusts is strongly controlled by the parent rock but duricrusts can contribute to geological mapping in that they apply a mineralogical 'fingerprint' of crust and rock properties (McFarlane, 1976; Dill et al., 2013).

15.6 Summary

This chapter places emphasis on the fact that duricrusts in southern Africa are often overlooked by conventional soil science. Conceptions of pedogenesis – mainly focusing on soils as a means for land use and agricultural production – tend to neglect the importance of duricrusts for understanding landscape evolution and in palaeoenvironmental and archaeological contexts. There are strong relationships between the spatial distribution of soils in southern Africa, weathering processes, *in situ* and redeposited soils and pedosediments, and these are relevant for hard crust (duricrust) formation (e.g. Dixon et al., 2009). Duricrusts are best formed under tropical and subtropical climates, provided there is a suitable dissolved mineral source, a mechanism to transfer the mineral in solution to a site of accumulation, and a means of triggering precipitation. Duricrusts are first-hand landscape markers for understanding past, present and possibly future weathering, soil and erosion dynamics in sensitive environments.

Acknowledgements

I express my gratitude to Joachim Eisenberg, Goethe Universität Frankfurt, who designed figures and maps for this chapter. My sincere thanks go to Robert A. Bean, University of Texas, Austin, and to Teresa Bulman, Portland State University, Portland, for checking and correcting the English.

References

Alexandre, J. (2002). *Les cuirasses latéritiques et autres formations ferrugineuses tropicales: exemple du Haute Katanga méridional.* Musée Royale de l'Afrique Centrale, Annales, Sciences Géologiques, 107, 118pp.

Atlhopheng, J. R. and Ekosse, G.-I. E. (2007). Mineralogical appraisal of sediments of duricrust suites and pans around Jwaneng area, Botswana. *Journal of Applied Science and Environmental Management*, 11, 51–56.

Blümel, W. (1982). Calcretes in Namibia and SE Spain: Relations to substratum, soil formation, and geomorphic factors. *Catena Supplement*, 1, 67–95.

Bond, G. (1964). Pleistocene environments in southern Africa. In *African Ecology and Human Evolution*, ed. F. C. Howell and F. Boulière. London: Methuen, pp. 308–314.

Botha, G. A. (1996). *The Geology and Palaeopedology of Late Quaternary Colluvial Sediments in Northern Kwazulu-Natal, South Africa.* Pretoria: Council for Geoscience, Memoir, 83, 165pp.

Burrough, S. L. and Thomas, D. S. G. (2008). Late Quaternary lake-level fluctuations in the Mababe Depression: Middle Kalahari palaeolakes and the role of Zambezi inflows. *Quaternary Research*, 69, 388–403.

Coltorti, M., Dramis, F. and Ollier, C. D. (2007). Planation surfaces in Northern Ethiopia. *Geomorphology*, 89, 287–296.

Dill, H. G., Weber, B. and Botz, R. (2013). Metalliferous duricrusts ("orecretes") – markers of weathering: A mineralogical and climatic-geomorphological approach to supergene Pb-Zn-Cu-Sb-P mineralization on different parent materials. *Neues Jahrbuch für Mineralogie - Abhandlungen (Journal of Mineralogy and Geochemistry)*, 190, 123–195.

Dixon, J. L., Heimsath, A. M. and Amundson, R. (2009). The critical role of climate and saprolite weathering in landscape evolution. *Earth Surface Processes and Landforms*, 34, 1507–1521.

Ehlen, J. (2005). Above the weathering front: Contrasting approaches to the study and classification of weathered mantle. *Geomorphology*, 67, 7–21.

FAO (2009). *Advances in the Assessment and Monitoring of Salinization and Status of Biosaline Agriculture*. Reports of expert consultation held in Dubai, United Arab Emirates, 26–29 November 2007. Rome: FAO, World Soil Resources Reports, 104, 82pp.

FAO-UNESCO (1977, 2012). *Soil Map of the World, 1:5.000.000, Africa, Volume VI*. Rome: FAO.

Fey, M. (2010). *Soils of South Africa*. Cambridge: Cambridge University Press, 287pp.

Fölster, H. (1964). Die Pedisedimente der südsudanesischen Pediplane, Herkunft und Bodenbildung. *Pédologie*, 14, 64–84.

Goudie, A. (1973). *Duricrusts in Tropical and Subtropical Landscapes*. Oxford: Clarendon Press, 192pp.

Grab, S. W. (2007). Rock surface temperatures of basalt in the Drakensberg alpine environment, Lesotho. *Geografiska Annaler*, 89A, 185–193.

Gupta, A. (2011). *Tropical Geomorphology*. Cambridge: Cambridge University Press, 400pp.

Hall, K. J. (1988). Weathering. In *The Geomorphology of Southern Africa*, ed. B. P. Moon and G. F. Dardis. Johannesburg: Southern Book Publishers, pp. 12–29.

Laker, M. C. (2000). Soil resources: Distribution, utilization, and degradation. In *The Geography of South Africa in a Changing World*, ed. R. Fox and K. Rowntree. Oxford: Oxford University Press, pp. 326–360.

Lewis, C. A. (1994). Protalus ramparts and the altitude of the local equilibrium line during the last glacial stage in Bokspruit, East Cape Drakensberg, South Africa. *Geografiska Annaler*, 76A, 37–48.

Lewis, C. A. (ed) (1996). *The Geomorphology of the Eastern Cape*. Grahamstown: Grocott and Sherry Publishers, 188pp.

McFarlane, M. J. (1976). *Laterite and Landscape*. New York: Academic Press, 164pp.

Meadows, M. E. (2001). The role of Quaternary environmental change in the evolution of landscapes: Case studies from southern Africa. *Catena*, 42, 39–57.

Moore, A. E., Cotterill, F. P. D. and Eckardt, F. D. (2012). The evolution and ages of Makgadikgadi Palaeo-lakes: Consilient evidence from Kalahari drainage evolution south-central Africa. *South African Journal of Geology*, 115, 385–413.

Nash, D. J., Coulson, S., Staurset, S., Ullyott, J. S., Babutsi, M., Hopkinson, L. and Smith, M. P. (2013). Provenancing of silcrete raw materials indicates long-distance transport to Tsodilo Hills, Botswana, during the Middle Stone Age. *Journal of Human Evolution*, 64, 280–288.

Nash, D. J., Shaw, P. A. and Thomas, D. S. G. (1994). Duricrust development and valley evolution: Process-landform links in the Kalahari. *Earth Surface Processes and Landforms*, 19, 299–317.

Nash, D. J. and McLaren, S. J. (2007). *Geochemical Sediments and Landscapes*. Chichester, New York: Wiley-Blackwell, 488pp.

Nash, D. J. and Ullyott, J. S. (2007). Silcrete. In *Geochemical Sediments and Landscapes*, ed. D. J. Nash and S. J. McLaren. Chichester, New York: Wiley-Blackwell, pp. 95–143.

Netterberg, F. (1978). Dating and correlation of calcretes and other pedocretes. *Transactions of the Geological Society of South Africa*, 81, 379–391.

Pain, C. F. and Ollier, C. D. (1995). Inversion of relief – a component of landscape evolution. *Geomorphology*, 12, 151–165.

Partridge, T. C., Wood, B. A. and deMenocal, P. B. (1995). The influence of global climatic change and regional uplift on large-mammalian evolution in east and southern Africa. In *Paleoclimate and Evolution with Emphasis on Human Origins*, ed. E. S. Vrba, G. H. Denton, T. C. Partridge and L. H. Burckle. New Haven, CT: Yale University Press, pp. 330–355.

Partridge, T. C. and Maud, R. R. (1987). Geomorphic evolution of Southern Africa since the Mesozoic. *South African Journal of Geology*, 90, 179–208.

Phillips, J. D. (2001). Inherited vs. acquired complexity in east Texas weathering profiles. *Geomorphology*, 40, 1–14.

Phillips, J. D. (2005). Weathering instability and landscape evolution. *Geomorphology*, 67, 255–272.

Ringrose, S., Harris, C., Huntsman-Mapila, P., Vink, B. W., Diskins, S., Vanderpost, C. and Matheson, W. (2009). Origins of strandline duricrusts around the Makgadikgadi Pans (Botswana Kalahari) as deduced from their chemical and isotope composition. *Sedimentary Geology*, 219, 262–279.

Rossetti, D. F. (2004). Paleosurfaces from northeastern Amazonia as a key for reconstructing paleolandscapes and understanding weathering products. *Sedimentary Geology*, 169, 151–174.

Runge, J. (1993). Lateritic crusts as climate-morphological indicators for the development of planation surfaces – possibilities and limits. *Zeitschrift für Geomorphologie, N.F., Supplementband*, 92, 201–216.

Runge, J. and Shikwati, J. (eds) (2011). *Geological Resources and Good Governance in Sub-Saharan Africa: Holistic Approaches to Transparency and Sustainable Development in the Extractive Sector*. Boca Raton: CRC Press, 292pp.

SCWG, Soil Classification Working Group (2012). Updating of the soil classification system for South Africa. Available from www.soils.org.za/scwg.html

Sobanski, R. and Valeton, I. (1994). Deep saprolitic weathering of the Miocene basalt from Triebendorf/Oberpfalz/S. Germany. In *Laterites, Palaeoweathering and Palaeosurfaces*, ed. B. J. Smith and P. A. Warke. Belfast: Eurolat, 94, pp. 47–48.

Stevenson, I. R. and McMillan, I. K. (2004). Incised valley fill stratigraphy of the Upper Cretaceous succession, proximal Orange Basin, Atlantic margin of southern Africa. *Journal of the Geological Society*, 161, 185–208.

Suttie, J. M., Reynolds, S. G. and Batello, C. (eds) (2005). *Grasslands of the World*. Rome: FAO, 495pp.

Thomas, M. F. (1994). *Geomorphology in the Tropics. A Study of Weathering and Denudation in Low Latitudes*. Chichester: John Wiley & Sons, 482pp.

Thomas, D. S. G. and Shaw, P. A. (1991). *The Kalahari Environment*. New York: Cambridge University Press, 300pp.

Turner, D. P. (2013). Perspectives on the principles and structure of the soil classification system in South Africa: Discussion and practical examples. *South African Journal of Plant and Soil*, 30, 61–68.

Twidale, C. R. (1988). Granite Landscapes. In *The Geomorphology of Southern Africa*, ed. B. P. Moon and G. F. Dardis. Johannesburg: Southern Book Publishers, pp. 198–230.

van de Graaff, W. J. E. (1983). Silcrete in Western Australia: geomorphological settings, textures, structures, and their genetic implications. In *Residual Deposits – Surface Related Weathering Processes and Materials*, ed. R. C. L. Wilson. London: Geological Society of London, Special Publications, 11, pp. 159–166.

van Huyssteen, C. W., Turner, D. P. and le Roux, P. A. L. (2013). Principles of soil classification and the future of the South African system. *South African Journal of Plant and Soil*, 30, 23–32.

Watson, A. and Nash, D. J. (1997). Desert crusts and varnishes. In *Arid Zone Geomorphology: Process, Form and Change in Drylands*, ed. D. S. G. Thomas. Chichester: John Wiley and Sons, pp. 69–107.

Wirthmann, A. (2000). *Geomorphology of the Tropics*. Heidelberg: Springer, 314pp.

Wright, P. W. (2007). Calcrete. In *Geochemical Sediments and Landscapes*, ed. D. J. Nash and S. J. McLaren. Chichester, New York: Wiley-Blackwell, pp. 10–45.

16

Karstic systems

KARIN HOLMGREN AND PAUL SHAW

Abstract

This chapter describes the karstic systems in southern Africa and explores their role in preserving records of past environmental and climatic conditions and archaeology. On longer glacial–interglacial timescales, multiproxy studies of clastic and organic cave deposits and calcium carbonate precipitates (speleothems) have provided information about the environment in which early humans lived. Speleothems also have potential to show millennial to annual-scale climate/environmental variability. The advantages of speleothems to provide precise ages, and the empirical relationships between speleothem chemistry and climate, are encouraging as we address the need for more globally dispersed terrestrial palaeoclimatic records. Future research should focus not only on obtaining more records, but also on improving the precision of data interpretation in terms of quantitative estimations of climate variables.

16.1 Introduction

Rainfall and temperature patterns over southern Africa respond to a variety of coupled ocean–atmosphere processes associated with both the Indian and Atlantic oceans. A better documentation of these patterns and changes over time is necessary for a better understanding of hydrological and geochemical processes on and beneath the land surface. Studies of deposits in southern African karstic systems, especially cave speleothems whose chemical composition is sensitive to changes in ocean–atmospheric processes, have significant potential to provide high-resolution palaeoclimatic records.

Karst topography is created by the process of solution, and is generally synonymous with areas that have calcium carbonate-rich rocks. Solution by weak carbonic acid, derived from rainfall, combined with atmospheric and soil carbon dioxide, etches downwards through fissures and joints in the bedrock to form a series of landforms, which includes dolines, karren and cave systems. A feature of all true karst is an absence of surface drainage, although streams may disappear and reappear in the landscape as changes take place in the level of phreatic water. Caves generally form in a subaquatic environment, and are then exposed to air circulation as water levels drop or land rises (Ford and Williams, 2007).

Caves provide an important repository of Cenozoic climate in southern Africa. The main elements of this record are:

1. Cave collapse, brecciation and sedimentation have preserved a number of world-famous hominid sites of palaeontological significance, such as Taung and the Sterkfontein complex (Sterkfontein/Swartkrans/Kromdraii) in South Africa;
2. Caves, both karst and non-karst, provided an important human habitat in the archaeological record, with habitation extending back to the Early Stone Age, as at Wonderwerk and Boomplas (South Africa), and Broken Hill (Zambia);
3. Caves are good archives of past climate, environmental and archaeological information through the preservation of clastic deposits that may contain pollen, bones, egg shells and other materials. The precipitation of calcium carbonate in the subaerial environment provides important palaeoenvironmental data such as oxygen and carbon isotopes, as in the case of tufa, found around springs, and in speleothems (a collective term for stalagmites, stalactites and flowstones found in caves).

The focus of this chapter is on the distribution of karst in southern Africa, and the use of caves as climatic, environmental and archaeological archives.

16.2 The distribution of karst in southern Africa

Carbonate rocks are widely distributed throughout the region (Fig. 16.1). A recent summary (Martini, 2006) identifies three main karst areas: the early Proterozoic Transvaal Supergroup extending into southeast Botswana; the late Proterozoic Cango Cave limestone of the Cape and Cenozoic limestones centred on Bredasdorp. In northern Namibia, Proterozoic limestones of Damara age, including the Otavi limestones, extend east into the Kgwebe Hills area of Botswana, while a Nama Group carbonate reef occupies the central spine of the country (Adams et al., 2004). Large areas of Zambia and parts of Mozambique are also underlain by Proterozoic limestones, which form important aquifers (De Waele and Follesa, 2003). In western Madagascar, well developed karst terrain occurs on limestone ranging in age from Jurassic to Miocene along the western length of the country.

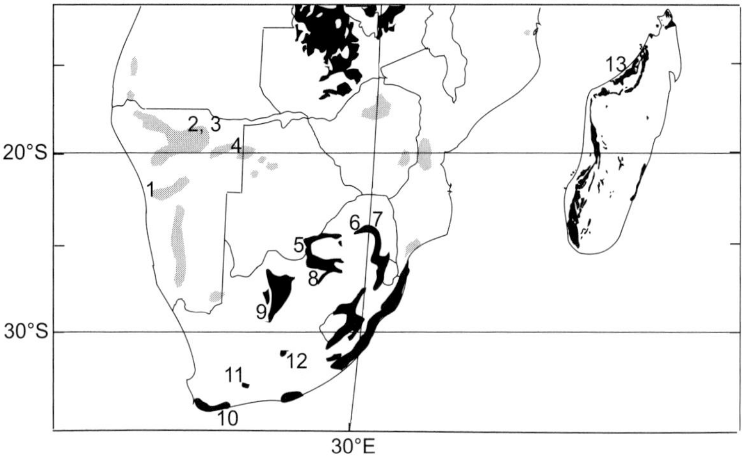

Fig. 16.1. Location of karst areas in southern Africa. Black areas: pure/continuous carbonate rocks; grey areas: impure/discontinuous carbonate rocks (after Williams, 2014). Numbers refer to caves mentioned in the text. 1: Rössing Cave; 2: Dante Cave; 3: Otavi cenotes; 4: Drotskys and Bone Caves; 5: Lobatse I, II Caves; 6: Cold Air Cave; 7: Wolkberg Cave; 8: Gladysvale Cave; 9: Gaap Escarpment, Wonderwerk Cave, Equus Cave; 10: Crevice Cave, Pinnacle Point caves; 11: Cango Cave; 12: Boomplaas Cave; 13: Anjohibe and Anjohikely Caves.

The development of karst landscapes is by no means equal across the carbonate outcrops. Martini (2006) estimates that approximately 750 caves, representing ~80% of the number of caves in South Africa, can be found in the Transvaal Supergroup. Although the number of recorded caves in Botswana, Namibia, Zambia and Mozambique is low, many are of major archaeological significance. In Madagascar, an expedition to Namokora recorded 16 caves in excess of 1 km length, with the longest, Ambatoharanana, reaching over 18 km (Laumanns and Gebauer, 1993). Surface karst features are also variable. Well-developed limestone pinnacles are a feature of Madagascar, and may be a form of palaeokarst originating from the wetter Mesozoic. In South Africa, the dominant karst topography of all three geological groups is more subdued, with features of deranged drainage such as dolines, poljes and blind valleys. Often, as in the Mpumalanga region, the relief may be covered by a thick weathered mantle (Wyberg, 1925). Cenotes (waterfilled sinkholes) are a dominant feature in northern Namibia, in particular lakes Otjikoto and Guinas, which have depths reaching 150 m and are recharged by groundwater.

16.3 Caves and speleothems as archives of the past

Caves, as closed systems, are often excellent preservers and recorders of evidence for past climate, environment and archaeology. Some of the oldest hominid specimens in the world have been found well preserved in calcified cave fillings of dolomitic bedrock, e.g. in Makapansgat Valley, along the Taung escarpment, and in the Sterkfontein complex, South Africa. These caves mostly formed through dissolution in the upper phreatic zone, and the first opening into the cave often took the form of a deep vertical shaft. Some caves in this vertical stage acted as death-traps for both humans and other fauna. Later, when the entrance was widened through further erosion and parts of the cave were filled with debris, sediments, biological remains and subsequent soil formation, the cave became suitable for human occupation (Partridge, 2000). Cave interiors tend to keep close to the regional annual average temperature, thus providing a comfortable living environment whilst the protective shelter that the cave provides, together with continuous subsurface karstification, also facilitates the preservation of human and other faunal remains.

Material deposited in caves is of three types: clastic sediments, organic deposits, and secondary calcium carbonate precipitates (speleothems). Studies of pollen, bone assemblages and stable light isotopes in organic cave deposits have provided records of past changes in moisture availability and vegetation (e.g. Van Zinderen Bakker, 1982; Beaumont et al., 1984; Scott, 1987; Scott and Thackeray, 1987; and references cited in Table 16.1). Chronological control of cave deposits and incorporated organic material is problematic in terms of resolution, continuity and time span, hampering reliable comparisons between studies and sites and therefore also hindering correlations and regional synthesis. Vrba (1974, 1975, 1995) used three phases of bovid evolution associated with *Australopithecine* remains as a temporal and ecological framework over 2 Ma. More recently, a range of radiometric dating techniques has been used, with varying degrees of success, to date both inorganic and organic deposits. At Wonderwerk Cave, palaeomagnetic, cosmogenic and optically stimulated luminescence (OSL) ages placed the onset of the Acheulian culture to ~1.6 Ma (Chazan et al., 2008), while electron spin resonance (ESR) dating of hominid teeth from Swartkrans suggests a maximum age of 2.1 Ma (Curnoe et al., 2001). The dating of the complete Sterkfontein *Australopithecene* fossil StW 573 (Little Foot) has been put back from ~2.2 Ma to ~3 Ma on the basis of stratigraphic re-interpretation of U-series-dated flowstones in the surrounding breccia, which itself has been strongly affected by post-depositional karstification (Bruxelles et al., 2014). Multiple dating techniques applied to beach, aeolian and human deposits and interspersed speleothems in cave remnants at Pinnacle Point, Mossel Bay (Western Cape), show that coastal caves suitable for humans existed prior to 1.1 Ma and that cycles of cave development, infilling and erosion reveal sea-level highstands at around 1.1 Ma and 0.4 Ma (Pickering et al., 2013).

By contrast to the clastic and organic cave deposits, speleothems can provide precisely U-series-dated continuous high-resolution records of climate and

Table 16.1 *Palaeoclimatic/environmental non-speleothem based cave studies published in the southern Africa region from 1991 onwards. Map ID refers to Fig. 16.1, LGM = last glacial maximum*

Country	Cave	Map ID	Coordinates	Climate/ Environmental Record	Dating	Source
South Africa	Wonderwerk Cave	9	S27°50'E23°33'	Stable isotopes in grazer tooth enamel indicate dominance of C_4 grassland throughout the Holocene and relative moist conditions at 4-2 kyr BP.	^{14}C-dates of Holocene age	Thackeray and Lee-Thorp, 1992; Lee-Thorp and Talma, 2000
South Africa	Equus Cave	9	S27°37'E24°37'	Pollen, isotopes in tooth enamels and bone assemblages show that parts of Pleistocene was cooler and possibly moister than Holocene. C_4 grasses were dominant over the last 30 kyr arguing against a total shift from summer to winter rainfall regime during the LGM.	^{14}C-dates of Pleistocene and Holocene age	Klein et al., 1991; Lee-Thorp and Beaumont, 1995
South Africa	Equus Cave	9	S27°37'E24°37'	Stable isotopes in ostrich egg shells indicate increasing moisture from 17–6 kyr BP and stable levels since then, minimum temperatures at 17–14 kyr BP and maximum temperatures during the late Holocene.	Amino acid racemisation and ^{14}C-dates	Johnson et al., 1997

South Africa	Boomplaas Cave	12	S31°E25°	Changes in ungulate assemblages indicate an expansion of grassland and increased winter rainfall from >65 kyr BP, with a peak around 24–18 kyr BP and a marked grassland/moisture decrease at the Pleistocene/Holocene transition.	^{14}C-dates of Pleistocene and Holocene age	Faith, 2013
Madagascar	Anjohibe and Anjohikely Caves	13	S15°33'E46°53'	Charcoal composition in clastic deposits indicates a decline in forest and increase in grassland in late Holocene.	^{14}C-dates of Holocene age	Burney et al., 1997

environments, albeit over shorter time spans (Fairchild and Baker, 2012). Speleothems are formed by calcite/aragonite precipitation from carbonate-rich seepage water entering dolomite or limestone caves (Figs. 16.2, 16.3), with formation generally favoured by humid and warm conditions, high soil biologic activity and a closed cave system. Speleothem growth can be interrupted by dry and/or cold conditions with limited soil biologic activity. Stalagmites are the most favourable speleothem archive on account of their structure. They are built up from the cave floor by precipitation from ceiling drip water and commonly form distinct growth layers, sometimes visible to the naked eye. The chemical composition and the thickness of the growth layers are dependent on climatic and environmental conditions, such as atmospheric circulation patterns, rainfall amount, temperature, vegetation type, soil thickness and soil biologic activity. These conditions can be elucidated by a number of different analyses of the stalagmite, such as stable oxygen ($\delta^{18}O$) and carbon ($\delta^{13}C$) isotope analysis, analysis of trace elements, luminescence, laminae thickness and calcite fabrics (Fairchild and Baker, 2012).

Fig. 16.2. Examples of structures within the Sterkfontein Caves, Cradle of Humankind (photos reproduced courtesy of Dominic Stratford). A black and white version of this figure will appear in some formats. For the colour version, please refer to the plate section.

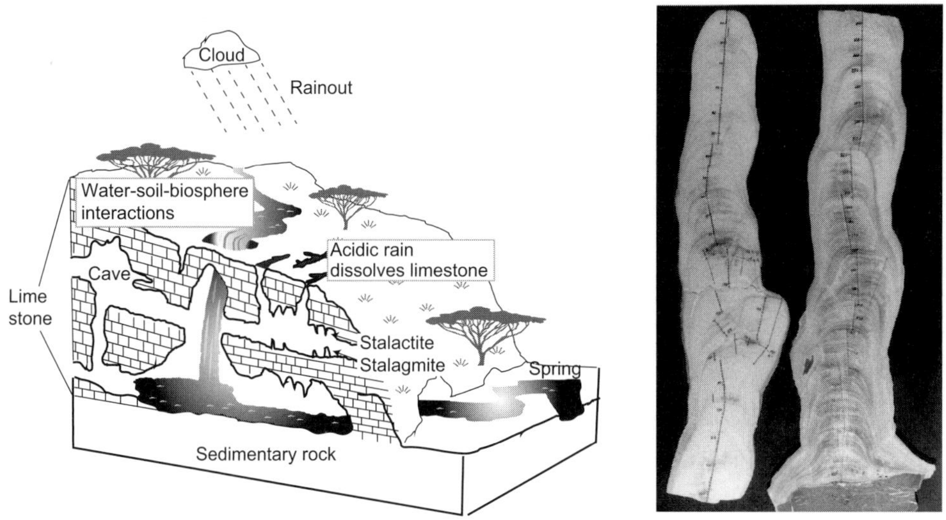

Fig. 16.3. Left: Principles behind cave and speleothem formation. Acidic rainwater percolates through cracks and fissures dissolving limestone bedrock before reaching the water table. Caves usually form in the saturated zone. If the seepage-water entering a cave in the aerated zone is enough rich in carbonate, the calcite may reprecipitate forming speleothems. Right: A typical stalagmite when sectioned has parts of clearly visible growth layers and parts with less visible layers. This 1 m long and 7000 year old stalagmite comes from Cold Air Cave, Makapansgat valley, South Africa.

The first attempt in southern Africa to extract palaeoclimatic data from speleothems was on Drotskys Cave, in the Kwihabe Hills of northwest Botswana (Cooke, 1975, 1984; Cooke and Verhagen, 1977), soon followed by a number of studies on ^{14}C-dated speleothem growth periods in caves across the region (e.g. Shaw and Cooke, 1986; Butzer et al., 1978; Brook, 1982; Burney et al., 1994, 1997; Brook et al., 1996, 1997, 1999a). At the time, these cave studies represented a revolutionary approach to palaeoenvironmental research, which had previously relied on lakes and other aquatic archives. The methodology was rapidly refined as the issues associated with ^{14}C dating became apparent, including open-system behaviour, limited time range and need for calibration (Bard et al., 1990). U/Th dating was widely adopted, with an extension of the records to ~300 kyr BP. However, the humid periods dated were a function of sampling strategy, not of climate, and frequently age groupings were produced by visual correlation, not by statistical analysis.

16.4 High-resolution palaeoenvironmental time-series from speleothems

The first stable isotope proxy temperature and vegetation record was based on ^{14}C-dated speleothems from Wolkberg Cave, South Africa (Talma et al., 1974). Subsequent stable oxygen and carbon isotopes studies rely on precise U/Th dating techniques (although the precision varies among the studies, depending on the quality of the material) and are now available from a number of caves in southern Africa. Other studies have focused on the identification of annual growth layers, micromorphology, mineralogy and trace elements. Table 16.2 summarises these studies and their major results.

Some longer timescale studies have utilised U-series-dated speleothem deposition in combination with isotopic analysis and/or sediment studies of clastic layers interspersed with speleothems. In dry Namibia, speleothem deposition took place during interglacials (marine isotope stages (MIS) 11, 9, 7 and 5), indicating global synchrony of these large-scale climatic events (Geyh and Heine, 2014). In Gladysvale Cave, in a relatively wetter region of South Africa, speleothem deposition occurred during both interglacials and glacials (Pickering et al., 2007) (Table 16.2).

Most speleothem records cover parts of the late Pleistocene and Holocene and provide evidence of regional changes in rainfall, temperature and vegetation regimes. For example, and by contrast to many other palaeoclimatic studies and syntheses from southern Africa (e.g. Talma and Vogel, 1992; Tyson and Preston-Whyte, 2000; Chase and Meadows, 2007; Faith, 2013), Pleistocene stalagmite records from Crevice Cave, South Africa, suggest that the summer rainfall zone expanded southwards during glacials (Bar-Matthews et al., 2010). This and other detailed speleothem studies have also contributed to discussion of the role of climate/environmental factors on both early hominid evolution (Hopley et al., 2007; Bar-Matthews et al., 2010) and historic societal developments (Tyson et al., 2002; Holmgren and Öberg, 2006; Sletten et al., 2013).

Speleothems from the Cold Air Cave (CAC) in Makapansgat Valley (Limpopo Province, South Africa) provided the first near-continuous information on climate/environmental evolution over the last 24 kyr (Lee-Thorp et al., 2001; Holmgren et al., 2003) (Table 16.2, Fig. 16.4a). By comparing the stalagmite stable oxygen isotope record with the fossil groundwater record, the regional mean temperature difference between the Holocene and Pleistocene was calculated to nearly 6ºC, which is in the same range as suggested by earlier studies (Talma and Vogel, 1992). Postglacial warming was initiated around 17 kyr BP; a hiatus between 12–10 kyr BP may indicate dry conditions followed by warm and mostly dry conditions in the early Holocene. Within the Holocene, the Little Ice Age period was clearly shown as an important cold and dry climate anomaly in southern Africa, with a peak at AD 1700 when temperatures were estimated to be ~1°C lower than the 1960–1990 average (Holmgren et al., 2001). Pronounced cycles of climate variability in the ranges of 4.0–2.5 kyr, 1.0 kyr, multidecadal, and interannual timescales, were identified and proposed to reflect solar cycles,

Table 16.2 *Detailed studies of speleothems published from southern Africa from 1994 onwards. Map ID refers to Fig. 16.1, MIS = marine isotope stage, LGM = last glacial maximum, LIA = Little Ice Age*

Country	Cave	Map ID	Coordinates	Record	Length of record (kyr)	Continuous record	Climate interpretation	Source
Namibia	Rössing Cave	1	S22°31' E14°47'	Growth periods, determined by Th/U-dates	420-121	no	Three (possibly four) wet periods identified by speleothem growth at 420-385 kyr; (ca 300 kyr); 207-229 kyr; 118-121 kyr, corresponding to MIS 11, 9, (7), and 5, respectively.	Geyh and Heine, 2014
Namibia	Dante Cave	2	S19°24' E17°53'	$\delta^{18}O$, $\delta^{13}C$, petrography, layer thickness	4.6	yes	Variable but generally wetter conditions inferred from depleted isotopes and fast growth at 4.6-3.3 kyr and 1.8 kyr-present. Drier conditions inferred by enriched isotopes and slow growth 3.3-1.8 kyr.	Sletten et al., 2013
Botswana	Drotskys Cave	4	S18°44' E21°42'	Petrography, annual layers	1.5	yes	Humidity conditions inferred from variable thickness of annual calcite/aragonite layers in a ^{14}C-dated stalagmite.	Railsback et al., 1994
Botswana	Lobatse II Cave	5	S25°13' E25°41'	$\delta^{18}O$, $\delta^{13}C$	51-21	no	Warmer, wetter conditions inferred by enriched isotopes at 51-43 kyr; drier, colder conditions by depleted isotopes: at 27-21 kyr.	Holmgren et al., 1995
South Africa	Cold Air Cave	6	S24°0' E25°36'	$\delta^{18}O$, $\delta^{13}C$, grey scale, trace elements, Temp-reconstruction for last 350 yrs	24.4	almost	LGM interpreted to be ~6°C cooler and LIA ~1.4°C cooler than present inferred from oxygen isotopes. Postglacial warming was initiated at 17 kyr. Early Holocene was warm and dry, mid-Holocene cooler, Medieval times variable but generally warmer, wetter and LIA cooler, drier.	Lee-Thorp et al., 2001; Finch et al., 2003; Holmgren et al., 2003; Sundqvist et al., 2013

Table 16.2 (*cont.*)

Country	Cave	Map ID	Coordinates	Record	Length of record (kyr)	Continuous record	Climate interpretation	Source
South Africa	Wolkberg Cave	7	S24°06' E29°53'	Mineralogy, $\delta^{18}O$, $\delta^{13}C$	58-46, at ca 40	no	Progressive drier and colder conditions between 58-46 kyr inferred from progressive enrichment in $\delta^{13}C$ and depletion in $\delta^{18}O$.	Holzkämper et al., 2009
South Africa	Gladysvale Cave	8	S25°70' E27°30'	Flow stones and clastic sediments: $\delta^{18}O$, $\delta^{13}C$	570-7	no	Wetter conditions inferred from flowstone growth at around 578, 393, 308, 255, 57, 43, 16-14 and 10-7 kyr, corresponding to both interglacial and glacial times; interspersed dry conditions are inferred from clastic deposits. Vegetation shifted from more C_4 vegetation, indicating drier conditions, in early Pleistocene to more C_3 vegetation and wetter in mid-Pleistocene.	Pickering et al., 2007
South Africa	Wonderwerk Cave	9	S27°50' E23°33'	Stalagmite and dung deposits; pollen, $\delta^{18}O$, $\delta^{13}C$, petrography	35	no	Age model is based on ^{14}C-datings of speleothems and must be viewed with caution. Wetter conditions inferred from pollen composition and depleted isotopes at 33, 23-17 and 4 kyr-present; drier conditions inferred from microstromatolitic carbonate at 17-13 kyr and depositional hiatuses at 33-23, and 13-4 kyr.	Brook et al., 2010

South Africa	Crevice Cave	10	S34°12' E22°05'	$\delta^{18}O$, $\delta^{13}C$	90-53	yes	Stronger winter rain, more C_3 grasses inferred from depleted isotopes at ca 90-84, and 67-50 kyr. Weaker winter rain, more C_4 grasses inferred by enriched isotopes at ca 83-67 kyr.	Bar-Matthews et al., 2010
South Africa	Cango Cave	11	S33°23' E22°13'	$\delta^{18}O$, $\delta^{13}C$	50	no	LGM interpreted to be ~6°C cooler than present inferred from oxygen isotopes in stalagmite and confined groundwater. Increased C_4 vegetation during late Holocene is inferred by enriched carbon isotopes.	Talma and Vogel, 1992
Madagascar	Anjohibe Cave	13	S16°44' E47°07'	Annual layers, petrography	Holocene, but only the last 450 years are interpreted	no	Rainfall amount is reflected by layer thickness. High rainfall inferred from AD 1550-2000 correlates with El Nino events.	Brook et al., 1999b

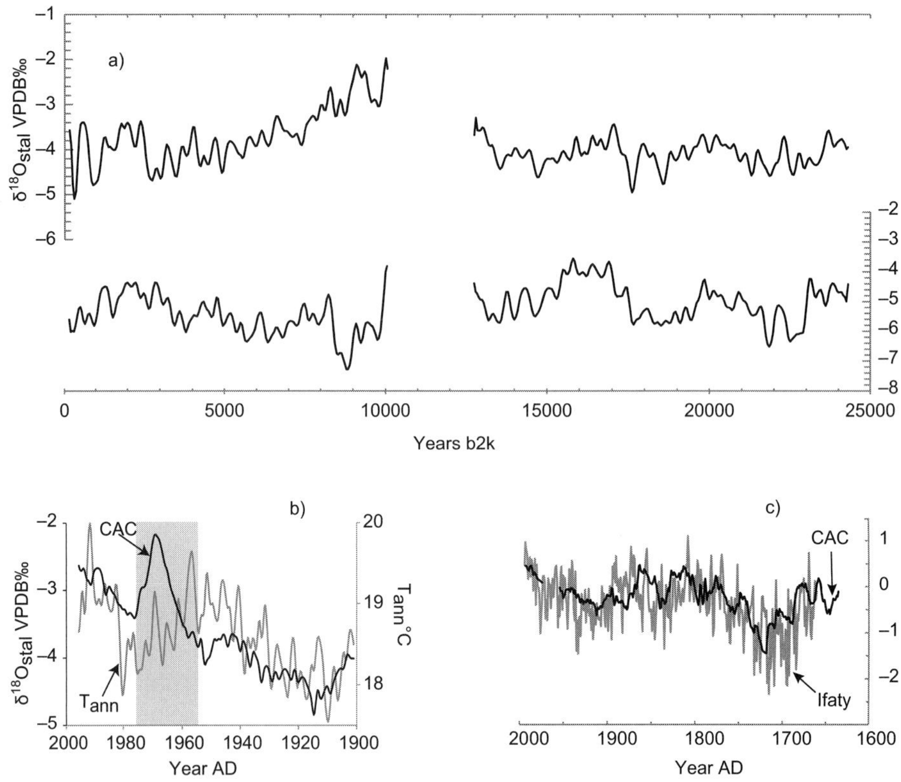

Fig. 16.4. The Cold Air Cave (CAC), Makapansgat Valley, stable isotope stalagmite record compared to air temperatures and Madagascar coral SST reconstruction. (a) Stalagmite T8 covers the last 24.4 kyr. Enriched $\delta^{18}O$ values are interpreted to indicate generally warmer climate and depleted values generally colder conditions. Enriched $\delta^{13}C$ values are interpreted to indicate good C_4 grass cover (generally wetter) and depleted values sparse C_4 grass cover (generally drier) (after Holmgren et al., 2003). (b) Stalagmite T7 high resolution $\delta^{18}O$ record compared with annual mean temperature (CRU TS3.1) for the period 1901–1995. Low correlation is observed at 1956–1976 (shaded interval) indicating that local factors affected the stalagmite T7 record at this time (after Sundqvist et al., 2013). (c) Stalagmite T7 high resolution air temperature reconstruction (1956–1976 excluded), covering the last 350 years (Sundqvist et al., 2013) compared with the Ifaty coral composite SST reconstruction (Zinke et al., 2014).

El Niño cycles, sea surface temperature (SST) changes and variations in the Indian Ocean Dipole (Tyson et al., 2002; Holmgren et al., 2003; Sundqvist et al., 2013), in turn affecting the Hadley circulation, the position of the intertropical convergence zone and El Niño-scale teleconnections.

Since the isotope composition in speleothems may be driven by various processes, alternative interpretations of the CAC record have also been proposed

(Brook et al., 2010; Chase et al., 2012; Stager et al., 2013). The hypothesis that the CAC $\delta^{18}O$ signal mainly reflects regional surface temperatures (and humidity to some extent) was supported in a recent study yielding a 350-year-long, near annually resolved $\delta^{18}O$ stalagmite record from CAC (Sundqvist et al., 2013). The temperature reconstruction, based on correlation between regional surface temperatures and stalagmite $\delta^{18}O$ variations, showed that the coldest period occurred

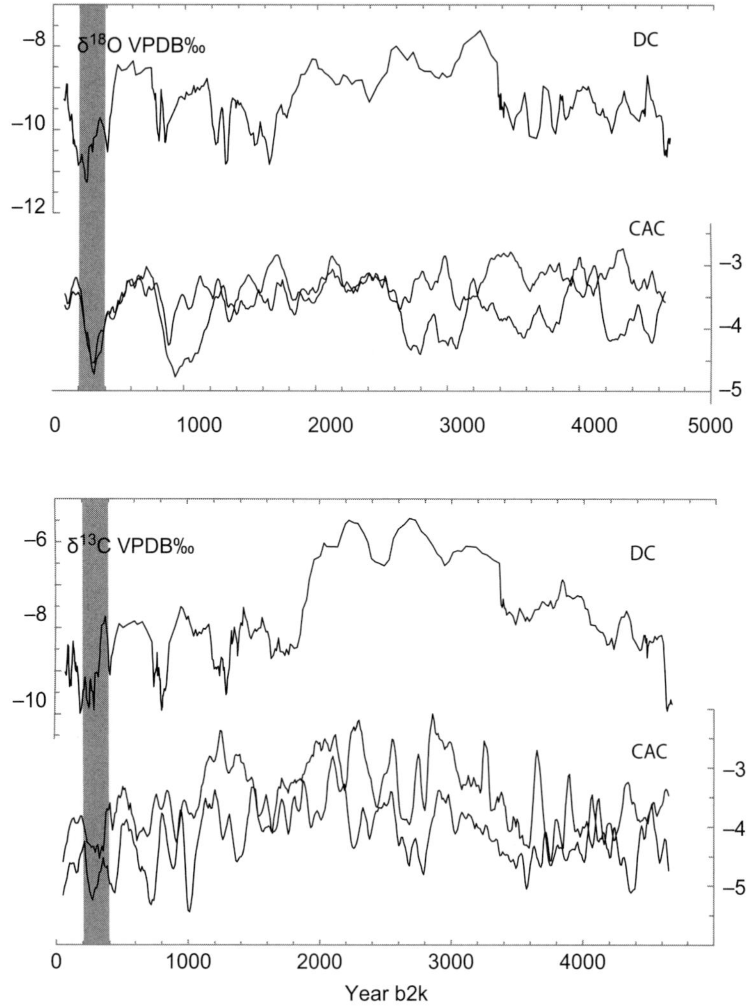

Fig. 16.5. The stable isotope record of Dante Cave (DC) (Sletten et al., 2013) compared with the two CAC stalagmite records available for the same time interval (Holmgren et al., 2003). Top panel: $\delta^{18}O$ record, bottom panel: $\delta^{13}C$ record. In the case of Dante Cave, enriched values indicate drier conditions, whereas in CAC enriched values indicate warmer, wetter conditions. Shaded area marks the time period corresponding to the Little Ice Age.

between AD 1690 and 1740, slightly lagging the solar Maunder Minimum (AD 1645–1710), and that parts of this period could have been as much as 1.4°C colder than today (Fig. 16.4b). This high-resolution CAC record agrees well with a coral-based temperature record southwest of Madagascar (Zinke et al., 2004, 2014) (Fig. 16.4c), which points at a dominant role of the Agulhas Current system for Indian Ocean SST multidecadal variability.

Similarly detailed studies carried out on a late Holocene stalagmite from Dante Cave in Namibia (Sletten et al., 2013), found that the stable carbon and oxygen isotope stalagmite record and changes in mineralogy were mainly driven by changes in humidity. Comparison of the isotopic signatures between Dante Cave and CAC shows both similarities and differences (Fig. 16.5). A similar isotopic signature can be found at around AD 1650–1800, corresponding to the Little Ice Age, indicating colder, drier conditions in northeastern South Africa, but wetter in Namibia. This result needs to be tested further and cautions against deterministic interpretations of $\delta^{18}O$ variations in low-latitude stalagmites, and suggests that the differences can be utilised for better understanding of regional climate driving processes and their large-scale teleconnections.

16.5 Summary

Karstic systems are important in preserving the history of climate, environment and cave-dwelling humans in southern Africa. Whilst these systems are fairly well documented and understood in South Africa, much remains to be done in neighbouring countries. The development of precise analytical techniques for dating and interpreting various proxies within speleothems has led to an enormous increase in studies globally, but much more can be done in southern Africa given the large number of caves with speleothems. However, there is also a need for better interpretation of the various proxies, preferably in terms of quantifiable climate variables. The combination of well-dated, high-resolution proxy data from stalagmites and isotope modelling can help to understand the environmental and climatic controls on speleothem formation, including long term precessional forcing as well as atmosphere-ocean coupling that give rise to SST variations and subsequent rainfall patterns.

References

Adams, E. W., Schroder, S., Grotzinger, J. P. and McCormick, D. S. (2004). Digital reconstruction and stratigraphic evolution of a microbial-dominated isolated carbonate platform (Terminal Proterozoic, Nama Group, Namibia). *Journal of Sedimentary Research*, 74, 479–497.

Bar-Matthews, M., Marean, C. W., Jacobs, Z., Karkanas, P., Fisher, E. C., Herries, A. I. R., Brown, K., Williams, H. M., Bernatchez, J., Ayalon, A. and Nilssen, P. J. (2010). A high resolution and continuous speleothem record of paleoclimate and paleoenvironment from 90 to 53 ka from Pinnacle Point on the south coast of South Africa. *Quaternary Science Reviews*, 29, 2131–2145.

Bard, E., Hamelin, B., Fairbanks, R. G. and Zindler, A. (1990). Calibration of the ^{14}C timescale over the past 30,000 years using mass spectrometric U–Th ages from Barbados corals. *Nature*, 345, 405–410.

Beaumont, P. B., Van Zinderen Bakker, E. M. and Vogel, J. C. (1984). Environmental changes since 32000 BP at Kathu Pan, northern Cape. In *Late Cainozoic Palaeoclimates of the Southern Hemisphere*, ed. J. C. Vogel. Rotterdam: Balkema, pp. 329–338.

Brook, G. A. (1982). Stratigraphic evidence of Quaternary climatic change at Echo Cave, Transvaal, and a paleoclimatic record for Botswana and northeastern South Africa. *Catena*, 9, 343–351.

Brook, G. A., Cowart, J. B., Brandt, S. A. and Scott, L. (1997). Quaternary climatic change in southern and eastern Africa during the last 300 ka: The evidence from caves in Somalia and the Transvaal region of South Africa. *Zeitschrift für Geomorphologie, Supplementband*, 108, 15–48.

Brook, G. A., Cowart, J. B. and Marais, E. (1996). Wet and dry periods in the southern African summer rainfall zone during the last 300 kyr from speleothem, tufa and sand dune age data. *Palaeoecology of Africa*, 24, 147–158.

Brook, G. A., Marais, E. and Cowart, J. B. (1999a). Evidence of wetter and drier conditions in Namibia from tufas and submerged speleothems. *Climbebasia*, 15, 29–39.

Brook, G. A., Rafter, M. A., Railsback, L. B., Shen, S.-W. and Lundberg, J. (1999b). A high-resolution proxy record of rainfall and ENSO since AD 1550 from layering in stalagmites from Anjohibe Cave, Madagascar. *The Holocene*, 9, 695–705.

Brook, G. A., Scott, L., Railsback, L. B. and Goddard, E. A. (2010). A 35 ka record of pollen and isotope record of environmental change along the southern margin of the Kalahari from a stalagmite and animal dung deposits in Wonderwerk Cave, South Africa. *Journal of Arid Environments*, 74, 870–884.

Bruxelles, L., Clarke, R. J., Maire, R., Ortega, R. and Stratford, D. (2014). Stratigraphic analysis of the Sterkfontein StW 573 Australopithecus skeleton and implications for its age. *Journal of Human Evolution*, 70, 36–48.

Burney, D. A., Brook, G. A. and Cowart, J. B. (1994). A Holocene pollen record for the Kalahari Desert of Botswana from a U-series dated speleothem. *The Holocene*, 4, 225–232.

Burney, D. A., James, H. F., Grady, F. V., Rafamantanantsoa, J.-G., Ramilisonina, Wright, H. T. and Cowart, J. B. (1997). Environmental change, extinction and human activity: Evidence from caves in NW Madagascar. *Journal of Biogeography*, 24, 755–767.

Butzer, K. W., Stuckenrath, R., Bruzewicz, A. and Helgren, D. M. (1978). Late Cenozoic paleoclimates of the Gaap Escarpment, Kalahari margin, South Africa. *Quaternary Research*, 10, 310–339.

Chase, B. and Meadows, M. E. (2007). Late Quaternary dynamics of southern Africa's winter rainfall zone. *Earth-Science Reviews*, 84, 103–138.

Chase, B. M., Scott, L., Meadows, M. E., Gil-Romera, G., Boom, A., Carr, A. S., Reimer, P. J., Truc, L., Valsecchi, V. and Quick, L. J. (2012). Rock hyrax middens: A palaeoenvironmental archive for southern African drylands. *Quaternary Science Reviews*, 56, 107–125.

Chazan, M., Ron, H., Matmon, A., Porat, N., Goldberg, P., Yates, R., Avery, M., Sumner, A. and Horwitz, L. K. (2008). Radiometric dating of the Earlier Stone Age sequence in Excavation 1 at Wonderwerk Cave, South Africa: Preliminary results. *Journal of Human Evolution*, 55, 1–11.

Cooke, H. J. (1975). The palaeoclimatic significance of caves and adjacent landforms in the Kalahari of western Ngamiland, Botswana. *Geographical Journal*, 141, 430–444.

Cooke, H. J. (1984). The evidence from northern Botswana of climate change. In *Late Cainozoic Palaeoclimates of the Southern Hemisphere*, ed. J. C. Vogel. Rotterdam: Balkema, pp. 265–278.

Cooke, H. J. and Verhagen. B. Th. (1977). The dating of cave development – an example from Botswana. *Proceedings of the 7th International Speleological Congress, Sheffield*. Sheffield, UK, pp. 122–124.

Curnoe, D., Grün, R., Taylor, L. and Thackeray, F. (2001). Direct ESR dating of a Pliocene hominin from Swartkrans. *Journal of Human Evolution*, 40, 379–391.

De Waele, J. and Follesa, R. (2003). Human impact on karst: The example of Lusaka (Zambia). *International Journal of Speleology*, 32, 71–83.

Fairchild, I. J. and Baker, A. (2012). *Speleothem Sciences: From Process to Past Environments*. Chichester: Wiley–Blackwell, 432pp.

Faith, J. T. (2013). Taphonomic and paleoecological change in the large mammal sequence from Boomplaas Cave, western Cape, South Africa. *Journal of Human Evolution*, 65, 715–730.

Finch, A. A., Shaw, P. A., Holmgren, K. and Lee-Thorp, J. (2003). Corroborated rainfall records from aragonitic stalagmites. *Earth and Planetary Science Letters*, 215, 265–273.

Ford, D. and Williams, P. (2007). *Karst Hydrology and Geomorphology* (3rd Ed). London: Unwin, 576pp.

Geyh, M. A. and Heine, K. (2014). Several distinct wet periods since 420 ka in the Namib Desert inferred from U-series dates of speleothems. *Quaternary Research*, 81, 381–391.

Holmgren, K., Karlén, W. and Shaw, P. A. (1995). Paleoclimatic significance of the stable isotopic composition and petrology of a late Pleistocene stalagmite from Botswana. *Quaternary Research*, 43, 320–328.

Holmgren, K., Lee-Thorp, J. A., Cooper, G. R. J., Lundblad, K., Partridge, T. C., Scott, L., Sithaldeen, R., Talma, A. S. and Tyson, P. D. (2003). Persistent millennial-scale climatic variability over the past 25,000 years in Southern Africa. *Quaternary Science Reviews*, 22, 2311–2326.

Holmgren, K., Moberg, A., Svanered, O. and Tyson, P. D. (2001). A preliminary 3000-year regional temperature reconstruction for South Africa. *South African Journal of Science*, 97, 49–51.

Holmgren, K. and Öberg, H. (2006). Climate change in Southern and East Africa during the past millennium and its implications for societal development. *Environment, Development and Sustainability*, 8, 185–195.

Holzkämper, S., Holmgren, K., Lee-Thorp, J., Talma, S., Mangini, A. and Partridge, T. (2009). Late Pleistocene stalagmite growth in Wolkberg Cave, South Africa. *Earth and Planetary Science Letters*, 282, 212–221.

Hopley, P. J., Marshall, J. D., Weedon, G. P., Latham, A. G., Herries, A. I. R. and Kuykendall, K. L. (2007). Orbital forcing and the spread of C_4 grasses in the late Neogene: Stable isotope evidence from South African speleothems. *Journal of Human Evolution*, 53, 620–634.

Johnson, B. J., Miller, G. H., Fogel, M. L. and Beaumont, P. B. (1997). The determination of late Quaternary palaeoenvironments at Equus Cave, South Africa, using stable isotopes and amino acid racemization in ostrich eggshell. *Palaeogeography, Palaeoclimatology, Palaeoecology*, 136, 121–137.

Klein, R. G., Cruz-Uribe, K. and Beaumont, P. B. (1991). Environmental, ecological, and paleoanthropological implications of the Late Pleistocene mammalian fauna from Equus Cave, northern Cape Province, South Africa. *Quaternary Research*, 36, 94–119.

Laumanns, M. and Gebauer, H. D. (1993). Namoroka 1992: Expedition to the karst of Namoroka and Narinda, Madagascar. *The International Caver*, 6, 30–36.

Lee-Thorp, J. A. and Beaumont, P. B. (1995). Vegetation and seasonality shifts during the Late Quaternary deduced from $^{13}C/^{12}C$ ratios of grazers at Equus Cave, South Africa. *Quaternary Research*, 43, 426–432.

Lee-Thorp, J. A. and Talma, A. S. (2000). Stable light isotopes and environments in the southern African Quaternary and late Pliocene. In *The Cenozoic of Southern Africa*, eds. T. C. Partridge and R. R. Maud. Oxford: Oxford University Press, pp. 236–251.

Lee-Thorp, J. A., Holmgren, K., Lauritzen, S.-E., Linge, H., Moberg, A., Partridge, T. C., Stevenson, C. and Tyson, P. D. (2001). Rapid climate shifts in the southern African interior throughout the mid to late Holocene. *Geophysical Research Letters*, 28, 4507–4510.

Martini, J. E. J. (2006). Karsts and Caves. In *The Geology of South Africa*, eds. M. R. Johnson, C. R. Anhaeusser and R. J. Thomas. Pretoria: Geological Society of South Africa/Council for Geoscience, pp. 661–668.

Partridge, T. C. (2000). Hominid-bearing cave and tufa deposits. In *The Cenozoic of Southern Africa*, eds. T. C. Partridge and R. R. Maud. Oxford: Oxford University Press, pp. 100–125.

Pickering, R., Hancox, P. J., Lee-Thorp, J. A., Grün, R., Mortimer, G. E., McCulloch, M. and Berger, L. R. (2007). Stratigraphy, U-Th chronology, and paleoenvironments at Gladysvale Cave: insights into the climatic control of South African hominin-bearing cave deposits. *Journal of Human Evolution*, 53, 602–619.

Pickering, R., Jacobs, Z., Herries, A. I. R., Karkanas, P., Bar-Matthews, M., Woodhead, J. D., Kappen, P., Fischer, E. and Marean, C. W. (2013). Paleoanthropologically significant South African sea caves dated to 1.1–1.0 million years using a combination of U–Pb, TT-OSL and palaeomagnetism. *Quaternary Science Reviews*, 65, 39–52.

Railsback, L. B., Brook, G. A., Chen, J., Kalin, R. and Fleisher, C. J. (1994). Environmental controls on the petrology of a late Holocene speleothem from Botswana with annual layers of aragonite and calcite. *Journal of Sedimentary Research, Section A*, 64, 147–155.

Scott, L. (1987). Pollen analysis of hyena coprolites and sediments from Equus Cave, Taung, southern Kalahari (South Africa). *Quaternary Research*, 28, 144–156.

Scott, L. and Thackeray, J. F. (1987). Multivariate analysis of late Pleistocene and Holocene pollen spectra from Wonderkrater, Transvaal, South Africa. *South African Journal of Science*, 83, 93–98.

Shaw, P. A. and Cooke, H. J. (1986). Geomorphic evidence for the Late Quaternary paleoclimates of the Middle Kalahari of northern Botswana. *Catena*, 13, 349–359.

Sletten, H. R., Railsback, L. B., Liang, F., Brook, G. A., Marais, E., Hardt, B. F., Cheng, H. and Edwards, R. L. (2013). A petrographic and geochemical record of climate change over the last 4600 years from a northern Namibia stalagmite, with evidence of abruptly wetter climate at the beginning of southern Africa's Iron Age. *Palaeogeography, Palaeoclimatology, Palaeoecology*, 376, 149–162.

Stager, J. C., Ryves, D. B., King, C., Madson, J., Hazzard, M., Neumann, F. H. and Maud, R. (2013). Late Holocene precipitation variability in the summer rainfall region of South Africa. *Quaternary Science Reviews*, 67, 105–120.

Sundqvist, H. S., Holmgren, K., Fohlmeister, J., Zhang, Q., Bar-Matthews, M., Spötl, C. and Körnich, H. (2013). Evidence of a large cooling between 1690 and 1740 AD in southern Africa. *Scientific Reports*, 3, 1767, doi:10.1038/srep01767.

Talma, A. S. and Vogel, J. C. (1992). Late Quaternary paleotemperatures derived from a speleothem from Cango Caves, Cape Province, South Africa. *Quaternary Research*, 37, 203–213.

Talma, A. S., Vogel, J. C. and Partridge, T. C. (1974). Isotopic contents of some Transvaal speleothems and their palaeoclimatic significance. *South African Journal of Science*, 70, 135–140.

Thackeray, J. F. and Lee-Thorp, J. A. (1992). Isotopic analysis of equid teeth from Wonderwerk Cave, northern Cape province, South Africa. *Palaeogeography, Palaeoclimatology, Palaeoecology*, 99, 141–150.

Tyson, P. D., Lee-Thorp, J., Holmgren, K. and Thackeray, J. F. (2002). Changing gradients of climate change in southern Africa during the past millennium: Implications for population movements. *Climatic Change*, 52, 129–135.

Tyson, P. D. and Preston-Whyte, R. A. (2000). *The Weather and Climate of Southern Africa*. Cape Town: Oxford University Press, 408pp.

Van Zinderen Bakker, E. M. (1982). Pollen analytical studies of the Wonderwerk Cave, South Africa. *Pollen et Spores*, 24, 235–250.

Vrba, E. S. (1974). Chronological and ecological implications of the fossil Bovidae at the Sterkfontein Australopithecine site. *Nature*, 250, 19–23.

Vrba, E. S. (1975). Some evidence of chronology and palaeoecology of Sterkfontein, Swartkrans and Kromdraii from the fossil Bovidae. *Nature*, 254, 301–304.

Vrba. E. S. (1995). The fossil record of African Antelopes (Mammalia, Bovidae) in relation to human evolution and palaeoclimate. In *Palaeoclimate and Evolution, with Emphasis on Human Origins*, eds. E. S. Vrba, G. H. Denton, T. C. Partridge and L. H. Burckle. New Haven CT: Yale University Press, pp. 385–424.

Williams, P. W. (2008). *World Heritage Caves and Karst: A Thematic Study*. World Heritage Convention, IUCN World Heritage Studies, no 2. Gland, Switzerland: IUCN, 57pp.

Wyberg, W. J. (1925). *Economic geology of Sable and Pilgrim's rest*. Memoirs, Geological Survey South Africa, 23, 124pp.

Zinke, J., Dullo, W. C., Heiss, G. A. and Eisenhauer, A. (2004). ENSO and Indian Ocean subtropical dipole variability is recorded in a coral record of southwest Madagascar for the period 1659 to 1995. *Earth and Planetary Science Letters*, 228, 177–194.

Zinke, J., Loveday, B. R., Reason, C. J. C., Dullo, W.-C. and Kroon, D. (2014). Madagascar corals track sea surface temperature variability in the Agulhas Current core region over the past 334 years. *Scientific Reports*, 4, 4393, doi:10.1038/srep04393.

17

Terrestrial ecosystem changes in the late Quaternary

MICHAEL E. MEADOWS AND LYNNE J. QUICK

Abstract

This chapter provides a perspective on vegetation changes (as a proxy for overall terrestrial ecosystem changes) within southern Africa's summer and winter rainfall zones. While climate is a major determinant on southern African ecosystems, topography, geology and fire also play critical roles in ecosystem pattern and processes. Late Quaternary climate changes impacted on southern African ecosystems but did not result in wholesale migration of these biomes. Instead, the palaeoenvironmental record reveals that community dynamics *within* biomes or at their ecotonal boundaries changed as individual species were favoured or disadvantaged by changing conditions (e.g. climate, fire regimes and/or human impacts) over time.

17.1 Introduction

In determining the impact that late Quaternary climate changes had on southern African ecosystems, it is essential to consider the range of different ecosystem drivers. Although climate is perceived as 'the principal dynamic component' (Schulze, 1997, p21) shaping vegetation patterns, especially at the biome scale (~10^4–10^7 km^2), ecosystem sensitivity to climate is mediated by other environmental variables and by human activities. For example, the impact of fire has varied through the Quaternary in response to climate and human influence (Daniau et al., 2013). Some environmental factors (fire and grazing) may be partially co-dependent on climate, whereas others (e.g. geology) are independent and more or less 'fixed' over the Quaternary timescales considered here. This is further complicated by the fact that ecosystems including soils are not necessarily in equilibrium with current climate, which is a problem that adds significantly to uncertainty in using evidence of past environmental changes to reconstruct climate, and *vice versa*.

The degree of correspondence between present-day southern African ecosystems and key environmental variables can be observed directly through consideration of subcontinental scale patterns of topography, underlying geology, rainfall and the distribution of major terrestrial biomes (Fig. 17.1A–D). Although there is a strong relationship between biome type and climate, in many cases the boundaries between major ecosystem units (ecotones) correspond to transitions between contrasting geological substrates. This is especially marked in the case of fynbos communities of the southwestern Cape, which have a strong affinity with particular geological substrates, in this case sandstones, which yield shallow and nutrient-impoverished soils. There is also a strong topographic component, as the Table Mountain Group sandstones generally form prominent mountainous terrain. Likewise, there are strong geomorphological controls on different types of savanna vegetation, e.g. valley-bottoms dominated by clay substrates support fine-leaved savannas, while broad-leaved savannas generally occur on sandy uplands (Scholes, 1997). Therefore, Quaternary climate change may not necessarily have resulted in the widespread redistribution of major ecosystems as it did elsewhere in Africa (Meadows and Chase, 2013). This does not mean that southern African ecosystems failed to respond rapidly to climate changes, but that biome-scale maps comparing two time periods would show fewer differences than for similar maps of North America or Europe where there was wholesale migration of biomes in response to glaciations. Since there was no widespread southern African Quaternary glaciation, the degree of landscape transformation would have likely been less than in localities experiencing major ice advance and retreat, or the development of land bridges by exposure of a shallow continental shelf. Thus, rather than involving substantial migration of terrestrial biomes, late Quaternary climate changes in southern Africa are expected to show changes in community dynamics *within* biomes or at their ecotonal boundaries as individual species were favoured or disadvantaged by changing conditions over time.

This chapter describes ecosystem responses to environmental change during the latest part of the Quaternary – the late Pleistocene – spanning the last glacial maximum (LGM), last glacial–interglacial transition (LGIT) and the Holocene. These ecosystem responses are considered within broader climate zones, *viz.* the summer rainfall zone (SRZ) (sub-divided where appropriate into arid/semi-arid and humid/sub-humid regions) and the winter rainfall zone (WRZ) (see Chase and Meadows, 2007). The focus is necessarily on vegetation as an indicator of overall ecosystem characteristics, due to the fact that information on other environmental parameters is limited, but geomorphic and other evidence may cast light on wider terrestrial environmental conditions and processes in specific instances. New approaches to reconstruct ecosystem changes that build on the traditional use of pollen records are now beginning to emerge, including botanical-climatological transfer functions (e.g. Truc et al., 2013) and lipid biomarker (leaf wax *n*-alkane) records (e.g. Schefuß et al.,

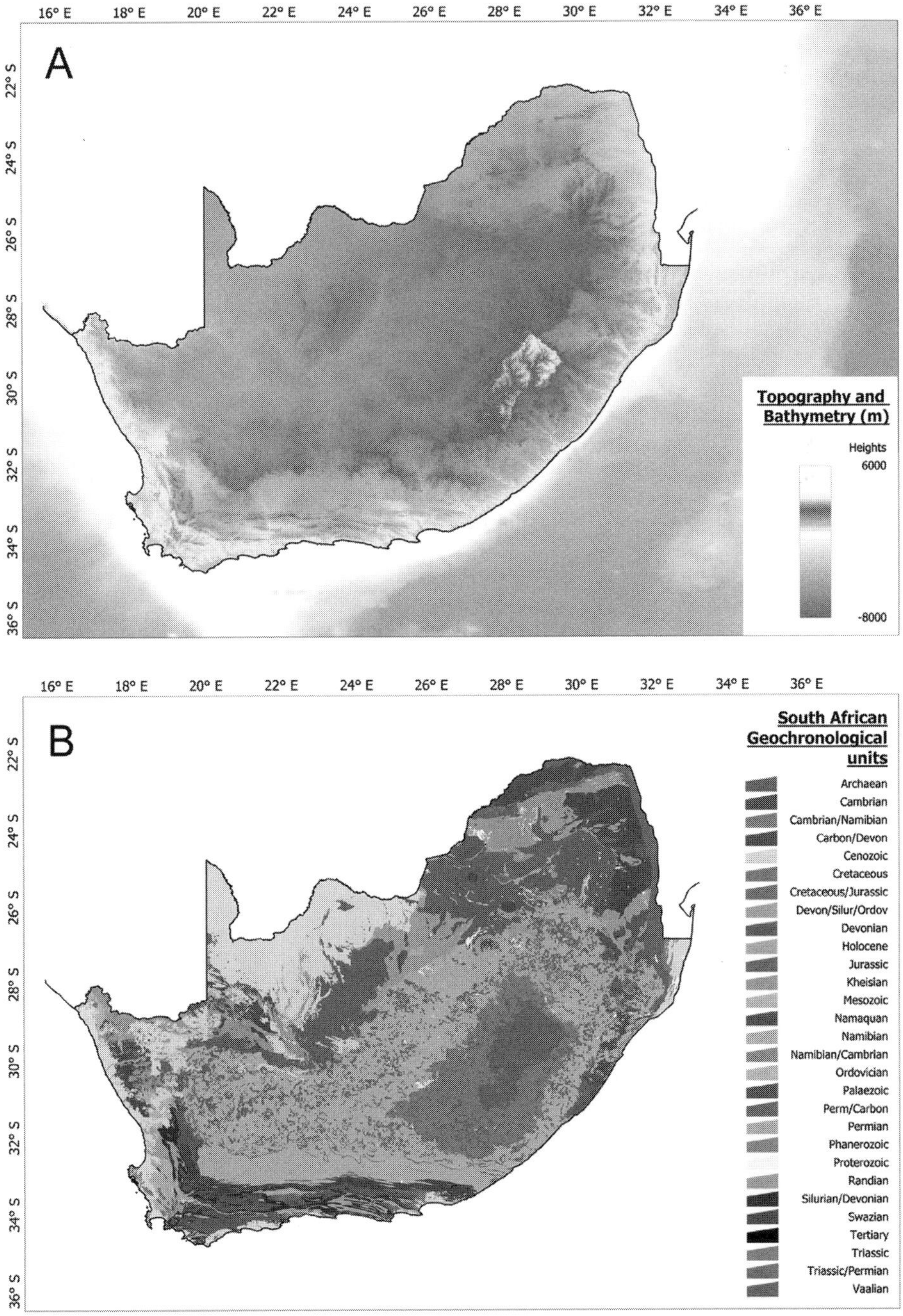

Fig. 17.1. (A) Topography and bathymetry of southern Africa (based on SRTM and ETOPO2 data); (B) geology of southern Africa depicted by geochronological units (data from the African Earth Observatory); (C) the distribution of mean annual rainfall over southern Africa and the approximate positions of the winter rainfall (WRZ), the year round (YRZ) and summer rainfall (SRZ) zones (after Hijmans et al., 2005); (D) the biomes of southern Africa (modified after Mucina and Rutherford, 2006). A black and white version of this figure will appear in some formats. For the colour version, please refer to the plate section.

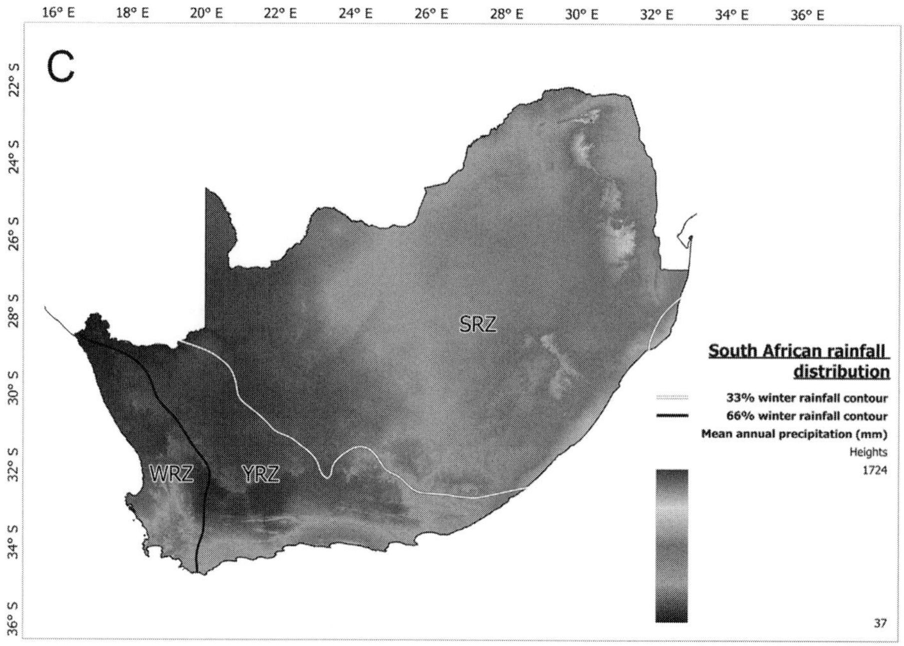
16° E
18° E
20° E
22° E
24° E
26° E
28° E
30° E
32° E
34° E
36° E
22° S
24° S
26° S
28° S
30° S
32° S
34° S
36° S
C
SRZ
WRZ
YRZ
South African rainfall distribution
33% winter rainfall contour
66% winter rainfall contour
Mean annual precipitation (mm)
Heights
1724
37

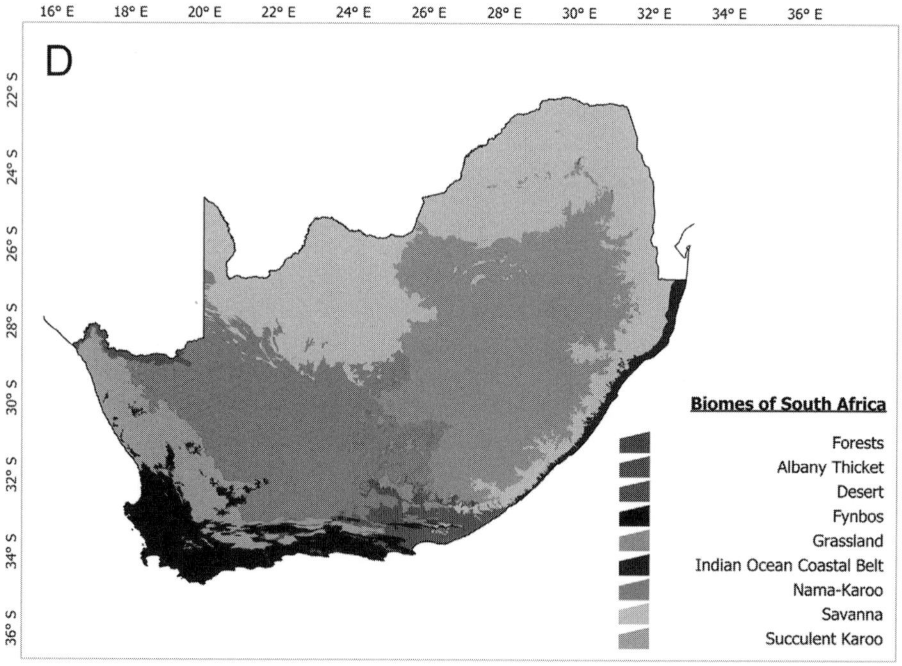
16° E
18° E
20° E
22° E
24° E
26° E
28° E
30° E
32° E
34° E
36° E
22° S
24° S
26° S
28° S
30° S
32° S
34° S
36° S
D
Biomes of South Africa
Forests
Albany Thicket
Desert
Fynbos
Grassland
Indian Ocean Coastal Belt
Nama-Karoo
Savanna
Succulent Karoo

Fig. 17.1. (*cont.*)

2011; Carr et al., 2014). These approaches are likely to play an increasingly significant role in improving our understanding of past ecosystem dynamics.

17.2 Late Pleistocene climate and ecosystems in southern Africa

17.2.1 Summer rainfall zone

It is still unclear how southern African savannas (spanning woodland to grassland) responded to late Pleistocene climate changes; vegetation responses are difficult to determine, in part because of uncertainty of the climate changes *per se* and also because there are few sites where ecosystem characteristics have been reconstructed.

There is more consistency in LGM temperature than in precipitation reconstructions (Chase and Meadows, 2007). For example, speleothem analyses at Cango Cave (Talma and Vogel, 1992) and palaeogroundwater records from the Letlhakeng (Kulongoski and Hilton, 2004), Stampriet (Stute and Talma, 1997) and Uitenhage (Heaton et al., 1983) aquifers all indicate that mean annual temperatures were 5–6°C cooler than present during the LGM. Vegetation modelling based on pollen records suggest 7°C cooler than present (Nash and Meadows, 2012). However, increased seasonality in temperature and precipitation may have been more significant (e.g. Wu et al., 2007; Truc et al., 2013) and may have impacted on the distribution of some plant species, especially given an increase in the number of frost days. For example, at Wonderkrater (northeast South Africa), heathland (fynbos) elements are more prominent around the LGM (Scott and Thackeray, 1987; Truc et al., 2013) and were much advantaged by cooler climates when compared with the more thermophilous savanna taxa that predominate today.

It is likely that more significant ecosystem responses would have been driven by changes in rainfall rather than temperature, although there is no regionally coherent pattern. Along the KwaZulu-Natal coast, Finch and Hill (2008) infer greater aridity around 24 cal kyr BP, causing the Mfabeni peatland and coastal forests to retreat, a conclusion that is also supported by geochemical evidence (Baker et al., 2014). The LGM itself is not recorded directly in the pollen record from the Tswaing Crater (Partridge et al., 1997), but variations in organic/inorganic carbon together with geochemical evidence suggest cooler and drier conditions around this time (Kristen et al., 2007), which would favour more open vegetation than the present savanna woodland. Cooler and drier LGM conditions are associated with the lowering of vegetation belts by up to 1000 m in altitude, as well as changes in the relative abundance of C_3 and C_4 grasses (Scott, 1999, 2002), although this may be a partial response to lower atmospheric carbon dioxide. Pollen records in sediments from the adjacent Atlantic Ocean indicate overall drier conditions between 21–17.5 cal kyr BP (Shi et al., 2000) and a correspondingly more open vegetation. In contrast, the dominance of Afromontane fynbos elements between ~18–13.5 cal kyr BP in the Mahwaqa record, a site within the present-day Grassland Biome, suggests cooler conditions and increased winter rainfall (Neumann et al., 2014). Periglacial and glacial geomorphology in the Drakensberg/Lesotho Highlands, combined with climate modelling, lends support to decreased temperatures and enhanced precipitation during the LGM (Mills et al., 2012). Altitudinal differences may be responsible for contrasting precipitation signals within the SRZ.

Fluctuations in precipitation and also windiness during the last glacial period are indicated by phases of aeolian activity in the Kalahari dunefields (Stokes et al., 1998; Burrough and Thomas, 2008) and highstands of Lake Makgadikgadi (Burrough et al., 2007, 2009). This suggests that a simple model of cooler and drier LGM climates driven by precessional forcing for the summer rainfall region as a whole is no longer applicable (Nash and Meadows, 2012). The lack of synchroneity in records across southern Africa may result from local, rather than regional, influences on climate (Burrough et al., 2007; Burrough and Thomas, 2013). Ecosystem responses are difficult to estimate with confidence, although maps depicting broad vegetation categories for Africa as a whole for several intervals during the late Quaternary have been produced (Adams and Faure, 1997; Meadows and Chase, 2013). The LGM situation indicates an expansion of drier savanna types across large areas of central and eastern southern Africa in response to presumed cooler and (largely) drier conditions, promoted by lower atmospheric carbon dioxide concentrations, which would have favoured C_4 grasses at the expense of C_3 trees and shrubs (Bond et al., 2003; Bragg et al., 2013).

17.2.2 Winter rainfall zone

Chase and Meadows (2007) reviewed the evidence for LGM climate conditions for the WRZ and, although the record is fragmented by region and time period, evidence suggests changing intensity and frequency of temperate, rain-bearing westerlies. Twelve sediment layers from Elands Bay Cave, Western Cape, dating from >40 ^{14}C kyr BP to ~12.5 cal kyr BP, show maximum values of woodland taxa around the LGM (Baxter, 1996). These findings are supported by the concurrent charcoal record that reveals the greater prominence of afromontane taxa during the LGM (Parkington et al., 2000). Some of these species (e.g. *Celtis africana, Grewia occidentalis*) are intolerant of prolonged summer drought and therefore suggest significantly extended winter rainfall during some periods of the last glaciation (Meadows and Baxter, 1999). Pollen evidence from core GeoB1711-4 off western Namibia indicates that the start of marine isotope stage (MIS) 2 at around 24 cal kyr BP was the wettest period in the region for the last 135 kyr (Shi et al., 2001) and that restio-dominated wetland ecosystems may have extended farther north than their current distribution. However, the GeoB1711-4 pollen data may reflect variations in windiness and pollen transport distance rather than climate at the coring location (Scott et al., 2004; Chase and Meadows, 2007).

Pollen and isotope evidence from rock hyrax middens in the nearby Cederberg (Scott and Woodborne, 2007; Chase et al., 2011; Quick et al., 2011) shows that MIS 2 ecosystems were subject to a cooler climate and a general lowering of vegetation belts, but the moisture signal is uncertain. Scott and Woodborne (2007) favour greater aridity associated with lower temperatures, whereas Chase and Meadows (2007) argue for more effective moisture, especially under lower atmospheric carbon dioxide which would have reduced plant available moisture. Thus, it can be said that the LGM in southwestern-most Africa appears to have been climatically complex and transitional (Nash and Meadows, 2012). This is difficult to depict at the biome scale (Fig. 17.2), but there may have been an extension of the fynbos vegetation northwards around this time (Meadows and Chase, 2013). Modern vegetation analogues of these different biomes are shown in Figure 17.3.

The highest resolution LGIT record obtained thus far comes from a large hyrax midden in the Cederberg (Chase et al., 2011; Quick et al., 2011; Valsecchi et al., 2013) where stable isotopes and pollen yield unequivocal evidence for a well-defined, colder and dry period corresponding to the Younger Dryas, and pointing to the influence of northern hemisphere climate drivers even as far south as 32°S. The impact on fynbos communities would have been through changes in fire regime suppression of moisture-sensitive elements in the vegetation (van Wilgen, 1992).

17.3 Holocene climate and ecosystems

As discussed above, records for the LGM and LGIT are rather rare in southern Africa. The Holocene, by comparison, is represented by more sites, although the extent to which such evidence translates into a clear picture of ecosystem conditions remains limited. Key global events of the Holocene include the 8.2 kyr BP Atlantic cold reversal, the mid-Holocene altithermal (HA), the Medieval Warm Period (MWP) and the Little Ice Age (LIA). There is evidence that southern African terrestrial ecosystems responded to many of these perturbations, although the temporal and spatial fragmentation of the record remains a constraint (Nash and Meadows, 2012). A key issue is the extent to which the so-called African Humid Period (AHP; 9–6 kyr BP) (deMenocal et al., 2000), widely recorded across Africa north of the equator and presumed a consequence of precessional forcing, was prevalent in southern Africa (Burrough and Thomas, 2013).

17.3.1 Summer rainfall zone

Holocene ecosystem change is recorded at a variety of sites across the region, with higher resolution records available through pollen analysis at Spitzkoppe (Chase et al., 2009), Austerlitz (Chase et al., 2010), Wonderkrater (Scott et al., 2003; Truc et al., 2013), Mfabeni (Finch and Hill, 2008), Sibaya (Neumann et al., 2008), Eteza (Neumann et al., 2010), Mahwaqa (Neumann et al., 2014) and Braamhoek

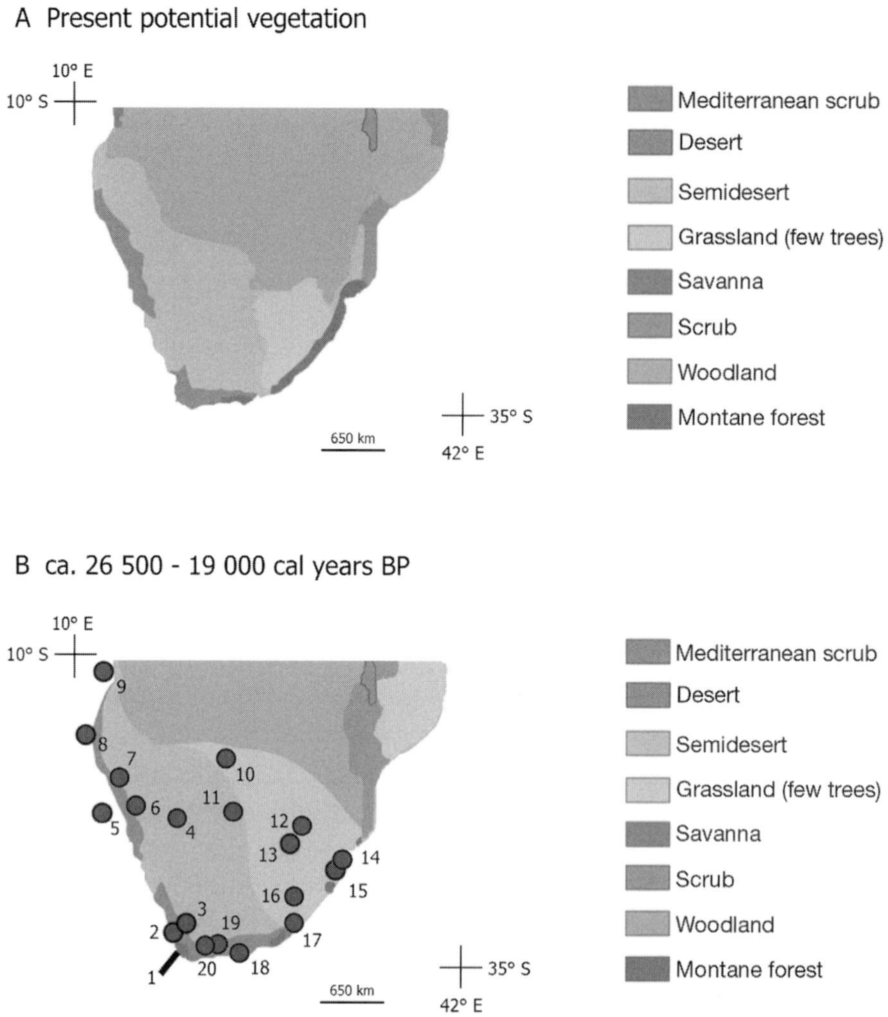

Fig. 17.2. (A) The present potential vegetation (the potential distribution of biomes before human alteration, which may be taken as an approximation for mid-Holocene biome distribution); (B) generalised perspective of LGM vegetation of southern Africa. Localities mention in the text: (1) Princess Vlei; (2) Elands Bay, Klaarfontein, Verlorenvlei; (3) Pakhuis Pass and De Rif, Cederberg; (4) Stampriet Aquifer; (5) GeoB 1711–4; (6) Spitzkoppe; (7) Austerlitz; (8) GeoB 1023–5; (9) ODP 1078C; (10) Makgadikgadi pans; (11) Letlhakeng; (12) Tswaing Crater; (13) Wonderkrater; (14) Lake Sibaya; (15) Mfabeni Peatland and Lake Eteza; (16) Braamhoek; (17) Maqhwa Mountain (18) Uitenhage; (19) Cango Cave; (20) Seweweekspoort (modified from Adams and Faure, 1997; Meadows and Chase, 2013).

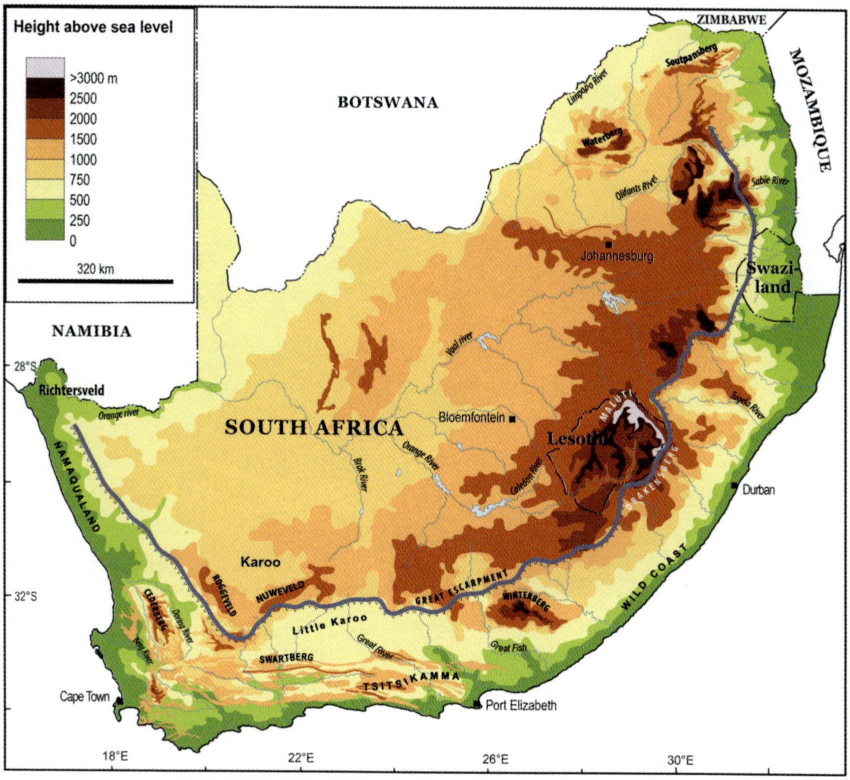

Fig. 1.1. Map of the topography of southern Africa (in m above sea level) and the location of the Great Escarpment (hatched line). Note that other key regional maps are presented elsewhere in this book: regional geology (Fig. 2.1), soils (Fig. 16.2) and ecosystems (Figs. 18.1, 18.2).

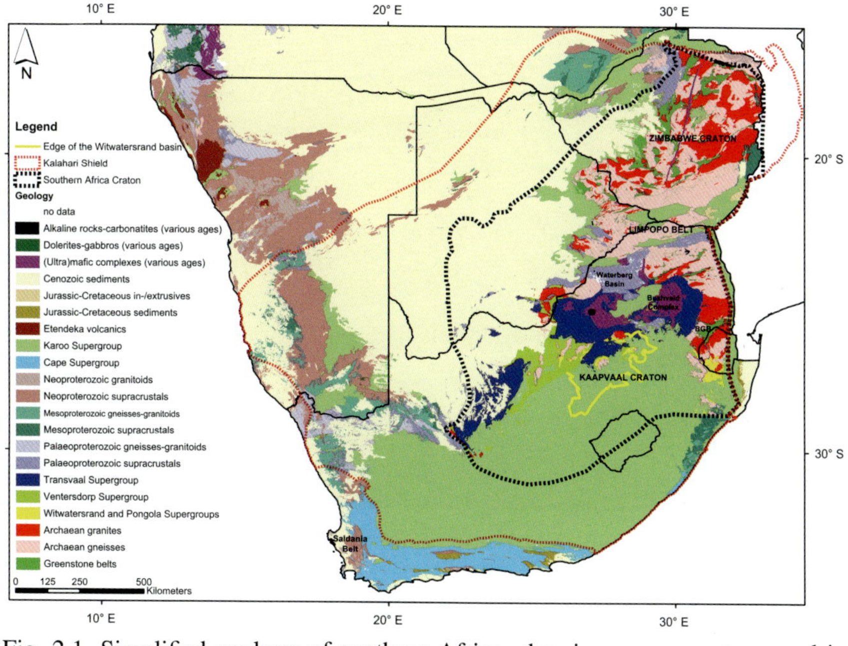

Fig. 2.1. Simplified geology of southern Africa, showing components named in the text (map compiled by Dr Hielke Jelsma).

Fig. 6.2. Features of the coastal plain are vertically exaggerated in this composite Landsat and radar topography image looking east from the Cape Peninsula, across the Cape Flats and False Bay to the mountains of the Cape Fold Belt. Beyond the CFB are the undulating hills of the Overberg with erosional outliers of the CFB that extend as rocky headlands separating sandy beaches along the coast (data from NASA/JPL-Caltech/NIMA).

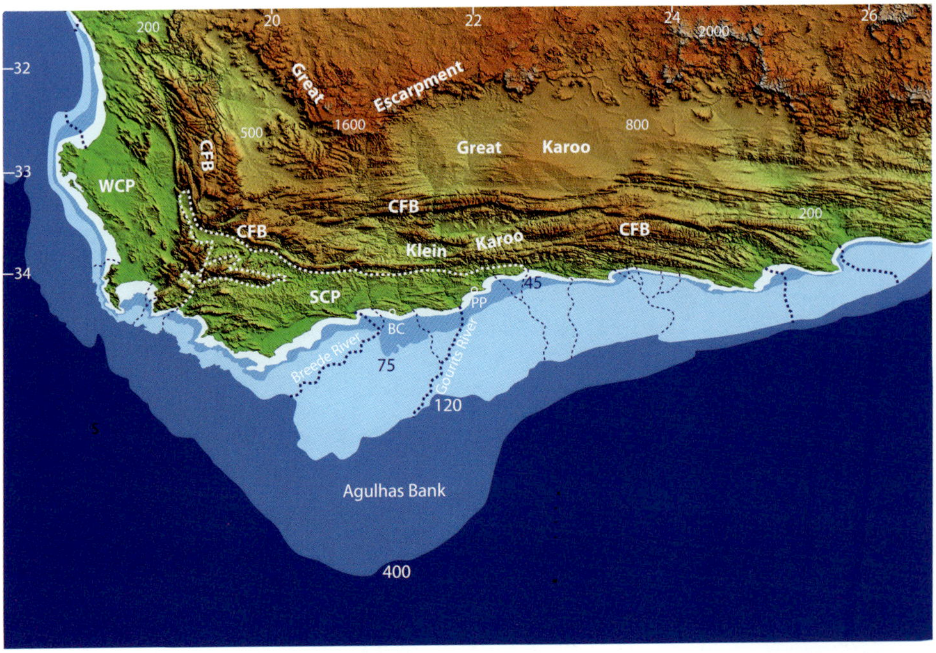

Fig. 6.4. Topography and bathymetry of the southern margin showing the Great Escarpment, Cape Fold Belt (CFB) and expansion of the western coastal plain (WCP) and southern coastal plain (SCP) when sea level is lowered by 45, 75 and 120 m (Compton, 2011). River courses are projected onto the glacial coastal plain, BC: Blombos Cave, PP: Pinnacle Point.

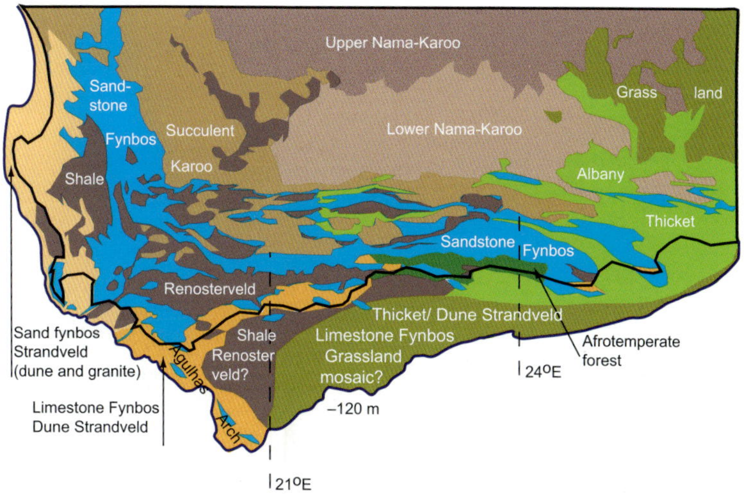

Fig. 6.6. Distribution of major vegetation biomes, simplified from Mucina and Rutherford (2006), and their proposed extrapolation onto the exposed coastal plain during periods of lowered sea level (after Compton, 2011).

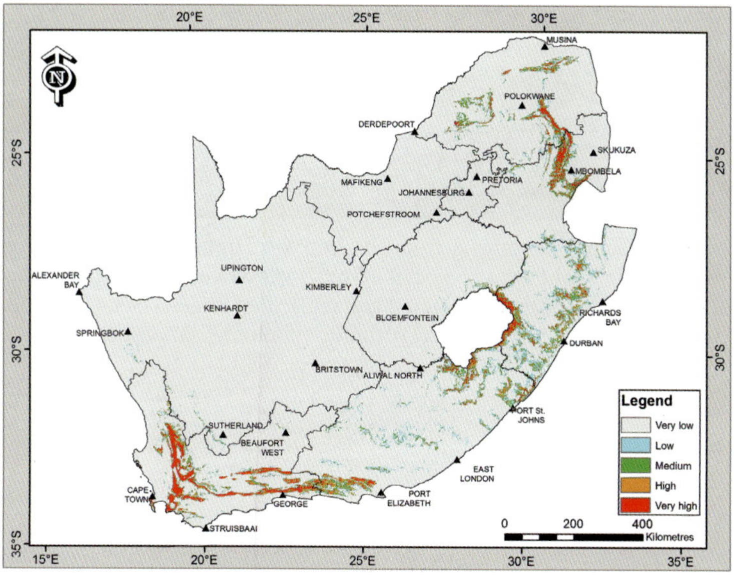

Fig. 9.1. Landslide susceptibility map for South Africa (adapted after Singh et al., 2011).

Fig. 8.4. Periglacial phenomena from the Lesotho Highlands. (A) Views of the blockstream near Sani Pass (as described by Boelhouwers et al., 2002) showing its well defined margins and weathered and reoriented large clasts; (B) stone bank lobes (arrows); (C) sorted circle; (D) thúfur (photos A,D: Jasper Knight; B,C: Stefan Grab).

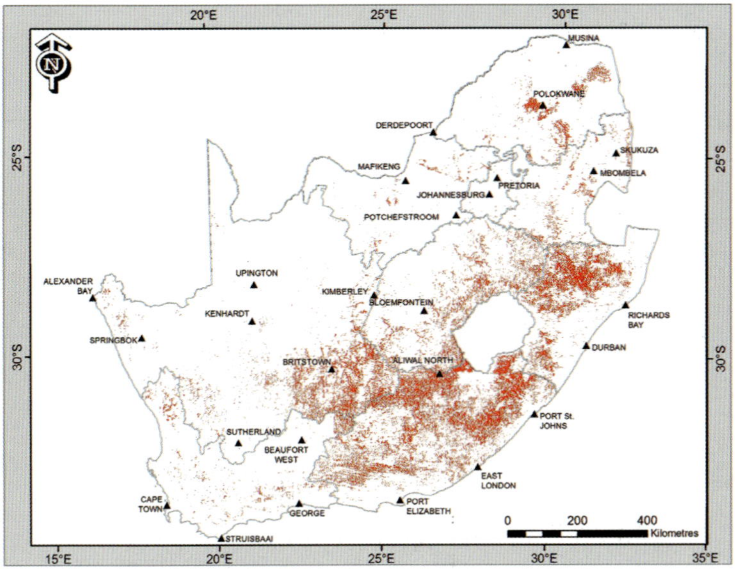

Fig. 9.3. The locations of gully systems in South Africa (adapted after le Roux, 2011). These are mainly cut into the Masotcheni Formation and equivalent colluvia.

Fig. 10.3. Different dune morphologies seen in the field. (a) Linear dunes in the SW Kalahari, which may support significant grass cover, especially in wet years; (b) SW Kalahari dunes exhibiting surface instability due to grazing pressure and drought; (c) sand ramp near Sossus Vlei, Namibia; (d) pan-fringing dunes at Witpan, Northern Cape (photos: David Thomas).

Fig. 9.4. Photos of depositional and erosional elements of gully systems in KwaZulu-Natal. (a) sedimentary succession of slope sediments and buried palaeosols (photo courtesy of Mary Evans), (b) road-cut at Keate's Drift showing bedrock-cut gullies which have then been infilled by slope sediments (I) (photo: Greg Botha).

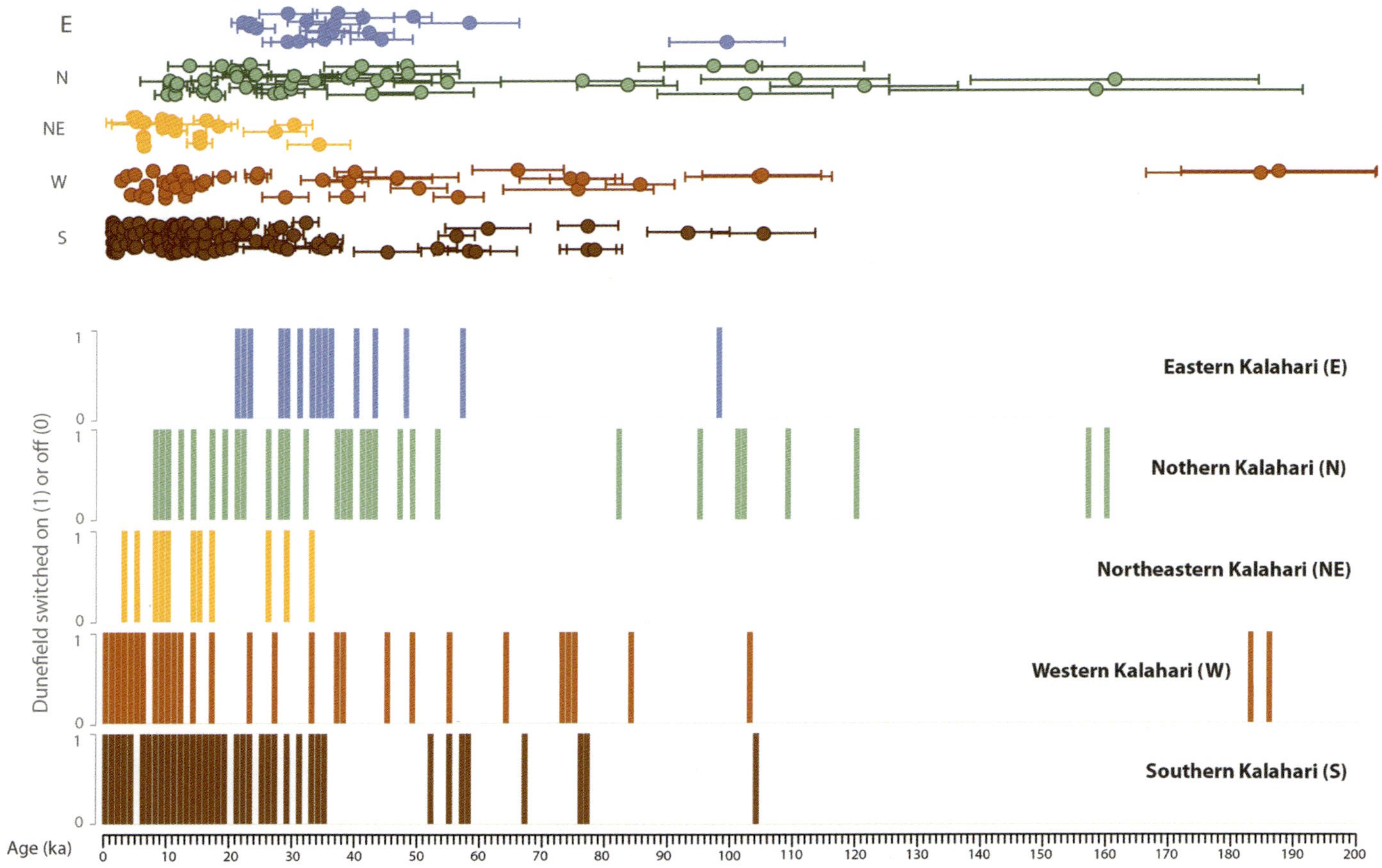

Fig. 10.4. Dated open field (linear dune and sand sheet) records from the Kalahari (adapted from Thomas and Burrough, 2014), see Table 10.2 and Figure 10.1 for locations. The data are presented by dunefield in two formats: (top) mean ages are each plotted with their respective one sigma errors; (bottom) mean ages only are plotted in a binary format in 1000-year bins. Empty bins mean no dune ages are recorded for that particular millennium, full bins indicate one or more ages are recorded for that millennium. Both approaches show near continuous dune accumulation through the last 30 kyr in presently drier southern and western areas, and a more nuanced accumulation record elsewhere, though high age errors (from older analyses) in northern dune samples precludes a clear record of development emerging.

Fig. 11.2. Photos of typical bedrock and alluvial features in southern Africa fluvial systems. (a) Ingrown meander of upper Orange River, incised into basalt, highlands of Lesotho. Flow direction from upper left to middle right; (b) terrace of the Modder River (site 4) with OSL ages (kyr BP) indicated. From bottom to top, the succession comprises: lower grey palaeosol (LGP), red palaeosol (RP), upper grey palaeosol (UGP), brown palaeosol (BP), and weathered silty sand (SC); (c) incised Schoonspruit (site 6). Flow direction from middle right to lower left (photo by T.S. McCarthy). OSL ages (kyr BP) are indicated; (d) Olifants River (site 8) showing the effects of extreme flooding in January 2012. Flow direction from left to right (photos a,b,d: Stephen Tooth).

Fig. 13.2. Examples of coastline types found on the southern Cape coast. (A) Cliffed coastlines in Table Mountain Group sandstone at Pondoland, (B) beachrock dominated foreshore with active back beach dunes at Reebok, (C) barrier dunes and backbarrier lagoons at Wilderness (photos: Hayley Cawthra).

Fig. 13.4. (A) Erosional gullies on a rocky shoreline, trending in a coast-perpendicular orientation, Durban Bluff, (B) multibeam echosounder data from Blood Reef showing submerged erosional gullies 17 m below sea level, of comparable dimensions to (A), (C) fractured joint sets on an outcrop of beachrock, Durban Bluff, (D) intertidal zone potholes, Reebok (Mossel Bay), (E-F) erosional potholes, now submerged 18 and 24 m below sea level respectively, Blood Reef, (G) wave-cut platform in the Msikaba Formation, Pondoland, (H) coastal sea-cliffs and incised overhands, Pondoland (photos: Hayley Cawthra).

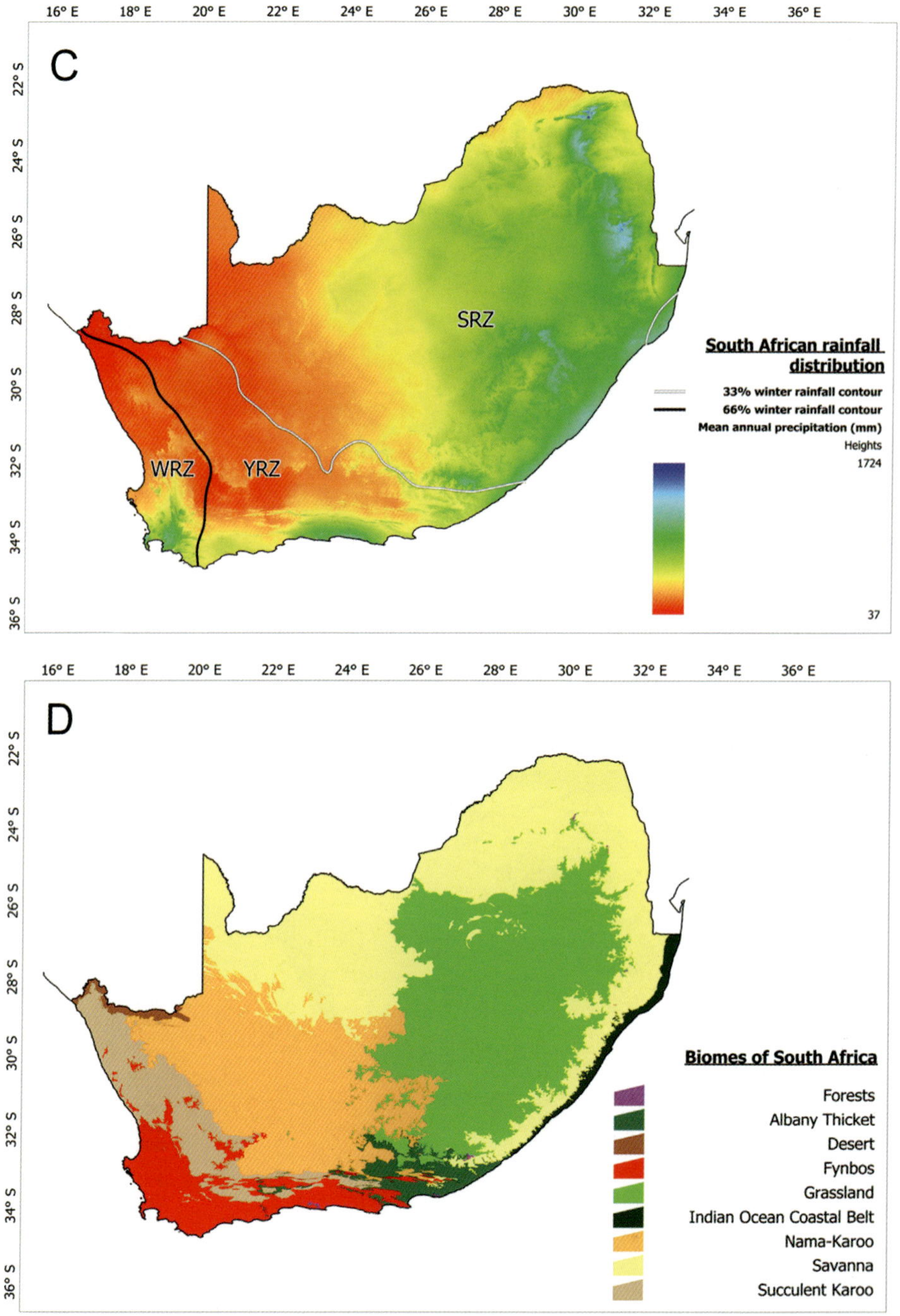

Fig. 17.3. Examples of the vegetation found within the biomes mapped in Figure 17.2. (A) Mediterranean Scrub: Mountain Fynbos, Western Cape, South Africa; (B) Desert: Namib dunes; (C) Semi-desert: Namibia; (D) Grassland: Southern Drakensberg Highland Grassland (Mucina and Rutherford, 2006, p367); (E) Savanna Biome: Angolan scarp savanna (Photo: www.eoearth.org/files/111001_111100/111082/375px-Angolansavanna.jpg); (F) Scrub: coastal scrub; (G) Woodland: Brachystegia woodland, Zimbabwe (Photo: www.vumba-nature.com/Msasa-spring-colour.jpg); (H) Montane Forest: Southern Cape Afromontane Forest, South Africa) (Photos A–C, F, H: Lynne Quick).

Fig. 19.3. Photomicrographs of charcoal and leaf remains from Wonderkrater and Sibudu. (A) *Phragmites* sp. viz. basal culm fragment with axillary bud (Wonderkrater ~15,167 cal yr BP), (B) *Phragmites* sp. viz. culm cross section showing vascular bundles between parenchyma cells (large light cells) (Wonderkrater), (C) grass culm cross section (Wonderkrater ~12,367 cal yr BP), (D) transverse section and (E) tangential longitudinal section of charcoal piece of *Acacia sieberiana* (Wonderkrater >32 kyr BP), (F) silicified leaf of *Cryptocarya woodii* (Sibudu rock shelter ~77 kyr BP), (G) detail of stomata from the leaf shown in (F), (H) transverse section and (I) tangential longitudinal section of charcoal of *Spirostachys africana* (Sibudu rock shelter ~38 kyr BP) (image: Marion Bamford).

Fig. 16.2. Examples of structures within the Sterkfontein Caves, Cradle of Humankind (photos reproduced courtesy of Dominic Stratford).

Fig. 23.3. Elands Bay Cave, Western Cape, South Africa. The view outwards to the Atlantic Ocean would have been dry land for much of the Pleistocene and the deposits here track the site's change from inland to coastal in location (photo reproduced with permission of G. Porraz).

Fig. 23.4. Rock painting near Maclear, Eastern Cape, South Africa (image RARI-RSA-SPR3-7D, reproduced with permission from Rock Art Research Institute, University of the Witwatersrand, South Africa).

Fig. 24.3. Balerno Main shelter, an aggregation site in the Shashe-Limpopo Confluence Area. The shelter itself (lower left) is 16–20 m wide (photo: Maria Schoeman).

Fig. 17.3. Examples of the vegetation found within the biomes mapped in Figure 17.2. (A) Mediterranean Scrub: Mountain Fynbos, Western Cape, South Africa; (B) Desert: Namib dunes; (C) Semi-desert: Namibia; (D) Grassland: Southern Drakensberg Highland Grassland (Mucina and Rutherford, 2006, p367); (E) Savanna Biome: Angolan scarp savanna (Photo: www.eoearth.org/files/111001_111100/111082/375px-Angolansavanna.jpg); (F) Scrub: coastal scrub; (G) Woodland: Brachystegia woodland, Zimbabwe (Photo: www.vumba-nature.com/Msasa-spring-colour.jpg); (H) Montane Forest: Southern Cape Afromontane Forest, South Africa) (Photos A–C, F, H: Lynne Quick). A black and white version of this figure will appear in some formats. For the colour version, please refer to the plate section.

(Norström et al., 2009). Burrough and Thomas (2013) synthesise a wide range of proxy evidence that suggests a more complex Holocene climate than inferred by a simple precession-driven AHP model alone. A simple orbital forcing model is also questioned by Chase et al. (2009) who argue, based on stable isotopes of hyrax midden material from Spitzkoppe in Namibia, for humid conditions in the first half of the Holocene being succeeded thereafter by significantly greater aridity. Although there is considerable heterogeneity of climate dynamics and ecosystem responses, a reconstruction of mid-Holocene biomes (Fig. 17.2A) suggests that a more humid and warmer climate at this time may have favoured savanna woodlands over grassland in much of the SRZ. Differences between eastern and western parts of the SRZ have also been suggested (Burrough and Thomas, 2013), although there is a lack of evidence to resolve this in detail.

17.3.2 Winter rainfall zone

Only a small number of high resolution pollen studies of Holocene ecosystems have been published from the WRZ, but the seasonality of rainfall has resulted in the intermittent accumulation of wetland sediments and thus a fragmented pollen record (Chase and Meadows, 2007), particularly in the early Holocene. The climatic progression from the cooler early Holocene to the HA was accompanied by increasing aridity and a reduced influence of westerly winds, certainly along the western margins of the WRZ (Chase and Meadows, 2007). For example, pollen sequences from 7.7–4.4 cal kyr BP at Pakhuis Pass, Western Cape, indicate a shift to warmer and drier climates (Scott and Woodbourne, 2007). Mesic thicket taxa begin disappearing from the Elands Bay Cave charcoal record at 11.3 cal kyr BP and are entirely absent by 4.7 cal kyr BP, having been replaced by a mosaic of asteraceous shrubs and xeric thicket (Parkington et al., 2000). Pollen from a core at Klaarfontein, 18 km inland from Elands Bay, indicates a dominance of Poaceae, Asteraceae and Chenopodiaceae from 7.5–4.0 cal kyr BP, reflecting a relatively arid environment (Meadows and Baxter, 2001). The hyrax midden stable isotope signal at Seweweekspoort, Western Cape, points to a strongly inverse relationship between temperature and moisture during the Holocene (Chase et al., 2013), supporting the concept of an arid HA in the WRZ.

Sequences covering the later Holocene allow for more detailed climate and ecosystem reconstructions for sites such as Princess Vlei (Neumann et al., 2011), Verlorenvlei (Meadows and Baxter, 1999; Stager et al., 2012) and Klaarfontein (Meadows and Baxter, 2001). Although much of the later Holocene appears to be drier than the HA, there is evidence for wetter phases such as 3.4–2.6 cal kyr BP at Princess Vlei (Neumann et al., 2011). A general warmer and drier MWP is followed by a cooler and wetter LIA which is well recorded in the diatom record from Verlorenvlei (Stager et al., 2012) and matched by evidence of greater moisture availability in the plant communities of the surrounding catchment (Meadows et al., 1996) around this time and prior to European colonisation.

17.4 Human impacts on terrestrial ecosystems

The southwestern Cape is likely the region of origin for behaviourally modern humans who were manipulating fire as long ago as 160 kyr BP (Brown et al., 2009). Despite the potential for such activities to modify landscapes, unequivocal evidence for significant human impact on the terrestrial ecosystems is unclear, although Neumann et al. (2011) tentatively attribute some vegetation changes at Princess Vlei within the last 2000 years to Khoi herders.

The arrival of colonial settlers from the 17th Century AD onwards had landscape-scale impacts that have been recorded in the sedimentary record. Meadows et al. (1996) and Meadows and Asmal (1996) document the impacts of farming on plant communities and sediment budgets respectively in the Verlorenvlei catchment. Here, a combination of vegetation clearance for crops, together with the local extinction of large herbivores (and carnivores), markedly altered the ecological balance in and around the coastal lake and substantially transformed the ecosystem. In the Princess Vlei record, there is evidence of the increasing scale and range of human impacts within the last two hundred years or so (Neumann et al., 2011). More widely, urbanisation and the commercialisation of agriculture under increasing human population pressure in southern Africa have transformed ecosystems irreversibly. Semi-arid ecosystems are especially vulnerable to disturbance and, as Thomas et al. (2005) demonstrate, the combined effects of climate change and inappropriate management have the potential to dramatically transform ecosystems at the landscape scale. Such changes may well exceed the amplitude of climate responses observed in the late Quaternary.

17.5 Summary

Widespread redistribution of southern African biomes does not appear to have taken place during the late Quaternary despite evidence of significant climate changes. Factors such as strong affiliations to geological substrate and fire regime are likely responsible for the lack of subcontinental-scale vegetation changes, but more localised changes are evident within biomes or at their ecotonal boundaries. While there is little doubt that temperatures during the LGM were lower than present, the distribution, amount and seasonality of precipitation is still uncertain. In general, periods of intensified winter precipitation (and decreased temperatures) promoted fynbos vegetation at the expense of savanna, grassland and thicket elements. The palaeoenvironmental record indicates that southern African terrestrial ecosystems responded to many global climate events within the Holocene, but the temporal and spatial fragmentation of the pollen record remains a major constraint. Warmer and wetter conditions during the mid-Holocene favoured savanna woodlands over grassland in much of the SRZ, while a more arid HA in the WRZ was likely associated with a reduction in mesic fynbos and the spread of more xeric (possibly karroid) communities.

References

Adams, J. and Faure, H. (1997). *Review and Atlas of Palaeovegetation: Preliminary Land Ecosystem Maps of the World since the Last Glacial Maximum*. Oak Ridge National Laboratory, TN, USA. Available from www.esd.ornl.gov/projects/qen/adams1.html.

Baker, A., Routh, J., Blaauw, M. and Roychoudhury, A.N. (2014). Geochemical records of palaeoenvironmental controls on peat forming processes in the Mfabeni peatland, Kwazulu Natal, South Africa since the Late Pleistocene. *Palaeogeography, Palaeoclimatology, Palaeoecology*, 395, 95–106.

Baxter, A. (1996). *Late Quaternary Palaeoenvironments of the Sandveld, Western Cape Province, South Africa*. Unpublished PhD Thesis, University of Cape Town, 710pp.

Bond, W. J., Midgley, G. F. and Woodward, F. I. (2003). The importance of low atmospheric CO_2 and fire in promoting the spread of grasslands and savannas. *Global Change Biology*, 9, 973–982.

Bragg, F. J., Prentice, I. C., Harrison, S. P., Eglinton, G., Foster, P. N., Rommerskirchen, F. and Rullkötter, J. (2013). Stable isotope and modelling evidence for CO_2 as a driver of glacial–interglacial vegetation shifts in southern Africa. *Biogeosciences*, 10, 2001–2010.

Brown, K. S., Marean, C. W., Herries, A. I. R., Jacobs, Z., Tribolo, C., Braun, D., Roberts, D. L., Meyer, M. C. and Bernatchez, J. (2009). Fire as an engineering tool of early modern humans. *Science*, 325, 859–862.

Burrough, S. L. and Thomas, D. S. G. (2008). Late Quaternary lake-level fluctuations in the Mababe Depression: Middle Kalahari palaeolakes and the role of Zambezi inflows. *Quaternary Research*, 69, 388–403.

Burrough, S. L. and Thomas, D. S. G. (2013). Central southern Africa at the time of the African Humid Period: A new analysis of Holocene palaeoenvironmental and palaeoclimate data. *Quaternary Science Reviews*, 80, 29–46.

Burrough, S. L., Thomas, D. S. G. and Bailey, R. M. (2009). Mega-Lake in the Kalahari: A late Pleistocene record of the Palaeolake Makgadikgadi system. *Quaternary Science Reviews*, 28, 1392–1411.

Burrough, S. L., Thomas, D. S. G., Shaw, P. A. and Bailey, R. M. (2007). Multiphase Quaternary highstands at Lake Ngami, Kalahari, northern Botswana. *Palaeogeography, Palaeoclimatology, Palaeoecology*, 253, 280–299.

Carr, A. S., Boom, A., Grimes, H. L., Chase, B. M., Meadows, M. E. and Harris, A. (2014). Leaf wax n-alkane distributions in arid zone South African flora: Environmental controls, chemotaxonomy and palaeoecological implications. *Organic Geochemistry*, 67, 72–84.

Chase, B. M., Boom, A., Carr, A. S., Meadows, M. E. and Reimer, P. J. (2013). Holocene climate change in southernmost South Africa: Rock hyrax middens record shifts in the southern westerlies. *Quaternary Science Reviews*, 82, 199–205.

Chase, B. M. and Meadows, M. E. (2007). Late Quaternary dynamics of southern Africa's winter rainfall zone. *Earth-Science Reviews*, 84, 103–138.

Chase, B. M., Meadows, M. E., Carr, A. S. and Reimer, P. J. (2010). Evidence for progressive Holocene aridification in southern Africa recorded in Namibian hyrax middens: Implications for African Monsoon dynamics and the "African Humid Period". *Quaternary Research*, 74, 36–45.

Chase, B. M., Meadows, M. E., Scott, L., Thomas, D. S. G., Marais, E., Sealy, J. and Reimer, P. J. (2009). A record of rapid Holocene climate change preserved in hyrax middens from southwestern Africa. *Geology*, 37, 703–706.

Chase, B. M., Quick, L. J., Meadows, M. E., Scott, L., Thomas, D. S. G. and Reimer, P. J. (2011). Late glacial interhemispheric climate dynamics revealed in South African hyrax middens. *Geology*, 39, 19–22.

Daniau, A.-L., Sánchez Goñi, M. F., Martinez, P., Urrego, D. H., Bout-Roumazeilles, V., Desprat, S. and Marlon, J. R. (2013). Orbital-scale climate forcing of grassland burning in southern Africa. *Proceedings of the National Academy of Sciences*, 110, 5069–5073.

deMenocal, P., Ortiz, J., Guilderson, T., Adkins, J., Sarnthein, M., Baker, L. and Yarusinsky, M. (2000). Abrupt onset and termination of the African Humid Period: Rapid climate responses to gradual insolation forcing. *Quaternary Science Reviews*, 19, 347–361.

Finch, J. M. and Hill, T. R. (2008). A late Quaternary pollen sequence from Mfabeni Peatland, South Africa: Reconstructing forest history in Maputaland. *Quaternary Research*, 70, 442–450.

Heaton, T. H. E., Talma, A. S. and Vogel, J. C. (1983). Origin and history of nitrate in confined groundwater in the western Kalahari. *Journal of Hydrology*, 62, 243–262.

Hijmans, R. J., Cameron, S. E., Parra, J. L., Jones, P. G. and Jarvis, A. (2005). Very high resolution interpolated climate surfaces for global land areas. *International Journal of Climatology*, 25, 1965–1978.

Kristen, I., Fuhrmann, A., Thorpe, J., Röhl, U., Wilkes, H. and Oberhänsli, H. (2007). Hydrological changes in southern Africa over the last 200 Ka as recorded in lake sediments from the Tswaing impact crater. *South African Journal of Geology*, 110, 311–326.

Kulongoski, J. T., Hilton, D. R. and Selaolo, E. T. (2004). Climate variability in the Botswana Kalahari from the late Pleistocene to the present day. *Geophysical Research Letters*, 31, L10204, doi:10.1029/2003GL019238.

Meadows, M. and Asmal, O. (1996). Chronology, sedimentology and geochemistry of sediments at Verlorenvlei (Western Cape Province, South Africa) as evidence of anthropogenically-induced land degradation. *Zeitschrift für Geomorphologie, Supplementband*, 107, 45–62.

Meadows, M. E. and Baxter, A. J. (1999). Late Quaternary palaeoenvironments of the southwestern Cape, South Africa: A regional synthesis. *Quaternary International*, 57–58, 193–206.

Meadows, M. E. and Baxter, A. J. (2001). Holocene vegetation history and palaeoenvironments at Klaarfontein Springs, Western Cape, South Africa. *The Holocene*, 11, 699–706.

Meadows, M. E., Baxter, A. J. and Parkington, J. (1996). Late Holocene environments at Verlorenvlei, Western Cape Province, South Africa. *Quaternary International*, 33, 81–95.

Meadows, M. E. and Chase, B. M. (2013). Pollen Records. In *Encyclopedia of Quaternary Science* (2nd Ed), ed. A. E. Scott. Amsterdam: Elsevier, pp. 9–17.

Mills, S. C., Grab, S. W., Rea, B. R., Carr, S. J. and Farrow, A. (2012). Shifting westerlies and precipitation patterns during the Late Pleistocene in southern Africa determined using glacier reconstruction and mass balance modelling. *Quaternary Science Reviews*, 55, 145–159.

Mucina, L. and Rutherford, M. C. (2006). *The Vegetation of South Africa, Lesotho and Swaziland, Strelitzia 19*. Pretoria: South African National Biodiversity Institute, 807pp.

Nash, D. J. and Meadows, M. E. (2012). Africa. In *Quaternary Environmental Change in the Tropics*, ed. S. E. Metcalfe and D. J. Nash. Chichester: Wiley & Sons, pp. 79–150.

Neumann, F. H., Botha, G. A. and Scott, L. (2014). 18,000 years of grassland evolution in the summer rainfall region of South Africa: evidence from Mahwaqa Mountain, KwaZulu-Natal. *Vegetation History and Archaeobotany*, 23, 665–681.

Neumann, F. H., Scott, L. and Bamford, M. K. (2011). Climate change and human disturbance of fynbos vegetation during the late Holocene at Princess Vlei, Western Cape, South Africa. *The Holocene*, 21, 1137–1149.

Neumann, F. H., Scott, L., Bousman, C. B. and van As, L. (2010). A Holocene sequence of vegetation change at Lake Eteza, coastal KwaZulu-Natal, South Africa. *Review of Palaeobotany and Palynology*, 162, 39–53.

Neumann, F. H., Stager, J. C., Scott, L., Venter, H. J. T. and Weyhenmeyer, C. (2008). Holocene vegetation and climate records from Lake Sibaya, KwaZulu-Natal (South Africa). *Review of Palaeobotany and Palynology*, 152, 113–128.

Norström, E., Scott, L., Partridge, T. C., Risberg, J. and Holmgren, K. (2009). Reconstruction of environmental and climate changes at Braamhoek wetland, eastern escarpment South Africa, during the last 16,000 years with emphasis on the Pleistocene–Holocene transition. *Palaeogeography, Palaeoclimatology, Palaeoecology*, 271, 240–258.

Parkington, J., Cartwright, C., Cowling, R. M., Baxter, A. and Meadows, M. (2000). Palaeovegetation at the Last Glacial Maximum in the Western Cape, South Africa: Wood charcoal and pollen evidence from Elands Bay Cave. *South African Journal of Science*, 96, 543–546.

Partridge, T. C., deMenocal, P. B., Lorentz, S. A., Paiker, M. J. and Vogel, J. C. (1997). Orbital forcing of climate over South Africa: A 200,000-year rainfall record from the Pretoria saltpan. *Quaternary Science Reviews*, 16, 1125–1133.

Quick, L. J., Chase, B. M., Meadows, M. E., Scott, L. and Reimer, P. J. (2011). A 19.5 kyr vegetation history from the central Cederberg Mountains, South Africa: Palynological evidence from rock hyrax middens. *Palaeogeography, Palaeoclimatology, Palaeoecology*, 309, 253–270.

Schefuß, E., Kuhlmann, H., Mollenhauer, G., Prange, M. and Pätzold, J. (2011). Forcing of wet phases in southeast Africa over the past 17,000 years. *Nature*, 480, 509–512.

Schulze, R. E. (1997). *South African Atlas of Agrohydrology and Climatology*. Pretoria: Water Resource Commission Report TT82/96, 276pp.

Scholes, R. J. (1997). Savanna. In *Vegetation of Southern Africa*, eds. R. M. Cowling, D. M. Richardson and S. M. Pierce. Cambridge: Cambridge University Press, pp. 258–277.

Scott, L. (1999). Vegetation history and climate in the Savanna biome South Africa since 190,000 ka: A comparison of pollen data from the Tswaing Crater (the Pretoria Saltpan) and Wonderkrater. *Quaternary International*, 57/58, 215–223.

Scott, L. (2002). Grassland development under glacial and interglacial conditions in southern Africa: Review of pollen, phytolith and isotope evidence. *Palaeogeography, Palaeoclimatology, Palaeoecology*, 177, 47–57.

Scott, L., Holmgren, K., Talma, A. S., Woodborne, S. and Vogel, J. C. (2003). Age interpretation of the Wonderkrater spring sediments and vegetation change in the

savanna biome, Limpopo Province, South Africa. *South African Journal of Science*, 99, 484–488.
Scott, L., Marais, E. and Brook, G. A. (2004). Fossil hyrax dung and evidence of Late Pleistocene and Holocene vegetation types in the Namib Desert. *Journal of Quaternary Science*, 19, 829–832.
Scott, L. and Thackeray, J. F. (1987). Multivariate analysis of Late Pleistocene and Holocene pollen spectra from Wonderkrater, Transvaal, South Africa. *South African Journal of Science*, 83, 93–98.
Scott, L. and Woodborne, S. (2007). Pollen analysis and dating of Late Quaternary faecal deposits (hyraceum) in the Cederberg, Western Cape, South Africa. *Review of Palaeobotany and Palynology*, 144, 123–134.
Shi, N., Dupont, L. M., Beug, H.-J. and Schneider, R. (2000). Correlation between vegetation in southwestern Africa and oceanic upwelling in the past 21,000 years. *Quaternary Research*, 54, 72–80.
Shi, N., Schneider, R., Beug, H.-J. and Dupont, L. M. (2001). Southeast trade wind variations during the last 135 kyr: Evidence from pollen spectra in eastern South Atlantic sediments. *Earth and Planetary Science Letters*, 187, 311–321.
Stager, J. C., Mayewski, P. A., White, J., Chase, B. M., Neumann, F. H., Meadows, M. E., King, C. D. and Dixon, D. A. (2012). Precipitation variability in the winter rainfall zone of South Africa during the last 1400 yr linked to the austral westerlies. *Climate of the Past*, 8, 877–887.
Stokes, S., Haynes, G., Thomas, D. S. G., Horrocks, J. L., Higginson, M. and Malifa, M. (1998). Punctuated aridity in southern Africa during the last glacial cycle: The chronology of linear dune construction in the northeastern Kalahari. *Palaeogeography, Palaeoclimatology, Palaeoecology*, 137, 305–322.
Stute, M. and Talma, A. S., (1997). Isotope techniques in the study of past and current environmental changes in the hydrosphere and the atmosphere. *IAEA Vienna Symposium 1997, Isotopic Techniques in the Study of Environmental Change*. Vienna: International Atomic Energy Agency, pp. 307–318.
Talma, A. S. and Vogel, J. C. (1992). Late Quaternary paleotemperatures derived from a speleothem from Cango Caves, Cape Province, South Africa. *Quaternary Research*, 37, 203–213.
Thomas, D. S. G., Knight, M. and Wiggs, G. F. S. (2005). Remobilization of southern African desert dune systems by twenty-first century global warming. *Nature*, 435, 1218–1221.
Truc, L., Chevalier, M., Favier, C., Cheddadi, R., Meadows, M. E., Scott, L., Carr, A. S., Smith, G. F. and Chase, B. M. (2013). Quantification of climate change for the last 20,000 years from Wonderkrater, South Africa: Implications for the long-term dynamics of the Intertropical Convergence Zone. *Palaeogeography, Palaeoclimatology, Palaeoecology*, 386, 575–587.
Valsecchi, V., Chase, B. M., Slingsby, J. A., Carr, A. S., Quick, L. J., Meadows, M. E., Cheddadi, R. and Reimer, P. J. (2013). A high resolution 15,600-year pollen and microcharcoal record from the Cederberg Mountains, South Africa. *Palaeogeography, Palaeoclimatology, Palaeoecology*, 387, 6–16.
van Wilgen, B. W., Richardson, D. M., Kruger, F. J. and van Hensbergen, H. J. (eds) (1992). *Fire in South African Mountain Fynbos: Ecosystem, Community, and Species Response at Swartboskloof.* Berlin: Springer-Verlag, 325pp.
Wu, H., Guiot, J., Brewer, S. and Guo, Z. (2007). Climatic changes in Eurasia and Africa at the last glacial maximum and mid-Holocene: Reconstruction from pollen data using inverse vegetation modelling. *Climate Dynamics*, 29, 211–229.

18

Faunal evidence for mid- and late Quaternary environmental change in southern Africa

JAMES S. BRINK

Abstract

Southern Africa is differentiated from other centres of aridity in Africa by the presence of an extended island of elevated, essentially treeless habitat in the central interior, known as the Highveld and the Karoo. This area coincides botanically with the Nama-Karoo and the Grassland Biomes. The large geographic extent of this habitat is unique to southern Africa, since it has no exact equivalent in modern-day east or north Africa. This uniqueness is reflected in the large herbivores of the central interior, the grazers and mixed feeders adapted to permanently available open habitat, which defines the endemic faunal character of the subregion. This contribution presents some of the faunal evidence for the appearance of permanently open habitat in central southern Africa, a process that formed part of a longer-term trend of faunal adaptation to aridification and global cooling that was initiated within the last 1 Ma, in a time known as the Cornelian Land Mammal Age (LMA). A secondary and overlapping theme deals with the appearance of lakes and wetlands on a subregional scale during the Florisian LMA, which lasted from c. 0.6 Ma to the end of the Pleistocene/early Holocene. The end of the Florisian LMA coincided with the regional extinction of wetland faunas in the interior and with the extinction of specialised grazing ungulates over the entire subregion, leading into the semi-arid conditions seen in the larger part of modern-day southern Africa.

18.1 Introduction

Climate change is often cited as a causal factor in biotic turnover (Vrba, 1995; Behrensmeyer, 2006). It has been suggested that in high latitude areas, temperature is the dominant climatic variable governing environmental change, but that precipitation may be more dominant in low latitudes (deMenocal and Bloemendal, 1995). The southern African faunal evidence appears to be in accord, pointing to a long-term trend towards drier climates throughout the Cenozoic (Pickford and Senut, 1999). Southern Africa shares with north and east Africa a semi-arid to arid climate, which gave rise to savannah grasslands, open grasslands and semi-deserts, providing the environmental basis for the evolution of characteristic large mammal faunas (Bigalke, 1978; Geraads, 1981; Skinner and Smithers, 1990; Dupont and Leroy, 1995; Kingdon, 1997). In the south, the appearance of aridity can be linked to the initiation during the mid-Cenozoic of the cold Benguela Current along the southwest coast of Africa, and to major tectonic uplift of the central plateau, following the separation of South America and Africa and the breakup of the Gondwana supercontinent (Axelrod and Raven, 1978, Pickford and Senut, 1999; Partridge and Maud, 2000; Tinker et al., 2008; McCarthy, 2013). This caused an island of essentially treeless habitat, an area commonly known as the Highveld and the Karoo, which botanically coincides with the Nama-Karoo and the Grassland Biomes (Mucina and Rutherford, 2006) (Fig. 18.1). The large herbivores of the central interior are mainly grazers and mixed feeders adapted to the permanently available open habitat of this region. They define the endemic faunal character of the subregion, and the marker taxa are the black wildebeest *Connochaetes gnou*, blesbok *Damaliscus pygargus*, and springbok *Antidorcas marsupialis* (Skinner and Smithers, 1990). To the north of the central interior, savannah grasslands are found, which contain some woody components (Fig. 18.1), but they are in structure and composition not very different from those of east Africa (Kingdon, 1997).

The aim of this chapter is to review the faunal evidence for the appearance of permanently open habitat in central southern Africa. This process was initiated during the Cornelian Land Mammal Age (LMA) at ~1 Ma, and formed part of a long-term trend of faunal adaptation to aridification and global cooling. Open grasslands became fully established during the Florisian LMA, which is the subsequent faunal stage (Tables 18.1–18.3). In addition, faunal evidence is provided for the appearance of wetlands on a subregional scale during the Florisian. The Florisian LMA lasted from ~0.6 Ma to the beginning of the Holocene and is defined by a number of now-extinct grazing ungulates, occurring in both the interior and coast (see below), and by a wetland faunal component in the interior. Wetlands are not unique to any given time, but their geographic extent during the mid- to late Quaternary was significant and provided a unique faunal signal that intersected the open habitat of the central interior and surrounding savannah, and excluded only the Cape montane and coastal areas (Fig. 18.1).

The thin strip of montane and coastal vegetation, which fringes the central plains to the south and east of southern Africa, has a distinctive mammalian fauna,

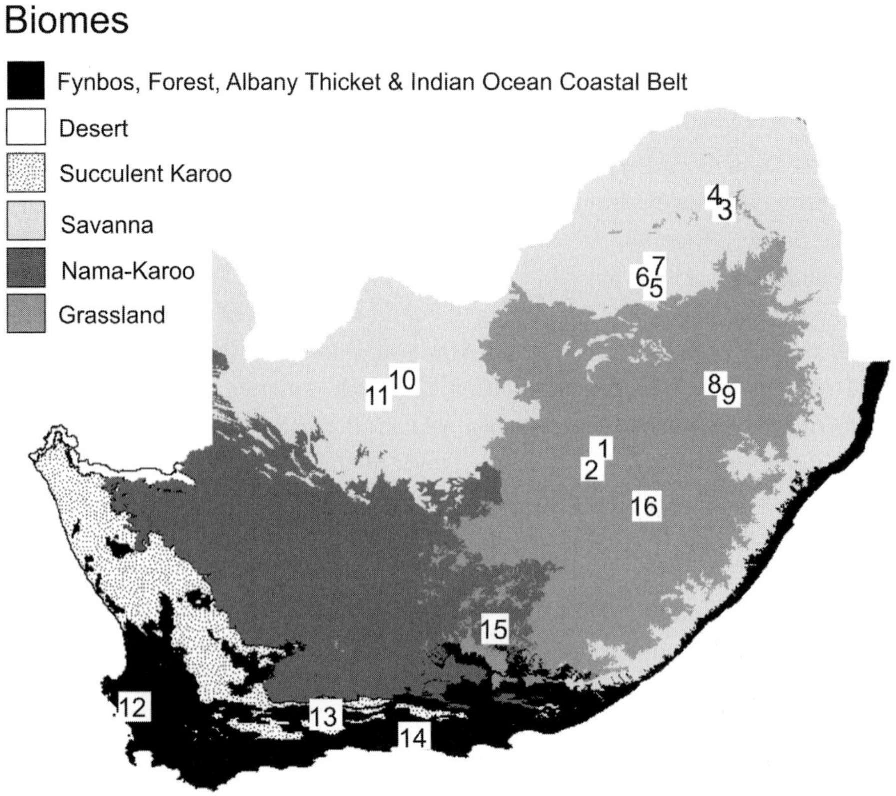

Fig. 18.1. Biomes of southern Africa (shaded), showing the fossil localities (numbered) referred to in the text and figures: 1. Florisbad, 2. Erfkroon, 3. Buffalo Cave, 4. Makapan Limeworks, 5. Gladysvale External Deposits, 6, Sterkfontein, 7. Swartkrans, 8. Cornelia-Uitzoek, 9. Cornelia-Mara, 10. Wonderwerk Cave, 11. Kathu Pan, 12. Elandsfontein Main, 13. Boomplaas Cave, 14. Klasies River, 15. Cradock Spring.

adapted to more closed, woody habitat, as seen in the Cape Fynbos, Albany Thicket and Indian Ocean Coastal Belt (Fig. 18.1). This area is biogeographically peripheral to the central plains, and particularly in the Fynbos Biome there is a unique endemic mammal fauna (Klein, 1983, 1984; Klein et al., 2007; Faith, 2013). Throughout the glacials of the mid- and late Quaternary, the Cape coastal zone periodically experienced incursions of plains game from the interior when the exposed continental margin provided additional habitat for grazing ungulates (Klein, 1983; Brink, 1993, 2005). Here, this area is referred to mainly for the insight that it provides in the early to mid-Quaternary, specifically the older Cornelian materials from Elandsfontein, known as 'Elandsfontein Main' (Klein et al., 2007). The Cape Florisian shares a number of open-habitat taxa with the

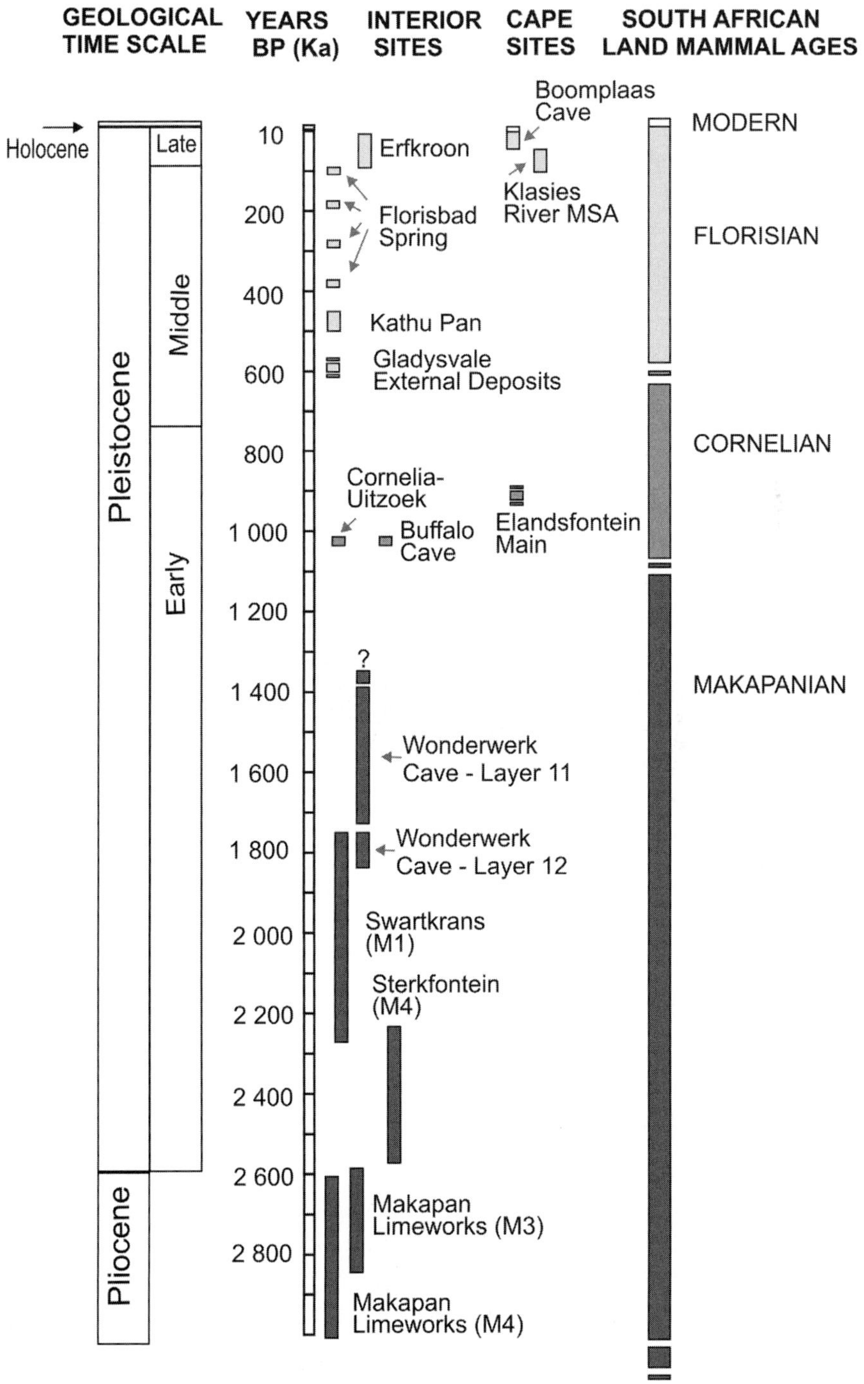

Fig. 18.2. Revised Pleistocene biochronology for southern Africa, including the name localities for the Land Mammal Ages (Makapan Limeworks, Cornelia-Uitzoek and Florisbad). Data from Hendey (1974), Grün et al. (1996), Lacruz et al. (2002), Herries et al. (2006, 2009); Chazan et al. (2008); Porat et al. (2010); Brink et al. (2012) and Braun et al. (2013).

interior Florisian, of which it is an impoverished version. These taxa, including an as yet unnamed caprine species, became extinct in the Cape ecozone at the end of the Pleistocene, coinciding with a postglacial, eustatic rise in sea level and a general reduction in grassland availability (Klein, 1983, 1984; Deacon et al., 1984; Brink, 1999, 2005; Faith, 2013, 2014).

18.2 Mammalian turnover and biogeography in southern Africa

The mammalian fossil record of southern Africa is divided into a sequence of Land Mammal Ages (LMA), which are periods of geologic time distinguished by their distinctive faunal character over a large area, and are defined by the presence or absence of time-sensitive taxa (Savage and Russel, 1983). In descending order of geological age and starting with the early Miocene, the southern African LMA scheme includes the Namibian, Langebaanian, Makapanian, Cornelian, Florisian and the Recent (Holocene) (Hendey, 1974; Klein, 1984) (Table 18.1). This scheme was first established during the mid-20th Century with attempts to create chronological and spatial order out of the growing complexity of the late Cenozoic fossil record (e.g. Cooke, 1952; Wells, 1962; Ewer and Cooke, 1964) (Table 18.1). The use of local southern African names reflects the difficulty of correlating with east Africa (Cooke, 1967), a trend that becomes more pronounced after 1 Ma (Brink et al., 2012). More recent studies of Quaternary fossil mammal studies in southern Africa have benefitted from Electron Spin Resonance (ESR), Optically Stimulated Luminescence (OSL) and palaeomagnetism dating techniques that allowed a major revision of the chronology of mid- to late Quaternary fossil mammal localities. It has now become possible to place the name localities of the LMA scheme and other important assemblages in a more accurate temporal order (Grün et al., 1996; Herries et al., 2009; Brink et al., 2012) (Fig. 18.2).

Table 18.1 *Southern African Land Mammal Ages in relation to approximate geological time periods, adapted from Hendey (1974).*

Land Mammal Age	Approximate geological age
Recent	Holocene
Florisian	Middle & late Pleistocene
Cornelian	End-early Pleistocene
Makapanian	Plio-Pleistocene
Langebaanian	Early Pliocene
Namibian	Miocene

The middle and late Pleistocene trend in southern African faunas towards biogeographic uniqueness is tempered by evidence for some faunal interchange with east Africa. This was first noted by zoologists (e.g. Balinsky, 1962) who proposed a southwest–northeast arid corridor linking the faunas of southwest Africa and northeast Africa at various times in the past. Wells (1962) applied this idea to the Pleistocene faunas of southern Africa. Previous biogeographic connections between southern and east Africa are evident also in historic and pre-Holocene distribution patterns of plains zebra *Equus quagga* subspp., hartebeest *Alcelaphus buselaphus* subspp, and blue wildebeest *Connochaetes taurinus* subspp. (Geraads, 1981; Kingdon, 1997). Such taxa provide some basis for correlation with east Africa during the mid- to late Quaternary. It is noteworthy that genetic data support periodic biogeographic connections between southern and east Africa. Changes between wetter and dryer climate periods may have allowed dispersal corridors to develop, facilitating the migration of large mammals adapted to varying palaeohabitats. In particular, the genetic data support the idea of evolutionary stability during the late Quaternary in southern Africa, as opposed to the biogeographic and evolutionary dynamism of east Africa at this time (Lorenzen et al., 2012).

18.3 The southern African faunal record

18.3.1 The end of the early Pleistocene: the Cornelian LMA

Biogeographic exchange between southern and east Africa is more evident in the earlier fossil record, such as the Makapanian assemblages (De Ruiter, 2003; Gilbert, 2008; Bernor et al., 2010; Gentry, 2010; Werdelin and Peigné, 2010; Reynolds and Kibii, 2011). The effects of these dispersals can be seen in the herbivores of the Cornelian assemblages. Taxa shared by southern and east Africa, but which became extinct in southern Africa before the Florisian, may be considered archaic in the Cornelian context. Examples of archaic forms from Cornelia-Uitzoek are the hipparion *Eurygnathohippus cornelianus*, the suid genera *Kolpochoerus* and *Metridiochoerus*, and the hippo *Hippopotamus gorgops* (Van Hoepen, 1932, 1947; Cooke, 1974; Harris and White, 1979; Bernor et al., 2010) (Table 18.2). At Elandsfontein, the archaic faunal component is more pronounced and includes *Theropithecus oswaldi*, *Megantereon whitei*, the genus *Kolpochoerus*, *Hippotragus gigas*, *Numidocapra arambourgi* and the genera *Sivatherium* and *Gazella* (Gentry and Gentry, 1978; Klein et al., 2007). *Numidocapra arambourgi* occurs at Buffalo Cave and Elandsfontein, but not at Cornelia-Uitzoek or at Cornelia-Mara (Brink et al., 2012). The presence of a gazelle and an extinct carnivore at Elandsfontein is particularly interesting, since

Table 18.2 *Taxonomic list of Cornelian faunas: a comparison between Cornelia-Uitzoek, Buffalo Cave and Elandsfontein Main. Extinct taxa also found in East Africa are underlined. Faunal lists modified and adapted from Herries et al. (2006), Klein et al. (2007) and Brink et al. (2012). † = Extinct*

	Cornelia-Uitzoek	Buffalo Cave	Elandsfontein Main
Rodentia			
Bathyergus suillus	-	-	X
Hystrix africaeasutralis	-	-	X
Lagomorpha			
Lepus capensis	-	-	X
Primates			
Homo sp.	X	-	X
Cercopithecidae indet.	-	X	-
Theropithecus oswaldi †	-	-	X
Pholidota			
Manis sp.	-	-	X
Carnivora			
Ictonyx striatus	-	-	X
Mellivora capensis	-	-	X
Suricata suricatta	-	-	X
Viverra civetta	-	-	X
Herpestes ichneumon	-	-	X
Atilax paludinosus	-	-	X
Canis mesomelas	-	-	X
Vulpes chama	-	-	X
Lycaon pictus	-	-	X
Crocuta crocuta	-	-	X
Parahyaena brunnea	-	-	X
Felis libyca	-	-	X
Felis caracal/serval	-	-	X
Panthera leo	X	cf.	X
Panthera pardus	-	-	X
<u>*Megantereon whitei*</u> †	-	-	X
Proboscidea			
Loxodonta atlantica zulu †		-	X
Indet.	X	-	-
Perissodactyla			
<u>*Eurygnathohippus cornelianus*</u> †	X	X	-
<u>*Hipparionini* indet.</u> †	-	X	-
Equus capensis †	X	-	X
Equus quagga	X	-	X
Equus sp.	X	-	-
Diceros bicornis	-	-	X
Ceratotherium simum	-	-	X
Rhinocerotidae indet.	X	-	-
Artiodactyla			
<u>*Hippopotamus gorgops*</u> †	X	-	-
Hippopotamus amphibius	-	-	X

Table 18.2 (*cont.*)

	Cornelia-Uitzoek	Buffalo Cave	Elandsfontein Main
Phacochoerus sp.	X	X	X
Metridiochoerus modestus †	X	-	-
Metridiochoerus compactus †	X	-	X
Kolpochoerus paiceae †	X	-	X
Sivatherium maurusium †	X	-	X
Taurotragus oryx	-	-	X
Tragelaphus strepsiceros	-	-	X
Syncerus antiquus †	X	?	X
Redunca arundinum	-	?	X
Hippotragus gigas †	-	?	X
Hippotragus leucophaeus	-	-	X
Damaliscus aff. lunatus	-	-	X
Damaliscus niro †	X	-	X
Damaliscus sp.	-	X	-
Numidocapra arambourgi †	-	X	X
Parmularius sp. †	-	X	cf.
Megalotragus eucornutus †	X	-	-
Megalotragus priscus †	-	-	X
Connochaetes gnou laticornutus	X	-	X
Connochaetes sp.	-	X	-
Budorcas makapaani †	-	X	-
Caprini indet. †	-	-	X
Gazella sp. †	-	-	X
Antidorcas recki †	X	-	X
Antidorcas australis †	-	-	X
Antidorcas bondi †	X	-	X
Aepyceros helmoedi †	X	-	-
Sylvicapra grimmia	X	-	-
Raphicerus melanotis	-	-	X
Raphicerus sp.	X	-	-

both are absent from Buffalo Cave and Cornelia-Uitzoek (Table 18.2). They may represent evolutionary remnants atypical of the Cornelian in general, and probably reflect the biogeographically distant position of the Cape coastal zone, where archaic forms are known to occur, as seen in the early Pliocene locality of Langebaanweg (Gentry, 1980). Taxa that may reflect early Pleistocene endemism in southern Africa are the giant wildebeest *Megalotragus eucornutus*, and the impala *Aepyceros helmoedi* (Van Hoepen, 1932, 1947; Brink, 2005; Faith, 2014).

Damaliscus niro is found at Olduvai (Gentry and Gentry, 1978), but at Cornelia-Uitzoek it is in an intermediate, Cornelian, stage of evolution (Thackeray and

Brink, 2004). *D. niro* became extinct in east Africa during the middle Pleistocene, but in the south it continued to evolve into a Florisian stage, before the lineage became extinct at the end-Pleistocene. An underived form of the extinct grazing springbok *Antidorcas bondi* is abundant at Cornelia-Uitzoek, but occurs in small numbers in the Elandsfontein Main assemblage and is so far not recorded at Buffalo Cave (Herries et al., 2006; Brink et al., 2012). Later, during the Florisian LMA, *A. bondi* evolved into a specialised grazer, characterised by extreme hypsodonty (Brink and Lee-Thorp, 1992; Brink et al., 2013). It is noteworthy that, with the exception of the long-horned buffalo *Syncerus antiquus*, none of the Cornelian herbivore taxa that have Florisian descendants and which became extinct at the end of the Florisian LMA are recorded in east Africa during the middle and late Pleistocene (Faith, 2014) (Table 18.3). This reflects the increasingly regionalised character of the large mammal faunas of southern Africa in relation to those of east Africa after 1 Ma.

The Cornelian LMA was a time of faunal transition and may have lasted from ~1 Ma to 0.6 Ma. The earliest-recorded full Florisian fauna is from the Gladysvale External Deposits, with a minimum (maximum) age of around 0.58 (0.78) Ma (Lacruz et al., 2002) (Fig. 18.2). Stable isotopes and the reduced taxonomic diversity in the faunal assemblages from Cornelia suggest that the interior Cornelian tended towards semi-arid conditions, whereas the evidence from Elandsfontein suggests that during the Cornelian the coastal areas may have been wetter than during the Florisian, in contrast to the situation in the interior (Codron et al., 2008; Brink et al., 2012; Braun et al., 2013).

18.3.2 Black wildebeest **Connochaetes gnou** *as a proxy for open habitat*

The black wildebeest is a southern African endemic with a remarkably complete local fossil record, which starts in the Cornelian LMA (Brink, 1993, 2005). It does not occur in east Africa during the time equivalent to the Cornelian LMA, or later (Gentry, 2010; Faith, 2014), although there are small-bodied wildebeest fossils from Olduvai, referred to *C. africanus*, which is considered to be similar to the black wildebeest (Gentry and Gentry, 1978).

The earliest black wildebeest specimens from Cornelia–Uitzoek are very close in morphology to contemporary blue wildebeest *C. taurinus prognu*, from which it evolved, and which had a pan-African distribution during the end-early Pleistocene and early-middle Pleistocene (Gentry and Gentry, 1978; Geraads, 1981; Gilbert, 2008). This evolutionary process was associated first with morphological changes in the skull, dentitions and horns that reflect adaptations to a specialised territorial reproductive behaviour; then by a reduction in body size and changes in body proportions (Brink, 1993, 2005) (Fig. 18.3a, b).

Table 18.3 *Taxonomic list of Florisian faunas: a comparison between Florisbad Spring, Erfkroon Last Interglacial levels (L/I) and the Gladysvale External Deposits. Faunal lists modified and adapted from Brink (1987, 1994), Churchill et al. (2000), Lacruz et al. (2002), and Brink et al. (2015). The extinct taxa also found in east Africa are underlined. †† = Extinct, † = Regionally extinct.*

	Florisbad Spring	Erfkroon L/I	Gladysvale External Deposits
Primates			
Homo helmei	X	-	-
Homo sp.	-	-	X
Papionini indet.	X	-	-
Carnivora			
Aonyx capensis	X	-	-
Cynictis penicillata	-	X	-
Galerella sanguinea	X	X	-
Atilax paludinosus	X	X	-
Canis mesomelas	X	X	X
Vulpes chama	-	X	-
Lycaon pictus	X	cf.	-
Crocuta crocuta	X	X	-
Panthera leo	X	cf.	X
Perissodactyla			
Equus capensis ††	X	X	X
Equus lylei ††	X	X	-
Equus quagga subsp.	X	X	X
Ceratotherium simum	X	X	-
Artiodactyla			
Hippopotamus amphibius	X	X	-
Phacochoerus africanus	X	-	-
Phacochoerus aethiopicus	X	-	-
Phacochoerus sp.	-	X	X
Giraffidae indet.	-	-	X
Taurotragus oryx	X	-	X
Tragelaphus streptsiceros	-	-	X
<u>*Syncerus antiquus*</u> ††	X	X	X
Kobus leche †	X	X	X
Kobus sp.†	X	X	-
Redunca arundimum	-	-	X
Redunca fulvorufula	-	-	X
Hippotragus sp. †	X	-	X
Damaliscus niro ††	X	X	-
Damaliscus pygargus	X	X	X
Alcelaphus buselaphus	X	X	-
Connochaetes gnou	X	X	-
Megalotragus priscus ††	X	X	-
Caprini indet. ††	-	-	X
Antidorcas bondi ††	X	X	X
Antidorcas marsupialis	X	X	-
Raphicerus campestris	X	-	-

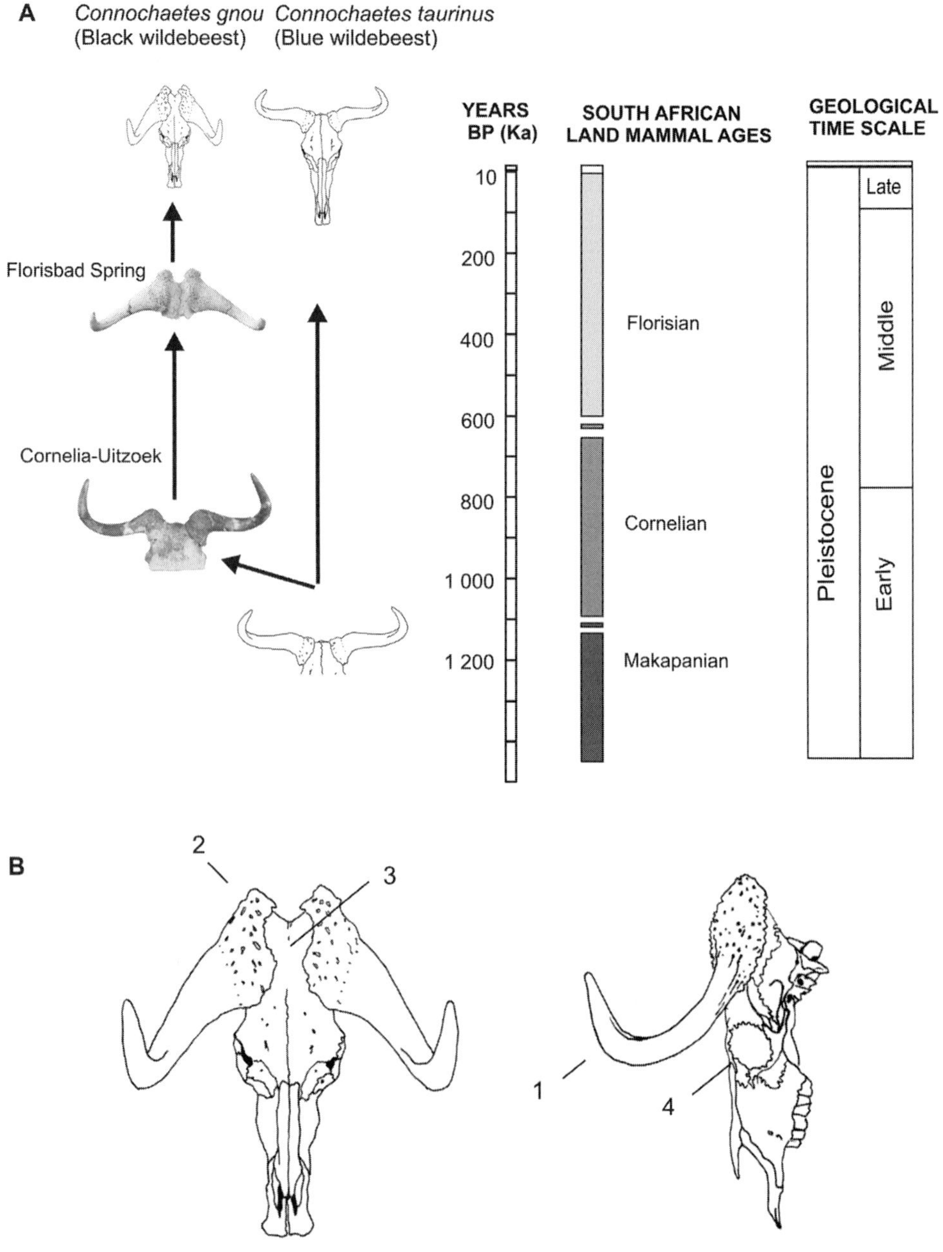

Fig. 18.3. (A) The temporal pattern of morphological change in the skulls of the black wildebeest as it evolved from a blue wildebeest ancestor. (B) The characteristic features of black wildebeest skulls are present in underived states in the earliest fossils from Cornelia-Uitzoek: 1) forward curving horns, 2) enlarged basal bosses 3) fused frontals' suture and 4) enlarged orbits. These features reflect the adaptation of black wildebeest to extreme territorial behaviour, which can only function in open, visually unobstructed habitat (adapted from Brink, 2005).

Black wildebeest are permanently territorial, and males patrol their territories by vision and defend them aggressively. This is seen in the increased size of the orbits, evidently linked to the greater need for vision in patrolling territories, and in the structural adaptations for increased stability of the skull, needed during head-to-head contact in the defence of territories. In contrast, its living ancestor, the blue wildebeest has a greater reliance on smell (Attwell, 1977; Brink, 2005). The morphological adaptations in black wildebeest skulls and dentition, therefore, do not reflect a trophic shift (Codron and Brink, 2007), but can be directly associated with adaptation towards greater territoriality in breeding behaviour, which can only function in open, unobstructed habitat. These properties point to the Highveld and Karoo regions of central southern Africa as the place of origin of this species.

The speciation of black wildebeest can be directly dated based on fossil evidence from Cornelia-Uitzoek, which has a palaeomagnetic age of around 1.0 Ma (Brink et al., 2012). This is supported independently by genetic studies (Corbet and Robinson, 1991; Corbet et al., 1994). The timing is significant and suggests a connection with increased global cooling and the initiation of the ~100 kyr eccentricity-driven glacial–interglacial cycle soon after 1.0 Ma. It is also likely that increased incidence of fire, which is related to the development of grasslands (Hoetzel et al., 2013), may have contributed towards greater openness in the Cornelian grasslands, and would have counteracted the propagation of woody plants (Bigalke and Willan, 1984). Regardless of causal factors, the evolution of the black wildebeest is evidently closely linked to the development of a Highveld-type open habitat. Thus, the presence of black wildebeest in the fossil record can be used as a proxy for the presence of open grassland habitats.

18.3.3 The middle and late Pleistocene: the Florisian LMA

Following the extinction of the Cornelian archaic component, an essentially modern-looking mammal fauna emerged (Table 18.3), with most of the lineages continuing into extant populations, excluding six specialised grazing ungulates that became extinct in southern Africa at the end of the Pleistocene. In addition, an as yet unnamed grazing caprine became extinct at the end of the late Pleistocene in the Cango valley and by the mid-Holocene in the northeastern Cape (Brink, 1999). The Cornelian–Florisian turnover brought about highly productive grasslands in the central interior with wetlands covering most of the interior of southern Africa (Brink, 1987; Brink and Lee-Thorp, 1992) (Fig. 18.4). It should be noted that a sivathere (?*Sivatherium maurusium*) was reported to occur at Florisbad and an extinct proboscidean (*Elephas recki*) probably survived for some time after the

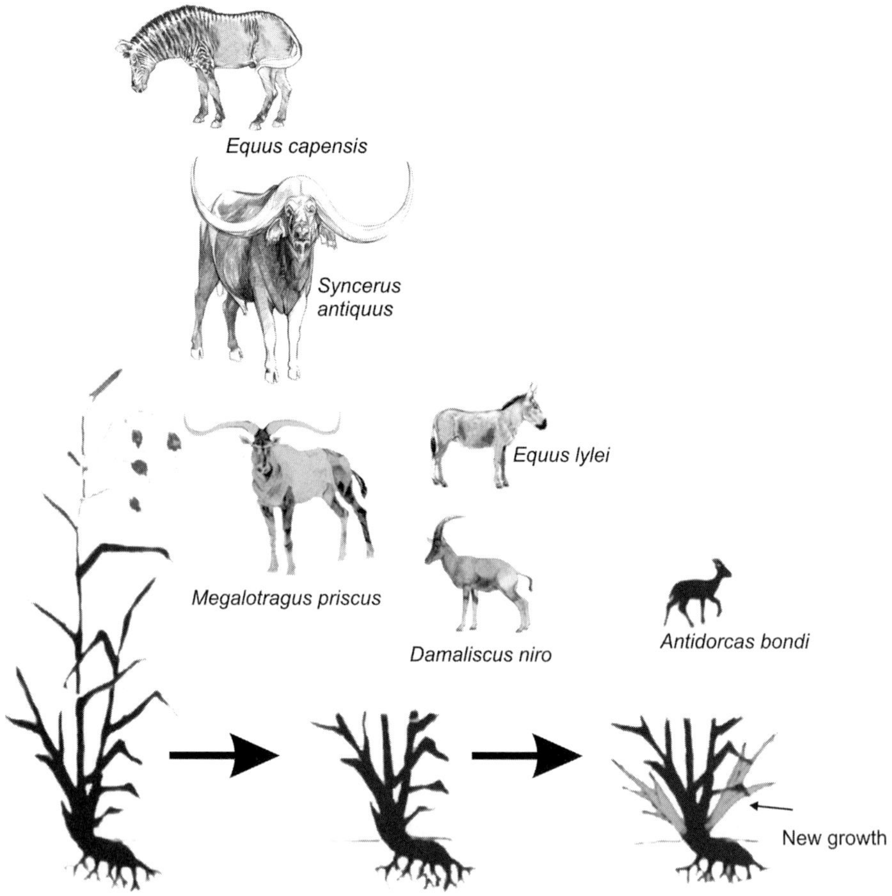

Fig. 18.4. A facilitating grazing succession suggested for the Florisian grasslands of the central interior of southern Africa, in which extinct and extant larger-bodied grazers prepared and maintained grasslands in a state of regrowth, providing niches for smaller-bodied grazers. The feeding niche of Bond's springbok *Antidorcas bondi* was that of the smallest of the specialised grazers and provided the basis for suggesting the existence of unusually productive grasslands during the Florisian that needed substantially increased precipitation compared to modern conditions (see Fig. 18.5) (adapted from Brink and Lee-Thorp, 1992).

Cornelian–Florisian transition (Klein, 1984; Porat et al., 2010). Also, an extinct grazing pig *Metridiochoerus compactus* has been recorded from Florisian contexts at Redcliff Cave. These occurrences are exceptional and not typical of the interior Florisian, and they are not included in Table 18.3.

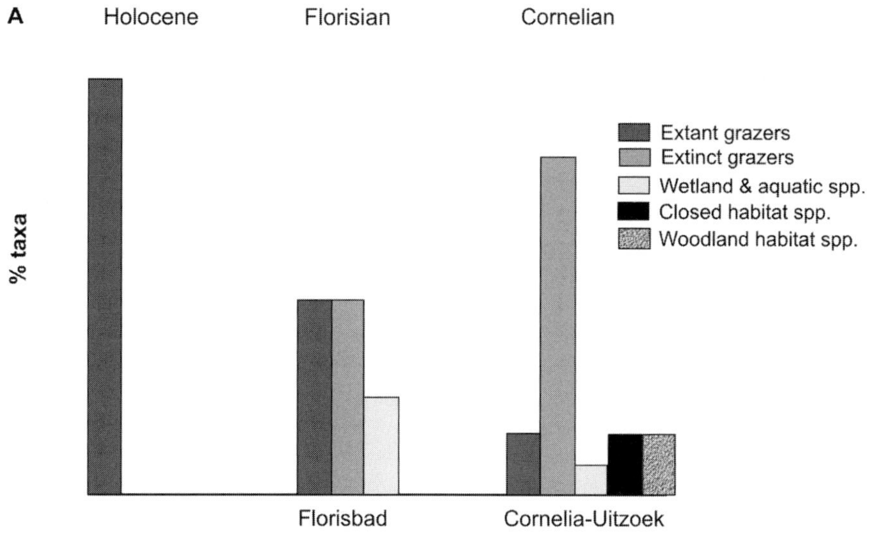

Fig. 18.5. A temporal trend in habitat types in the central interior of southern Africa (A). Cornelian grasslands containing some woody and closed-habitat component gave way to open grasslands and wetlands during the Florisian LMA. The Florisian wetlands terminated at the end of the late Pleistocene due to widespread aridification of the central interior of southern Africa with the local extinction of the wetland forms and the extinction of the six specialised grazers (see Fig. 18.4). The Florisian grasslands were linked to an extensive palaeolake system, which today survives as dry pans, of which an example is given from near Florisbad (B). The fossil presence of Bond's springbok and lechwe *Kobus leche* reflects the existence of palaeolakes and wetlands across the landscape during the Florisian LMA (adapted from Grobler and Loock, 1988; Brink, 2005).

18.3.4 Florisian wetlands and the end-Pleistocene extinction

Florisian grasslands existed between 0.6 and 0.01 Ma (Fig. 18.2) and are primarily indicated by the association of an evolved form of Bond's springbok *Antidorcas bondi* with a range of wetland and aquatic taxa, such as waterbuck (*Kobus ellipsiprymnus*), lechwe (*K. leche*) and hippo (*Hippopotamus amphibius*). During the Florisian LMA, Bond's springbok reached the climax of its adaptation as a specialised small-bodied grazer, and it has been suggested that this niche was facilitated by larger-bodied grazers, both extinct and extant, and by markedly increased precipitation (Brink and Lee-Thorp, 1992; Brink, 2005; Codron et al., 2008) (Fig. 18.4). The Florisian wetland and aquatic taxa were closely associated with a palaeolake system that extended across central southern Africa into eastern Zimbabwe, southern Zambia and northern Botswana, where today a Florisian remnant wetland fauna survives. The Florisian palaeolake system is today visible in the pan veld of the central and western interior, such as around Florisbad (Loock and Grobler, 1988) (Fig. 18.5).

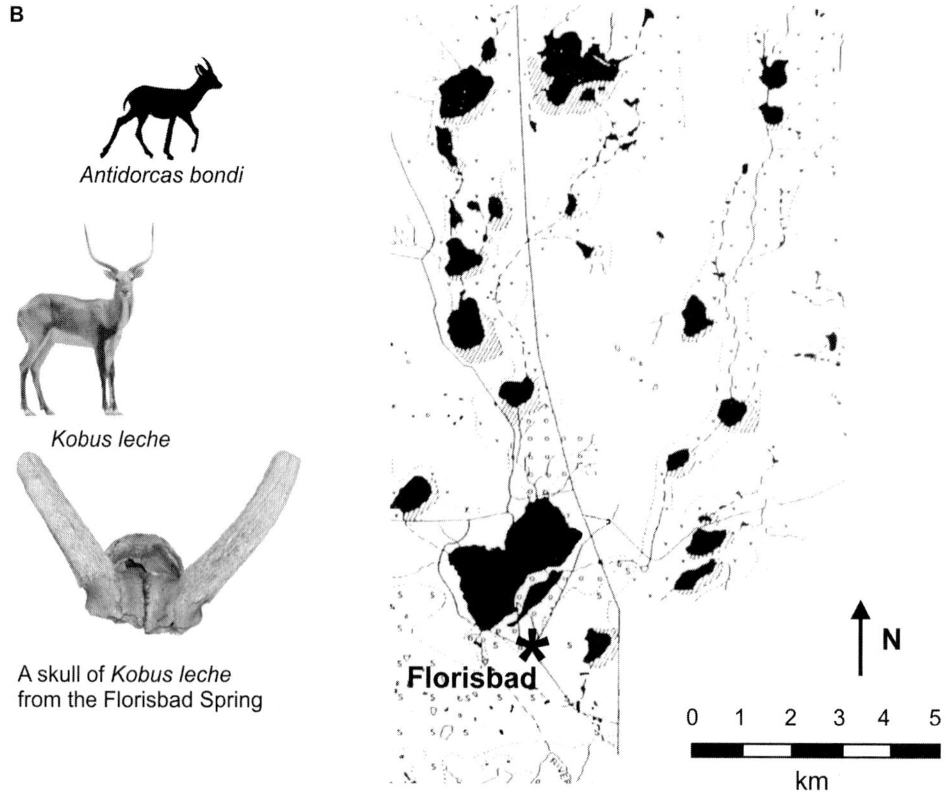

Fig. 18.5. (*cont.*)

It is also present in the Florisbad spring mound record, where high lake levels are recorded until immediately before the start of the Holocene (Visser and Joubert, 1990). Based on the occurrence of lechwe and Bond's springbok, it is clear that the Florisian wetland system extended over much of the width of southern Africa and from Cradock in the south to southern Zimbabwe in the north (Fig. 18.6). These wetlands were anomalous and a temporary departure from the long-term trend towards aridification as seen in the sub-Saharan fossil record (Pickford and Senut, 1999; Bobe, 2006). The cause for the Florisian wetlands is still unclear, but in spite of palaeoclimatic fluctuations, which would have followed the glacial–interglacial cycles, conditions remained sufficiently stable to allow taxonomic stasis and the persistence of the wetland indicator taxa in this region over a period exceeding 0.5 Ma (Figs. 18.2, 18.5; Table 18.3). The stability in the Florisian record is also reflected in ungulate phylogeographic data, suggesting a stable, long-standing southern refuge (Lorenzen et al., 2012).

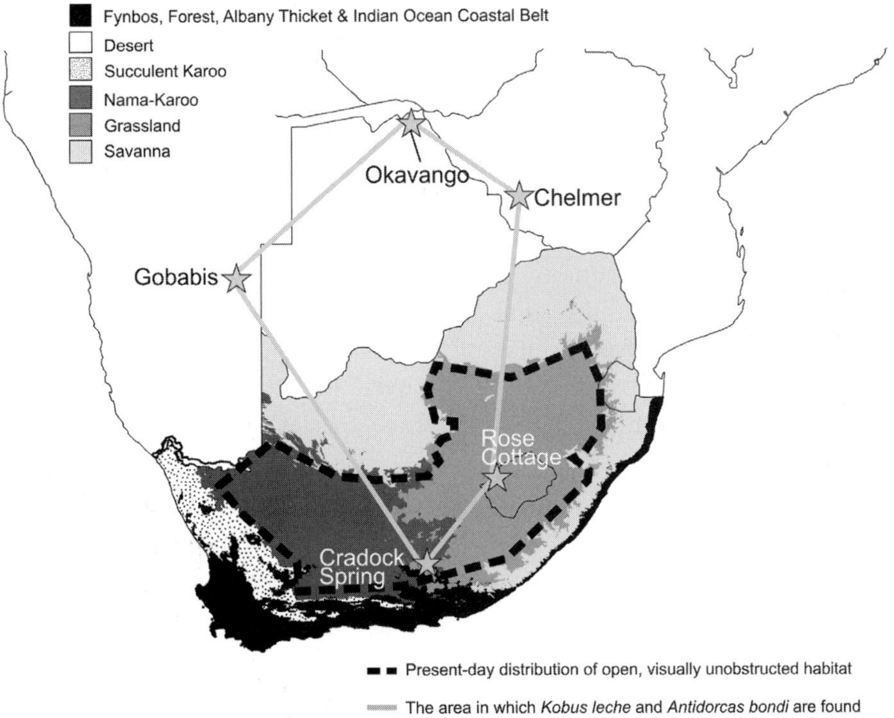

Fig. 18.6. The extent of late Quaternary fossil localities with Bond's springbok (*Antidorcas bondi*) and lechwe (*Kobus leche*). This distribution overlaps with modern open habitat (Grassland and Nama-Karoo Biomes) and the savanna grasslands (Savanna Biome) and gives a conservative impression of the geographic extent of the Florisian palaeolakes and wetlands.

The Florisian ecosystem was disrupted towards the end of the Pleistocene by intense aridification, when primary productivity of the grasslands was diminished and the palaeo-lakes dried up to be transformed into the modern semi-arid grassland and pan system (Loock and Grobler, 1988; Brink and Lee-Thorp, 1992). The interior fossil assemblages dating to around the Last Glacial Maximum record this transformation by a decline in frequency and eventual disappearance of wetland and aquatic fauna, such as Bond's springbok, lechwe and hippopotamus (Klein et al., 1991; Plug and Engela, 1992; Brink, 2005). The extinction of the Florisian grazing ungulates coincided with the disappearance of wetland elements at the end of the late Pleistocene/early Holocene (Klein, 1984; Thackeray, 1984; Brink, 2005). Extinction affected only the more specialised grazers, those with extremely large (*Equus capensis*, *Syncerus antiquus* and *Megalotragus priscus*), or very small body size (*Antidorcas bondi*), or those with close competitors (*D. niro* and *E. lylei*) (Brink and Lee-Thorp, 1992). The pattern of extinction is consistent with aridification and disruption of Florisian ecosystems through a marked reduction in primary productivity of interior grasslands. Grassland reduction and impoverishment also coincided with extinction in the Cape ecozone (Klein, 1983; Deacon et al., 1984; Faith, 2013, 2014). The clear evidence for major regional ecological shifts at the late Pleistocene–Holocene boundary argues against a human role in the end-Pleistocene extinctions. Also, the stability in the southern African archaeological record at this time (*vide* Deacon and Deacon, 1999) gives further support to this being a natural process (see also Faith, 2014). By the end of the Pleistocene and early Holocene, the process of extinction was complete in both the interior and in the Cape coastal zone.

18.4 Summary

Aridification provided the environmental basis for the evolution of large mammal faunas that are adapted to arid and semi-arid conditions in southern Africa. By the end of the early Pleistocene, southern endemism became more pronounced in Cornelian assemblages through extinction, dispersal and *in loco* evolution. Black wildebeest evolved locally from a blue wildebeest ancestor, and its presence since the Cornelian LMA reflects the appearance of open, Highveld-type grasslands as a stable landscape component. During the Florisian LMA, an essentially modern-looking fauna emerged, which included six extinct specialised grazing ungulates and an as yet unnamed caprine species in montane grassland areas. Florisian wetlands extended beyond modern biome boundaries into the wooded grassland areas, covering most of southern Africa, but excluded the coastal zones. Today some of the Florisian wetland forms survive as biogeographic remnants in the Okavango Delta of northern Botswana and in southern Zambia. The Florisian faunas were disrupted by extreme aridification towards the end of the late Pleistocene and early Holocene. This resulted in the extinction of the specialised grazers, the local extinction of the wetland forms and the emergence by the early Holocene of mainly arid and semi-arid adapted faunas, as seen historically over most of southern Africa.

References

Atwell, C. A. M. (1977). *Reproduction and population ecology of the blue wildebeest* (Connochaetes taurinus taurinus) *in Zululand*. Unpublished PhD dissertation, University of Natal, 303pp.

Axelrod, D. I. and Raven, P. H. (1978). Late Cretaceous and Tertiary vegetation history of Africa. In *Biogeography and Ecology of Southern Africa*, ed. M. J. A. Werger. The Hague: W. Junk, pp. 77–130.

Balinsky, B. I. (1962). Patterns of animal distribution on the African continent. *Annals of the Cape Provincial Museums*, 2, 299–310.

Behrensmeyer, A. K. (2006). Climate change and human evolution. *Science*, 311, 476–478.

Bernor, R., Armour-Chelu, M. J., Gilbert, H., Kaiser, T. M. and Schultz, E. (2010). Equidae. In *Cenozoic Mammals of Africa*, eds. L. Werdelin and W. J. Saunders. Berkeley: University of California Press, pp. 685–722.

Bigalke, R. C. (1978). Present-day mammals of Africa. In *Evolution of African Mammals*, eds. V. Maglio and H. S. B. Cooke. Cambridge: Harvard University Press, pp. 1–16.

Bigalke, R. C. and Willan, K. (1984). Effects of Fire Regime on Faunal Composition and Dynamics. In *Ecological Effects of Fire in Southern African Ecosystems,* eds. P. de V. Booysen and N. M. Tainton. Berlin: Springer-Verlag, pp. 255–271.

Bobe, R. (2006). The evolution of arid ecosystems in East Africa. *Journal of Arid Environments*, 66, 564–584.

Braun, D. R., Levin, N. E., Stynder, D., Herries, A. I. R., Archer, W., Forrest, F., Roberts, D. L., Bishop, L. C., Matthews, T., Lehmann, S. B., Pickering, R. and Fitzsimmons, K. E. (2013). Mid-Pleistocene Hominin occupation at Elandsfontein, Western Cape, South Africa. *Quaternary Science Reviews*, 82, 145–166.

Brink, J. S. (1987). *The Archaeozoology of Florisbad, Orange Free State*. Bloemfontein: Memoirs van die Nasionale Museum, 24, 151pp.

Brink, J. S. (1993). Postcranial evidence for the evolution of the black wildebeest, *Connochaetes gnou*: an exploratory study. *Palaeontologia Africana*, 30, 61–69.

Brink, J. S. (1994). An ass, *Equus* (*Asinus*) sp., from late Quaternary mammalian assemblages of Florisbad and Vlakkraal, central Southern Africa. *South African Journal of Science*, 90, 497–500.

Brink, J. S. (1999). Preliminary report on a caprine from the Cape mountains, South Africa. *Archaeozologia*, 10, 11–26.

Brink, J. S. (2005). *The Evolution of the Black Wildebeest (*Connochaetes gnou*) and Modern Large Mammal Faunas of Central Southern Africa*. Unpublished DPhil dissertation, University of Stellenbosch, 485pp.

Brink, J. S. and Lee-Thorp, J. A. (1992). The feeding niche of an extinct springbok, *Antidorcas bondi* (Antilopini, Bovidae), and its palaeoenvironmental meaning. *South African Journal of Science*, 88, 227–229.

Brink, J. S., Herries, A. I. R., Moggi-Cecchi, J., Gowlett, J. A. J., Bousman, C. B., Hancox, J. P., Grün, R., Eisenmann, V., Adams, J. W. and Rossouw, L. (2012). First hominine remains from a ~1.0 million year old bone bed at Cornelia-Uitzoek, Free State Province, South Africa. *Journal of Human Evolution,* 63, 527–535.

Brink, J. S., de Beer, F., Hoffman, J. and Bam, L. (2013). The evolutionary meaning of *Raphicerus*-like morphology in the dentitions and postcrania of *Antidorcas bondi* (Antilopini). *Zitteliana,* 31 (Series B), 21.

Brink, J. S., Bousman, C. B. and Grün, R. (2015). A reconstruction of the skull of *Megalotragus priscus* (Broom, 1909), based on a find from Erfkroon, Modder River, South Africa, with notes on the chronology and biogeography of the species. *Palaeoecology of Africa*, 33, 71–94.

Chazan, M., Ron, H., Matmon, A., Porat, N., Goldberg, P., Yates, R., Avery, M., Sumner, A., Horwitz, L. K. (2008). Radiometric dating of the Earlier Stone Age sequence in

Excavation I at Wonderwerk Cave, South Africa: preliminary results. *Journal of Human Evolution*, 55, 1–11.
Codron, D. and Brink, J. S. (2007). Trophic ecology of two savanna grazers, blue wildebeest *Connochaetes taurinus* and black wildebeest *Connochaetes gnou*. *European Journal of Wildlife Research*, 53, 90–99.
Codron, D., Brink, J. S., Rossouw, L. and Clauss, M. (2008). The evolution of ecological specialization in southern African ungulates: Competition- or physical environmental turnover? *Oikos*, 117, 344–353.
Cooke, H. B. S. (1952). Mammals, ape-men and stone age men in Southern Africa. *South African Archaeological Bulletin*, 7 (26), 59–69.
Cooke, H. B. S. (1967). The Pleistocene sequence in South Africa and problems of correlation. In *Background to evolution in Africa*, eds. W. W. Bishop and J. D. Clark. Chicago: University of Chicago Press, pp. 175–184.
Cooke, H. B. S. (1974). The fossil mammals of Cornelia, O.F.S., South Africa. *Memoirs van die Nasionale Museum, Bloemfontein*, 9, 63–84.
Corbet, S. W. and Robinson, T. J. (1991). Genetic divergence in South African wildebeest: comparative cytogenetics and analysis of mitochondrial DNA. *Journal of Heredity*, 82, 447–452.
Corbet, S. W., Grant, W. S. and Robinson, T. J. (1994). Genetic divergence in South African wildebeest: analysis of allozyme variability. *Journal of Heredity*, 85, 479–183.
De Ruiter, D. (2003). Revised faunal lists for Members 1–3 of Swartkrans, South Africa. *Annals of the Transvaal Museum*, 40, 29–41.
Deacon, H. J. and Deacon, J. (1999). *Human Beginnings in South Africa*. Cape Town: David Philip, 224pp.
Deacon, H. J., Deacon, J., Scholtz, A., Thackeray, J. F., Brink, J. S. and Vogel, J. C. (1984). Correlation of palaeoenvironmental data from the Late Pleistocene and Holocene deposits at Boomplaas Cave, southern Cape. In *Late Cainozoic Palaeoclimates of the Southern Hemisphere*, ed. J. C. Vogel. Rotterdam: Balkema, pp. 339–352.
deMenocal, P. B. and Bloemendal, J. (1995). Plio-Pleistocene climatic variability in subtropical Africa and the paleoenvironment of hominid evolution: A combined data-model approach. In *Paleoclimate and Evolution, with Emphasis on Human Origins*, eds. E. S. Vrba, G. H. Denton, T. C. Partridge and L. H. Burckle. New Haven: Yale University Press, pp. 262–288.
Dupont, L. M. and Leroy, S. A. G. (1995). Steps towards drier climatic conditions in northwestern Africa during the upper Pliocene. In *Paleoclimate and Evolution, with Emphasis on Human Origins*, eds. E. S. Vrba, G. H. Denton, T. C. Partridge and L. H. Burckle. New Haven: Yale University Press, pp. 289–298.
Ewer, R. F. and Cooke, H. S. B. (1964). The Pleistocene mammals of southern Africa. *Monographiae Biologicae*, 14, 35–38.
Faith, J. T. (2013). Taphonomic and paleoecological change in the large mammal sequence from Boomplaas Cave, western Cape, South Africa. *Journal of Human Evolution,* 65, 715–730.
Faith, J. T. (2014). Late Pleistocene and Holocene mammal extinctions on continental Africa. *Earth-Science Reviews*, 128, 105–121.
Gentry, A. W. (1980). Fossil bovidae (Mammalia) from Langebaanweg, South Africa. *Annals of the South African Museum*, 79, 213–337.
Gentry, A. W. (2010). Bovidae. In *Cenozoic Mammals of Africa*, eds. L. Werdelin and W. J. Saunders. Berkeley: University of California Press, pp. 741–796.

Gentry, A. W. and Gentry, A. (1978). Fossil Bovidae (Mammalia) of Olduvai Gorge, Tanzania. Part I. *Bulletin of the British Museum (Natural History)*, 29, 290–446.

Geraads, D. (1981). Bovidae et Giraffidae (Artiodactyla, Mammalia) du Pléistocène de Ternifine (Algérie). *Bulletin Muséum national d'Histoire naturelle, Paris*, 4 (3), 47–86.

Gilbert, W. H. (2008). Introduction. In *Homo Erectus: Pleistocene Evidence from the Middle Awash, Ethiopia,* eds. W. H. Gilbert and B. Asfaw. Berkeley and Los Angeles: University of California Press, pp. 1–12.

Grobler, N. J. and Loock, J. C. (1988). Morphological development of the Florisbad deposit. *Palaeoecology of Africa,* 19, 163–168.

Grün, R., Brink, J. S., Spooner, N. A., Taylor, L., Stringer, C. B., Franciscus, R. B. and Murray, A. (1996). Direct dating of the Florisbad hominid. *Nature*, 382, 500–501.

Harris, J. M. and White, T. (1979). Evolution of the Plio-Pleistocene African Suidae. *Transactions of the American Philosophical Society*, 69, 1–128.

Hendey, Q. B. (1974). Faunal dating of the late Cenozoic of southern Africa, with special reference to the Carnivora. *Quaternary Research*, 4, 149–161.

Herries, A. I. R., Reed, K. E., Kuykendall, K. L. and Latham, A. G. (2006). Speleology and magnetobiostratigraphic chronology of the Buffalo Cave fossil site, Makapansgat, South Africa. *Quaternary Research*, 66, 233–245.

Herries, A. I. R., Hopley, P. J., Adams, J. W., Curnoe, D. and Maslin, M. A. (2009). Geochronology and palaeoenvironments of the South African early hominin bearing sites – a reply to Wrangham et al., 2009: "Shallow-Water Habitats as Sources of Fallback Foods for Hominins". *American Journal of Physical Anthropology*, 143, 640–646.

Hoetzel, S., Dupont, L., Schefuß, E., Rommerskirchen, F. and Wefer, G. (2013). The role of fire in Miocene to Pliocene C_4 grassland and ecosystem evolution. *Nature Geoscience*, 6, 1027–1030.

Kingdon, J. (1997). *The Kingdon Field Guide to African Mammals*. London: Academic Press, 496pp.

Klein, R. G. (1983). Paleoenvironmental implications of Quaternary large mammals in the Fynbos region. In *Fynbos Palaeoecology: A Preliminary Synthesis*, eds. H. J. Deacon, Q. B. Hendey and J. J. N. Lambrechts. Pretoria: South African National Scientific Programmes Report 75, pp. 116–138.

Klein, R. G. (1984). The large mammals of southern Africa: Late Pliocene to recent. In *Southern African Prehistory and Palaeoenvironments*, ed. R. G. Klein. Rotterdam: A.A. Balkema, pp. 107–146.

Klein, R. G., Cruz-Uribe, K. and Beaumont, P. B. (1991). Environmental, ecological, and paleoanthropological implications of the late Pleistocene mammalian fauna from Equus Cave, northern Cape Province, South Africa. *Quaternary Research*, 36, 94–119.

Klein, R. G., Avery, G., Cruz-Uribe, K. and Steele, T. E. (2007). The mammalian fauna associated with an archaic hominin skullcap and later Acheulean artifacts at Elandsfontein, Western Cape Province, South Africa. *Journal of Human Evolution*, 52, 164–186.

Lacruz, R. S., Brink, J. S., Hancox, P. J., Skinner, A. R., Herries, A., Schmid, P. and Berger, L. R. (2002). Palaeontology and geological context of a Middle Pleistocene faunal assemblage from the Gladysvale Cave, South Africa. *Palaeontologia Africana*, 38, 99–114.

Loock, J. C. and Grobler, N. J. (1988). The Regional Geology of Florisbad. *Navorsinge van die Nasionale Museum, Bloemfontein*, 5, 489–497.

Lorenzen, E. D., Heller, R. and Siegismuld, H. R. (2012). Comparative phylogeography of African savannah ungulates. *Molecular Ecology*, 21, 3656–3670.

McCarthy, T. S. (2013). The Okavango delta and its place in the geomorphological evolution of southern Africa. *South African Journal of Geology*, 116, 1–54.

Mucina, L. and Rutherford, M. C. (2006). *The vegetation of South Africa, Lesotho and Swaziland. Strelitzia 19*. Pretoria: South African National Biodiversity Institute, 807pp.

Partridge, T. C. and Maud, R. R. (2000). Macro-scale geomorphic evolution of southern Africa. In *The Cenozoic of Southern Africa*, eds. T. C. Partridge and R. R. Maud. Oxford: Oxford University Press, pp. 3–18.

Pickford, M. and Senut, B. (1999). *Geology and Palaeobiology of the Namib Desert*. Windhoek: Memoir of the Geological Survey of Namibia, 18, 155pp.

Plug, I. and Engela, R. (1992). The macrofaunal remains from recent excavations at Rose Cottage Cave, Orange Free State. *South African Archaeological Bulletin*, 47, 16–25.

Porat, N., Chazan, M., Grün, R., Aubert, M., Eisenmann, V. and Horwitz, L. K. (2010). New radiometric ages for the Fauresmith industry from Kathu Pan, southern Africa: Implications for the Earlier to Middle Stone Age transition. *Journal of Archaeological Science*, 37, 269–283.

Reynolds, S. C. and Kibii, J. M. (2011). Sterkfontein at 75: Review of palaeoenvironments, fauna, dating and archaeology from the hominin site of Sterkfontein (Gauteng Province, South Africa). *Palaeontologia Africana*, 46, 59–88.

Savage, D. E. and Russel, D. E. (1983). *Mammalian Paleofaunas of the World*. Reading (Massachusetts): Addison Wesley, 462pp.

Skinner, J. and Smithers, R. H. (1990). *The Mammals of the Southern African Subregion*, 2nd Ed. Pretoria: University of Pretoria, 768pp.

Thackeray, J. F. (1984). *Man, Animals and Extinctions: The Analysis of Holocene Faunal Remains from Wonderwerk Cave, South Africa*. Unpublished PhD dissertation, Yale University, 293pp.

Thackeray, J. F. and Brink, J. S. (2004). *Damaliscus niro* horns from Wonderwerk Cave and other Pleistocene sites: Morphological and chronological considerations. *Palaeontologia Africana*, 40, 89–93.

Tinker, T., de Wit, M. J. and Brown, R. (2008). Mesozoic exhumation of the southern Cape, South Africa, quantified using apatite fission track themochronology. *Tectonophysics*, 455, 77–93.

Van Hoepen, E. C. N. (1932). Voorlopige beskrywing van Vrystaatse soogdiere. *Paleontologiese navorsinge van die Nasionale Museum, Bloemfontein*, 2, 63–65.

Van Hoepen, E. C. N. (1947). A preliminary description of new Pleistocene mammals of South Africa. *Paleontologiese navorsinge van die Nasionale Museum, Bloemfontein*, 2, 103–106.

Visser, J. J. N. and Joubert, A. (1991). Cyclicity in the late Pleistocene to Holocene spring and lacustrine deposits at Florisbad, Orange Free State. *South African Journal of Geology*, 94, 123–131.

Vrba, E. S. (1995). On the connections between paleoclimate and evolution. In *Paleoclimate and Evolution, with Emphasis on Human Origins*, eds. E. S. Vrba, G. H. Denton, T. C. Partridge and L. H. Burckle. New Haven: Yale University Press, pp. 24–45.

Wells, L. H. (1962). Pleistocene faunas and the distribution of mammals in Southern Africa. *Annals of the Cape Provincial Museums*, 2, 37–40.
Werdelin, L. and Peigné, S. (2010). Carnivora. In *Cenozoic Mammals of Africa*, eds. L. Werdelin and W. J. Saunders. Berkeley: University of California Press, pp. 603–658.

19

Pollen, charcoal and plant macrofossil evidence of Neogene and Quaternary environments in southern Africa

MARION K. BAMFORD, FRANK H. NEUMANN AND LOUIS SCOTT

Abstract

Pollen and macrofossils are an integral part of palaeoenvironmental reconstruction. Here we discuss palaeobotanical evidence for vegetation and climate changes since the origins of modern biomes in southern Africa during the Miocene, and through the Pleistocene and Holocene. Examples of palaeobotanical records are provided from different biomes in different climate zones across southern Africa. These examples show that different biomes responded in different ways to climate changes throughout the Neogene and Quaternary, and that these environmental changes are also recorded in different ways though pollen, charcoal and macrofossils. In the latter part of the record, biome composition also reflects the impact of human activity.

19.1 Introduction

Pollen and macrofossils such as seeds, leaves, wood and charcoal play an integral part in palaeoenvironmental reconstruction. Here we discuss palaeobotanical evidence for vegetation and climate changes since the origins of modern biomes in southern Africa during the Miocene (23.0–5.3 Ma), and through the Pleistocene (2.6 Ma to 11,700 yr BP) and Holocene (last 11,700 yr). Archives where fossil plant material may be preserved include lakes (Scott, 1999a, b), springs or swamps (van Zinderen Bakker, 1957; Coetzee, 1967; Scott, 1982; Neumann et al., 2014), dung deposits, speleothems (Scott, 1987, 1994; Scott and Bousman, 1990; Brook et al., 2010; Chase et al., 2012; Valsecchi et al., 2013) and marine cores (Dupont et al., 2005). Constraints in interpreting these records include problems with dating (contamination, reservoir effects) and palynological precision of identification,

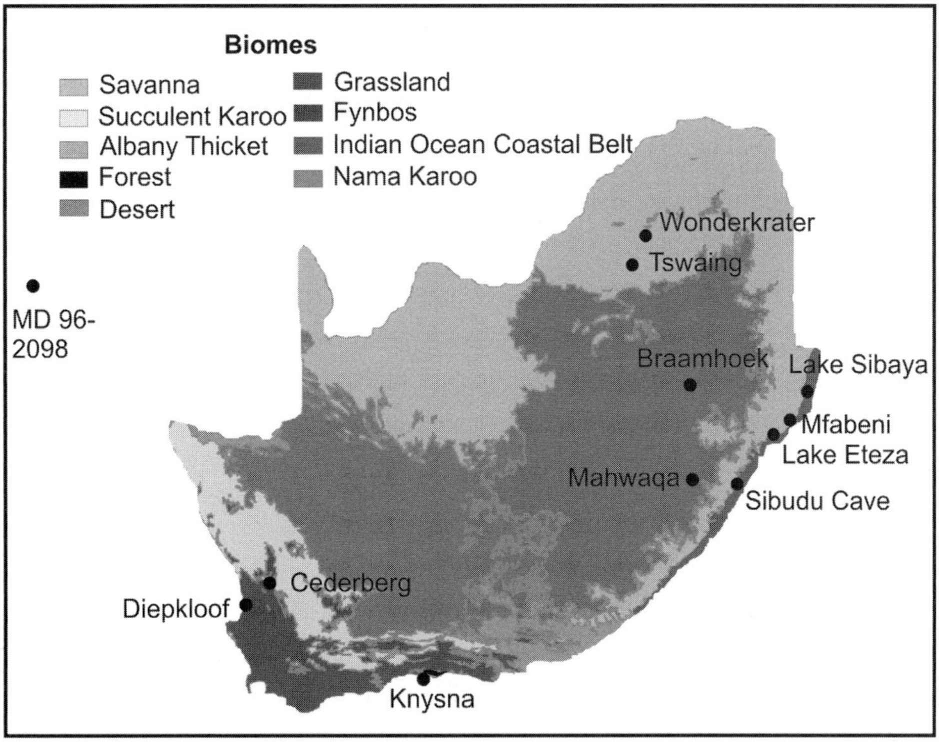

Fig. 19.1. Biome map of South Africa with sites mentioned in the text, including the position of Atlantic core MD 96-2098.

which is rarely better than genus level and often only possible to family level (e.g. with Poaceae, Asteraceae). Taphonomic problems include differential pollen production and transport, and pollen preservation (Hamilton, 1972; Jarzen and Nichols, 1996). The uses of macrofossils are limited if they are scarce or too fragmentary for identification.

This chapter describes selected palaeobotanical case studies, in particular the use of pollen, charcoal and macro-plant remains from Wonderkrater, Mahwaqa and Tswaing Crater, respectively (Figs. 19.1–19.3). Due to the scarcity of these deposits, studies of hyena and hyrax dung in cave deposits or rock shelters have gradually become more important as palaeoenvironmental records (Scott, 1987; Carrión et al., 1999, 2000; Scott and Woodborne, 2007a, b; Quick et al., 2011; Chase et al., 2012; Valsecchi et al., 2013).

Both micro- and macro-charcoal records have been used for vegetation and climate reconstructions. Many records of micro-charcoal remains, including some from South Africa (e.g. Scott, 2002), have been incorporated into large databases such as the Global Palaeofire Working Group (Power et al., 2008). Analysis shows that rising temperatures resulted in an increase in global fire activity for the last

21 kyr, but that moderate moisture levels increase fire activity through fuel productivity. A core off the Namaqualand coast, South Africa, showed that burning increased when the climate was wetter and cooler (Daniau et al., 2013). Length to width ratios of micro-charcoal pieces show shifts in fires of grassland or woody vegetation, where longer pieces are from grasses. The increase in burning has been linked to the expansion of C_4 grasslands in the Miocene and Pliocene (Hoetzel et al., 2013).

19.1.1 The major biomes of South Africa during the Neogene

The biomes in southern Africa evolved during the Neogene (see Scott et al., 1997). Due to an arid climate and the erosional character of the interior, most fossiliferous sediments at this time accumulated along the coasts and offshore (Scott, 1995), which makes marine records particularly invaluable (van Zinderen Bakker, 1984; Dupont et al., 2005). Palaeogeographical changes, such as the opening of the Drake passage between Antarctica and South America and the Tasmanian gateway between Australia and Antarctica (see Hassold et al., 2009), led to the establishment of the circum-Antarctic current and decreasing temperatures following the Middle Miocene climatic optimum (Coetzee, 1978; van Zinderen Bakker, 1984; Majewski and Bohaty, 2010).

Palynomorphs from lignites near Knysna, on the southern coast of South Africa, point to wetland conditions and a mixed forest with (sub)-tropical affinity (Thiergart et al., 1962; Thiergart, 1964; Helgren and Butzer, 1977; Coetzee et al., 1983; Carr et al., 2010). The similarity of the palynoflora to that of other pollen sequences in the southwestern Cape at Langebaanweg, Noordhoek and Rondeberg (Coetzee, 1978; Coetzee and Rogers, 1982; Coetzee et al., 1983, Coetzee and Muller, 1984; Coetzee and Praglowski, 1984; Scott, 1995; Sciscio et al., 2013; Roberts et al., 2013), suggests a similar age which might even postdate the Middle Miocene (Carr et al., 2010). In contrast, other workers suggest an early-middle Miocene age for the deposits at Langebaanweg, Noordhoek and Rondeberg (Roberts et al., 2011, 2013). Pollen of palms, which are now extinct in the Cape, were identified, along with *Podocarpus* pollen, Poaceae, Restionaceae, Proteaceae, Sapotaceae and others (Fig. 19.2). Palynofloras from the southwestern Cape suggest a humid, subtropical climate (Scott et al., 1997; Carr et al., 2010; Roberts et al., 2013). Several families, extinct in the Cape but still dwelling in Madagascar or Australia (e.g. Casuarinaceae, Chloranthacaeae, Sarcolaenaceae and Winteraceae), exist in the palynological record up to the mid-Miocene (Coetzee and Muller, 1984; Coetzee and Praglowski, 1984; Goldblatt, 1996). Based on phytoliths, at the end of the Miocene, C_4 grasslands were spreading in the southwestern Cape, whereas C_3 plants (trees, shrubs, forbs, some grasses) were

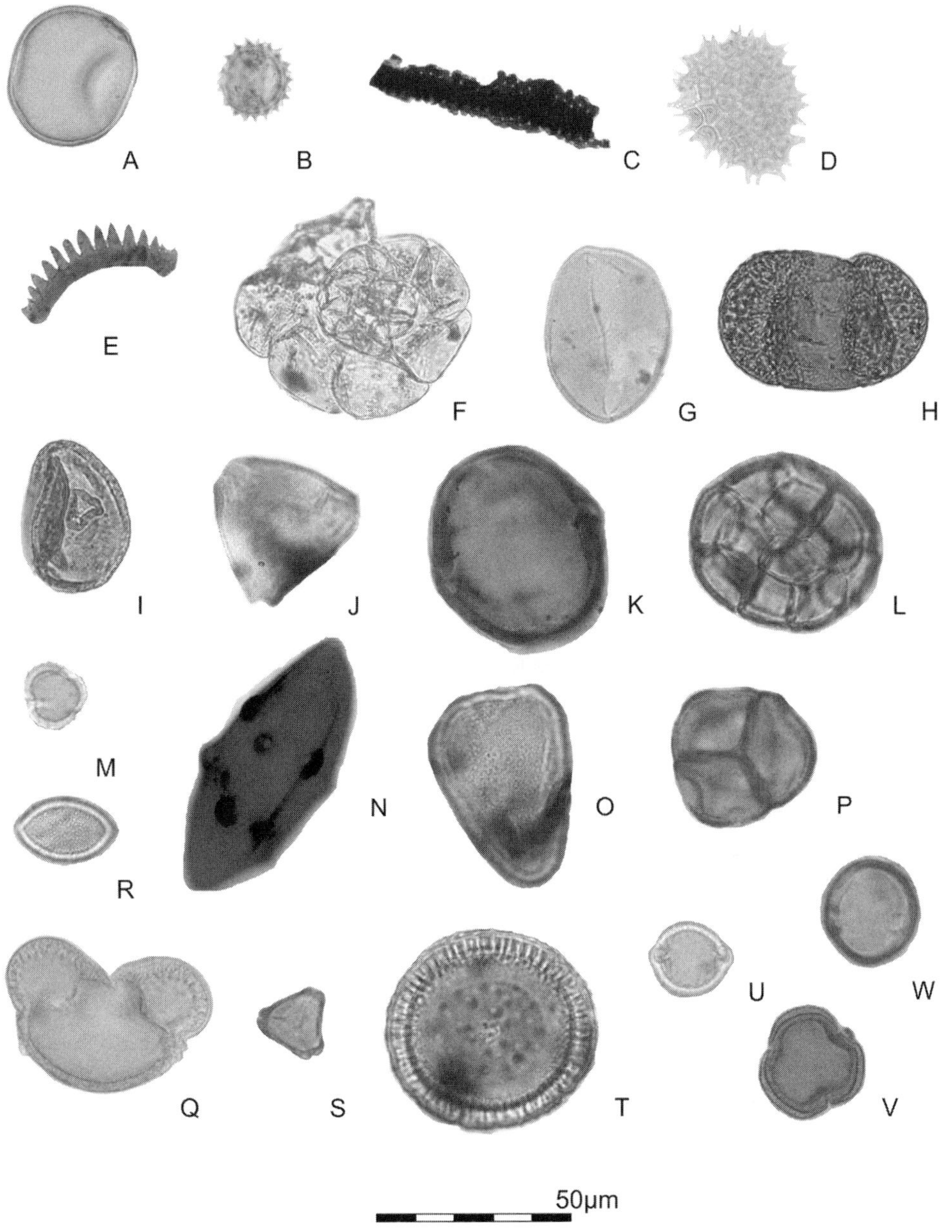

Fig. 19.2. Important palynomorphs including microcharcoal and pollen from terrestrial sites in South Africa. (A) Poaceae (Princessvlei, Holocene), (B) Asteraceae (Langebaanweg, Miocene), (C) micro-charcoal (Gosberg valley, recent), (D) *Pediastrum* sp. (Verlorenvlei, sub-recent, with chlorophyll), (E) Scolecodont (Eteza, Holocene), (F) microforaminiferal lining (Blood Reef, Holocene), (G) *Arecipites* sp. (Saldanha Bay, Miocene), (H) *Podocarpidites* sp. (Saldanha Bay, Miocene), (I) *Restioniidites* sp. (Saldanha Bay, Miocene), (J) *Proteacidites* sp. (Langebaanweg, Miocene), (K) *Sapotaceoidaepollenites* sp. (Langebaanweg, Miocene), (L) *Acacia* sp. (Sibaya, Holocene), (M) *Stoebe*-type sp. (Princessvlei, Holocene), (N) *Mediaverrunitis* sp. (Mahwaqa, Holocene), (O) Cyperaceae (Braamhoek, Holocene), (P) Ericaceae (Princessvlei, Holocene), (Q) *Pinus* sp. (Mahwaqa, Holocene), (R) *Phoenix reclinata* (Sibaya, Holocene), (S) *Eucalyptus* sp. (Princessvlei, Holocene), (T) *Isoglossa woodii* (Eteza, Holocene), (U) *Clif-fortia* sp. (Princessvlei, Holocene), (V) *Anthospermum* sp. (Verlorenvlei, Holocene), (W) *Dodonaea* sp. (Princessvlei, Holocene).

diminished at this time (Rossouw et al., 2009). Open fynbos vegetation, already established during the early Pliocene, in association with subtropical C_3 vegetation, suggests a more seasonal climate with a stronger Benguela Current which was established at about 14 Ma (Hendey, 1984; Scott, 1995; Roberts et al., 2011).

Palynological and palaeobotanical data from the contemporary interior savanna and grassland regions of southern Africa are sparse. The Sterkfontein region is important for understanding human evolution, and data on vertebrate fossils and pollen from caves permit some insights on Neogene vegetation dynamics (see Scott and Bonnefille, 1986; Scott et al., 1997). Based on work by Vrba (1985), it was suggested that at ~2.5 Ma (Pliocene–Pleistocene boundary), vegetation changed from mesic and wooded to open, xeric environments. Pollen samples from Sterkfontein and Kromdraai indicate an open *Protea* savanna at the Plio-Pleistocene boundary, while fossil wood from Member 4 in the Sterkfontein Cave indicates the existence of a local gallery forest (Bamford, 1999).

The evolution of the Desert, Succulent Karoo and Nama Karoo biomes can be reconstructed from marine cores off southwestern Africa (Dupont et al., 2005) and fossil wood and pollen from Namaqualand (van Zinderen Bakker, 1984; Scott et al., 1997; Bamford, 2003). The middle Miocene was characterised by subtropical and humid conditions (Bamford and de Wit, 1993); Miocene clays from sites in Namaqualand show a transition from humid woodland with *Olea* and *Podocarpus* and pollen spectra similar to the southwestern Cape, to open karroid vegetation where fynbos elements also occur (Scott, 1995). Increased aridity along the west coast of southern Africa since the late Miocene is attested by the predominance of grasses, chenopods and Asteraceae, which point to a Karroid shrubland. A marine core off the coast of Namibia shows a late Miocene–Pliocene (8–3 Ma) spread of C_4 grasses in an environment of shrinking grasslands and expanding deserts (Hoetzel et al., 2013).

19.2 Quaternary case studies

Selected terrestrial sites from various biomes (Fig. 19.1) allow us to address floristic changes during the Quaternary, but also problems with palaeobotanical interpretations associated with dating, pollen taphonomy, and sedimentary hiatuses.

19.2.1 Tswaing (I & II) and Wonderkrater: Savanna Biome

Situated in the Savanna Biome and characterised by a co-dominance of grasses and woody plants (e.g. *Acacia*), Wonderkrater and Tswaing have the longest Quaternary terrestrial pollen records for this region (Scott, 1982, 1999a, b). The Middle and Late Pleistocene sections of these cores are discontinuous and the older parts suffer from anomalous dating results and gaps in the pollen record, probably due to oxidation of palynomorphs (Scott, 1999b). Summaries of these records have been provided by Scott and Thackeray (1987), Scott (1999a, b), and Scott et al. (2003, 2012). Chase et al. (2010) and Truc et al. (2013) reinterpreted palaeoclimatic conditions of the Holocene part of Wonderkrater sequence, questioning previous interpretations of dry conditions during the early Holocene. Dryness during the early Holocene is supported by the palynology of a new core (Tswaing II) (Metwally et al., 2014), by independent proxies of the crater lake sediments (Kristen et al., 2010), and is in line with interpreted conditions at Wonderkrater (Scott and Thackeray, 1987; Scott et al., 2003, 2012). A new interpretation of the Wonderkrater sequence based on radiocarbon and luminescence dating of archaeological excavations, ~20 m distance from the original cores, also contributes to the vegetation history of the Savanna Biome (McCarthy et al., 2010; Backwell et al., 2014). This understanding can potentially be advanced for the Pleistocene section at Tswaing if the chronology is improved and the palynology, phytoliths and cuticles are studied at high resolution.

19.2.2 Braamhoek and Mahwaqa: Grassland Biome

An 18 kyr pollen record from a wetland in the Mahwaqa mountains, an outlier of the Drakensberg mountain range and within the grassland biome (Neumann et al., 2014), is compared here to a 16 kyr palynological archive at Braamhoek on the Great Escarpment (Norström et al., 2009) (Fig. 19.1). Micro-charcoal, fungal remains and wetland elements, especially Cyperaceae, are abundant (Fig. 19.2). The Mahwaqa record shows a predominance of fynbos vegetation (e.g. Ericaceae and Restionaceae), between 18–13 cal kyr BP. This vegetation was presumably influenced by westerlies with higher rainfall during winter and assumed glaciers in the Lesotho highlands (Mills et al., 2012). During a transitional period up to 8.5 cal kyr BP, grasses gradually displaced fynbos elements. At Braamhoek, fynbos vegetation made up about 10% of the terrestrial pollen sum until ~10 cal kyr BP (Norström et al., 2009). Mahwaqa was thus more strongly affected by the cooling effects of the Drakensberg range than was Braamhoek. The grassland at both sites became drier in the mid-Holocene altithermal, shown by an increase of Asteraceae especially at 4.6–3.5 cal kyr BP at Mahwaqa, and 6–3 cal kyr BP at Braamhoek (Scott et al., 2012). Enhanced humidity with increasing *Podocarpus* pollen percentages characterised the late Holocene at both sites (Norström et al., 2009; Scott et al., 2012, Neumann et al., 2014). At Mahwaqa, neophytes such as *Pinus* mark increased anthropogenic impact by European settlers. Both records feature periods of very low sedimentation (e.g. between 17–8 cal kyr BP at Mahwaqa) where hiatuses cannot be excluded.

19.2.3 Mfabeni peatland, Lakes Sibaya and Eteza: Indian Ocean Coastal Belt Biome (IOCB)

The IOCB is characterised by subtropical trees, palms, lianas and epiphytes (Mucina and Rutherford, 2006). The longest continuous Quaternary record in the region is the Mfabeni peatland (Fig. 19.1) close to Lake St. Lucia, but its comparison to other sites is hampered by the chronology that consists of only six radiocarbon dates (Finch and Hill, 2008). The record starts at > 46,950 cal yr BP. Shorter coastal sequences at Lake Sibaya and Lake Eteza (25 km inland) have well-constrained chronologies (Neumann et al., 2008, 2010). The Eteza record shows a humid environment characterised by *Phoenix* palms (at 10.2–6.8 cal kyr BP) which gives way to drier grasslands (Neumann et al., 2010). This change coincided with rising sea surface temperatures (SST) and sea levels in the adjacent Mozambique Channel (Ramsay, 1995). Between 6.8–3.6 cal kyr BP, tree pollen percentages, especially of *Podocarpus*, are high and mangrove and swamp forest elements indicate high sea levels and proximity to the coast. During the Holocene Altithermal, the Mfabeni record is characterised by moist, warm environments and forest expansion (Finch and Hill, 2008). At Eteza, after 3.6 cal kyr BP, tree pollen decreased whereas grassland associated with *Phoenix* increased. The Sibaya record is affected by a hiatus of > 5 kyr, separating a short pollen sequence at ~7 cal kyr BP, which indicates a moist forest environment similar to Eteza, from a section covering the last 1.5–1.3 kyr (including a radiocarbon age error of 50–550 yr; Neumann et al., 2008). Wet and dry cycles are evident at Sibaya for the last ~1800 years from a well dated diatom record (Stager et al., 2013), showing good agreement with similar climatic fluctuations in the Karkloof *Podocarpus* tree ring record (Hall, 1976). The Iron Age at Lake Sibaya is characterised by high *Podocarpus* pollen percentages which decrease probably due to human (burning) activities. *Zea mays* pollen is detected during the last 300 years (Neumann et al., 2008, Walther and Neumann, 2011). Recent tree plantations lead to the dominance of *Pinus* and other neophyte pollen such as *Eucalyptus* at the top of the Sibaya profile (Neumann et al., 2008), consistent with recent anthropogenic increases in total carbon, nitrogen and phosphorus in Lake Sibiya (Humphries and Benitez-Nelson, 2013). Vegetation changes during the last 700 years at Eteza included a decline of *Podocarpus* and an increase of grasses and neophytes. In concordance with the pollen record at Lake Sibaya, human disturbances by Iron Age farmers and then by Europeans caused vegetation changes (Neumann et al., 2010). An alternative explanation is high sea levels at ~1.5–1.2 cal kyr BP at the Kariega Estuary, followed by sea-level fall at ~1.1 cal kyr BP (Strachan et al., 2014). At Eteza, the dominant understorey plants in the coastal forest, *Podocarpus* and *Isoglossa*, decrease sharply at ~1 cal kyr BP, which may reflect shoreline retreat.

19.2.4 Cederberg: Fynbos Biome

In the Cederberg, a fynbos biome record is preserved in the pollen content of hyrax middens (Quick et al., 2011; Valsecchi et al., 2013). Potential problems of using midden records are discussed in Scott (1990) and Chase et al. (2012). An advantage is that they can be found in dry areas and can provide a more direct representation of slope and hillside pollen production than wetlands. In contrast, wet climate causes the decay of hyrax middens. The high resolution hyrax sequences from Pakhuis Pass and De Rif in the Cederberg (Valsecchi et al., 2013) show that there were more significant fluctuations in the late Quaternary vegetation than is reported by Meadows and Sugden (1990) for this mountain range. From 15.6–13.3 cal kyr BP, the De Rif sequence shows that Ericaceae was gradually replaced by Restionaceae, together with high proportions of microscopic charcoal that gradually declined (Valsecchi et al., 2013). The record resumes at ~11.2 cal kyr BP with high Proteaceae and *Dodonaea* pollen, reflecting a warmer and relatively humid climate. Proteaceae, which peak at the start of this phase, declined sharply by 10 cal kyr BP and is not prominent during the remainder of this phase. The period 7.7–4.2 cal kyr BP shows a more karroid vegetation with small shrubs, succulents and an increase in microscopic charcoal. With some minor fluctuations, the final phase of the sequence shows pollen spectra comparable with today's vegetation.

19.2.5 Macrobotanical remains: Fynbos, Savanna and IOCB

Sites with macrobotanical remains are rarer than sites with microfossils; the most common preservation type is that of charcoal fragments associated with human occupation and fire activities. In most cases the woods were used by humans primarily for fuel but also for insecticidal properties, medicines and rituals (Wadley et al., 2011). Nonetheless, the woods selected would reflect local vegetation and thus an indication of climate. Here, to provide a broad perspective, sites discussed cover a range of biomes and climate regimes: Diepkloof (southwestern Cape coast), Wonderkrater (central interior), and Sibudu (northeast coast of KwaZulu Natal) (Fig. 19.1).

The Diepkloof cave shelter Middle Stone Age record spans the pre-Still Bay, Still Bay, and Howiesons Poort lithic technologies, but the absolute dates are in question, ranging from 70–58 kyr BP (Jacobs et al., 2008; Tribolo et al., 2009) and ~109–52 kyr BP (Guérin et al., 2013; Tribolo et al., 2013). Identified charcoal from Diepkloof Rock Shelter indicates a shift of Afromontane forest taxa and proteoid fynbos over time, and an increasing diversity of taxa during the Howiesons Poort phase (Cartwright, 2013). Some common Afromontane taxa are *Cunonia capensis, Olinia ventosa, Rapanea melanophloeos*, *Leucodendron* spp., *Gymnosporium buxifolia, Podocarpus* sp. and *Widdringtonia* sp. Woodland and thicket taxa include *Kiggelaria africana, Myrsine africana, Maytenus* spp.,

Searsia spp., *Leucospermum* spp., *Dodonaea viscosa* ssp. *angustifolia* and *Olea europaea* ssp. *africana,* and *Euclea* spp.

From the central interior, Wonderkrater has a good pollen record and some macro-charcoal remains dominated by the wetland reed *Phragmites australis*, other sedges, grasses and some woody pieces (Backwell et al., 2014) (Fig. 19.3). Around 45 cal kyr BP there are four species of *Acacia* and one species of *Combretum*. Charcoal from ~30 cal kyr BP represents grasses, flowering plants and trees. The lower sand unit at ~12 cal kyr BP contains *Phragmites* and grass alternating with *Phragmites*, grass and a wood (*Diospyros austro-africana*). The macro-flora does not correspond with the much more detailed pollen record, which contains long-distance wind-transported elements.

Sibudu rock shelter on the Thongati River contains fine sediments, ash, hearths, charcoal, faunal and lithic material. From the oldest layer (~77 kyr BP), there is evidence of bedding in the form of layers of sedge culms overlain by leaves of the pesticidal *Cryptocarya woodii* (Wadley et al., 2011; Miller and Sievers, 2012) (Fig. 19.3). Bedding from younger layers has been burned but the charcoal and charcoalified seeds have been preserved. Charcoal from the occupational layers and hearths indicates a change in wood selection and vegetation over time (Allott, 2006). In the layers older than 60 kyr, the presence of *Podocarpus* sp., *Buxus* sp. *Brachylaena* sp., *Sapium/Spirostachys*, *Kirkia* sp. and *Ptaeroxylon obliquum* point to an evergreen forest but with warm woodland savanna not too far away (e.g. *Kirkia* sp.). Around 50 kyr BP, there are fewer evergreen forest elements and more bushveld taxa (*Acacia* spp., Fabaceae and fewer *Erica* spp.), which indicate a drier climate. The woody vegetation at ~37 kyr BP was similar to today's, with *Kirkia* sp., *Brachylaena* sp., *Acacia* spp., *Erica* spp., *Clerodendrum glabrum, Xylotheca kraussiana, Mystroxylon aethiopicum* and *Phoenix reclinata.*

A variety of seeds and sedge nutlets has also been recovered and identified from Sibudu (Wadley, 2004; Sievers, 2006), but the pollen record is poor (Renaut and Bamford, 2006). The combined macrobotanical data were subjected to GIS-based co-existence analysis by Bruch et al. (2012). Results show that during the Howiesons Poort industry, winters were slightly colder and drier but summers were similar to present; during post-Howiesons Poort, this climate regime continued but summer rainfall decreased; and during the late Middle Stone Age, winter temperatures were higher than previously, and summer rainfall increased, resulting in denser vegetation (thus, the climate was similar to that of today).

19.3 Future research directions and potential

To extend the long term palaeobotanical record in southern Africa is challenging, in particular given the scarcity of records in the arid interior, but despite this some

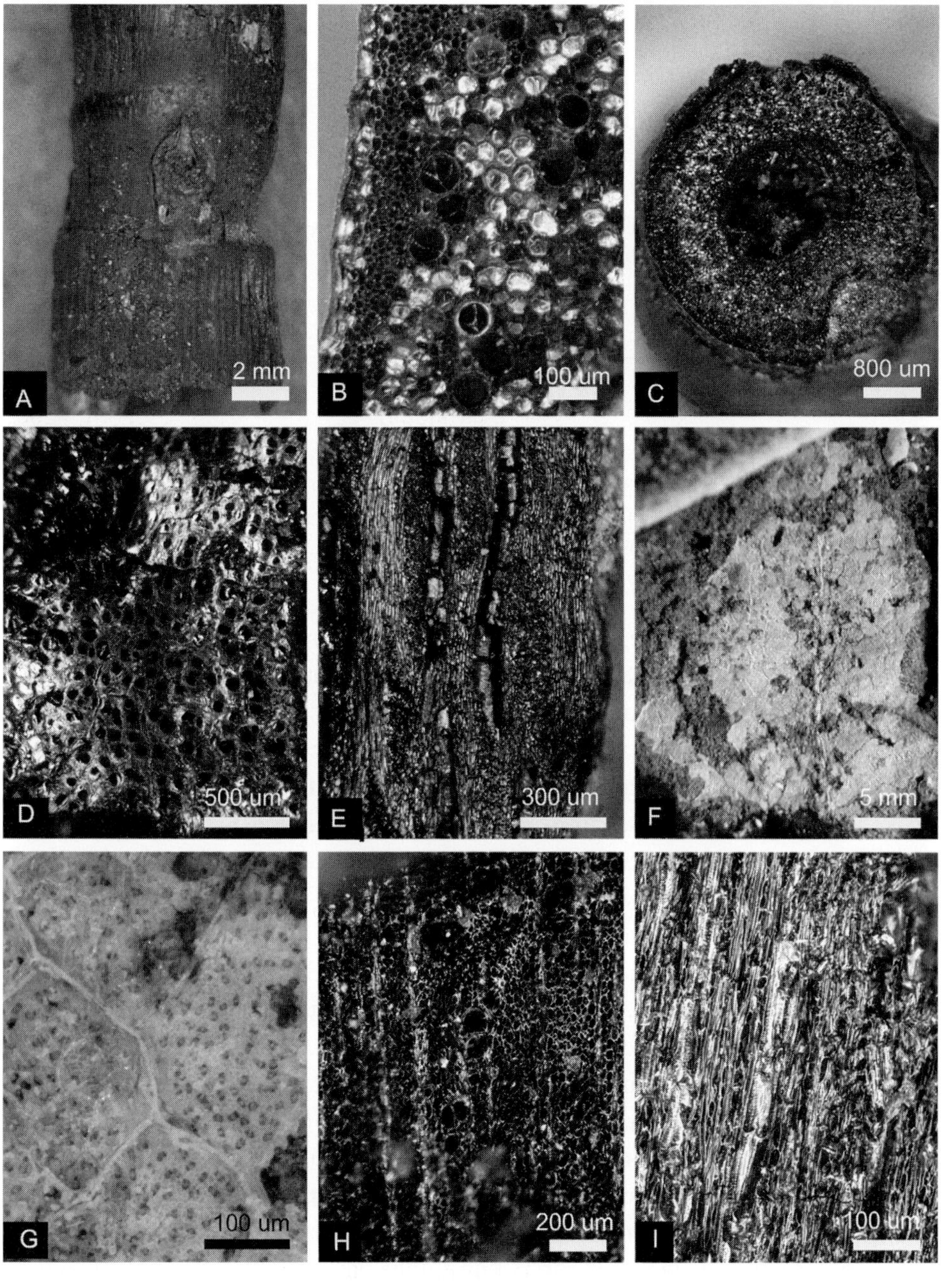

Fig. 19.3. Photomicrographs of charcoal and leaf remains from Wonderkrater and Sibudu. (A) *Phragmites* sp. viz. basal culm fragment with axillary bud (Wonderkrater ~15,167 cal yr BP), (B) *Phragmites* sp. viz. culm cross section showing vascular bundles between parenchyma cells (large light cells) (Wonderkrater), (C) grass culm cross section (Wonderkrater ~12,367 cal yr BP), (D) transverse section and (E) tangential longitudinal section of charcoal piece of *Acacia sieberiana* (Wonderkrater >32 kyr BP), (F) silicified leaf of *Cryptocarya woodii* (Sibudu rock shelter ~77 kyr BP), (G) detail of stomata from the leaf shown in (F), (H) transverse section and (I) tangential longitudinal section of charcoal of *Spirostachys africana* (Sibudu rock shelter ~38 kyr BP) (image: Marion Bamford). A black and white version of this figure will appear in some formats. For the colour version, please refer to the plate section.

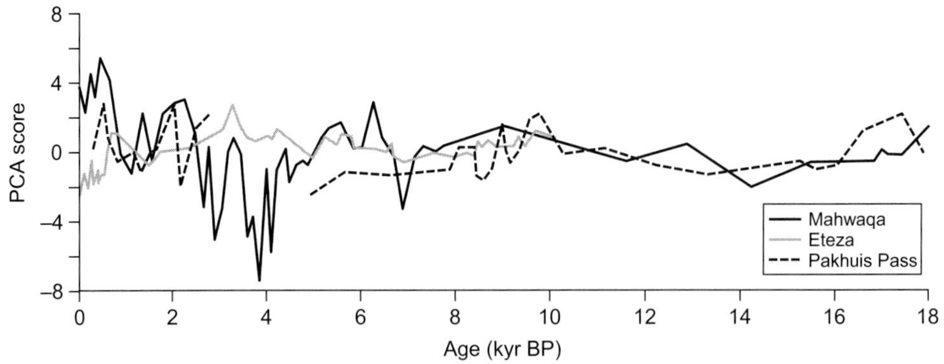

Fig. 19.4. Variations in moisture-related principal component scores (vertical axis) over time (horizontal axis) from three major late Quaternary sites in South Africa: Mahwaqa (Grassland Biome), Eteza (Indian Ocean Coastal Belt Biome) and Pakhuis Pass (Fynbos Biome).

patterns are emerging. Here, we adapted a multivariate method employed originally by Scott and Thackeray (1987), generating moisture-related principal components of selected late Quaternary pollen indicators to compare three key sites (Mahwaqa, Eteza, Pakhuis Pass) that have long and continuous records, reasonable time resolution and chronological control, and representing different biomes (Fig. 19.4). Principal component analyses from Pakhuis Pass (Cederberg, Fynbos Biome; Scott and Woodborne, 2007b; Scott et al., 2012) and Mahwaqa (Grassland Biome; Neumann et al., 2014) show similar fluctuations until ~10 cal kyr BP. From 10–5 cal kyr BP, moisture at Pakhuis Pass declines, indicating less winter rain (Scott et al., 2012). In the summer-rainfall region at Mahwaqa and Eteza (Indian Ocean Coastal Belt Biome; Neuman et al., 2010), lower humidity occurs ~7 cal kyr BP but these trends diverge at ~5.0–2.5 cal kyr BP. Between ~6.8–3.5 cal kyr BP, conditions at Eteza were very humid with a spread of subtropical trees, whereas the Drakensberg region and grassland biome experienced dry conditions. Scott et al. (2012) proposed that the proportion of moisture, brought from the Agulhas Current to the coastal regions and by low-pressure cells to the interior in summers, may have varied. The last 2500 years present non-uniform patterns at all three sites but with increased moisture at Mahwaqa and drier conditions at the coast.

It is necessary to understand if differences between regional observations reflect underlying forcing patterns or incorrect interpretations (Chase et al., 2010; Truc et al., 2013). To do this, palaeobotanical research should be complemented by faunal, geological, geochemical, palaeoclimatic and modelling studies. Southern Africa has a large number of potential environmental archives such as coastal wetlands (e.g. Kosi Bay, Pondoland), and bogs/mires in the Drakensberg range (Fisher et al., 2013), that have not yet been exploited (see Fitchett et al., this volume). In addition, quantitative climate reconstruction, local/regional vegetation-pollen relationships, and modern (including seasonal) pollen monitoring should receive attention. This may also include maintaining and extending pollen and charcoal reference collections, such as the one at the University of the Free State, and building new reference materials for grass cuticle analysis (e.g. Palmer, 1976; Wooller et al., 2000).

19.4 Summary

Pollen and macrofossil records in southern Africa reflect local vegetation composition but are also influenced by regional to global climate changes through the Neogene to Holocene. Discussion of pollen records from sites located in different contemporary biomes shows spatial and temporal variations in both vegetation (biome) composition, and in patterns of reconstructed climate. This shows the complexity of landscape and vegetation dynamics, particularly in the latter part of the records when human disturbance is a more significant factor. Macrofossils are not as common as microfossils in preserved records, but are useful in that they show the local vegetation, whereas pollen may be wind-transported from the wider region. The woods and plant materials selected by past occupants of caves and shelters are a subset of the flora growing in the vicinity. When macrofossils are combined with microfossil remains, a much more complete picture of the vegetation and its changes can be developed.

References

Allott, L. F. (2006). Archaeological charcoal as a window on palaeovegetation and wood-use during the Middle Stone Age at Sibudu Cave. *Southern African Humanities*, 18, 173–201.

Backwell, L. R., McCarthy, T. S., Wadley, L., Henderson, Z., Steininger, C. M., deKlerk, B., Barré, M., Lamothe, M., Chase, B. M., Woodborne, S., Susino, G. J., Bamford, M. K., Sievers, C., Brink, J. S., Rossouw, L., Pollarolo, L., Trower, G., Scott, L. and d'Errico, F. (2014). Multiproxy record of late Quaternary climate change and Middle Stone Age human occupation at Wonderkrater, South Africa. *Quaternary Science Reviews*, 99, 42–59.

Bamford, M. (1999). Pliocene fossil woods from an early hominid cave deposit, Sterkfontein, South Africa. *South African Journal of Science*, 95, 231–237.

Bamford, M. K. (2003). Fossil woods from Auchas and their palaeoenvironment. Geology and Palaeobiology of the central and southern Namib Desert, southwestern Africa. *Geological Survey of Namibia, Memoirs*, 19, 23–34.

Bamford, M. K., and de Wit, M. C. J. (1993). Taxonomic descriptions of fossil wood from Cainozoic Sak River terraces near Brandvlei, South Africa. *Palaeontologia Africana*, 30, 71–80.

Brook, G. A., Scott, L., Railsback, L. B. and Goddard, E. A. (2010). A 35 ka pollen and isotope record of environmental change along the southern margin of the Kalahari from a stalagmite and animal dung deposits in Wonderwerk Cave, South Africa. *Journal of Arid Environments*, 74, 870–884.

Bruch, A. A., Sievers, C. and Wadley, L. (2012). Quantification of climate and vegetation from the southern African Middle Stone Age sites – an application using Late Pleistocene plant material from Sibudu, South Africa. *Quaternary Science Reviews*, 45, 7–17.

Carr, A. S., Boom, A. and Chase, B. M. (2010). The potential of plant biomarker evidence derived from rock hyrax middens as an indicator of palaeoenvironmental change. *Palaeogeography, Palaeoclimatology, Palaeoecology*, 285, 321–330.

Carrión, J. S., Scott, L. and Vogel, J. C. (1999). Twentieth century changes in montane vegetation in the eastern Free State, South Africa, derived from palynology of hyrax dung middens. *Journal of Quaternary Science*, 14, 1–16.

Carrión, J. S., Scott, L., Huffman, T. and Dreyer, C. (2000). Pollen analysis of Iron Age cow dung in southern Africa. *Vegetation History and Archaeobotany*, 9, 239–249.

Cartwright, C. R. (2013). Identifying the woody resources of Diepkloof Rock Shelter (South Africa) using scanning electron microscopy of the MSA wood charcoal assemblages. *Journal of Archaeological Science*, 40, 3463–3474.

Chase, B. M., Meadows, M. E., Carr, A. S. and Reimer, P. J. (2010). Evidence for progressive Holocene aridification in southern Africa recorded in Namibian hyrax middens: Implications for African monsoon dynamics and the "African Humid Period". *Quaternary Research*, 74, 36–45.

Chase, B. M., Scott, L., Meadows, M. E., Gil-Romera, G., Boom, A., Carr, A. S., Reimer, P. J., Truc, L., Valsecchi, V. and Quick, L. J. (2012). Rock hyrax middens: A palaeoenvironmental archive for southern African drylands. *Quaternary Science Reviews*, 56, 107–125.

Coetzee, J. A. (1967). Pollen analytical studies in east and southern Africa. *Palaeoecology of Africa*, 3, 1–146.

Coetzee, J. A. (1978). Climatic and biological changes in southwestern Africa during the Late Cenozoic. *Palaeoecology of Africa and the Surrounding Islands*, 10, 13–29.

Coetzee, J. A. and Muller, J. (1984). The phytogeographic significance of some extinct Gondwana pollen types from the Tertiary of the southwestern Cape (South Africa). *Annals of the Missouri Botanical Garden*, 71, 1088–1099.

Coetzee, J. A. and Praglowski, J. (1984). Pollen evidence for the occurrence of *Casuarina* and *Myrica* in the Tertiary of South Africa. *Grana*, 23, 23–41.

Coetzee, J. A. and Rogers, J. (1982). Palynological and lithological evidence for the Miocene palaeoenvironment in the Saldanha region (South Africa). *Palaeogeography, Palaeoclimatology, Palaeoecology*, 39, 71–85.

Coetzee, J. A., Scholtz, A. and Deacon, H. J. (1983). Palynological studies and the vegetation history of the fynbos. In *Fynbos Palaeecology: A Preliminary Synthesis*, ed. H. J. Deacon, Q. B. Hendey and J. J. N. Lambrechts. Pretoria: South African National Scientific Programmes Report 75, pp. 156–173.

Daniau, A.-L., Sánchez Goni, M. F., Martinez, P., Ureggo, D. H., Bout-Roumazeilles, V., Desprat, S. and Marlon, J. R. (2013). Orbital-scale climate forcing of grassland burning in southern Africa. *Proceedings of the National Academy of Sciences*, 110, 5069–5073.

Dupont, L. M., Donner, B., Vidal, L., Pérez, E. M. and Wefer, G. (2005). Linking desert evolution and coastal upwelling: Pliocene climate change in Namibia. *Geology*, 33, 461–464.

Finch, J. M. and Hill, T. R. (2008). A late Quaternary pollen sequence from Mfabeni Peatland, South Africa: Reconstructing forest history in Maputaland. *Quaternary Research*, 70, 442–450.

Fisher, E. C., Albert, R.-M., Botha, G., Cawthra, H. C., Esteban, E., Harris, J., Jacobs, Z., Jerardino, A., Marean, C. W., Neumann, F. H., Pargeter, J., Poupart, M. and Venter, J. (2013). Archaeological reconnaissance for Middle Stone Age sites along the Pondoland coast, South Africa. *Journal of Palaeoanthropology*, 2013, 104–137.

Goldblatt, P. (1996). Floristic diversity in the Cape Flora of South Africa. *Biodiversity and Conservation*, 6, 359–377.

Guérin, G., Murray, A. S., Jain, M., Thomsen, K. J. and Mercier, N. (2013). How confident are we in the chronology of the transition between Howieson's Poort and Still Bay? *Journal of Human Evolution*, 64, 314–317.

Hall, M. J. (1976). Dendroclimatology, rainfall and human adaptation in the Later Iron Age of Natal and Zululand. *Annals of the Natal Museum*, 22, 693–703.

Hamilton, A. C. (1972). The interpretation of pollen diagrams from highland Uganda. *Palaeoecology of Africa*, 7, 45–149.

Hassold, N. J. C., Rea, D. K., van der Pluijm, B. A. and Parés, J. M. (2009). A physical record of the Antarctic Circumpolar Current: Late Miocene to recent slowing of abyssal circulation. *Palaeogeography, Palaeoclimatology, Palaeoecology*, 275, 28–36.

Helgren, D. M. and Butzer, K. W. (1977). Palaeosols of the Southern Cape coast, South Africa: implications for laterite definition, genesis and age. *Geographical Review*, 67, 430–445.

Hendey, Q. B. (1984). Southern African late Tertiary vertebrates. In *Southern African Prehistory and Palaeoenvironments*, ed. R. G. Klein. Rotterdam: Balkema, pp. 81–106.

Hoetzel, S., Dupont, L., Schefuss, E., Rommerskirchen, F. and Wefer, G. (2013). The role of fire in Miocene to Pliocene C_4 grassland and ecosystem evolution. *Nature Geoscience*, 6, 1027–1030.

Humphries, M. S. and Benitez-Nelson, C. R. (2013). Recent trends in sediment and nutrient accumulation rates in coastal, freshwater Lake Sibaya, South Africa. *Marine and Freshwater Research*, 64, 1087–1099.

Jacobs, Z., Roberts, R. G., Galbraith, R. F., Deacon, H. J., Grün, R., Mackay, A., Mitchell, P., Vogelsang, R. and Wadley, L. (2008). Ages for the Middle Stone Age of southern Africa: implications for human behavior and dispersal. *Science*, 322, 733–735.

Jarzen, D. M. and Nichols, D. J. (1996). Pollen. In *Palynology: Principles and Applications. Vol. 1. Principles*, ed. J. Jansonius and D. C. McGregor. Salt Lake City: American Association of Stratigraphic Palynologists Foundation, pp. 261–292.

Kristen, I., Wilkes, H., Vieth, A., Zink, K.-G., Plessen, B., Thorpe, J., Partridge, T. C. and Oberhänsli, H. (2010). Biomarker and stable carbon isotope analyses of sedimentary organic matter from Lake Tswaing: Evidence for deglacial wetness and early Holocene drought from South Africa. *Journal of Palaeolimnology*, 44, 143–160.

Majewski, W. and Bohaty, S. M, (2010). Surface-water cooling and salinity decrease during the Middle Miocene climate transition at Southern Ocean ODP site 747 (Kerguelen Plateau). *Marine Micropaleontology*, 74, 1–14.

McCarthy, T. S., Ellery, W. N., Backwell, L., Marren, P., de Klerk, B., Tooth, S., Brandt, D. and Woodborne, S. (2010). The character, origin and palaeoenvironmental significance of the Wonderkrater spring mound, South Africa. *Journal of African Earth Sciences*, 58, 115–126.

Meadows, M. E. and Sugden, J. M. (1990). Late Quaternary vegetation history of the Cederberg, southwestern Cape. *Palaeoecology of Africa*, 21, 269–282.

Metwally, A. A., Scott, L., Neumann, F. H., Bamford, M. K. and Oberhänsli, H. (2014). Holocene palynology and palaeoenvironments in the Savanna Biome at Tswaing Crater, central South Africa. *Palaeogeography, Palaeoclimatology, Palaeoecology*, 402, 125–135.

Miller, C. E. and Sievers, C. (2012). An experimental micromorphological investigation of bedding construction in the Middle Stone Age of Sibudu, South Africa. *Journal of Archaeological Science*, 39, 3039–3051.

Mills, S. C., Grab, S. W., Rea, B. R., Carr, S. J. and Farrow, A. (2012). Shifting westerlies and precipitation patterns during the Late Pleistocene in southern Africa determined using glacier reconstruction and mass balance modelling. *Quaternary Science Reviews*, 55, 145–159.

Mucina, L. and Rutherford, M. C. (2006). *The Vegetation of South Africa, Lesotho and Swaziland. Strelitzia 19*. Pretoria: South African National Biodiversity Institute, 807pp.

Neumann, F. H., Stager, J. C., Scott, L., Venter, H. J. T. and Weyhenmeyer, C. (2008). Holocene vegetation and climate records from Lake Sibaya, KwaZulu-Natal (South Africa). *Review of Palaeobotany and Palynology*, 152, 113–128.

Neumann, F. H., Scott, L., Bousman, C. B. and van As, L. (2010). A Holocene sequence of vegetation change at Lake Eteza, coastal KwaZulu-Natal, South Africa. *Review of Palaeobotany and Palynology*, 162, 39–53.

Neumann, F. H., Botha, G. A. and Scott, L. (2014). 18,000 years of grassland evolution in the summer rainfall region of South Africa: evidence from Mahwaqa Mountain, KwaZulu-Natal. *Vegetation History and Archaeobotany*, 23, 665–681.

Norström, E., Scott, L., Partridge, T. C., Risberg, J. and Holmgren, K. (2009). Reconstruction of environmental and climate changes at Braamhoek wetland, eastern escarpment South Africa, during the last 16,000 years with emphasis on the Pleistocene–Holocene transition. *Palaeogeography, Palaeoclimatology, Palaeoecology*, 271, 240–258.

Palmer, P. G. (1976). Grass cuticles: A new paleoecological tool for East African lake sediments. *Canadian Journal of Botany*, 54, 1725–1734.

Power, M. J., Marlon, J., Ortiz, N., Bartlein, P.J., Harrison, S. P., Mayle, F. E., Ballouche, A., Bradshaw, R. H. W., Carcaillet, C., Cordova, C., Mooney, S., Moreno, P. I., Prentice, I. C., Thonicke, K., Tinner, W., Whitlock, C., Zhang, Y., Zhao, Y., Ali, A. A., Anderson, R. S., Beer, R., Behling, H., Briles, C., Brown, K. J., Brunelle, A., Bush, M., Camill, P., Chu, G. Q., Clark, J., Colombaroli, D., Connor, S., Daniau, A.-L., Daniels, M., Dodson, J., Doughty, E., Edwards, M. E., Finsinger, W., Foster, D., Frechette, J., Gaillard, M.-J., Gavin, D. G., Gobet, E., Haberle, S., Hallett, D. J., Higuera, P., Hope, G., Horn, S., Inoue, J., Kaltenrieder, P., Kennedy, L., Kong, Z. C., Larsen, C., Long, C. J., Lynch, J., Lynch, E. A., McGlone, M., Meeks, S., Mensing, S., Meyer, G., Minckley, T., Mohr, J., Nelson, D. M., New, J., Newnham, R., Noti, R., Oswald, W., Pierce, J., Richard, P. J. H., Rowe, C., Sanchez Goñi, M. F., Shuman, B. N., Takahara, H., Toney, J., Turney, C., Urrego-Sanchez, D. H., Umbanhowar, C., Vandergoes, M., Vanniere, B., Vescovi, E., Walsh, M., Wang, X., Williams, N., Wilmshurst, J. and Zhang, J. H. (2008). Changes in fire regimes since the Last Glacial Maximum: An assessment based on a global synthesis and analysis of charcoal data. *Climate Dynamics*, 30, 887–907.

Quick, L. J., Chase, B. M., Meadows, M. E., Scott, L. and Reimer, P. J. (2011). A 19.5 kyr vegetation history from the central Cederberg Mountains, South Africa: Palynological evidence from rock hyrax middens. *Palaeogeography, Palaeoclimatology, Palaeoecology*, 309, 253–270.

Ramsay, P. J. (1995). 9000 years of sea-level change along the southern African coastline. *Quaternary International*, 31, 71–75.

Renaut, R. and Bamford, M. K. (2006). Results of a preliminary palynological analysis at Sibudu Cave. *Southern African Humanities*, 18, 235–240.

Roberts, D. L., Matthews, T., Herries, A. I. R., Boulter, C., Scott, L., Musekiwa, C., Mthembi, P., Browning, C., Smith, R. M. H., Haarhoff, P. and Bateman, M. D. (2011). Regional and global context of the Late Cenozoic Langebaanweg (LBW) palaeontological site: West Coast of South Africa. *Earth-Science Reviews*, 106, 191–214.

Roberts, D. L., Sciscio, L., Herries, A. I. R., Scott, L., Bamford, M. K., Musekiwa, C. and Tsikos, H. (2013). Miocene fluvial systems and palynofloras at the southwestern tip of Africa: Implications for regional and global fluctuations in climate and ecosystems. *Earth-Science Reviews*, 124, 184–201.

Rossouw, L., Stynder, D. D. and Haarhof, P. (2009). Evidence for opal phytolith preservation in the Langebaanweg 'E' Quarry Varswater Formation and its potential for palaeohabitat reconstruction. *South African Journal of Science*, 105, 223–227.

Sciscio, L., Neumann, F. H., Roberts, D., Tsikos, H., Scott, L. and Bamford, M. (2013). Fluctuations in Miocene climate and sea levels along the south-western South African coast: Inferences from biogeochemistry, palynology and sedimentology. *Palaeontologia Africana*, 48, 2–18.

Scott, L. (1982). A late Quaternary pollen record from the Transvaal bushveld, South Africa. *Quaternary Research*, 17, 341–370.

Scott, L. (1987). Pollen analysis of hyena coprolites and sediments from Equus Cave, Taung, Southern Kalahari (South Africa). *Quaternary Research*, 28, 144–156.

Scott, L. (1990). Hyrax (Procaviidae) and dassie rat (Petromuridae) middens in palaeoenvironmental studies in Africa. In *Packrat Middens: The Last 40,000 Years of Biotic Change*, ed. J. L. Betancourt, T. R. van Devender and P. S. Martin. Tucson: University Arizona Press, pp. 408–427.

Scott, L. (1994). Palynology of late Pleistocene hyrax middens, southwestern Cape Province, South Africa: A preliminary report. *Historical Biology*, 9, 71–81.

Scott, L. (1995). Pollen evidence for vegetational and climatic change in southern Africa during the Neogene and Quaternary. In *Paleoclimate and Evolution, with Emphasis on Human Origins*, ed. E. S. Vrba, G. H. Denton, T. C. Partridge and L. H. Burckle. New Haven: Yale University Press, pp. 65–76.

Scott, L. (1999a). Palynological analysis of the Pretoria Saltpan (Tswaing Crater) sediments and vegetation history of the Bushveld Savanna biome, South Africa. In *Tswaing – Investigations into the Origin, Age and Palaeoenvironments of the Pretoria Saltpan*, ed. T. C. Partridge. Pretoria: Council for Geoscience, pp. 143–166.

Scott, L. (1999b). Vegetation history and climate in the Savanna biome South Africa since 190,000 ka: A comparison of pollen data from the Tswaing crater (the Pretoria Saltpan) and Wonderkrater. *Quaternary International*, 57/58, 215–223.

Scott, L. (2002). Microscopic charcoal in sediments: Quaternary fire history of the grassland and savanna regions in South Africa. *Journal of Quaternary Science*, 17, 77–86.

Scott, L., Anderson, H. M. and Anderson, J. M. (1997). Vegetation history. In *Vegetation of Southern Africa*, ed. R. M. Cowling, D. M. Richardson and S. M. Pierce. Cambridge: Cambridge University Press, pp. 62–84.

Scott, L. and Bonnefille, R. (1986). Search for pollen from the hominid deposits of Kromdraai, Sterkfontein and Swartkrans: Some problems and preliminary results. *South African Journal of Science*, 82, 380–382.

Scott, L. and Bousman, C. B. (1990). Palynological analysis of hyrax middens from Southern Africa. *Palaeogeography, Palaeoclimatology, Palaeoecology*, 76, 367–379.

Scott, L., Holmgren, K., Talma, A. S., Woodborne, S. and Vogel, J. C. (2003). Age interpretation of the Wonderkrater spring sediments and vegetation change in the Savanna Biome, Limpopo Province, South Africa. *South African Journal of Science*, 99, 484–488.

Scott, L. and Thackeray, J. F. (1987). Multivariate analysis of late Pleistocene and Holocene pollen spectra from Wonderkrater, Transvaal, South Africa. *South African Journal of Science*, 83, 93–98.

Scott, L., Neumann, F. H., Brook, G. A., Bousman, C. B., Norström, E. and Metwally, A. A. (2012). Terrestrial fossil-pollen evidence of climate change during the last 26 thousand years in Southern Africa. *Quaternary Science Reviews*, 32, 100–118.

Scott, L. and Woodborne, S. (2007a). Pollen analysis and dating of Late Quaternary faecal deposits (hyraceum) in the Cederberg, Western Cape, South Africa. *Review of Palaeobotany and Palynology*, 144, 123–134.

Scott, L. and Woodborne, S. (2007b). Vegetation history inferred from pollen in Late Quaternary faecal deposits (hyraceum) in the Cape winter-rain region, and its bearing on past climates in South Africa. *Quaternary Science Reviews*, 26, 941–953.

Sievers, C. (2006). Seeds from the Middle Stone Age layers at Sibudu cave. *Southern African Humanities*, 18, 203–222.

Stager, J. C., Ryves, D. B., King, C., Madson, J., Hazzard, M., Neumann, F. H. and Maud, R. (2013). Late Holocene precipitation variability in the summer rainfall region of South Africa. *Quaternary Science Reviews*, 67, 105–120.

Strachan, K. L., Finch, J. M., Hill, T. and Barnett, R. L. (2014). A late Holocene sea-level curve for the east coast of South Africa. *South African Journal of Science*, 110, doi:10.1590/ sajs.2014/20130198.

Thiergart, F. (1964). Ein Vergleich der flözbildenden Elemente des niederrheinischen Tertiärs mit mehr oder weniger gleichaltrigen Südafrikas und Asiens auf Grund pollenanalytischer Untersuchungen. *Fortschritte der Geologie Rheinland und Westfalen*, 12, 105–112.

Thiergart, F., Frantz, U. and Raukopf, K. (1962). Palynologische Untersuchungen von Tertiärkohlen und einer Oberflächenprobe nahe Knysna, Südafrika. *Advancing Frontiers in Plant Science*, 4, 151–178.

Tribolo, C., Mercier, N., Douville, E., Joron, J.-L., Reyss, J.-L., Rufer, D., Cantin, N., Lefrais, Y., Miller, C. E., Porraz, G., Parkington, J., Rigaud, J.-Ph. and Texier, P.-J. (2013). OSL and TL dating of the Middle Stone Age sequence at Diepkloof Rock Shelter (South Africa): A clarification. *Journal of Archaeological Science*, 40, 3401–3411.

Tribolo, C., Mercier, N., Valladas, H., Joron, J. L., Guibert, P., Lefrais, Y., Selo, M., Texier, P.-J., Rigaud, J.-Ph., Porraz, G., Poggenpoel, C., Parkington, J., Texier, J.-P. and Lenoble, A. (2009). Thermoluminescence dating of a Stillbay–Howiesons Poort sequence at Diepkloof Rock Shelter (Western Cape, South Africa). *Journal of Archaeological Science*, 36, 730–739.

Truc, L., Chevalier, M., Favier, C., Cheddadi, R., Meadows, M. E., Scott, L., Carr, A. S., Smith, G. F. and Chase, B. M. (2013). Quantification of climate change for the last 20,000 years from Wonderkrater, South Africa: Implications for the long-term dynamics of the Intertropical Convergence Zone. *Palaeogeography, Palaeoclimatology, Palaeoecology*, 386, 575–587.

Valsecchi, V., Chase, B. M., Slingsby, J. A., Carr, A. S., Quick, L. J., Meadows, M. E., Cheddadi, R. and Reimer, P. J. (2013). A high resolution 15,600-year pollen and microcharcoal record from the Cederberg Mountains, South Africa. *Palaeogeography, Palaeoclimatology, Palaeoecology*, 387, 6–16.

van Zinderen Bakker, E. M. (1957). A pollen analytical investigation of the Florisbad deposits (South Africa). In *Proceedings of the Third-Pan-African Congress on Prehistory, Livingstone, 1955*, ed. J. D. Clarke. London: Chatto and Windus, pp. 56–67.

van Zinderen Bakker, E. M. (1984). Palynological evidence for Late Cenozoic arid conditions along the Namibia coast from holes 532 and 530A, Leg 57. *Deep Sea Drilling Project. Initial Reports of the Deep Sea Drilling Project*, 75, 763–768.

Vrba, E. S. (1985). Early hominids in southern Africa: Updated observations on chronological and ecological background. In *Hominid Evolution*, ed. P. V. Tobias. New York: Alan R. Liss, pp. 195–200.

Wadley, L. (2004). Vegetation changes between 61 500 and 26 000 years ago: The evidence from seeds in Sibudu Cave, KwaZulu-Natal. *South African Journal of Science*, 100, 167–173.

Wadley, L., Sievers, C., Bamford, M., Goldberg, P., Berna, F. and Miller, C. (2011). Middle Stone Age bedding construction and settlement patterns at Sibudu, South Africa. *Science*, 334, 1388–1391.

Walther, S. and Neumann, F. H. (2011). Sedimentology, isotopes and palynology of late Holocene cores from Lake Sibaya and the Kosi Bay system (KwaZulu-Natal, South Africa). *South African Geographical Journal*, 93, 133–153.

Wooller, M. J., Street-Perrott, F. A. and Agnew, A. D. Q. (2000). Late Quaternary fires and grassland palaeoecology of Mount Kenya, East Africa: Evidence from charred grass cuticles in lake sediments. *Palaeogeography, Palaeoclimatology, Palaeoecology*, 164, 207–230.

20

Minerogenic microfossil records of Quaternary environmental change in southern Africa

JENNIFER M. FITCHETT, JASPER KNIGHT AND STEFAN W. GRAB

Abstract

Minerogenic microfossils are abundantly preserved in sedimentary sequences from a wide range of aquatic environments, including shallow and deep ocean basins, lakes, wetlands and estuaries, and in environments with a range of pH, temperature, salinity and nutrient loads. In southern Africa, pollen is used more commonly as a palaeoenvironmental proxy than are minerogenic microfossils, despite the wider range of environmental variables to which minerogenic micro-organisms respond. Palaeoenvironmental reconstructions in southern Africa that have utilised some of these microfossils demonstrate their value, particularly in multi-proxy analyses, when comparing microfossil community changes with those represented by pollen, charcoal and stable isotopes. This chapter outlines the minerogenic microfossils that are most commonly examined globally, and discusses some specific case studies from southern Africa that demonstrate the utility of microfossils in reconstructing Quaternary palaeoenvironments. We argue that efforts should be made to expand the use of minerogenic microfossils in southern African palaeoenvironmental studies, given the valuable information they provide, both as proxies and through facilitating isotope analysis and dating.

20.1 Introduction

Minerogenic microfossils, hereafter referred to as microfossils, are the fossil remains (exo- or endoskeletons or preserved body parts) of micro-organisms with a size range of ~10–100 μm (Armstrong and Brasier, 2005). These fossils are abundant in a range of marine, freshwater and terrestrial environments, and many of them are well preserved within sediments deposited in these environments

(De Wever et al., 2002). Due to their abundance, microfossils can be chemically isolated for analysis from even small amounts of sediment (<1 g). This can facilitate the high-resolution investigation of sediment cores, using multiple proxies (Battarbee, 2000; Armstrong and Brasier, 2005). Micro-organisms from which many microfossils are derived, often have short life spans, ranging from days to weeks, and have specific environmental and habitat preferences (Stevenson and Pan, 2004). This means that micro-organism assemblages are potentially highly responsive to environmental changes, which indicates their potential importance as palaeoenvironmental proxies, providing high-resolution records of coeval and rapid environmental change (Smol and Stoermer, 2010). Micro-organisms tend to reflect micro-environments rather than broader regional environments, as they live within clearly defined environmental niches and are deposited locally as subfossils, immediately below their living habitat (Round, 1984). Aquatic microfossils are deposited *in situ* within water bodies, and are thus unlikely to be transported away from their locality of origin by wind, when compared with terrestrial proxies such as pollen and spores. When transport does occur, via river water flow and ocean currents, the provenance can be more easily traced due to the more specific environmental attributes of each species (Round, 1984; Jones, 2006).

Following the development of high magnification microscopes at the turn of the 20th Century and increased concerns of water quality, the analysis and understanding of microfossils has rapidly increased, facilitating the use of these proxies in palaeoenvironmental work (Pike and Kemp, 1996; Battarbee et al., 2001). Many of these proxies can usefully supplement more commonly used palaeoenvironmental proxies, both in their capacity to corroborate evidence on warm-cool, wet-dry gradients, but also by providing additional environmental information, including water quality, pH and anthropogenic influence through pollution and nutrient loads (Battarbee, 2000; Rörig, 2013; Lotter, 2014). The shells and valves of certain microfossils can also be used to more directly reconstruct water temperatures through the extraction of oxygen isotopes (Leng and Barker, 2006). Microfossil work offers great potential for an improved understanding of rapid environmental and climate changes that took place during the Pleistocene–Holocene transition, and of anthropogenic environmental changes during the Holocene (Finné et al., 2010). From the perspective of anthropogenic change, phytoliths in particular are a useful additional proxy in combination with pollen analysis to explore human-induced shifts from forest to grassland ecosystems, and to determine the timing and rate of such change (Scott, 2002). Despite these numerous applications of microfossils in palaeoenvironmental work, such research has predominantly been centred in East Africa, due to the prevalence of lake systems in the region (e.g. Haberyan, 1987; Stager, 1988; Barker et al., 2003, 2013; Stager et al., 2005; Pasche et al., 2010; Roubiex et al., 2014). The lack of southern African palaeoenvironmental

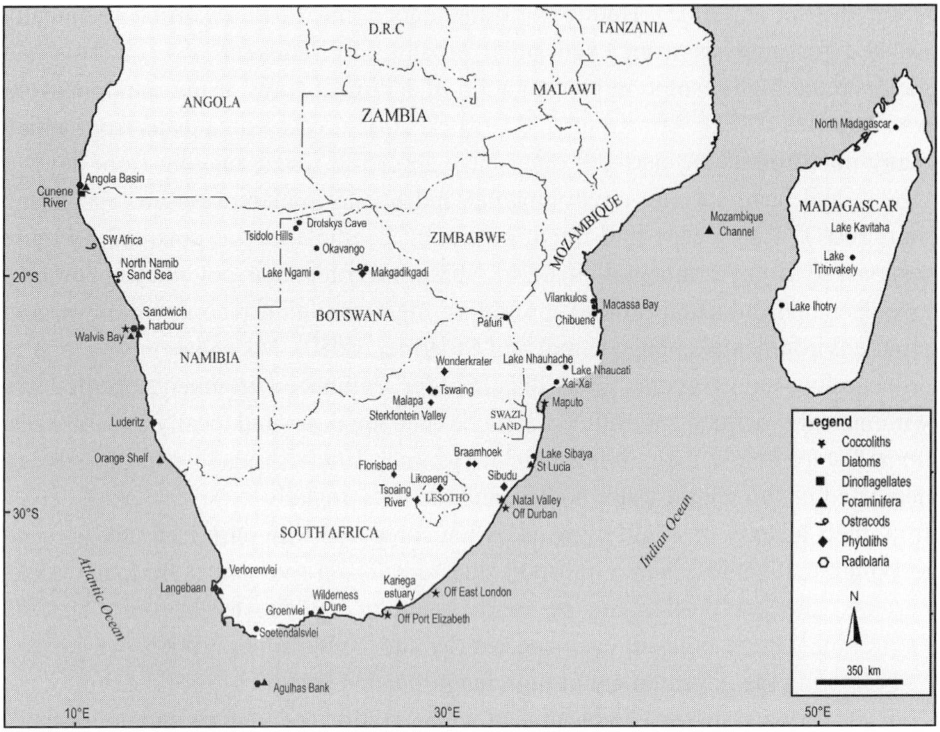

Fig. 20.1. Distribution of key sites in southern Africa (Namibia, Botswana, South Africa, Lesotho, Swaziland, Mozambique, Madagascar) where palaeoenvironmental reconstructions using minerogenic microfossils have taken place.

reconstructions that are based on microfossil evidence (Fig. 20.1) is due more to the relatively late inception of all proxy based palaeoenvironmental work in the region, and the associated scarcity of skills, rather than the absence of microfossils in the sediment record.

This chapter first outlines the different types of microfossils recorded in southern African records, and then presents four case studies illustrating their use with respect to environmental reconstruction. Some of the advantages and limitations of microfossils as palaeoenvironmental proxies are then discussed.

20.2 Microfossil types and their environmental significance

The classification of microfossils is based on laboratory methods rather than biological differences, requiring isolation from the surrounding sediments and the use of microscopy for analysis (Armstrong and Brasier, 2005). Microfossils can be broadly separated into organisms that produce minerogenic shells or valves, and those that have organic cell walls (Armstrong and Brasier, 2005) (Table 20.1).

Table 20.1 *Microfossil categorisation (after Armstrong and Brasier, 2005), and examples of their use at key palaeoenvironmental sites in southern Africa. ZA – South Africa, MG – Madagascar, LS – Lesotho, BW – Botswana*

Microfossil classification	Microfossil	Location of study	Period of reconstruction	Reference
Organic	Pollen	Braamhoek, ZA	16 kyr–present	Norström et al., 2009
		Tswaing, ZA	200 kyr–present	Scott, 1999
		Kruger National Park, ZA	1280–1950 AD	Ekblom et al., 2012
		Lake Tritrivakely, MG	40 kyr–present	Gasse & Van Campo, 1998
		Wonderkrater, ZA	>45 kyr–present	Backwell et al., 2014
		Verlorenvlei, ZA	5.5 kyr–present	Baxter and Meadows, 1999
		Mfabeni, ZA	44 kyr–present	Finch & Hill, 2008
		Groenvlei, ZA	8 kyr–present (inferred)	Martin, 1959, 1968
		Cederberg (hyrax middens), ZA	19.5–7.3 kyr	Chase et al., 2011
		Lake Sibaya, ZA	7 kyr–present	Neumann et al., 2008
		Lake Nhaucati, MZ	1.6 kyr–present	Ekblom and Stabell, 2008
		Tsoaing River Basin, LS	12–4 kyr	Grab et al., 2005
	Fungal spores	Kruger National Park, ZA	1300–2000 AD	Ekblom et al., 2012
Minerogenic				
Siliceous	Phytoliths	Braamhoek, ZA	16 kyr–present	Finné et al., 2010
		Tswaing, ZA	200 kyr–present	McLean and Scott, 1999
		Wonderkrater, ZA	>45 kyr–present	Backwell et al., 2014
		Makgadikgadi, BW	140 kyr–present	Burrough et al., 2012
		Tsoaing River Basin, LS	12–4 kyr	Grab et al., 2005
	Diatoms	Braamhoek, ZA	16 kyr–present	Finné et al., 2010
		Tswaing, ZA	200 kyr–present	Metcalfe, 1999
		Kruger National Park, ZA	1280–1950 AD	Ekblom et al., 2012

Table 20.1 (*cont.*)

Microfossil classification	Microfossil	Location of study	Period of reconstruction	Reference
		Lake Tritrivakely, MG	40 kyr–present	Gasse and Van Campo, 1998
		Groenvlei, ZA	8 kyr–present (inferred)	Martin, 1967
		Lake Sibaya, ZA	6.5 kyr–present	Neumann et al., 2008; Stager et al., 2013
		Makgadikgadi, BW	40.5 kyr–present	Huntsman-Mapila et al., 2006
		Lake Nhaucati, MZ	1.6 kyr–present	Ekblom and Stabell, 2008
	Radiolaria	Benguela Upwelling Zone, NA	350 kyr–present	Jacot Des Combes and Abelman, 2007
Calcareous	Ostracods	Makgadikgadi, BW	50 kyr–present	Riedel et al., 2014
	Coccoliths	Benguela Upwelling Zone, NA	750 kyr–present	Baumann and Freitag, 2004
	Foraminifera	Benguela Upwelling Zone, NA	160 kyr–present	Little et al., 1997
		Kariega Estuary, ZA	1.2 kyr–present	Strachan et al., 2014

The latter group includes pollen, which has already been discussed by Bamford et al. (this volume), as well as non-pollen palynomorphs including spores, bryophytes and algal cell remains (Armstrong and Brasier, 2005). This chapter focuses on minerogenic microfossils, which can be classified on the basis of their dominant mineral composition. Siliceous microfossils include diatoms and phytoliths, which have been used more frequently in Quaternary environmental studies in southern Africa. Calcareous microfossils have been used less commonly and include foraminifera, radiolaria and ostracods (Table 20.1). A third and rarer class, phosphatic microfossils, is not discussed in this chapter, but include conodonts, scolecodonts and ichthyoliths (Burnett and Riggs, 2006). Testate amoebae are also minerogenic microfossils, but as their tests can be siliceous, calcareous or formed through agglutination of sediments, they cannot be further categorised within this group (Mitchell et al., 2008). A final microfossil type that is not included in this chapter, but which can provide useful palaeoenvironmental information, is the fossil remains of crustaceans and insects with chitin skeletons, including cladocera, coleoptera, diptera and chironomids, as chitin is composed of long-chain sugars rather than minerals (Hofmann, 1998). Minerogenic microfossils can also be categorised as plants or animals, but given that many of the plant microfossils are motile and do not rely exclusively on photosynthesis, such distinctions become difficult and add little value to microfossil classification (Armstrong and Brasier, 2005).

Of these different types of microfossils, palaeoenvironmental studies in southern Africa have predominantly focused on the use of diatoms and phytoliths, largely because these are most abundant in many different environmental settings. Many microfossils, including testate amoebae, cocodonts, scolecodonts, cladocera, coleoptera and chironomids have not been used extensively in palaeoenvironmental reconstructions for the region, and these represent an under-exploited potential set of palaeoenvironmental proxies. Based on the classification discussed, a brief introduction to the commonly-used siliceous and calcareous microfossils is presented, with specific reference to their use in southern African palaeoenvironmental reconstructions.

20.2.1 Siliceous microfossils

Phytoliths are one of the most common minerogenic microfossils used in southern African palaeoenvironmental work, at sites including Tswaing Crater (Scott, 2002), Florisbad (Scott and Rossouw, 2005), Braamhoek wetland (Finné et al., 2010), the Kalahari (Burrough et al., 2012) and Wonderkrater (Backwell et al., 2014). Formed by plants through the uptake of silica from the soil, phytoliths are rigid silica structures found within plant tissue (Piperno, 2006) (Fig. 20.2). Due to the resilience of silica, phytoliths retain their shape after the decay of the plant, and

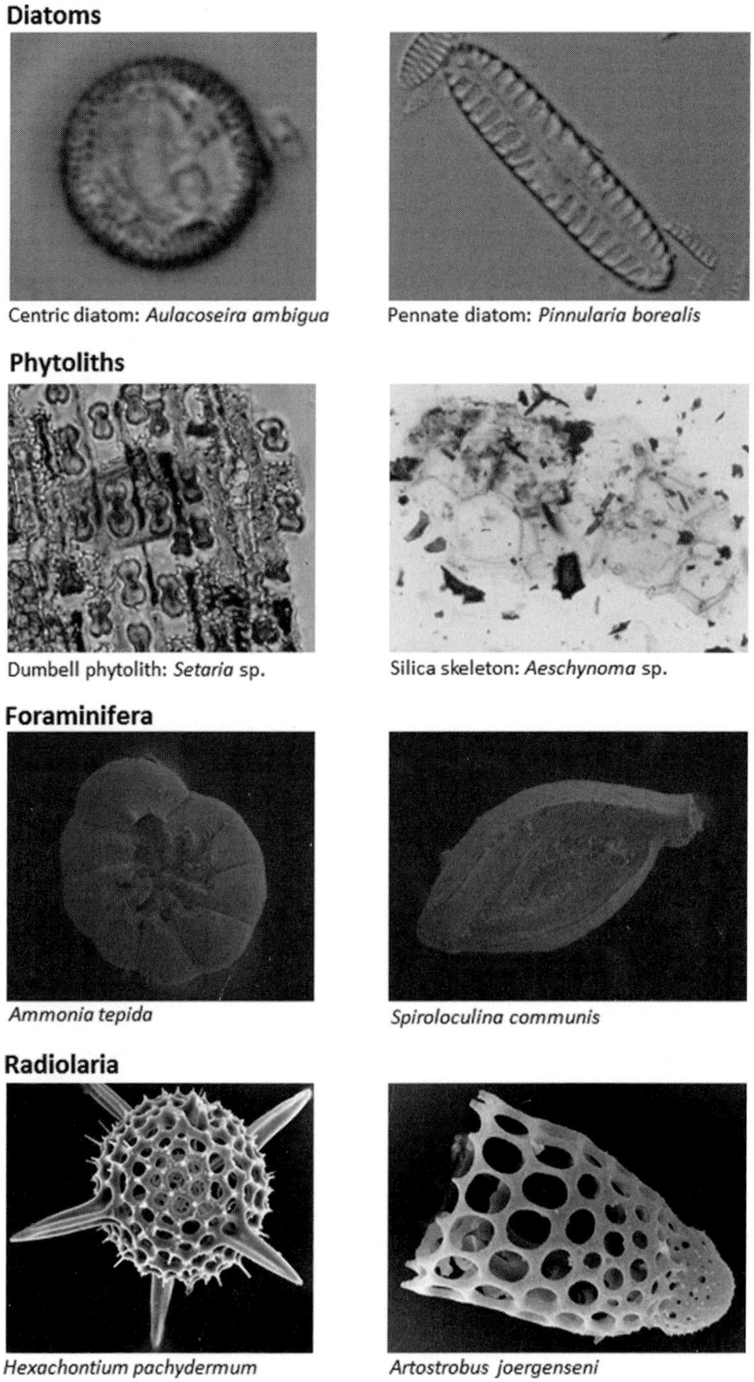

Fig. 20.2. Light microscope and SEM photographs of the siliceous and calcareous minerogenic microfossils discussed in this chapter. Diatoms, pennate and centric (photos: Jennifer Fitchett), phytoliths, dumbbell and silica skeleton (photos reproduced by permission of Marion Bamford), foraminifera species (photos reproduced by permission of Kate Leigh Strachan), radiolaria species (photos reproduced by permission of Jane Dolven).

are well preserved in sediment records, even in rocks of late Devonian age (Smol and Stoermer, 2010) and in southern Africa of Mio-Pliocene age (Rossouw et al., 2009). A wide range of different terrestrial plants produce phytoliths, but the most common are from the family Poaceae (grass) (Finné et al., 2010; Cordova, 2013). This is particularly beneficial to palaeoenvironmental studies, as it facilitates higher resolution grass taxonomy than fossil pollen is able to offer, which can also enable differences in C_3 and C_4 grass composition to be determined (Scott, 2002). Phytolith studies have consequently been used together with pollen analysis in grassland regions of southern Africa, and key palaeoenvironmental study sites are increasingly being revisited to include phytolith analysis (e.g. McLean and Scott, 1999; Scott, 2002; Burrough et al., 2012; Chazan et al., 2012; Backwell et al., 2014). In addition to Poaceae phytoliths, other plant groups in southern Africa produce identifiable, ecologically significant phytoliths, including Restionaceae and Cyperaceae (Cordova, 2013), and tree species (Mercader et al., 2009). The limitations in phytolith analysis involve both their production and taxonomy. Various parts of a single plant may produce phytoliths, resulting in over-counting for that population, and different plants are able to produce morphologically identical phytoliths, which can result in problems for environmental attribution (Tsartsidou et al., 2007; Cordova and Scott, 2011; Cordova, 2013).

Diatoms are now a commonly used palaeoenvironmental proxy in southern Africa. These are single cell algae of the class Bacillariophyceae, which produce siliceous tests, unique to species level (Mackay et al., 2003; Smol and Stoermer, 2010) (Fig. 20.2). The siliceous remains of diatoms are present in the fossil records of marine and freshwater environments as far back as the Lower Cretaceous, and are abundant in Quaternary sedimentary sequences (Battarbee et al., 2001; Smol and Stoermer, 2010). Diatoms are particularly valuable as they have a short life span (approximately a week) and individual species have very specific environmental tolerances, whilst the group itself is found in a range of environments including marine, riverine, lacustrine and shallow wetlands (Mackay et al., 2003; Smol and Stoermer, 2010). Consequently, changes in diatom communities are good indicators of changing environmental, aquatic and climate conditions, and the ecosystem sensitivity to changing environmental thresholds (Smol and Stoermer, 2010). Diatoms are responsive to, and thus when found in the fossil record are indicative of, changes in water depth, temperature, pH, salinity, dissolved oxygen content and nutrient loading (Battarbee et al, 2001; Rörig, 2013). Unlike terrestrial plants to which phytoliths can be confidently attributed, our understanding of the habitats, survival strategies and autecology of diatoms remains incomplete, and represents the greatest challenge to their use in palaeoenvironmental work (Mackay et al., 2003; Bere, 2014). Although research initially focused only on contemporary diatom assemblages and their role as water quality indicators (Schoeman, 1972;

Taylor et al., 2005), they are increasingly used for southern African environmental reconstructions, including establishing sea-level curves (Martin, 1959), moisture gradients (Huntsman-Mapila et al., 2006; Stager et al., 2012), and as multi-proxies to complement findings from foraminifera and phytolith analysis (Kristen et al., 2007; Finné et al., 2010). Much work is still required to establish locally based diatom transfer functions. At present, the most appropriate transfer functions are arguably from East Africa (Stager et al., 2013), although continued work on the environmental attributes of contemporary diatom assemblages may prove more valuable (Harding and Taylor, 2011).

Radiolaria represent the third commonly used siliceous microfossil group globally, but their use in southern African palaeoenvironmental studies remains limited. Radiolaria are marine protozoa that produce intricate siliceous skeletons around a central capsule (Sanfilippo et al., 1989; De Wever et al., 2002) (Fig. 20.2). Living species are restricted to marine environments, but are presently distributed across the world's oceans and throughout the water column, with a wide range of environmental tolerances, and their skeletal remains comprise the largest component of siliceous ooze on the ocean floor. Radiolaria are some of the oldest microfossils, extending back to the start of the Cambrian, with 80% of the fossil species extinct, which limits their use as environmental proxies (Sanfilippo et al., 1990; Feng et al., 2007). Due to their highly resistant silica skeletons, radiolaria are particularly valuable within ocean floor sediments located below the carbonate compensation depth (CCD), and for the reconstruction of ocean temperatures, but they can also be used to infer changes in salinity, dissolved oxygen and levels of nitrates and phosphates (De Wever et al., 2002). The use of radiolaria in southern African environmental reconstructions is limited to investigations into upwelling of the Benguela and Agulhas currents (Rau et al., 2002; Jacot Des Combes and Abelman, 2007; Lazarus et al., 2008).

20.2.2 Calcareous microfossils

Ostracods are microscopic crustaceans with calcareous bivalve shells, occupying a broad range of aquatic habitats in marine, freshwater and terrestrial environments (Boomer et al., 2003; Martens et al., 2008; Rodriguez-Lazaro and Ruiz-Muñoz, 2012) (Fig. 20.2). Ostracods are useful palaeoenvironmental indicators as their species composition within a given environment is dependent on temperature, salinity, conductivity and nutrient loads (Zhu et al., 2010; Horne et al., 2012). They are also sensitive to water depth, water transparency and water retention time and overturning (Battarbee, 2000; Zhu et al., 2010). On the basis of absence/presence of a particular species, palaeo-air temperatures can be estimated using the concept of the Mutual Ostracod Temperature Range (Horne, 2007). Ostracods have a fossil record extending from the Ordovican to present, but are limited in marine environments by the CCD (Armstrong and Brasier, 2005; Rodriguez-Lazaro and Ruiz-Muñoz, 2012). While surveys of ostracods in southern Africa have been undertaken (Dingle et al., 1989; Martens, 2007; Siziba, 2013), very little palaeoenvironmental work in the region has utilised this proxy. This is predominantly as a result of the scarcity of funding, expertise and facilities in southern Africa to presently investigate wide selections of proxies (Chase and Meadows, 2007).

Coccoliths are comprised of calcium carbonate plates formed by single cell algae (coccolithophores) and are widespread in the photic zone of most marine habitats (Billard and Inouye, 2004). Coccoliths are relatively young in the geological record, first appearing in the Jurassic (De Vargas et al., 2007). The coccolith exoskeleton (coccosphere) is well preserved due to its low magnesium content. Coccospheres can be examined by Scanning Electron Microscopy (SEM) and identified to species level based on plate microstructures (Samtleben and Bickert, 1990; Tyrrell and Young, 2009) (Fig. 20.2). As with radiolaria, the use of coccoliths in environmental reconstruction in southern Africa has been limited to examining ocean upwelling systems (Mitchell-Innes and Winter, 1987; Summerhayes et al., 1995; Baumann and Freitag, 2004).

Foraminifera are the most commonly used calcareous microfossil for southern African palaeoenvironmental reconstructions. These are a group of exclusively marine amoeboid protists identified through the preservation of their tests (Murray, 2006; Jones, 2013) (Fig. 20.2). These tests are predominantly comprised of $CaCO_3$, but identification to species level is sometimes hampered by agglutinated sediment and organic particles, and silica-test genus variants (Jones, 2013). The earliest foraminifera date back to the Cambrian, but they have been morphologically consistent since the Permian (Murray, 2006). In southern Africa, foraminifera provide evidence for sea-level change and changes in sedimentary conditions (McMillan, 2003; Franceschini et al., 2005; Strachan et al., 2014, 2015), upwelling processes (Giraudeau, 1993; Little et al., 1997), and for temperature fluctuations and changes associated with El Niño and the Little Ice Age (Herbert, 1987). There has also been considerable research into the contemporary ecologies of living foraminifera in the Benguela Upwelling Zone, the Mozambique Channel and the Agulhas Bank (Schmiedl et al., 1997; Licari and Mackensen, 2005; Leiter and Altenback, 2010).

20.3 Southern African microfossils: Case studies

20.3.1 Braamhoek wetland, South Africa

A multi-proxy palaeoenvironmental reconstruction for the late Pleistocene to present is available at Braamhoek wetland, within the temperate grassland biome of South Africa (Fig. 20.3), using pollen and charcoal (Norström, 2009), and

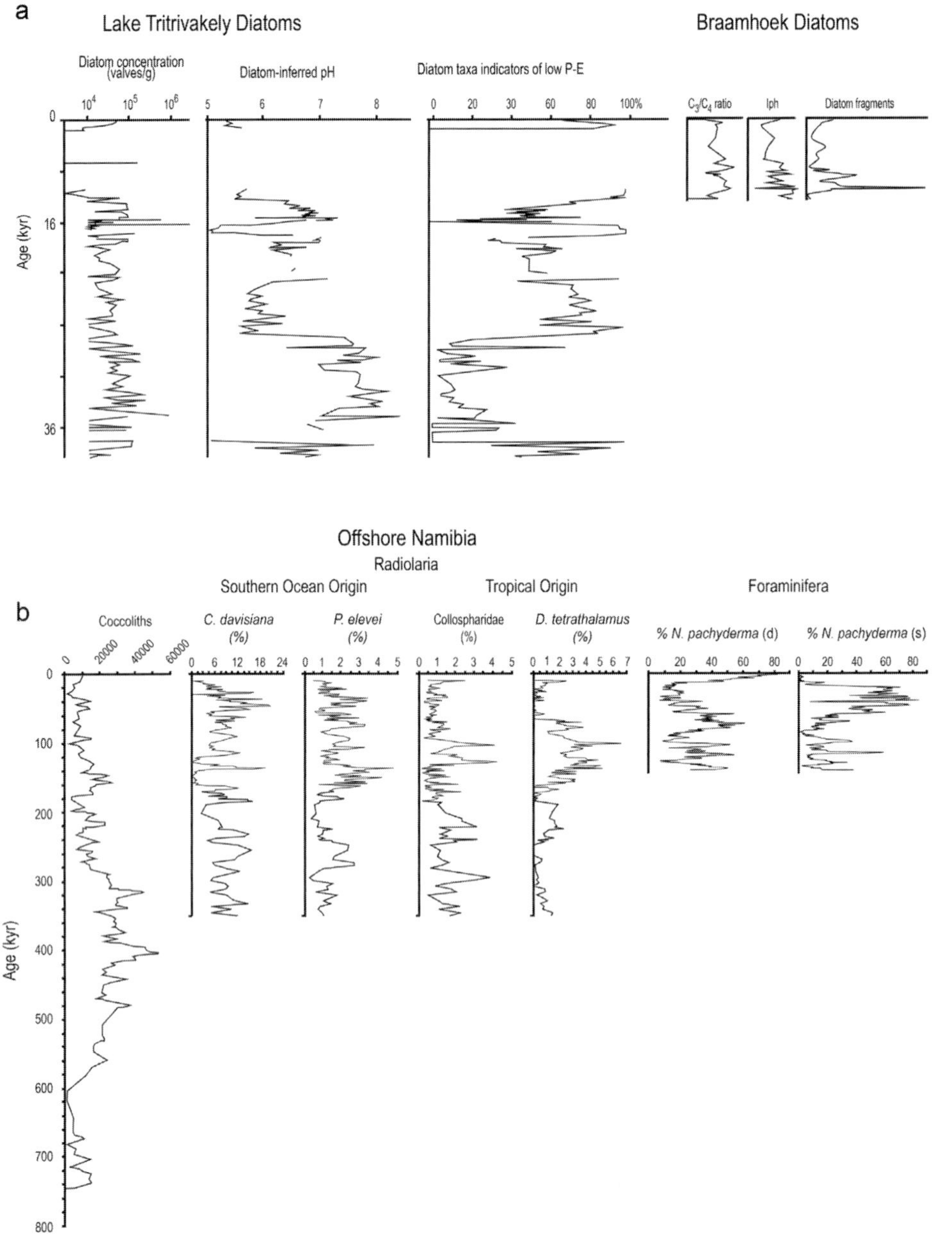

Fig 20.3. Composite stratigraphic diagrams demonstrating proxy evidence for palaeoenvironmental change at (a) Lake Tritrivakely (Gasse and Van Campo, 1998), Braamhoek wetland (Finné et al., 2010), and (b) the Benguela Upwelling Zone offshore from the Namibian coast (Little et al., 1997; Baumann and Freitag, 2004; Jacot Des Combes and Abelman, 2007).

phytoliths and diatoms (Finné et al., 2010). A 3 m continuous core was extracted from the wetland, spanning the last 16 kyr (Finné et al., 2010). The study simultaneously extracted and counted the diatoms and phytoliths, with phytoliths representing between 80.0–99.4% of the total microfossil count throughout the sediment core (Finné et al., 2010). While accounting for the majority of microfossils counted, C_3 species dominated the phytolith assemblage and no clear relationships between C_3 and C_4 grasses could be inferred (Finné et al., 2010). The diatoms, which represented a very small proportion of the microfossils counted, demonstrated fluctuations between higher diatom counts paired with higher proportions of planktonic species indicative of deeper water and wetter conditions, and low diatom counts combined with proportionally more benthic and aerophilic species, which suggest periods of drying (Finné et al., 2010). The planktonic species, dominated by *Aulacoseira ambigua, Aulacoseira distans, Cyclotelly cyclopuncta* and *Discotella* sp., reach maximum concentrations at the Pleistocene-Holocene transition, with particularly wet periods inferred at 13.6, 11.3 and 10.4–10.0 kyr BP (Finné et al., 2010). This supports evidence from pollen analysis at the site, from which wet periods are inferred for the periods 13.7–12.8, 10.5–9.5 and 1.5 kyr BP, based on relative changes in the concentration of Cyperaceae (Norström et al., 2009). The study raises two important considerations. The first is the use of C_3:C_4 plant ratios to reconstruct temperature changes in mountain environments. Such reconstructions require the study site to be located within an altitudinal range in which both groups of plants could potentially have thrived under past temperature fluctuations (Roberts et al., 2013). The altitude of the Braamhoek wetland is likely too high for plant community shifts to have spanned the C_3:C_4 divide. The second consideration relates to the importance of exploring proxies individually. By using a combined count of all siliceous microfossils, the more notable changes in diatoms throughout the sequence are less statistically robust, given the low total counts of diatom valves analysed. Independent counts of diatoms from the sediment profile would provide a more accurate representation of species composition changes through time, from which more accurate environmental inferences could be made.

20.3.2 Lake Tritrivakely, Madagascar

A diatom-based environmental reconstruction for a much longer time period is available for Lake Tritrivakely, in the central highlands of Madagascar (Fig. 20.3). A 40 m long core was extracted from the maar lake, presently situated in a crater 19 m deep and 55 m wide, which, due to changes in water level, presently fluctuates between a shallow lake and an oligotrophic bog (Gasse and Van Campo, 2001). AMS ^{14}C dates from the upper 13 m of the core constrain it to 0–41 cal kyr BP, while the age of the remainder of the core has been interpolated by comparison between the Vostok δD isotopic record and the pollen-based temperature inferences for the core. This method estimates a basal age for the core of 150 kyr BP (Gasse and Van Campo, 2001). The diatom-based reconstructions discussed are only for the period from 41 kyr BP to present (Gasse and Van Campo, 1998, 2001). Similar to Braamhoek, environmental inferences from diatoms are made largely on the basis of the ratio between planktonic and aerophilic species, and indicate a gradual transition from a permanent slightly alkaline lake dominated by the planktonic *A. ambigua* and *Aulacoseira granulata,* to acidic subaerial conditions dominated by aerophilic *Eunotia* and *Pinnularia* spp. together with acidophilous *Cymbella* and *Navicula* spp., all of which demonstrated dissolution (Gasse and Van Campo, 1998). An intermediate phase, with species indicative of circumneutral, freshwater swamps, was dominated by *Cymbella* and *Stauroneis* spp. together with *Anomoeoneis brachysira* and *Nitzschia linearis* (Gasse and Van Campo, 1998). These shifts in diatom species assemblages are concurrent with a decrease in Ericaceae pollen and the emergence of Cyperaceae with the transition into bog conditions (Gasse and Van Campo, 1998). The multi-proxy reconstructions for the lake suggest cool, dry conditions at the Last Glacial Maximum, with postglacial warming and wet conditions occurring relatively suddenly at 17 and 15 kyr BP, respectively, and bog development throughout the last 4 kyr (Gasse et al., 1994; Gasse and Van Campo, 1998, 2001). The value of using a multi-proxy approach here is in the ability to reconstruct and differentiate between changes in both the regional environment and the succession of the water body. The simultaneous shifts in pollen and diatom communities are of interest, particularly when considering the differences in geographic area represented by the two proxies.

20.3.3 Makgadikgadi, Botswana

Desert environments are unusual locations for palaeoenvironmental reconstructions based on minerogenic microfossils, as their existence alone tends to suggest aquatic or marine conditions and thus pluvial climates. The Makgadikgadi salt pan is thus a particularly interesting case study, as it is a palaeolake located southeast of the Okavango Delta, entirely surrounded by the Kalahari Desert (Thomas and Shaw, 1991). While the region is more commonly recognised for its Quaternary-age sand dunes and palaeolake shorelines (Thomas and Shaw, 1991; Burrough et al., 2007; Moore et al., 2012), palaeoenvironmental reconstructions have also been undertaken based on diatoms, phytoliths and ostracods. Studies on ostracods focus on their presence/absence, as lake conditions would be required for their growth (Riedel et al., 2014). While diatom studies include inferences made on the basis of concentration, with diatomite beds themselves indicating palaeolake shores, further analyses explore more intricate environmental changes based on shifts in species assemblage composition (Saw, 2010; Riedel et al., 2014). Changes in lake level are inferred from *Campylodiscus, Cyclotella* and *Synedra* spp. during lacustrine conditions, whereas higher energy fluvial systems are indicated by *Fragilaria* and *Rhopalodia* spp. (Shaw et al., 1997). Diatoms are also used here to infer pH changes from slightly alkaline to neutral, with neutral conditions indicated by increasing numbers of *Nitzschia, Navicula, Melosira* and *Amphora* spp. (Shaw et al., 1997). A more recent analysis of the diatoms in the adjacent Lake Ngami reports shifts from the lower section of the core, dominated by *A. ambigua, Denticula kuetzingii* and *Fragilaria construens*, suggesting well mixed conditions in deeper water, whereas species from more recent sediments include *Achnanthes minutissima* and *Cymbella silesiaca* which both live in shallower water (Huntsman-Mapila et al., 2006). Phytolith analysis of sandy shoreline deposits indicates significant shifts in the savannah vegetation that occurs concurrent with lake level changes (Shaw, 1997; Huntsman-Mapila et al., 2006), with highest lake levels associated with grassland vegetation (Burrough et al., 2012). A progressive shift towards mesic and C_3 taxa is observed during lake events after 36 kyr BP, with samples dominated by an abundance of Pooid phytoliths, and occasional re-establishment of xeric C_4-dominated vegetation (Burrough et al., 2012). However, these results are still debated, as the dominant Pooid phytoliths identified in this study are inconsistent with the present day vegetation (Cordova et al., 2013), emphasising the need for detailed calibration studies and sufficient understanding of the contemporary local and regional environments. The discovery of microfossil sequences in the Kalahari suggests that they are considerably more widespread across the southern African landscape than previously realised, and are not limited to moist environments. This finding has implications for facilitating an improved spatial-resolution of southern Africa's palaeoenvironmental record.

20.3.4 Offshore Namibia

The previous three case studies demonstrate the application of microfossil palaeoenvironmental reconstructions to aquatic environments at inland locations, controlled in part by habitats in the surrounding terrestrial regions. This case study explores the use of minerogenic microfossils to reconstruct palaeoceanographic changes in upwelling strength of the Benguela Current, located off the coast of Namibia. This marine core record covers significantly longer time periods (from 750–130 kyr BP; Little et al., 1997; Baumann and Freitag, 2004) but at lower resolution than terrestrial records. The earliest of the three studies on palaeo-upwelling strength involves the use of foraminifera, with a comparison of the counts of cold water species *Neogloboquadrina pachyderma* and the warmer-water species *Globorotalia inflata*, from 160 kyr BP to present (Little et al., 1997) (Fig. 20.3). Peaks in *N. pachyderma* are indicative of periods with increased upwelling, which are demonstrated for rapidly changing but discrete periods throughout Marine Isotope Stage (MIS) 2 and 3 (Little et al., 1997). These pulses of strong upwelling do not correlate with glacial–interglacial cycles, but are posited to rather reflect shifts in the position of the Angola-Benguela front (Little et al., 1997). A study of radiolaria in this region based on a core record of 350 kyr length similarly relies on the use of indicator species for tracing water provenance (Jacot Des Combes and Abelman, 2007). This study uses *Didymorcyrtis tetrathalamus* and Collosphaeridae as indicators of tropical waters, and *Phorticium clevei* and *Cyladophora davisiana* for cold, Southern Ocean waters (Fig. 20.3). These findings by contrast show that strongest upwelling intensity corresponds to interglacial periods (MIS 3, 5 and 8), and weakest upwelling to glacial periods (MIS 2, 4 and 6) (Jacot Des Combes and Abelman, 2007). However, within MIS 4 they note the presence of species indicative of cold Antarctic waters, and in MIS 5 find species of tropical provenance (Jacot Des Combes and Abelman, 2007). A 750 kyr upwelling reconstruction for the region based on coccoliths shows elements in support of both these studies, suggesting changes in upwelling strength driven by both glacial–interglacial cycles on long-time scales, and more rapid shifts in the Angola-Benguela front (Baumann and Freitag, 2004). The study relies predominantly on coccolith concentrations rather than tracing counts of particular species, as coccolith species reach highest concentrations in warmer waters (Baumann and Freitag, 2004). Coccolith concentrations were lowest in glacial periods, and reached a maximum between 280–560 kyr BP (Baumann and Freitag, 2004) (Fig. 20.3). Whilst their concentrations broadly tracked glacial–interglacial cycles, smaller fluctuations in concentration map shifts in the Angola-Benguela front, which is a finding supported through the use of individual species to determine the depth of upwelling. Given the significant impact of the Benguela Current and upwelling on aridity over southwestern Africa, such longer-term reconstructions of upwelling strength are of value to climate modelling. Offshore records are of further interest as they represent systems with considerably higher energy than aquatic, terrestrial environments. This has implications for the spatial scale of environmental niches for different microfossil species, and for the temporal variability of environmental conditions during that micro-organism's lifespan.

20.4 Discussion

Minerogenic microfossils are valuable palaeoenvironmental proxies for three primary reasons. First, they are associated with conditions which are either aquatic/marine (diatoms, ostracods, radiolaria, foraminifera) or terrestrial (phytoliths). Thus, their presence in a sediment profile, and associated relative changes in their concentration, may indicate change in wet/dry or cold/warm periods (Riedel et al., 2014), as was discussed for ostracods at Makgadikgadi and coccoliths in the Benguela Current. Second, these microfossils can be identified to high levels of taxonomic resolution, and have very specific environmental preferences at such resolution (Stevenson and Pan, 2004; Smol and Stoermer, 2010). Species assemblages can therefore indicate a potential early-warning of environmental change, and can be used to identify critical environmental thresholds in the palaeo-record. Most southern African palaeoenvironmental work using microfossils adopts this approach. Third, the siliceous and calcareous shells or valves of these microfossils are suitable for isotope analysis and radiometric dating (Smol and Stoermer, 2010). Recent isotope work includes using $^{16}O/^{18}O$ ratios and Mg/Ca ratios in ostracods and foraminifera as proxies of hydrological cycle vigour, ice volume and temperature (Lamb et al., 2005; Gray et al., 2014). In addition, $\delta^{18}O_{diatom}$ values can be used to reconstruct past temperatures and the relationships between evaporation and precipitation rates (Leng and Barker, 2006; Wilson et al., 2014). The short life span of micro-organisms means that high-resolution (seasonal to annual) changes in environmental variables may be reconstructed, which is significantly higher than those from surrounding bulk sediments. Despite this potential for high-resolution analyses, they have not yet been applied in many cases, presenting an important aspect for future microfossil research (Horne et al., 2012).

As with all palaeoenvironmental proxies, limitations exist in their ability to reflect past conditions. The primary limitations are common to palaeoenvironmental studies, and involve making some general assumptions: first, that sediments and microfossils contained therein have a consistent stratigraphy and that no reworking has occurred since deposition; second, that sediments and microfossils originate from and are representative of a small geographic region and third, that present processes or ecological communities can be used as an analogue for those observed in the fossil record (Soon et al., 2003; Horne et al., 2012). In reality, these assumptions are not always valid. Significant reworking occurs in most environments from a combination of factors which include geological (e.g. seismic or flood events), geomorphological (e.g. weathering, infiltration and denudation), biological (e.g. bioturbation), and increasingly anthropogenic (e.g. human site disturbance) (Loubere, 1989; Maslin et al., 2014). To account for this, a high temporal resolution of age-dates through a sediment sequence is required, and caution should be used when relying on the stratigraphic context alone (Loubere, 1989). It is important to understand the local geographic context, as geomorphology, hydrology and climate often affect the region from which sediments and microfossils originate, and these may have changed over time (Soon et al., 2003). Further caution is required when using the Principle of Uniformitarianism, particularly during the Anthropocene, as levels of human influence on natural environments prevent any true analogues from existing (Knight and Harrison, 2014; Meadows, 2014). While these concerns apply to all fields of palaeoenvironmental study, they are particularly problematic when dealing with microfossils, as it is significantly more difficult to identify such issues from anomalies in their species composition alone (Smol and Stoermer, 2010; Birks, 2014). Unlike larger terrestrial animals and plants for which there is considerable knowledge regarding their identification, classification and habitat, our understanding of microfossils is relatively new and thus incomplete. New micro-organism species are discovered regularly, and the environmental preferences of these and many extant species remain unknown, and even for species with well-defined environmental preferences, specific environmental tolerances have not yet been tested (Birks, 2014). Consequently, there is a risk of misidentification of species, and misattribution of their ecological preferences from which palaeoenvironments are inferred (Cordova et al., 2013). This has necessitated ongoing efforts to develop and improve statistical microfossil-based transfer functions (Gasse et al., 1995; Juggins et al., 2015), and experimental calibration- and training-sets to test both the transfer functions and the relationship between fossils and their living communities (Bigler and Hall, 2002; Mackay et al., 2003; Seppä et al., 2004).

In southern Africa, suitable archives preserving palaeoenvironmental proxies remain relatively sparse (Chase and Meadows, 2007; Scott et al., 2012). While a large number of northern hemisphere lakes with deep sediment profiles have been examined for microfossils, fewer such lakes exist in southern Africa, largely due to its aridity (Scott et al., 2012). Likewise, few nearshore marine cores have been recovered from the Atlantic, Southern and Indian oceans; hence records from radiolaria, coccoliths, foraminifera and marine diatoms are still uncommon (Shi et al., 2001). Southern African microfossil records have nevertheless assisted with the reconstruction of past environments, in corroborating inferences made from other proxies, and in providing information on a broader range of environmental variables, including quantifying precipitation and lake level changes (e.g. Holmgren et al., 2012; Stager et al., 2013). There are also other micro-organisms that can be used for palaeoenvironmental reconstruction (e.g. coleoptera, chironomids, testate amoebae and cladocera), but these are very poorly understood in southern Africa. To ensure accuracy in palaeoenvironmental reconstructions made from these varied proxies, future efforts should include contemporary regional calibration datasets for each proxy (Cordova et al., 2013).

20.5 Summary

Minerogenic microfossils are widely preserved in sediments of different types, usually in large numbers, and from a wide variety of habitats and environments. This makes them particularly valuable environmental proxies for southern Africa, a region where considerable contemporary spatial environmental variability exists, and for which past climate and environmental changes (including the extent of sea-level rise, the age of palaeolakes and Pleistocene glaciation in highlands) remains uncertain. Due to their abundance within samples and high stratigraphic resolution, there is potential for good palaeoenvironmental information from microfossils. This is particularly valuable for southern Africa where many potential sites have not yet been cored and examined. Inferences can be made from the presence of these microfossils, their species composition, and from isotope analysis and dating. Additional value would be derived from their inclusion in multiproxy studies. The paucity of studies using minerogenic microfossils in southern Africa does not imply a lack of appropriate study sites, but rather that such proxies have not yet been fully appreciated in terms of their potential to improve local and regional palaeoenvironmental reconstructions.

References

Antoniades, D., Douglas, M. S. V. and Smol, J. P. (2005). Benthic diatom autecology and inference model development from the Canadian High Arctic Archipelago. *Journal of Phycology*, 41, 30–45.

Armstrong, H. A. and Brasier, M. D. (2005). *Microfossils*, 2nd ed. Chichester: Blackwell, 306pp.

Backwell, L., McCarthy, T. S., Wadley, L., Henderson, Z., Steininger, C. M., de Klerk, B., Barré, M., Lamothe, M., Chase, B. M., Woodborne, S., Susino, G. J., Bamford, M. K., Sievers, C., Brink, J. S., Rossouw, L., Pollarolo, L., Trower, G., Scott, L. and d'Errico, F. (2014). Multiproxy record of late Quaternary climate change and Middle Stone Age human occupation at Wonderkrater, South Africa. *Quaternary Science Reviews*, 99, 42–59.

Barker, P. A., Hurrell, E. R., Leng, M. J., Plessen, B., Wolff, C., Conley, D. J., Keppens, E., Milne, I., Cumming, B. F., Laird, K. R., Kendrick, C. P., Wynn, P. M. and Verschuren, D. (2013). Carbon cycling within an East African lake revealed by the carbon isotope composition of diatom silica: A 25-ka record from Lake Challa, Mt. Kilimanjaro. *Quaternary Science Reviews*, 66, 55–63.

Barker, P., Williamson, D., Gasse, F. and Gibert, E. (2003). Climatic and volcanic forcing revealed in a 50,000-year diatom record from Lake Massoko, Tanzania. *Quaternary Research*, 60, 368–376.

Battarbee, R. W. (2000). Palaeolimnological approaches to climate change, with special regard to the biological record. *Quaternary Science Reviews*, 19, 107–124.

Battarbee, R. W., Jones, V. J., Flower, R. J., Cameron, N. G., Bennion, H., Carvalho, L. and Juggins, S. (2001). Diatoms. In *Tracking Environmental Change Using Lake Sediments: Terrestrial, Algal and Siliceous Indicators*, ed. J. P. Smol, H. J. B. Birks, W. M. Last, R. S. Bradley and K. Alverson. Berlin: Kluwer Academic Publishers, pp. 155–202.

Baumann, K.-H. and Freitag, T. (2004). Pleistocene fluctuations in the northern Benguela Current system as revealed by coccolith assemblages. *Marine Micropalaeontology*, 52, 195–215.

Baxter, A. J. and Meadows, M. E. (1999). Evidence for Holocene sea level change at Verlorenvlei, Western Cape, South Africa. *Quaternary International*, 56, 65–79.

Bere, T. (2014). Ecological preferences of benthic diatoms in a tropical river system in Sao Carlos-SP, Brazil. *Tropical Ecology*, 55, 47–61.

Bigler, C. and Hall, R. I. (2002). Diatoms as indicators of climatic and limnological change in Swedish Lapland: A 100-lake calibration set and its value for paleoecological reconstructions. *Journal of Paleolimnology*, 27, 97–115.

Billard, C. and Inouye, I. (2004). What is new in coccolithophore biology? In *Coccolithophores: From molecular processes to global impact*, ed. H. R. Thierstein and J. R. Young. Berlin: Springer, pp. 1–30.

Birks, H. J. B. (2014). Quantitative palaeoenvironmental reconstructions from Holocene biological data. In *Global Change in the Holocene*, ed. J. Birks, R. Battarbee, A. Mackay and I. J. Slipper. Oxford: Routledge, pp. 107–123.

Boomer, I., Horne, D. J. and Slipper, I. J. (2003). The use of ostracods in palaeoenvironmental studies, or what can you do with an ostracod shell? *Palaeontological Society Papers*, 9, 153–179.

Burnett, W. C. and Riggs, S. R. (eds) (2006). *Phosphate Deposits of the World: Volume 3, Neogene to Modern Phosphorites*, 4th Ed. Cambridge: Cambridge University Press, 484pp.

Burrough, S. L., Thomas, D. S. G., Shaw, P. A. and Bailey, R. M. (2007). Multiphase Quaternary highstands at Lake Ngami, Kalahari, northern Botswana. *Palaeogeography, Palaeoclimatology, Palaeoecology*, 253, 280–299.

Burrough, S. E., Breman, E. and Dodd, C. (2012). Can phytoliths provide an insight into past vegetation of the Middle Kalahari palaeolakes during the Quaternary? *Journal of Arid Environments*, 82, 156–164.

Chase, B. M. and Meadows, M. E. (2007). Late Quaternary dynamics of southern Africa's winter rainfall zone. *Earth-Science Reviews*, 84, 103–138.

Chase, B. M., Quick, L. J., Meadows, M. E., Scott, L., Thomas, D. S. G. and Reimer, P. J. (2011). Late glacial interhemispheric climate dynamics revealed in South African hyrax middens. *Geology*, 39, 19–22.

Chazan, M., Avery, D. M., Bamford, M. K., Berna, F., Brink, J., Fernandez-Jalvo, Y., Goldberg, P., Holt, S., Matmon, A., Porat, N., Ron, H., Rossouw, L., Scott, L. and Horwitz, L. K. (2012). The Oldowan horizon in Wonderwerk Cave (South Africa): Archaeological, geological, paleontological and paleoclimate evidence. *Journal of Human Evolution*, 63, 859–866.

Cordova, C. E. (2013). C_3 Poaceae and Restionaceae phytoliths as potential proxies for reconstructing winter rainfall in South Africa. *Quaternary International*, 287, 121–180.

Cordova, C. E., Chase, B. M. and Smith, G. F. (2013). Comment on "Burrough, S.E., Breman, E., and Dodd, C., 2012. Can phytoliths provide an insight into past vegetation of the Middle Kalahari paleolakes during the Quaternary? Journal of Arid Environments 82, 156–164". *Journal of Arid Environments*, 92, 113–116.

Cordova, C. E. and Scott, L. (2010). The potential of Poaceae, Cyperaceae, and Restionaceae phytoliths to reflect past environmental conditions in South Africa. In *African Palaeoenvironmental and Geomorphic Landscape Evolution, Palaeoecology of Africa, 30*, ed. J. Runge. Boca Raton, Florida: CRC Press, pp. 107–133.

De Vargas, C., Aubry, M.-P., Probert, I. and Young, J. (2007). Origin and Evolution of Coccolithophores: From coastal hunters to organic farmers. In *Evolution of Primary Producers of the Sea*, ed. P. G. Falkowski and A. H. Knoll. Boston: Academic Press, pp. 251–285.

De Wever, P., Dumitrica, P., Caulet, J. P., Nigrini, C. and Caridroit, M. (2002). *Radiolarians in the Sedimentary Record.* Amsterdam: CRC Press, 533pp.

Dingle, R. V., Lord, A. R. and Boomer, I. D. (1989). Ostracod faunas and water masses across the continental margin off southwestern Africa. *Marine Geology*, 87, 323–328.

Ekblom, A. and Stabell, B. (2008). Paleohydrology of Lake Nhaucati (southern Mozambique), ~400 AD to present. *Journal of Paleolimnology*, 40, 1127–1141.

Ekblom, A., Gillson, L., Risberg, J., Holmgren, K. and Chidoub, Z. (2012). Rainfall variability and vegetation dynamics of the lower Limpopo Valley, Southern Africa, 500 AD to present. *Palaeogeography, Palaeoclimatology, Palaeoecology*, 363–364, 69–78

Feng, Q., He, W., Gu, S., Meng, Y., Jin, Y. and Zhang, F. (2007). Radiolarian evolution during the latest Permian in South China. *Global and Planetary Change*, 55, 177–192.

Finch, J.M. and Hill, T.R. (2008). A late Quaternary pollen sequence from Mfabeni Peatland, South Africa: Reconstructing forest history in Maputaland. *Quaternary Research*, 70, 442–450.

Finné, M., Norström, E., Risberg, J. and Scott, L. (2010). Siliceous microfossils as late-Quaternary paleo-environmental indicators at Braamhoek wetland, South Africa. *The Holocene*, 20, 747–760.

Franceschini, G., McMillan, I. K. and Compton, J. S. (2005) Foraminifera of Langebaan Lagoon salt marsh and their application to the interpretation of late Pleistocene depositional environments at Monwabisi, False Bay coast, South Africa. *South African Journal of Geology*, 108, 285–296.

Gasse, F., Cortijo, E., Disnar, J.-R., Ferry, L., Gilbert, E., Kissel, C., Laggoun-Défarge, F., Lallier-Vergès, E., Miskovsky, J.-C., Ratsimbazafy, B., Ranaivo, F., Robison, L., Tucholka, P., Saos, J-L.., Sifeddine, A., Taieb, M., Van Campo, E. and Williamson, D. (1994). A 36 ka environmental record in the southern tropics: Lake Tritrivakely (Madagascar). *Comptes Rendus de l'Académie des Sciences*, 318, 1513–1519.

Gasse, F., Juggins, S. and Ben Khelifa, L. (1995). Diatom-based transfer functions for inferring past hydrochemical characteristics of African Lakes. *Palaeogeography, Palaeoclimatology, Palaeoecology*, 117, 31–54.

Gasse, F. and Van Campo, E. (1998). A 40 000 year pollen and diatom record from Lake Tritrivakely, Madagascar, in the southern Tropics. *Quaternary Research*, 49, 299–311.

Gasse, F. and Van Campo, E. (2001). Late Quaternary environmental changes from a pollen and diatom record in the southern tropics (Lake Tritrivakely, Madagascar). *Palaeogeography, Palaeoclimatology, Palaeoecology*, 167, 287–308.

Giraudeau, J. (1993). Planktonic foraminiferal assemblages in surface sediments from the southwest African continental margin. *Marine Geology*, 110, 47–62.

Grab, S., Scott, L., Rossouw, L. and Meyer, S. (2005). Holocene palaeoenvironments inferred from a sedimentary sequence in the Tsoaing River Basin, western Lesotho. *Catena*, 61, 49–62.

Gray, W., Holmes, J. and Shevenell, A. (2014). Evaluation of foraminiferal trace element cleaning protocols on the Mg/Ca of marine ostracod genus Krithe. *Chemical Geology*, 382, 14–23.

Haberyan, K. A. (1987). Fossil diatoms and the paleolimnology of Lake Rukwa, Tanzania. *Freshwater Biology*, 17, 429–436.

Harding, W. R. and Taylor, J. C. (2011). *The South African Diatom Index (SADI) – A preliminary index for indicating water quality in rivers and streams in southern Africa*. WRC, Pretoria: Water Research Commission Report No. 1701/1/11.

Herbert, R. S. (1987). *Late Holocene Climatic Change: The Little Ice Age and El Niño from planktonic foraminifera in Sediments off Walvis Bay, South West Africa*. Cape Town; Bulletin No. 18, Marine Geosciences Unit, Department of Geology, University of Cape Town, 45pp.

Hofmann, W. (1998). Cladocerans and chironomids as indicators of lake level changes in north temperate lakes. *Journal of Paleolimnology*, 19, 55–62.

Holmgren, K., Risberg, J., Freudendahl, J., Achimo, M., Ekblom, A., Mugabe, J., Norström, E. and Sitoe, S. (2012). Water-level variations in Lake Nhauhache, Mozambique, during the last 2,300 years. *Journal of Paleolimnology*, 48, 311–322.

Horne, D. J. (2007). A Mutual Temperature Range method for Quaternary palaeoclimatic analysis using European nonmarine Ostracoda. *Quaternary Science Reviews*, 26, 1398–1415.

Horne, D. J., Holmes, J. A., Rodriguez-Lazaro, J. and Viehlberg, F. A. (2012). Ostracoda as proxies for Quaternary climate change: Overview and future prospects. In *Ostracoda as proxies for Quaternary climate change*, ed. D. Horne, J. Holmes, F. Viehberg and J. Rodriguez-Lazaro. Amsterdam: Elsevier, pp. 305–324.

Huntsman-Mapila, P., Ringrose, S., Mackay, A. W., Downey, W. S., Modisi, M., Coetzee, S. H., Tiercelin, J.-J., Kampunzu, A. B. and Vanderpost, C. (2006). Use of the geochemical and biological sedimentary record in establishing palaeo-environments and climate change in the Lake Ngami basin, NW Botswana. *Quaternary International*, 148, 51–64.

Jacot Des Comes, H. and Abelman, A. (2007). A 350-ky radiolarian record of Lüderitz, Namibia – evidence for changes in the upwelling regime. *Marine Micropaleontology*, 62, 194–210.

Jones, R. W. (2013). *Foraminifera and Their Applications*. Cambridge: Cambridge University Press, 401pp.

Jones, V. (2006). Diatom Introduction. In *Encyclopedia of Quaternary Science*, ed. S. A. Elias. London: Elsevier, pp. 476–484.

Juggins, S., Simpson, G. L. and Telford, R. J. (2015). Taxon selection using statistical learning techniques to improve transfer function prediction. *The Holocene*, 25, 130–136.

Knight, J. and Harrison, S. (2014). Limitations of uniformitarianism in the Anthropocene. *Anthropocene*, 5, 71–75.

Lamb, A. L., Leng, M. J., Sloane, H. J. and Telford, R. J. (2005). A comparison of the palaeoclimate signals from diatom oxygen isotope ratios and carbonate oxygen isotope ratios from a low latitude crater lake. *Palaeogeography, Palaeoclimatology, Palaeoecology*, 223, 290–302.

Lazarus, D., Bittniok, B., Diester-Haass, L., Billups, K., Ogawa, Y., Takahashi, K. and Meyers, P. (2008). Radiolarian and sedimentologic paleoproductivity proxies in late Pleistocene sediments of the Benguela Upwelling System, ODP Site 1084. *Marine Micropalaeontology*, 68, 223–235.

Leiter, C. and Altenbach, A. V. (2010). Benthic foraminifera from the diatomaceous mud belt off Namibia: characteristic species for severe anoxia. *Palaeontologia Electronica*, 13, Issue 2, 11A, 19pp. http://palaeo-electronica.org/2010_2/188/index.html

Leng, M. J. and Barker, P. A. (2006). A review of the oxygen isotope composition of lacustrine diatom silica for palaeoclimate reconstruction. *Earth-Science Reviews*, 75, 5–27.

Licari, L. and Mackensen, A. (2005). Benthic foraminifera off West Africa (1°N to 32°S): Do live assemblages from the topmost sediment reliably record environmental variability? *Marine Micropalaeontology*, 55, 205–233.

Little, M. G., Schneider, R. R., Kroon, D., Price, B., Bickert, T. and Wefer, G. (1997). Rapid palaeoceanographic change in the Benguela Upwelling System for the last 160,000 years as indicated by abundances of planktonic foraminifera. *Palaeogeography, Palaeoclimatology, Palaeoecology*, 130, 135–161.

Lotter, A. F. (2014). Multi-proxy climatic reconstructions. In *Global Change in the Holocene*, ed. J. Birks, R. Battarbee, A. Mackay and F. Oldfield. Oxford: Routledge, pp. 373–383.

Loubere, P. (1989). Bioturbation and sedimentation rate control of benthic microfossil taxon abundances in surface sediments: A theoretical approach to the analysis of species microhabitats. *Marine Micropaleontology*, 14, 317–325.

Mackay, A. W., Jones, V. J. and Battarbee, R. W. (2003). Approaches to Holocene climate reconstructions using diatoms. In *Global Change in the Holocene*, ed. J. Birks, R. Battarbee, A. Mackay and F. Oldfield. Oxford: Routledge, pp. 294–309.

Martens, K. (2007). On a new species and genus in the Cypridini (Crustacea, Ostracoda, Cyprididae) from South Africa, with a phylogenetic analysis of the tribe and a discussion on the genus concept in this group. *Journal of Natural History*, 41, 381–399.

Martens, K., Schön, I., Meisch, C. and Horne, D. J. (2008). Global diversity of ostracods (Ostracoda, Crustacea) in freshwater. *Hydrobiologia*, 595, 185–193.

Martin, A. R. H. (1959). The stratigraphy and history of Groenvlei, a South African coastal fen. *Australian Journal of Botany*, 7, 142–167.

Martin, A. R. H. (1968). Pollen analysis of Groenvlei Lake sediments, Knysna (South Africa). *Review of Palaeobotany and Palynology*, 7, 107–144.

Maslin, M., Pike, J., Stickley, C. and Ettwein, V. (2014). Evidence of Holocene climate variability in marine sediments. In *Global Change in the Holocene*, ed. J. Birks, R. Battarbee, A. Mackay and F. Oldfield. Oxford: Routledge, pp. 185–209.

McLean, B. and Scott, L. (1999). Phytoliths in sediments of the Pretoria Saltpan (Tswaing Crater) and their potential as indicators of the environmental history at the site. In *Tswaing – Investigations into the Origin, Age and Paleoenvironments of the Pretoria Saltpan*, ed. T. C. Partridge. Pretoria: Council for Geosciences, pp. 167–171.

McMillan, I. K. (2003). Foraminiferally defined biostratigraphic episodes and sedimentation pattern of the Cretaceous drift succession (Early Barremian to Late Maastrichtian) in seven basins of on the South African and southern Namibian continental margin. *South African Journal of Science*, 99, 537–576.

Meadows, M. E. (2014). Recent methodological advances in Quaternary palaeoecological proxies. *Progress in Physical Geography*, 38, 807–817.

Mercader, J., Bennett, T., Esselmont, C., Simpson, S. and Walde, D. (2009). Phytoliths in woody plants from the Miombo woodlands of Mozambique. *Annals of Botany*, 104, 91–113.

Metcalfe, S. E. (1999). Diatoms from the Pretoria Saltpan – a record of lake evolution and environmental change. In *Tswaing – Investigations into the Origin, Age and Paleoenvironments of the Pretoria Saltpan*, ed. T. C. Partridge. Pretoria: Council for Geosciences, pp. 172–192.

Mitchell-Innes, B. A. and Winter, A. (1987). Coccolithophores: A major phytoplankton component in mature upwelled waters off the Cape Peninsula, South Africa in March, 1983. *Marine Biology*, 95, 25–30.

Mitchell, E. A. D., Charman, D. J. and Warner, B. G. (2008). Testate amoebae analysis in ecological and paleoecological studies of wetlands: Past, present and future. *Biodiversity Conservation*, 17, 2115–2137.

Moore, A. E., Cotterill, F. P. D. and Eckhardt, F. D. (2012). The evolution and ages of Makgadikgadi palaeo-lakes: consilient evidence from Kalahari drainage evolution south-central Africa. *South African Journal of Geology*, 115, 385–413.

Murray, J. W. (2006). *Ecology and Applications of Benthic Foraminifera*. Cambridge: Cambridge University Press, 440pp.

Neumann, F. H., Stager, J. C., Scott, L., Venter, H. J. T. and Weyhenmeyer, C. (2008). Holocene vegetation and climate records from Lake Sibaya, KwaZulu-Natal (South Africa). *Review of Palaeobotany and Palynology*, 152, 113–128.

Norström, E., Scott, L., Partridge, T. C., Risberg, J. and Holmgren, K. (2009). Reconstruction of environmental and climate changes at Braamhoek wetland, eastern escarpment South Africa, during the last 16,000 years with emphasis on the Pleistocene – Holocene transition. *Palaeogeography, Palaeoclimatology, Palaeoecology*, 271, 240–258.

Noström, E., Risberg, J., Gröndahl, H., Holmgren, K., Snowball, I., Mugabe, J. A. and Sitoe, S. R. (2012). Coastal paleo-environment and sea-level change at Macassa Bay, southern Mozambique, since c 6600 cal BP. *Quaternary International*, 260, 153–163.

Pasche, N., Alunga, G., Mills, K., Muvundja, F., Ryves, D. B., Schurter, M., Wehrli, B., Schmid, M. (2010). Abrupt onset of carbonate deposition in Lake Kivu during the 1960s: Response to recent environmental changes. *Journal of Paleolimnology*, 44, 931–946.

Pike, J. and Kemp, A. E. S. (1996). Preparation and analysis techniques for studies of laminated sediments. In *Palaeoclimatology and Palaeoceanography from Laminated Sediments*, ed. A. E. S. Kemp. London: Geological Society of London Special Publication 116, pp. 37–48.

Piperno, D. R. (2006). *Phytoliths: A Comprehensive Guide for Archaeologists and Paleocologists*. Lanham, MD: AltaMira Press, 304pp.

Rau, A. J., Rogers, J., Lutjeharms, J. R. E., Giraudeau, J., Lee-Thorp, J. A., Chen, M.-T. and Waelbroek, C. (2002). A 450-kyr record of hydrological conditions on the western Agulhas Bank Slope, south of Africa. *Marine Geology*, 180, 183–201.

Riedel, F., Henderson, A. C. G., Heußner, K.-U., Kaufmann, G., Kossler, A., Leipe, C., Shemang, E. and Taft, L. (2014). Dynamics of a Kalahari long-lived mega-lake system: hydromorphological and limnological changes in the Makgadikgadi Basin (Botswana) during the terminal 50 ka. *Hydrobiologia*, 739, 25–53.

Roberts, P., Lee-Thorp, J. A., Mitchell, P. J. and Arthur, C. (2013). Stable carbon isotopic evidence for climate change across the late Pleistocene to early Holocene from Lesotho, southern Africa. *Journal of Quaternary Science*, 28, 360–369.

Rodriguez-Lazaro, J. and Ruiz-Muñoz, F. (2012). A general introduction to ostracods: Morphology, distribution, fossil record, and applications. In *Ostracoda as Proxies for Quaternary Climate Change*, ed. D. Horne, J. Holmes, F. Viehberg and J. Rodriguez-Lazaro. Amsterdam: Elsevier, pp. 1–14.

Rörig, L. F. (2013). Algal Analysis. In *Handbook of Water Analysis*, 3rd Ed, ed. L. M. L. Nollet and L. S. P. De Gelder. Boca Raton: CRC Press, pp. 163–188.

Rossouw, L., Stynder, D. D. and Haarhof, P. (2009). Evidence for opal phytolith preservation in the Langebaanweg 'E' Quarry Varswater Formation and its potential for palaeohabitat reconstruction. *South African Journal of Science*, 105, 223–227.

Roubeix, V., Chalie, F. and Gasse, F. (2014). The diatom Thalassiosira faurri (Gasse) Hasle in the Ziway-Shala lakes (Ethiopia) and implications for paleoclimatic reconstructions: Case study of the Glacial–Holocene transition in East Africa. *Palaeogeography, Palaeoclimatology, Palaeoecology*, 402, 104–112.

Round, F. E. (1984). *The Ecology of Algae*. Cambridge: Cambridge University Press, 664pp.

Samtleben, C. and Bickert, T. (1990). Coccoliths in sediment traps from the Norwegian Sea. *Marine Micropaleontology*, 16, 39–64.

Sanfilippo, A., Westberg-Smith, M.J. and Riedel, W.R. (1989). Cenozoic radiolarian. In *Plankton Stratigraphy: Volume 2, Radiolaria, Diatoms, Silicoflagellates,*

Dinoflagellates and Ichthyoliths, ed. H. M. Bolli, J. B. Saunders and K. Perch-Nielsen. Cambridge: Cambridge University Press, pp. 631–712.

Schmiedl, G., Mackensen, A. and Müller, P. J. (1997). Recent benthic foraminifera from the eastern South Atlantic Ocean: Dependence on food supply and water masses. *Marine Micropaleontology*, 32, 249–287.

Schoeman, F. R. (1972). A further contribution to the diatom flora of sewage enriched waters in southern Africa. *Phycologia*, 11, 239–245.

Scott, L. (1999). Vegetation history and climate in the Savanna biome South Africa since 190,000 ka: A comparison of pollen data from the Tswaing Crater (the Pretoria Saltpan) and Wonderkrater. *Quaternary International*, 57–58, 215–223.

Scott, L. (2002). Grassland development under glacial and interglacial conditions in southern Africa: A review of pollen, phytolith and isotope evidence. *Palaeogeography, Palaeoclimatology, Palaeoecology*, 177, 47–57.

Scott, L., Neumann, F. H., Brook, G. A., Bousan, C. B., Norström, E. and Metwally, A. A. (2012). Terrestrial fossil-pollen evidence of climate change during the last 26 thousand years in Southern Africa. *Quaternary Science Reviews*, 32, 100–118.

Scott, L. and Rossouw, L. (2005). Reassessment of botanical evidence for palaeoenvironments at Florisbad, South Africa. *South African Archaeological Bulletin*, 60, 96–102.

Seppä, H., Birks, H. J. B., Odland, A., Poska, A. and Veski, S. (2004). A modern pollen – climate calibration set from northern Europe: Developing and testing a tool for palaeoclimatological reconstructions. *Journal of Biogeography*, 31, 251–267.

Shaw, P. A., Stokes, S., Thomas, D. S. G., Davies, F. B. M. and Holmgren, K. (1997). Palaeoecology and age of a Quaternary high lake level in the Makgadikgadi Basin of the Middle Kalahari, Botswana. *South African Journal of Science*, 93, 273–276.

Shaw, P. A. (2010). The use of diatoms as palaeoenvironmental indicators in Botwana: A review and species checklist. *Botswana Notes and Records*, 42, 154–162.

Shi, N., Schneider, R. R., Beug, H.-J. and Dupont, L. M. (2001). Southeast trade wind variations during the last 135 kyr: Evidence from pollen spectra in eastern South Atlantic sediments. *Earth and Planetary Science Letters*, 187, 311–321.

Siziba, N., Chimbari, M. J., Masundire, H., Mosepele, K. and Ramberg, L. (2013). Variation in assemblages of small fishes and microcrustaceans after inundation of rarely flooded wetlands of the lower Okavango Delta, Botswana. *Environmental Management*, 52, 1386–1399.

Smol, J. P. and Stoermer, E. F. (Eds) (2010). *The Diatoms: Applications for the Environmental and Earth Sciences*. Cambridge: Cambridge University Press, 686pp.

Soon, W., Baliunas, S., Idso, C., Idso, S. and Legates, D. R. (2003). Reconstructing Climatic and Environmental Changes of the Past 1000 Years: A Reappraisal. *Energy and Environment*, 14, 233–296.

Stager, J. C. (1988). Environmental Changes at Lake Cheshi, Zambia since 40,000 Years B.P. *Quaternary Research*, 29, 54–65.

Stager, J. C., Ryves, D. B., King, C., Madson, J., Hazzard, M., Neumann, F. H. and Maud, R. (2013). Late Holocene precipitation variability in the summer rainfall region of South Africa. *Quaternary Science Reviews*, 67, 105–120.

Stager, J. C., Westwood, J., Grzesik, D. and Cumming, B. F. (2005). A 5500-year environmental history of Lake Nabugabo, Uganda. *Palaeogeography, Palaeoclimatology, Palaeoecology*, 218, 347–354.

Strachan, K. L., Finch, J. M., Hill, T. and Barnett, R. L. (2014). A late Holocene sea-level curve for the east coast of South Africa. *South African Journal of Science*, 110(1/2), Art. #2013-0198, 9pp, doi:10.1590/sajs.2014/20130198.

Strachan, K. L., Hill, T. R., Finch, J. M. and Barnett, R. L. (2015). Vertical zonation of foraminifera assemblages in Galpins salt marsh, South Africa. *Journal of Foraminiferal Research*, 45, 29–41.

Stevenson, R. J. and Pan, Y. (2004). Assessing environmental conditions in rivers and streams with diatoms. In *The Diatoms: Applications for the Environmental and Earth Science*, ed. E. F. Stoermer and J. P. Smol. Cambridge: Cambridge University Press, pp. 11–40.

Summerhayes, C. P., Kroon, D., Rosell-Melé, A., Jordan, R. W., Schrader, H.-J., Hearn, R., Villanueva, J., Grimalt, J. O. and Eglinton, G. (1995). Variability in the Benguela Current upwelling system over the past 70,000 years. *Progress in Oceanography*, 35, 207–251.

Taylor, J. C., Harding, W. R., Archibald, C. G. M and van Rensberg, L. (2005). Diatoms as indicators of water quality in the Jukskei-Crocodile river system in 1956 and 1957, a re-analysis of diatom count data generated by BJ Cholnoky. *Water SA*, 31, 237–246.

Thomas, D. S. G. and Shaw, P.A. (1991). *The Kalahari Environment*. Cambridge: Cambridge University Press, 298pp.

Tsartsidou, G., Lev-Yadun, S., Albert, R.-M., Miller-Rosen, A., Efstratiou, N. and Weiner, S. (2007). The phytolith archaeological record: Strengths and weaknesses evaluated based on a quantitative modern reference collection from Greece. *Journal of Archaeological Science*, 34, 1262–1275.

Tyrrell, T. and Young, J. R. (2009). Coccolithophores. In *Elements of Physical Oceanography*, ed. J. H. Steele, S. A. Thorpe and K. K. Turekian. London: Academic Press, pp. 407–415.

Wilson, K. E., Leng, M. J. and Mackay, A. W. (2014). The use of multivariate statistics to resolve multiple contamination signals in the oxygen isotope analysis of biogenic silica. *Journal of Quaternary Science*, 29, 641–649.

Zhu, L., Peng, P., Xie, M., Wang, J., Frenzel, P., Wrozyna, C. and Schwalb, A. (2010). Ostracod-based environmental reconstruction over the last 8,400 years of Nam Co Lake on the Tibetan plateau. *Hydrobiologia*, 648, 157–174.

21

Development of the archaeological record in southern Africa during the Earlier Stone Age

KATHLEEN KUMAN

Abstract

The Earlier Stone Age record is rich in South Africa, thanks to a long history of research in the country. It also has examples of every stage of chronological development, from the Oldowan to the final Earlier Stone Age (i.e. from ~2.0 to 0.3 Ma). This chapter discusses the details of this lithic and cultural chronology, using those sites in better preserved context that also have dates or relative age estimates.

21.1 Introduction

The African Earlier Stone Age (ESA) includes cultural traditions that are grouped both chronologically and technologically as Oldowan from 2.6 Ma, Acheulean from 1.76 Ma, and late to terminal Acheulean/ESA from ~0.4 Ma to the appearance of the Middle Stone Age ~0.3 Ma. The southern African archaeological record documents all of these industries at a number of sites, beginning from at least 2 Ma. It is clear that ESA hominids were widespread in southern Africa, even though the restrictive conditions for good site preservation and the small number of active researchers have limited our record for these earliest archaeological traces (Klein, 2000; Kuman, 2007).

21.2 The earliest archaeology

21.2.1 Oldowan (2.6–1.76 Ma)

Plummer (2004) lists 17 Oldowan localities across Africa, comprising 37 sites, with all but four located in East Africa. In southern and south central Africa, there have been suggestions that Oldowan sites exist in Angola south of the Zambezi,

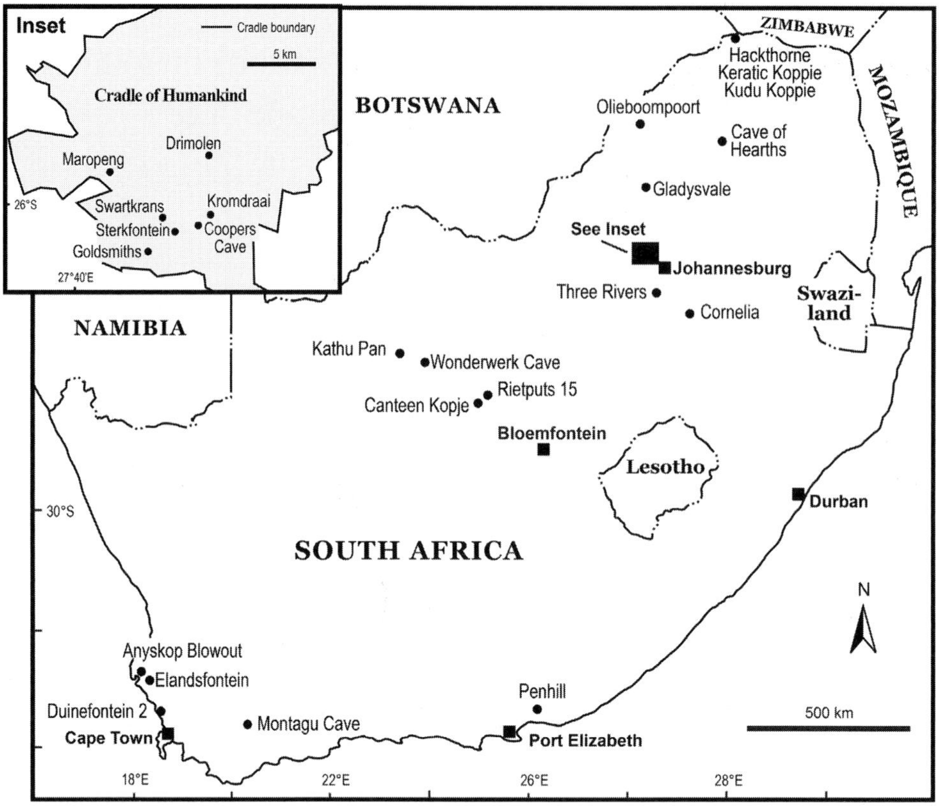

Fig. 21.1. Locations of selected Earlier Stone Age sites discussed in the text.

and in southern Zambia in the Luangwa Valley, but these are undated occurrences (Barham and Mitchell, 2008), and the latter may be no older than 1 Ma (Barham et al., 2011). Until recently, Sterkfontein in South Africa's Cradle of Humankind (Gauteng Province) was the only well documented southern African example of an Oldowan industry (see Fig. 21.1 for all site locations). However, a second assemblage is now identified at the adjacent site of Swartkrans (Lower Bank of Member 1; Sutton, 2012), and the nearby site of Kromdraai B has two artefacts of Oldowan age (Kuman et al., 1997). The Oldowan also exists in the interior of the country at Wonderwerk Cave, where Stratum 12 has 30 artefacts ~1.8 Ma (Chazan et al., 2008, 2012; Matmon et al., 2012) (Table 21.1). Contrasting with the numerous open-air sites of East Africa, most Oldowan material is found in underground karstic cave infills where artefacts have washed in from surface occupations close to steep cave shafts (see Holmgren and Shaw, this volume). However, the first open-air Oldowan site in the country has recently been announced by

Table 21.1. *Major South African ESA sites and their ages and contexts. Sites are marked on Fig. 21.1. For additional sites, see Sampson (1974) and Kuman (2007).*

Site Name	Approximate age (Ma)	Stratigraphic context	Basis for age estimate	Cultural industry	Selected references
Canteen Kopje	>2	Alluvial gravels, Pit 4	Cosmogenic nuclide burial dating	Oldowan under excavation	Gibbon et al. (2013); Leader et al. (in prep.)
Sterkfontein M5E	~2	Karst cave infill, Member 5 East	Fauna; cosmogenic burial dating	Oldowan	Kuman and Clarke (2000); Kuman and Field (2009); Granger et al. (2015)
Swartkrans M1	~2	Karst cave infill, Member 1 Lower Bank	Fauna; cosmogenic burial dating; U-Pb on speleothems	Oldowan	Pickering et al. (2012); Sutton (2012); Gibbon et al. (2014)
Kromdraai B	>1.9	Karst cave infill	Fauna; palaeomagnetism	2 or 3 artefacts/ Oldowan	Kuman et al. (1997); Thackeray et al. (2002); L. Bruxelles (pers. comm., 2014)
Wonderwerk Cave Stratum 12	~1.85±0.23	Cave site	Cosmogenic nuclide burial dating, magnetostratigraphy	30 artefacts/Oldowan	Chazan et al. (2008); Matmon et al. (2012)
Rietputs 15	1.7 to 1.3	Alluvial gravels with several dated pits; artefacts throughout sequence	Cosmogenic nuclide burial dating	Early Acheulean	Gibbon et al. (2009); Leader (2009)
Sterkfontein M5W	~1.6	Karst cave infill, Member 5 West	Fauna	Early Acheulean	Kuman and Clarke (2000)
Three Rivers	~1.6	Alluvial gravels	Artefact technology	Early Acheulean	Mason (1962)
Wonderwerk Cave Stratum 11	~1.6	Cave deposit	Palaeomagnetism	Early Acheulean	Chazan et al. (2008); Chazan (2015)
Drimolen	1.5–2.0	Karst cave infill, decalcified	Fauna	Several artefacts	artefacts unpublished

Table 21.1. (*cont.*)

Site Name	Approximate age (Ma)	Stratigraphic context	Basis for age estimate	Cultural industry	Selected references
Canteen Kopje	>1.5	Alluvial gravels, Pit 6, two lowest layers	Cosmogenic nuclide burial dating	Early Acheulean	Leader (2013)
Swartkrans M2	~1.5	Karst cave infill, Member 2	Fauna	Early Acheulean	Brain (1993)
Kromdraai A	~1.5–1.0	Karst cave infill and associated dump	Fauna	Early Acheulean	Kuman et al. (1997)
Coopers D	1.4/1.5	Karst cave infill	U-Pb	50 artefacts, undiagnostic	G. Hall (unpub. ms.)
Canteen Kopje	>1.0	Alluvial gravels, Pit 6, top layer	Cosmogenic nuclide burial dating	Victoria West Acheulean	McNabb and Beaumont (2011a, b); Leader (2013)
Maropeng	>1.0	Coarse colluvium	Artefact technology	Early Acheulean	Pollarolo et al. (2010)
Goldsmith's	>1.0/1.5?	Karst cave infill disturbed/dump	Fauna	Early Acheulean	Jacoby et al. (2013)
Swartkrans M3	~1.0	Karst cave infill, Member 3	Fauna; ESR	Early Acheulean	Brain (1993)
Wonderwerk Cave Stratum 10	0.99–1.07	Cave deposits	Palaeomagnetism	Acheulean	Chazan et al. (2008); Berna et al. (2012); Chazan (2015)
Gladysvale	>0.780	Karst cave infill, decalcified	Paleomagnetism, ESR	Early Acheulean	Hall et al. (2006)
Cornelia	~1.0	Mottled yellow clay, palaeosurface in abandoned channel	Fauna	Acheulean	Brink et al. (2012)
Penhill, Sundays River Valley	<1.34	Debris flow fill within dated fine alluvium	Cosmogenic nuclide burial dating (tba)	Acheulean	Erlanger et al. (2012); Lotter et al. (2014)
Elandsfontein	~0.6	Thin palimpsests in dune sands on palaeo-surface near water hole	Fauna, artefacts	Later Acheulean	Klein et al. (2007); Klein (2009: 344)

Kathu Pan Stratum 4a	~0.464/0.542	Pan margin, silty to gravelly sands with spring vents	OSL/U-series, ESR	Acheulean/ Fauresmith	Porat et al. (2010)
Montagu Cave	<0.6	Cave strata	Artefacts	Later Acheulean	Keller (1973)
Cave of Hearths Beds I-III	~0.6–0.4	Cave breccias	Palaeomagnetism, artefacts	Later Acheulean	McNabb (2009)
Wonderwerk Cave Stratum 6	>0.35	Cave deposit	U-series	Later Acheulean	Beaumont and Vogel (2006); Chazan et al. (2008)
Wonderwerk Cave Strata 3-4	0.286–0.276	Cave deposit	U-series on stalagmites	Acheulean/ Fauresmith	Beaumont and Vogel (2006)
Wonderwerk Cave	>0.2	Cave deposit	U-series	Late Acheulean/ Fauresmith	Beaumont and Vogel (2006)
Olieboompoort	<0.6	Rockshelter deposits with a basal rubble	Artefacts	Later Acheulean	Mason (1962); Sampson (1974)
Canteen Kopje	<0.6	Base of aeolian (Hutton) sands	Artefacts	Acheulean/ Fauresmith	McNabb and Beaumont (2011a); Kuman (2014)
Kudu Koppie	~0.3	Koppie rubble with refits confirming integrity	Artefacts	Final ESA	Pollarolo et al. (2010); Wilkins et al. (2010); Sumner (2013); Sumner and Kuman (2014)
Hackthorne	~0.3	Deflated lag sealed by younger sands	Artefacts	Final ESA	Kuman et al. (2005); Kempson (2007); Le Baron et al. (2010)
Keratic Koppie	~0.3	Deflated lag sealed by younger sands	Artefacts	Final ESA	Kuman et al. (2005); Kempson (2007); Le Baron et al. (2010)
Duinefontein 2, Horizon 2	>0.29	Buried land surface near a marsh or pond	OSL on capping sands, fauna, artefacts, U-series	Late Acheulean	Cruz-Uribe et al. (2003)

Table 21.1. (*cont.*)

Site Name	Approximate age (Ma)	Stratigraphic context	Basis for age estimate	Cultural industry	Selected references
Duinefontein 2, Horizon 4	>0.27	Buried land surface near a marsh or pond	OSL on capping sands, fauna, artefacts, U-series	Late Acheulean	Cruz-Uribe et al. (2003)
Anyskop Blowout	>0.182	Artefacts exposed on calcrete following deflation	OSL on sands 0.5 m above the calcrete	Late Acheulean	Kandel and Conard (2012)
Wonderwerk Cave Excavation 6	>0.187±0.08	Cave deposit	U-Th on stalagmite fragment in deposit	Acheulean/ Fauresmith	Chazan and Horwitz (2010); Chazan (2015)

Gibbon et al. (2013) at Canteen Kopje, in coarse alluvium dated by the cosmogenic nuclide burial method to >2 Ma. With the Oldowan now at both Wonderwerk and Canteen Kopje, it is clear that the small number of early sites is due to limited research and not to the absence of these hominids.

Due to good fossil preservation, faunal age estimates have been used for the Cradle karst sites, as southern Africa lacks volcanic deposits for absolute dating. However, cosmogenic burial dating is improving the chronology of early sites and the earliest cultural record. The Sterkfontein Oldowan (Member 5 East) has a faunal age of 1.7–2.0 Ma (Kuman and Clarke, 2000), but Granger et al. (2015) now confirm a cosmogenic age of 2.18±0.21 Ma for this period. At Swartkrans, the generalised faunal age was 1.7/1.8 Ma for the Lower Bank of Member 1 (Brain, 1993), but new cosmogenic dates for this sub-member are now 2.19±0.08 and 1.80±0.09 Ma (Pickering et al., 2011; Gibbon et al., 2014). The fauna from Kromdraai B are considered to be older than those from Swartkrans (Vrba, 1982, 1985), but new stratigraphic and dating work is in progress, and a palaeomagnetic age of ca 1.9 Ma (Thackeray et al., 2002) applies to only a part of the deposits.

The Oldowan (Fig. 21.2) is a simple core and flake industry in which flaking is opportunistic, core technology is simple, and flake tools are the main goal. Freehand flaking of cores dominates, but the 'bipolar' hammer and anvil technique is also established, particularly for assemblages in quartz. Despite its simple organisation, however, there is considerable evidence for cognitive and manual skills in many Oldowan assemblages, and the South African material is no exception. At Sterkfontein, quartzite and chert are the most common rock types in gravels close to the site, but vein quartz was overwhelmingly selected for tools (Kuman and Field, 2009). The large assemblage (n=3500) is dominated by small flakes and flaking debris, and the focus for tools is on sharp edges for light-duty cutting tasks. Vein quartz is brittle and easy to flake, yielding chunks that can be further cycled into cores. Preliminary work on the Swartkrans Oldowan suggests that tool production at this site, only 1.5 km from Sterkfontein, was more opportunistic and chert was more heavily used due to its ready availability (Sutton, 2012). As the two assemblages are of similar age, this lithological difference probably relates to different types of activities. Swartkrans seems to have provided easy access to rocks that were tested, while Sterkfontein received more imported material. This is also suggested by the quartzite lithics at Sterkfontein, which form a very small component of the Oldowan, mainly as medium-sized flakes that were brought into the site. However, Swartkrans also reveals hominid selectivity for better quality rocks over what was generally available in the gravels and surrounding landscape, as the archaeological lithics contain fewer flaws and are of better quality than random samples (Sherwood, 2013). The fauna associated with the Sterkfontein Oldowan show little evidence of hominid interaction and accumulated

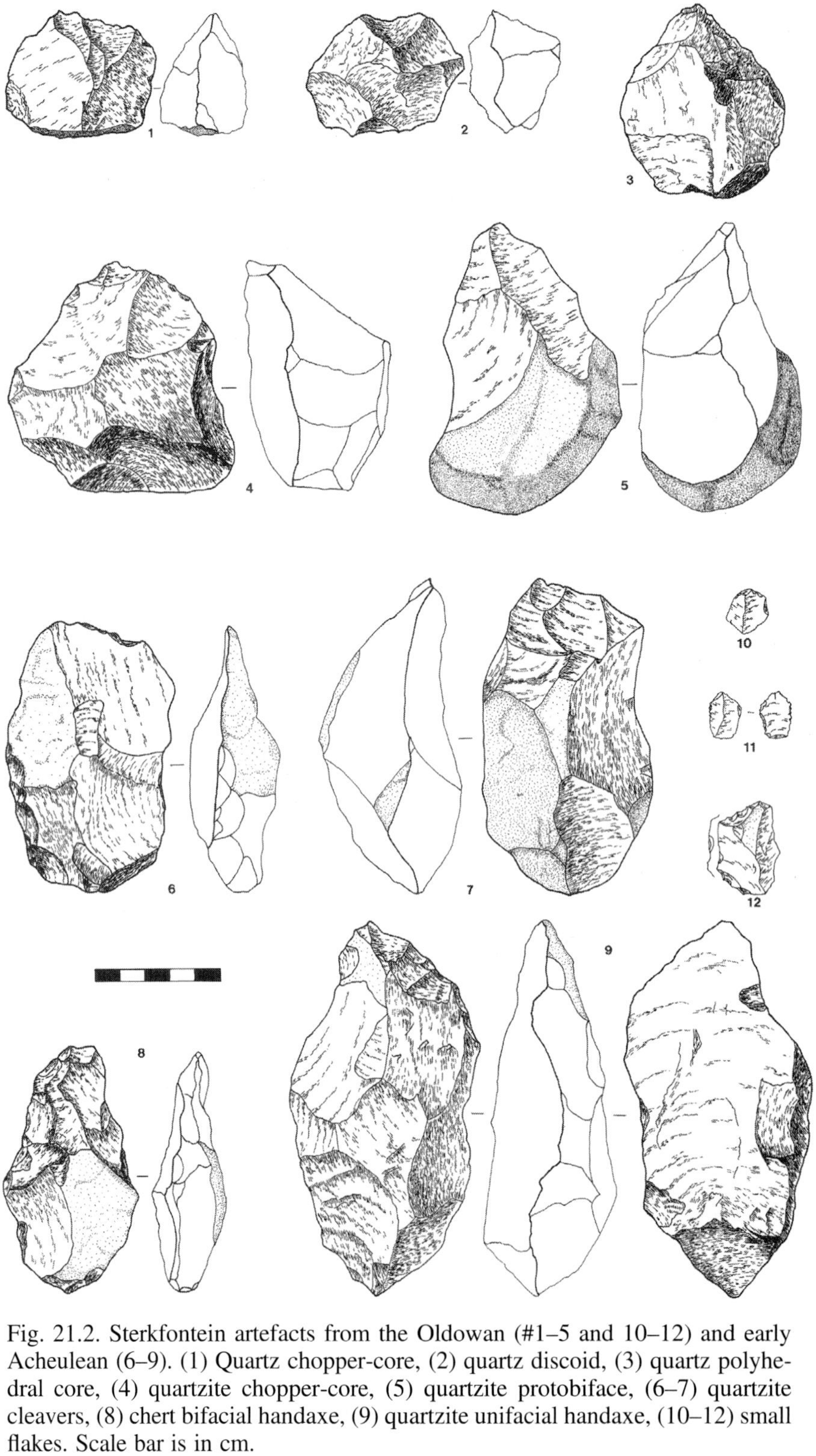

Fig. 21.2. Sterkfontein artefacts from the Oldowan (#1–5 and 10–12) and early Acheulean (6–9). (1) Quartz chopper-core, (2) quartz discoid, (3) quartz polyhedral core, (4) quartzite chopper-core, (5) quartzite protobiface, (6–7) quartzite cleavers, (8) chert bifacial handaxe, (9) quartzite unifacial handaxe, (10–12) small flakes. Scale bar is in cm.

mainly as a deathtrap assemblage, with some additional elements entering through slopewash (Pickering, 1999). At Swartkrans, the associated fauna were accumulated by carnivores, and to a lesser extent, by hominids who butchered parts of large animal carcasses on site (Pickering et al., 2008). Overall, the modes of accumulation for fauna in the early underground karst cave sites are dominated by natural and biological agents, and only at Swartkrans are there good signs of some hominid involvement, which makes the site rather unique.

In South Africa, the hominid species associated with Oldowan assemblages are *Paranthropus robustus* at Sterkfontein and Swartkrans and *Homo habilis* at Swartkrans. However, *Paranthropus* became extinct about 1 Ma and so is unlikely to have had the flexible, adaptive behaviour of a toolmaker. A juvenile cranium of *H. habilis* (SK 27) derives from Swartkrans Member 1 (Clarke, 1977; Sutton et al., 2012) and is a more likely species responsible for the artefacts. The environment at the time of the Sterkfontein Oldowan has been reconstructed, based on faunal carbon isotope ratios, as a medium-density woodland (Luyt and Lee-Thorp, 2003). The Swartkrans environment was also a medium-density woodland, with bushland and patches of grassland (Reed, 1997), in agreement with Watson (1993) who interprets a savannah woodland with riparian woodland and reed beds. There are no water-dependent species at Sterkfontein as there are at Swartkrans, which could explain the differences in site activities discussed above.

21.2.2 Early Acheulean (1.76–>1 Ma)

The Acheulean industry sees the appearance of purposefully shaped handaxes and cleavers as heavy-duty tools, in tandem with more advanced hominid species and more open environments. The karstic hominid sites in Gauteng have until recently been the only (datable) Early Acheulean sites in southern Africa. Sterkfontein Member 5 is the best such site and includes 14 handaxes and cleavers (Kuman, 1994; Kuman and Clarke, 2000). The faunal age is ~1.6 Ma, but the assemblage could range between 1.7–1.4 Ma (Kuman and Clarke, 2000). The best archaeological context at Sterkfontein occurs in Member 5 West where an assemblage of 701 pieces is dominated by cores and manuports, showing that the site was used consistently for an extended period of time (Kuman, 1998; Field, 1999). Most small flaking debris was eroded from the site as the habitat was considerably more open and drier than during the Oldowan (Luyt and Lee-Thorp, 2003). At Swartkrans, there are two small assemblages, in Member 2 at ~1.5 Ma and Member 3 at 1 Ma (Field, 1999; Gibbon et al., 2014). Several handaxes and cleavers are present but were retrieved from lime miners' breccia dumps. The size range of the assemblages is varied but shows an incomplete capture of the lithic assemblages from surface occupations. Both Member 2 and 3 habitats were more open than

Member 1: a moderately open savannah or wooded grassland with wetlands for Member 2, whereas for Member 3 there was open grassland associated with a river or stream (Vrba, 1975; Reed, 1997). Swartkrans is somewhat closer to the Blaauwbank River than Sterkfontein and water-dependent species continue in these members, again contrasting with Sterkfontein only just over a kilometre away.

At Swartkrans, a partial cranium of *Homo ergaster* (SK 847) derives from a younger part of Member 1 (the Hanging Remnant), which unfortunately lacks artefacts. This fossil was the first discovery made of this species (Groves and Mazák, 1975; Clarke, 1977, 1994, 2012). The East and South African *H. ergaster* fossils are contemporary, both appearing at >1.7 Ma in the earliest Acheulean (Pickering et al., 2012). A second key specimen from Swartkrans, the SK 15 mandible (Robinson, 1953), derives from Member 2 at ~1.5 Ma (Brain, 1993). At Sterkfontein, a partial mandible of *H. ergaster*, StW 80, is very similar to SK 15 and was found in direct association with a cleaver in the Member 5 West Early Acheulean assemblage (Kuman and Clarke, 2000). Also in the Cradle is a third Early Acheulean assemblage at Kromdraai A, ~1.0–1.5 Ma, with just over 100 artefacts very like those at Sterkfontein and Swartkrans (Kuman et al., 1997). The habitat has been reconstructed as wooded grasslands or scrub woodland with grassland (Reed, 1997), and open savannah based on bovid evidence (Vrba, 1975).

The fauna of Swartkrans Members 2 and 3 are taphonomically similar, with carnivore feeding activities responsible for the bulk of the assemblages, but hominids also contributing either through active hunting or aggressive scavenging, as they had early access to large ungulates (Pickering et al., 2004, 2007, 2008). In contrast, the Sterkfontein Acheulean fauna accumulated through more natural processes such as erosion, slopewash, and collecting by carnivores and porcupines (Pickering, 1999, 2002).

Two other karst cave sites have undated handaxes of Early Acheulean type. Gladysvale has only one artefact >0.78 Ma (Hall et al., 2006), while Goldsmith's has handaxes within a large assemblage of artefacts derived from miners' dumps, with fauna present but out of context (Jacoby et al., 2013). Also in undatable contexts is a large collection of Early Acheulean lithics from the Maropeng open-air site just west of the dolomite caves, where artefacts accumulated on an eroded colluvial surface (Pollarolo et al., 2010a). At Three Rivers in Gauteng, but outside of the Cradle, Mason (1962) excavated a collection of Early Acheulean artefacts very similar to those at Sterkfontein from an alluvial gravel. The contexts of these latter two collections are poor, but they demonstrate that hominids were living more widely in the region than just at the cave sites.

Beyond Gauteng, a new region for Early Acheulean discoveries has been identified at Rietputs 15 in alluvial deposits of the Vaal River at Windsorton, Northern Cape (Fig. 21.1). Although the archaeological context is not ideal, the application of cosmogenic burial dating has provided good dates of 1.7–1.3 Ma (Gibbon et al., 2009; Leader, 2009; the younger end of the published age range is suggested based on new model outputs—R. Gibbon, pers. comm.). Collections were made from two diamond mining pits, but artefacts occur in other pits as well (Gibbon et al., 2009). Thus, the Early Acheuelean appears in South Africa as early as in East Africa, where its first appearance is dated at Konso in Ethiopia to ~1.7 Ma (Beyene et al., 2013) and at Kokiselei in Kenya to ~1.76 Ma (Lepre et al., 2011). Also in the Northern Cape, a small assemblage of artefacts is present in Stratum 11 at Wonderwerk Cave, aged between ~1.1 and 1.5 Ma (Chazan, 2015), and three stratified Acheulean assemblages are excavated at Canteen Kopje (Barkly West) in coarse alluvium and dated by cosmogenics, two of them to >1.5 Ma and one at >1 Ma (Gibbon et al., in prep.).

The youngest of the three Canteen Kopje assemblages belongs to the Victoria West Industry, one of the most interesting developments of the Acheulean (McNabb and Beaumont, 2011a, b; Leader, 2013). The reduction strategies employed here include a component of prepared core technology, which takes the standardised form of large handaxe-shaped cores, and is worked asymmetrically. The upper surface is prepared to strike off a single, large, side-struck flake that is then used as a blank for handaxes, and probably occasionally for cleavers or as simple knives. There are a few other independently developed prepared core technologies elsewhere in the world which arose through the concept of shaping of Acheulean bifaces, but the Victoria West Industry at Canteen Kopje is now the oldest such development at >1.0 Ma (Beaumont and Vogel, 2006; Gibbon et al. in prep.).

21.2.3 Assessment of the early archaeology record

The foregoing sections demonstrate that southern Africa's earliest archaeology is substantial, if limited by a paucity of researchers and an absence of sites in fine sediments in datable open-air contexts. Although hominid presence from ~2–1 Ma appears to have been geographically widespread, archaeological visibility is restricted to localities close to river gravels where lithic raw materials are abundant, or close to rock outcrops in the case of Maropeng. Some of Gauteng's dolomitic cave sites, such as Kromdraai and Goldsmith's, are relatively close to gravels but artefacts are preserved in smaller numbers, perhaps because they were not geographically favoured for human activity (Kuman, 2003). Two sites at a distance from the Blauubank gravels, Gladysvale and Drimolen, have yielded only a handful of artefacts between them, and other karst sites have none at all. Gauteng's earliest archaeology must therefore be read from 'keyhole' sites in secondary contexts, and it is not yet possible to study behaviour across dated landscapes, as is done in East Africa. The open-air sites in the Vaal River basin at Windsorton and Barkly West are also in secondary contexts. However, with careful analysis of site formation processes and lithic size profiles, all of the foregoing sites are valuable archives of the technology, chronology and distribution of early tool-making hominids in southern Africa. They show that the technology and overall pattern of hominid behaviour in the Oldowan and Early Acheulean are comparable to the better preserved record in East Africa. At Sterkfontein and Swartkrans, the presence of *Homo habilis* and *H. ergaster* fossils in deposits with Oldowan and Early Acheulean artefacts is a most valuable record of the respective tool makers.

21.3 The Acheulean from ~1 Ma

The controlled use of fire is demonstrated at two South African sites by 1 Ma. In Swartkrans Member 3, burnt bones show histological changes consistent with sustained burning at high temperatures in camp fires (Brain and Sillen, 1988; Brain, 1993; Hanson and Cain, 2007). As the karst cave shafts are steep and the caves not usually inhabitable by hominids, it is most probable that the bones washed in from camp fires made on the surface. The Member 3 artefact assemblage is also small and its size profile does not suggest a cave occupation. A different type of evidence comes from Wonderwerk Cave, where *in situ* burning of vegetation is reported in an Acheulean occupation layer at ~1 Ma (Berna et al., 2012). Swartkrans and Wonderwerk thus provide some of the world's earliest evidence for the controlled use of fire. Although the use of fire has been proposed for certain open-air sites as early as 1.6–1.5 Ma, this is debated (see James, 1989, for discussion). In East Africa, patches of reddened sediments are argued to be caused by camp fires, and the patterned spatial distribution of artefacts and bones is used as supporting evidence for hearths that have not survived (Barbetti, 1986; Bellomo, 1993). A range of techniques is applied in this research, including studies of burnt artefacts and bone, charred plant remains, phytoliths, thermoluminescence analysis and palaeomagnetism. By 0.79 Ma, the evidence from the Israeli site of Gesher Benot Ya'aqov is clearly apparent (Alperson-Afil et al., 2007; Alperson-Afil, 2008; Alperson-Afil and Goren-Inbar, 2010).

Sites in southern Africa of ~1 Ma are rare, although our inability to date potential occurrences influences this observation. However, Cornelia in the Free State is dated by palaeomagnetism and biostratigraphy to the Jaramillo subchron (1.07–0.99 Ma), and recent work has documented a site formed on the palaeosurface of an abandoned channel at Cornelia (Brink et al., 2012). The carnivore-accumulated fauna reflect open grasslands typical of interior South Africa. Also present are one hominid tooth, and handaxes, cleavers and lithic elements suggesting their manufacture. These large cutting tools (LCTs) are similar to a larger collection studied by Clark (1974). Both are characterised by an improved flaking technology compared with the Early Acheulean, but with bold scars and less secondary trimming than is found in Later Acheulean assemblages. Wonderwerk Strata 11–10 also have small assemblages dated to the Jaramillo subchron (0.99–1.07 Ma) (Chazan et al., 2008; Berna et al., 2012), with handaxes shaped largely by non-invasive removals (Chazan, 2015). Another site with the potential to fill this time gap is Penhill, which is currently under excavation in the Sundays River valley, Eastern Cape (Lotter et al., 2014). Here, a large assemblage is currently dated to <1.34 Ma (Erlanger et al., 2012). More precise cosmogenic dating of Penhill is being attempted.

Sites that are ~0.6 Ma and younger belong to the Later Acheulean. They are the most numerous and generally show more refined flaking methods (Klein, 2000; Clark, 2001a). Elandsfontein is a major fossil and archaeological locality of ~0.6 Ma (Klein et al., 2007; Klein, 2009), located on the west coast of South Africa where ponds and marshes once existed. It has been the venue of decades of collecting but only one excavation has been published (Singer and Wymer, 1968). The site is well known for the skullcap of an early *H. sapiens*, its Later Acheulean artefacts, and for faunal studies that document most bones as deriving from natural deaths (Klein and Cruz-Uribe, 1991). A new systematic excavation programme with emphasis on understanding site formation began in 2008 and promises to provide behavioural information from those sites with the best integrity (Braun et al., 2013). This work has identified several specimens at one locality with butchery cutmarks, and the project has some landscape archaeology potential (*ibid.*). The environment is reconstructed as a grassy moist habitat during an interglacial period (Klein and Cruz-Uribe, 1991), with isotopic results indicating a grassy bushveld that is unusual for this area today (Luyt et al., 2000).

ESA occupation of other west coast open-air sites also corresponds with periods of favourable climates in this low rainfall region, with sites characterised by low artefact densities, reflecting ephemeral use of the landscape and high population mobility. Where lithic raw materials are not available, some rocks have been transported up to 25 km (Luyt et al., 2000). Duinefontein 2 is a late Acheulean site that has two buried horizons on the margin of a large pond or marsh, with Optically Stimulated Luminescene (OSL) dates on capping deposits at >290 and >270 kyr BP; occupation was under more favourable climatic conditions than today because grass and broad-leafed bush were then more prominent (Klein et al., 1999; Cruz-Uribe et al., 2003). ESA artefacts have also been recorded at the Anyskop Blowout site east of the Langebaan Lagoon (Kandel and Conard, 2012). Here, wind has removed sands overlying a calcrete layer on which generally fresh artefacts are resting. In contrast with Duinefontein and Elandsfontein, where artefacts accumulated on the margin of a body of water, Anyskop is located on a hilltop and seemingly formed as a short-term butchery site. Here, 24 finished handaxes that were not flaked on site are among the assemblage of 193 artefacts at this site (Kandel and Conard, 2012).

There are three cave occupation sites and one rock shelter with Later Acheulean artefacts, all of which probably are <0.6 Ma: Wonderwerk Cave (Northern Cape Province), Montagu Cave (Western Cape Province), Cave of Hearths (Limpopo Province) and Olieboompoort Shelter (Gauteng Province). Unfortunately the lithic assemblages are only extensive at two of the sites. Montagu Cave on the southern coast was largely used as a factory site where beach cobbles provided abundant lithic materials (Keller, 1973), and flaking debris, rough-outs and broken LCTs are common (McNabb et al., 2004). Both dating and technological analysis are needed at this site. Cave of Hearths is a breccia site with Acheulean levels in the range of 0.6–0.4 Ma (McNabb and Sinclair, 2009). Most of the excavations were conducted in the 1950s by Mason (1962, 1988). McNabb (2009) re-analysed the 2212 Acheulean artefacts and concluded that there is no *in situ* knapping profile. The LCTs were brought into the cave as finished tools, but they are not highly standardised and show a least-effort approach to shaping (McNabb et al., 2004; McNabb, 2009). However, in comparison with earlier Acheulean assemblages, they have more extensive and often finer flaking, and some LCTs are very well made. The faunal assemblage was largely accumulated by hominids, although carnivores and porcupines had a role in altering the bones (Ogola, 2009).

The Northern Cape Province holds a rich record of Later Acheulean Fauresmith Industry sites which are particularly interesting because they combine smaller handaxes, prepared core technology, blades and convergent flakes (points). With these more advanced features and the paucity of good dates, the Fauresmith was previously considered a late Acheulean development (Underhill, 2011). However, dating advances now demonstrate a longer age range for the Fauresmith. Combined OSL and Electron Spin Resonance results at Kathu Pan place Fauresmith Stratum 4a at ~400–500 kyr BP (Porat et al., 2010). Not only is blade production well developed at this site (Wilkins and Chazan, 2012), but impact fractures and basal shaping of the points indicate that these tools were used as hafted spearheads (Wilkins et al., 2012, 2015). Evidence from Kathu Pan now suggests that the Fauresmith is an industry that extends from the later to the final Acheulean. Canteen Kopje also has Fauresmith assemblages excavated by different researchers from the base of aeolian sands that unconformably cap the Victoria West Industry gravels (McNabb and Beaumont, 2011b; Leader, 2013; Lotter et al., 2016). Whilst our own OSL efforts on these sands in Pit 6 did not succeed, Chazan et al. (2013) dated the oldest grains in their samples to ~0.3 Ma and suggest that the aeolian sands began to accumulate by this time. The implication is that the Canteen Kopje Fauresmith is a final Acheulean development. Blades occur but are not as common as at Kathu Pan, and core rejuvenation elements suggest some sophisticated flaking strategies.

In northern Limpopo Province, close to the Limpopo River, three sites belong to a final ESA industry transitional to the Middle Stone Age (Kuman et al., 2005; Le Baron et al., 2010; Pollarolo et al., 2010b; Wilkins et al., 2010). Here, the lithic industries have more in common with the cultural sequence of Zimbabwe than with South Africa (Kempson, 2007). The assemblages could not be dated but the artefacts compare most closely with the transitional Charaman Industry in Zimbabwe (Sampson, 1974), with occupation during a warm, moist phase of the Pleistocene probably ±0.3 Ma. The best preserved site is Kudu Koppie where large numbers of conjoining artefacts indicate a context with good integrity (Sumner, 2013; Sumner and Kuman, 2014). Artefacts at the Hackthorne site are in a deflated open-air context on an escarpment overlooking the Limpopo River, where they appear to have accumulated adjacent to a game trail leading down to the river (Kuman et al., 2005). Artefacts at Keratic Koppie also occur in a deflated lag context but on sandstone, adjacent to a large rock outcrop, a context where lithic concentrations are always greatest (Le Baron et al., 2010).

Beyond South Africa, the best understood and well published late ESA is the Sangoan Industry at Kalambo Falls, Zambia (Clark, 2001b) which, like the Charaman, follows the Acheulean in time. The industry is characterised by small scrapers and heavy-duty picks and core axes said to relate to the woodland resources that dominated this region, even during drier periods of the Pleistocene (*ibid.*). Because some Acheulean style handaxes persist in the Sangoan, some researchers see it as a variant of the late Acheulean. The most likely toolmaker of this period is represented by the Kabwe cranium from Broken Hill, Zambia, an example of archaic *Homo sapiens* that is undated but probably at least 0.4 Ma.

21.4 Summary

From 0.6–0.5 Ma, the number of archaeological sites in southern Africa increases (Clark, 2001a), not merely because younger sites are more likely to be preserved, but because of the adaptability of the evolving populations of early *Homo sapiens*, exemplified by fossil hominids from Kabwe and Elandsfontein. It is plausible that these later ESA sites reflect a slower, more modern rate of maturation in our genus, which allowed for the evolution of increasingly flexible and adaptable behaviour. The ability of the first Acheulean hominids to detach large flakes (>10 cm) and to shape handaxes and cleavers was a milestone in human evolution. However, the crude or simple appearance of most Early Acheulean LCTs suggests that *H. ergaster* juveniles matured relatively quickly and did not have the kind of prolonged adolescence that allowed modern *H. sapiens* the time and freedom to experiment and innovate. Support for this view is found in a juvenile male skeleton of *H. ergaster* from Nariokotome (West Turkana, Kenya) who died at about 8 or 9 years of age and is argued to have been much closer to his adult height than modern boys of this age (Graves et al., 2010). Like the Oldowan, the Acheulean is a highly sophisticated development that allowed early *Homo* to adapt and survive well, but it is widely considered a conservative tradition that does not reflect the kind of complexity typical of modern *H. sapiens* (e.g. Sharon et al., 2011). The most variable and skilled assemblages of the ESA are found in its final phases and in the transition to the Middle Stone Age, which is characterised by a faster pace of change and greater complexity of tool traditions and cultural behaviour in general.

References

Alperson-Afil, N. (2008). Continual fire-making by Hominins at Gesher Benot Ya'aqov, Israel. *Quaternary Science Reviews*, 27, 1733–1739.

Alperson-Afil, N. and Goren-Inbar, N. (2010). *The Acheulian Site of Gesher Benot Ya'aqov, Vol. II, Ancient Flames and Controlled Use of Fire*. New York and London: Springer, 120pp.

Alperson-Afil, N., Richter, D. and Goren-Inbar, N. (2007). Phantom hearths and the use of fire at Gesher Benot Ya'aqov, Israel. *PaleoAnthropology*, 2007, 1–15.

Barbetti, M. (1986). Traces of fire in the archaeological record, before one million years ago? *Journal of Human Evolution*, 15, 771–781.

Barham, L., Phillips, W. M., Maher, B. A., Karloukovski, V., Duller, G. A. T., Jain, M. and Wintle, A. G. (2011). The dating and interpretation of a Mode 1 site in the Luangwa Valley, Zambia. *Journal of Human Evolution*, 60, 549–570.

Barham, L. and Mitchell, P. (2008). *The First Africans: African Archaeology from the Earliest Toolmakers to Most Recent Foragers*. Cambridge: Cambridge University Press, 622pp.

Bellomo, R. V. (1993). A methodological approach for identifying archaeological evidence of fire resulting from human activities. *Journal of Archaeological Science*, 20, 525–553.

Beaumont, P. B. and Vogel, J. C. (2006). On a timescale for the past million years of human history in central South Africa. *South African Journal of Science*, 102, 217–228.

Berna, F., Goldberg, P., Horowitz, L. K., Brink, J., Holt, S., Bamford, M. and Chazan, M. (2012). Microstratigraphic evidence of in situ fire in the Acheulean strata of Wonderwerk Cave, Northern Cape province, South Africa. *Proceedings of the National Academy of Sciences,* 109, E1215–E1220.

Beyene, Y., Katoh, S., WoldeGabriel, G., Hart, W. K., Uto, K., Sudo, M., Kondo, M., Hyodo, M., Renee, P. R., Suwa, G. and Asfaw, B. (2013). The characteristics and chronology of the earliest Acheulean at Konso, Ethiopia. *Proceedings of the National Academy of Sciences*, 110, 1584–1591.

Brain, C. K. (ed) (1993). *Swartkrans, a Cave's Chronicle of Early Man*. Pretoria: Transvaal Museum Monograph, No 8, 270pp.

Brain, C. K. and Sillen, A. (1988). Evidence from the Swartkrans Cave for the earliest use of fire. *Nature*, 336, 464–466.

Braun, D. R., Levin, N. E., Roberts, D., Stynder, D., Forrest, F., Herries, A. I., Matthews, T., Bishop, L., Archer, W. and Pickering, R. (2013). Initial investigations of Acheulean hominin behaviour at Elandsfontein. In *The Archaeology of the West Coast of South Africa*, ed. A. Jerardino, A. Malan and D. Braun. Cambridge: Cambridge Monographs in African Archaeology 84, pp. 10–23.

Brink, J. S., Herries, A. I. R., Moggi-Cecchi, J., Gowlett, J. A. J., Bousman, C. B., Hancox, J. P., Grün, R., Eisenmann, V., Adams, J. W. and Roussouw, L. (2012). First hominine remains from a ~1.0 million year old bone bed at Cornelia-Uitzoek, Free State Province, South Africa. *Journal of Human Evolution*, 63, 527–535.

Chazan, M. (2015). Technological trends in the Acheulean of Wonderwerk Cave, South Africa. *African Archaeological Review*, 32, 701–728.

Chazan, M., Avery, D. M., Bamford, M. K., Berna, F., Brink, J., Fernandez-Jalvo, Y., Goldberg, P., Holt, S., Matmon, A., Porat, N., Ron, H., Rossouw, L., Scott , L. and Horwitz, L. K. (2012). The Oldowan horizon in Wonderwerk Cave (South Africa): Archaeological, geological, paleontological and paleoclimatic evidence. *Journal of Human Evolution*, 63, 859–866.

Chazan, M., Hagai, R. Matmon, A., Porat, N., Goldberg, P., Yates, R., Avery, M., Sumner, A. and Horwitz, L. K. (2008). Radiometric dating of the Earlier Stone Age sequence in Excavation I at Wonderwerk Cave, South Africa: preliminary results. *Journal of Human Evolution*, 55, 1–11.

Chazan, M. and Horwitz, L. K. (2010). Milestones in the development of symbolic behaviour: A case study from Wonderwerk Cave, South Africa. *World Archaeology*, 41, 521–539.

Chazan, M., Porat., N. Sumner, T. A. and Horwitz, L. K. (2013). The use of OSL dating in unstructured sands: The archaeology and chronology of the Hutton Sands at Canteen Kopje (Northern Cape Province, South Africa). *Archaeological and Anthropological Sciences*, 5, 351–363.

Clark, J. D. (1974). The stone artifacts from Cornelia, O.F.S., South Africa. *Memoirs van die Nasionale Museum (Bloemfontein)*, 9, 33–62.

Clark, J. D. (2001a). Variability in primary and secondary technologies of the Later Acheulian in Africa. In *A Very Remote Period Indeed: Papers on the Palaeolithic presented to Derek Roe*, ed. S. Milliken and J. Cook. Oxford: Oxbow Books, pp. 1–18.

Clark, J. D. (2001b). *Kalambo Falls Prehistoric Site, Vol. III, The Earlier Cultures: Middle and Earlier Stone Age (Clark: Kalambo Falls Prehistoric Site)*. Cambridge: Cambridge University Press, 704pp.

Clarke, R. J. (1977). A juvenile cranium and some adult teeth of early *Homo* from Swartkrans, Transvaal. *South African Journal of Science*, 73, 46–50.

Clarke, R. J. (1994). The significance of the Swartkrans *Homo* to the *Homo erectus* problem. *Courier Forschungs-Institut Senckenberg*, 171, 185–193.

Clarke, R. J. (2012). The history of research in human evolution in Africa and what lessons have been learned. In *Human Origins Sites and the World Heritage Convention in Africa*, ed. N. Sanz. New York: UNESCO, World Heritage Papers 33, pp. 44–67.

Cruz-Uribe, K., Klein, R. G., Avery, G., Avery, M., Halkett, D., Hart, T., Milo, R. G., Sampson, C. G. and Volman, T. P. (2003). Excavation of buried Late Acheulean (Mid-Quaternary) land surfaces at Duinefontein 2, Western Cape Province, South Africa. *Journal of Archaeological Science*, 30, 559–575.

Erlanger, E. D., Granger, D. E. and Gibbon, R. J. (2012). Rock uplift rates in South Africa from isochron burial dating of fluvial and marine terraces. *Geology*, 40, 1019–1022.

Field, A. S. (1999). *An Analytical and Comparative Study of the Earlier Stone Age Archaeology of the Sterkfontein Valley*. Unpublished MSc dissertation, University of the Witwatersrand, Johannesburg, 2 vol.

Gibbon, R. J., Granger, D., Kuman, K., Leader, G., Lotter, M. and Forssman, T. (2013). Isochron burial dating of the Earlier Stone Age deposits at Canteen Kopje, South Africa. *Paper presented at the Association of Southern African Professional Archaeologists Biennial Conference, Gaborone, Botswana, 3–7 July 2013.*

Gibbon, R. J., Granger, D. E., Kuman, K. and Partridge, T. C. (2009). Early Acheulean technology in the Rietputs Formation, South Africa, dated with cosmogenic nuclides. *Journal of Human Evolution*, 56, 152–160.

Gibbon, R. J., Pickering, T. R, Sutton, M. B., Heaton, J. L., Kuman, K., Clarke, R. J., Brain, C. K. and Granger, D. E. (2014). Cosmogenic nuclide burial dating of hominin-bearing Pleistocene cave deposits at Swartkrans, South Africa. *Quaternary Geochronology*, 24, 10–15.

Granger, D. E., Gibbon, R. J., Kuman, K., Clarke, R. J., Bruxelles, L. and Caffee, M. W. (2015). New cosmogenic burial ages for Sterkfontein Member 2 *Australopithecus* and Member 5 Oldowan. *Nature*, 522, 85–88.

Graves, R. R., Lupo, A. C., McCarthy, R. C., Wescott, D. J. and Cunningham, D. L. (2010). Just how strapping was KNM-WT 15000? *Journal of Human Evolution*, 59, 542–554.

Groves, C. P. and Mazák, V. (1975). An approach to the taxonomy of the Hominidae: gracile Villafranchian hominids of Africa. *Casopis pro Mineralogii a Geologii*, 20, 225–246.

Hall, G., Pickering, R., Lacruz, R., Hancox, J., Berger, L. R. and Schmid, P. (2006). An Acheulean handaxe from Gladysvale Cave Site, Gauteng, South Africa. *South African Journal of Science*, 102, 103–105.

Hanson, M. and Cain, C. R. (2007). Examining histology to identify burned bone. *Journal of Archaeological Science*, 34, 1902–1913.

Jacoby, B., Phillips, N., Clarke, R. and Kuman, K. (2013). New research at the Goldsmith's Pleistocene site, Cradle of Humankind World Heritage Site (Gauteng Province, South Africa). *Poster presented at the Association of Southern African Professional Archaeologists Biennial Conference, Gaborone, Botswana, 3–7 July 2013.*

James, S. R. (1989). Hominid use of fire in the Lower and Middle Pleistocene. A review of the evidence. *Current Anthropology*, 30, 1–26.

Kandel, A. W. and Conard, N. J. (2012). Settlement patterns during the Earlier and Middle Stone Age around Langebaan Lagoon, Western Cape (South Africa). *Quaternary International*, 270, 15–29.

Keller, C. M. (1973). *Montagu Cave in Prehistory: A Descriptive Analysis*. Berkeley and London: University of California Press, Anthropological Records No 28, 150pp.

Kempson, H. (2007). *Late Earlier Stone Age sites in the Mapungubwe National Park, South Africa: A technological Study*. Unpublished MSc Dissertation, University of the Witwatersrand, Johannesburg, 110pp.

Klein, R. G. (2000). The Earlier Stone Age of southern Africa. *South African Archaeological Bulletin*, 55, 107–122.

Klein, R. G. (2009). *The Human Career: Human Biological and Cultural Origins* (3rd ed.). Chicago and London: The University of Chicago Press, 1024pp.

Klein, R. G., Avery, G., Cruz-Uribe, K., Halkett, D., Hart, T., Milo, R. G. and Volman T. P. (1999). Duinefontein 2: an Acheulean site in the Western Cape Province of South Africa. *Journal of Human Evolution*, 37, 153–190.

Klein, R. G., Avery, G., Cruz-Uribe, K. and Steele, T. E. (2007). The mammalian fauna associated with an archaic hominin skullcap and later Acheulean artifacts at Elandsfontein, Western Cape Province, South Africa. *Journal of Human Evolution*, 52, 164–186.

Klein, R. G. and Cruz-Uribe, K. (1991). The bovids from Elandsfontein, South Africa, and their implications for the age, palaeoenvironment, and origins of the site. *African Archaeological Review*, 9, 21–79.

Kuman, K. (1994). The archaeology of Sterkfontein – past and present. *Journal of Human Evolution*, 27, 471–495.

Kuman, K. (1998). The earliest South African industries. In *Early Human Behavior in Global Context: The Rise and Diversity of the Lower Palaeolithic Record*, ed. M. Petraglia and R. Korisettar. London: Routledge Press, pp. 151–186.

Kuman, K. (2003). Site formation in the early South African Stone Age sites and its influence on the archaeological record. *South African Journal of Science*, 99, 251–254.

Kuman, K. (2007). The Earlier Stone Age in South Africa: site context and the influence of cave studies. In *Breathing Life into Fossils: Taphonomic Studies in Honor of C.K. (Bob) Brain*, ed. T. R. Pickering, K. Schick and N. Toth. Gosport, IN: Stone Age Institute Press, pp. 181–198.

Kuman, K. (2014). The Acheulean Industrial Complex. In *Encyclopedia of Global Archaeology, Vol. 8*, ed. C. Smith. New York: Springer, pp. 7–18.

Kuman, K. and Clarke, R. J. (2000). Stratigraphy, artefact industries and hominid associations for Sterkfontein, Member 5. *Journal of Human Evolution,* 38, 827–847.

Kuman, K. and Field, A. S. (2009). The Oldowan Industry from Sterkfontein Caves, South Africa. In *The Cutting Edge: New Approaches to the Archaeology of Human Origins*, ed. K. Schick and N. Toth. Gosport, Indiana: Stone Age Institute Press, pp. 151–169.

Kuman, K., Field, A. S. and Thackeray, J. F. (1997). Discovery of new artefacts at Kromdraai. *South African Journal of Science*, 93, 187–193.

Kuman, K., Gibbon, R. J., Kempson, H., Langejans, G., Le Baron, J. C., Pollarolo, L. and Sutton, M. (2005). Stone Age signatures in northernmost South Africa: Early archaeology of the Mapungubwe National Park and vicinity. In *From Tools to Symbols, From Early Hominids to Modern Humans*, ed. F. D'Errico and L. Backwell. Johannesburg: Witwatersrand University Press, pp. 163–182.

Leader, G. M. (2009). *The Early Acheulean in the Vaal River Basin, Rietputs Formation, Northern Cape Province, South Africa*. Unpublished MSc dissertation, University of the Witwatersrand, Johannesburg, 96pp.

Leader, G. M. (2013). *A Techno-Typological Analysis of the Earlier Acheulean Assemblages at Canteen Kopje, Northern Cape Province, South Africa, with a New Interpretation of the Victoria West Core Phenomenon*. Unpublished PhD thesis, University of the Witwatersrand, Johannesburg.

Le Baron, J. C., Kuman, K. and Grab, S. W. (2010). The landscape distribution of Stone Age artefacts on the Hackthorne Plateau, Limpopo River Valley, South Africa. *South African Archaeological Bulletin*, 65, 123–131.

Lepre, C. J., Roche, H., Kent, D. V., Harmand, S., Quinn, R. L., Brugal, J.-P., Texier, P.-J., Lenoble, A. and Feibel, C. S. (2011). An earlier origin for the Acheulian. *Nature*, 477, 82–85.

Lotter, M. G., Gibbon, R. J., Kuman, K., Leader, G. M., Forssman, T. and Granger, D. E. (2016). A geoarchaeological study of the Middle and Upper Pleistocene levels at Canteen Kopje, Northern Cape Province, South Africa. *Geoarchaeology: An International Journal,* in press.

Lotter, M., Kuman, K., Gibbon, R. and Granger, D. (2014). The archaeology of the lower Sundays River Valley, Eastern Cape Province, South Africa: An assessment of Earlier Stone Age alluvial terrace sites. *Paper presented at the 14th Congress of the Pan African Archaeological Association for Prehistory and Related Studies and 22nd*

Biennial Meeting of the Society of Africanist Archaeologists. University of the Witwatersrand, 14–18 July 2014.

Luyt, C. J. and Lee-Thorp, J. A. (2003). Carbon isotope ratios of Sterkfontein fossils indicate a marked shift to open environments c. 1.7 Myr ago. *South African Journal of Science*, 99, 271–273.

Luyt, J., Lee-Thorp, J. A. and Avery, G. (2000). New light on Middle Pleistocene west coast environments from Elandsfontein, Western Cape Province, South Africa. *South African Journal of Science*, 96, 399–403.

Mason, R. J. (1962). *Prehistory of the Transvaal: A Record of Human Activity*. Johannesburg: Witwatersrand University Press, 498pp.

Mason, R. J. (1988). *Cave of Hearths, Makapansgat, Transvaal.* Johannesburg: University of the Witwatersrand, Archaeological Research Unit, Occasional Paper No 21, 713pp.

Matmon, A., Ron, H., Chazan, M., Porat, N. and Horwitz, L. K. (2012). Reconstructing the history of sediment deposition in caves: A case study from Wonderwerk Cave, South Africa. *Geological Society of America Bulletin*, 124, 611–625.

McNabb, J. (2009). The ESA stone tool assemblage from the Cave of Hearths, Beds 1-3. In *The Cave of Hearths: Makapan Middle Pleistocene Research Project*, ed. J. McNabb and A. Sinclair. Oxford: Archaeopress, University of Southampton Series in Archaeology No 1, pp. 75–104.

McNabb, J. and Beaumont, P. (2011a). *A Report on the Archaeological Assemblages from Excavations by Peter Beaumont at Canteen Koppie, Northern Cape, South Africa.* Oxford: BAR International Series 2275, University of Southampton Series in Archaeology No 4, 86pp.

McNabb, J. and Beaumont, P. (2011b). Excavations in the Acheulean levels of the Earlier Stone Age site of Canteen Koppie, Northern Cape Province, South Africa. *Proceedings of the Prehistory Society*, 78, 51–71.

McNabb, J., Binyon, F. and Hazelwood, L. (2004). The large cutting tools from the South African Acheulean and the question of social traditions. *Current Anthropology*, 45, 653–677.

McNabb, J. and Sinclair, A. (eds) (2009). *The Cave of Hearths: Makapan Middle Pleistocene Research Project*. Oxford: Archaeopress, University of Southampton Series in Archaeology No 1, 193pp.

Ogola, C. (2009). The taphonomy of the Cave of Hearths Acheulean faunal assemblage. In *The Cave of Hearths: Makapan Middle Pleistocene Research Project*, ed. J. McNabb and A. Sinclair. Oxford: Archaeopress, University of Southampton Series in Archaeology No 1, pp. 65–74.

Pickering, R., Kramers, J. D., Hancox, P. J., de Ruiter, D. and Woodhead, J. (2011). Contemporary flowstone development links early hominin bearing cave deposits in South Africa. *Earth and Planetary Science Letters*, 306, 23–32.

Pickering, T. R. (1999). *Taphonomic Interpretations of the Sterkfontein Early Hominid Site (Gauteng, South Africa) Reconsidered in Light of Recent Evidence*. Unpublished PhD thesis, University of Wisconsin, Madison, 1114pp.

Pickering, T. R. (2002). Reconsideration of criteria for differentiating faunal assemblages created by hyenas and hominids. *International Journal of Osteoarchaeology*, 12, 127–141.

Pickering, T. R., Domínguez-Rodrigo, M., Egeland, C. P. and Brain, C. K. (2004). New data and ideas on the foraging behaviour of Early Stone Age hominids at Swartkrans Cave, South Africa. *South African Journal of Science*, 100, 215–219.

Pickering, T. R., Domínguez-Rodrigo, M., Egeland, C. P. and Brain, C. K. (2007). Carcass foraging by early hominids at Swartkrans Cave (South Africa): A new investigation of the zooarchaeology and taphonomy of Member 3. In *Breathing Life into Fossils: Taphonomic Studies in Honor of C.K. (Bob) Brain*, ed. T. R. Pickering, K. Schick and N. Toth. Gosport, Indiana: Stone Age Institute Press, pp. 233–253.

Pickering, T. R., Egeland, C. P., Domínguez-Rodrigo, M., Brain, C. K. and Schnell, A. G. (2008). Testing the "shift in the balance of power" hypothesis at Swartkrans, South Africa: Hominid cave use and subsistence behavior in the Early Pleistocene. *Journal of Anthropological Archaeology*, 27, 30–45.

Pickering, T. R., Sutton, M. B., Heaton, J. L., Clarke, R. J., Brain, C. K. and Kuman, K. (2012). New stratigraphic interpretations and hominid fossils from Swartkrans Member 1 (South Africa). *Journal of Human Evolution*, 62, 618–628.

Plummer, T. (2004). Flaked stones and old bones: Biological and cultural evolution at the dawn of technology. *Yearbook of Physical Anthropology*, 47, 118–164.

Pollarolo, L., Susino, G., Kuman, K. and Bruxelles, L. (2010a). Acheulean artefacts at Maropeng in the Cradle of Humankind World Heritage Site, Gauteng Province, South Africa. *South African Archaeological Bulletin*, 65, 3–12.

Pollarolo, L., Wilkins, J., Kuman, K. and Galletti, L. (2010b). Site formation at Kudu Koppie: A late Earlier and Middle Stone Age site in northern Limpopo Province, South Africa. *Quaternary International*, 216, 151–161

Porat, N., Chazan, M., Grün, R., Aubert, M., Eisenmann, V. and Horwitz, L. K. (2010). New radiometric ages for the Fauresmith industry from Kathu Pan, southern Africa: Implications for the Earlier to Middle Stone Age transition. *Journal of Archaeological Science*, 37, 269–283.

Reed, K. E. (1997). Early hominid evolution and ecological change through the African Plio-Pleistocene. *Journal of Human Evolution*, 32, 289–322.

Robinson, J. T. (1953). *Telanthropus* and its phylogenetic significance. *American Journal of Physical Anthropology*, 11, 445–501.

Sampson, C. G. (1974). *The Stone Age Archaeology of Southern Africa*. New York: Academic Press, 518pp.

Sharon, G., Alperson-Afil, N. and Goren-Inbar, N. (2011). Cultural conservatism and variability in the Acheulian sequence of Gesher Benot Ya'aqov. *Journal of Human Evolution*, 60, 387–397.

Sherwood, N. (2013). *Lithic Raw Material Procurement Through Time at Swartkrans: Earlier to Middle Stone Age*. Unpublished MSc Dissertation, University of the Witwatersrand, Johannesburg, 111pp.

Singer, R. and Wymer, J. J. (1968). Archaeological investigations at the Saldanha skull site in South Africa. *South African Archaeological Bulletin*, 25, 63–74.

Sumner, T. A. (2013). A refitting study of Late Early to Middle Stone Age lithic assemblages from the site of Kudu Koppie, Limpopo Province, South Africa. *Journal of African Archaeology*, 11, 133–153.

Sumner, T. A. and Kuman, K. (2014). Refitting evidence for the stratigraphic integrity of the Kudu Koppie Early to Middle Stone Age site, northern Limpopo Province, South Africa. *Quaternary International*, 343, 169–178.

Sutton, M. (2012). *The Archaeology of Swartkrans: Members 1 and 4*. Unpublished PhD thesis, University of the Witwatersrand, Johannesburg, 227pp.

Sutton, M., Kuman, K., Clarke, R. J., Pickering, T. R., Heaton, J. L. and Brain, C. K. (2012). The Oldowan and *Homo habilis* at Swartkrans. *PaleoAnthropology 2012: A1-A39. Abstracts of the Paleoanthropology Society 2012 Meeting, Tennessee, USA.*

Thackeray, J. F., Kirschvink, J. L. and Raub, T. D. (2002). Palaeomagnetic analyses of calcified deposits from the Plio-Pleistocene hominid site of Kromdraai, South Africa. *South African Journal of Science,* 98, 537–540.

Underhill, D. (2011). The study of the Fauresmith: A review. *South African Archaeological Bulletin*, 66, 15–26.

Vrba, E. S. (1975). Some evidence of chronology and palaeoecology of Sterkfontein, Swartkrans, and Kromdraai from the fossil Bovidae. *Nature*, 254, 301–304.

Vrba, E. S. (1982). Biostratigraphy and chronology, based particularly on Bovidae, of southern hominid associated assemblages: Makapansgat, Sterkfontein, Taung, Kromdraai, Swartkrans; also Elandsfontein (Saldanha), Broken Hill (now Kabwe) and Cave of Hearths. *Congrès International de Paleéontologie Humaine, 1er Congrès, Tome 2*. Nice: CNRS, pp. 707–752.

Vrba, E. S. (1985). Early hominids in southern Africa: Updated observations on chronological and ecological background. In *Hominid Evolution. Past, Present and Future*, ed. P. V. Tobias. New York: Alan R. Liss, pp. 195–200.

Watson, V. (1993). Composition of the Swartkrans bone accumulations, in terms of skeletal parts and animals represented. In *Swartkrans, a Cave's Chronicle of Early Man*, ed. C. K. Brain. Pretoria: Transvaal Museum Monograph No 8, pp. 35–74.

Wilkins, J. and Chazan, M. (2012). Blade production ~500 thousand years ago at Kathu Pan 1, South Africa: Support for a multiple origins hypothesis for early Middle Pleistocene blade technologies. *Journal of Archaeological Science*, 39, 1883–1900.

Wilkins, J., Pollarolo, L. and Kuman, K. (2010). Prepared core reduction at the site of Kudu Koppie in northern South Africa: Temporal patterns across the Earlier and Middle Stone Age boundary. *Journal of Archaeological Science*, 37, 1279–1292.

Wilkins, J., Schoville, B. J., Brown, K. S. and Chazan, M. (2012). Evidence for early hafted hunting technology. *Science*, 338, 942–946.

Wilkins, J., Schoville, B. J., Brown, K. S. and Chazan, M. (2015). Kathu Pan 1 points and the assemblage-scale, probabilistic approach: a response to Rots and Plisson, "Projectiles and the abuse of the usewear method in a search for impact". *Journal of Archaeological Science*, 54, 294–299.

22

Development of the archaeological record during the Middle Stone Age of South Africa

SARAH WURZ

Abstract

It is in the context of the Middle Stone Age (MSA) that characteristic modern human morphotypes and behaviours appeared. In this chapter, MSA technologies, subsistence and site maintenance behaviours and their cognitive and behavioural implications are discussed in broad terms. The earliest MSA stone tool technology already shows behavioural complexity, but current evidence also indicates that bone tools, pyrotechnology, ochre technology, jewellery and engraved objects – which are all conventional markers for complex and modern behaviours – only appear in notable quantities from marine isotope stage (MIS) 5 onwards (from 130 to 85 kyr BP), with a discernible surge in MIS 4. Two of the MSA aspects that remain largely obscure due to a lack of fine grained information, however, are the regional extent of bone and stone tool technologies and the degree to which climatic events and environmental factors shaped human evolution and behaviour.

22.1 Introduction and background

The Middle Stone Age (MSA), a period associated with shifting temporal and geographical boundaries, existed between ~300 and 22 kyr BP. Since the 1990s, an increase in research focus on the MSA by several local and international groups has resulted in a remarkable diversity of perspectives on the MSA in South Africa, albeit predominantly on MIS 4 phenomena, which is discussed in some detail in this chapter.

In southern Africa, there are relatively few sites that record the Early Middle Stone Age (EMSA), dating to between ~300 and 130 kyr BP and within marine isotope stages (MIS) 8–6 (Table 22.1). Of these sites, few have been examined in

Table 22.1 *The South African and Lesotho Stone Age sequence (after Lombard et al., 2012)*

Approximate time period (kyr BP)	Marine Isotope Stage (MIS) association	Associated techno-complex	Alternative nomenclature
40–18	3–2	Early Later Stone Age	Late Pleistocene microlithic
40–20	3–2	Final Middle Stone Age	MSA IV at Klasies River, MSA 4 generally
60–45	3	Sibudu	late MSA; post-Howiesons Poort, MSA III at Klasies, Sibudan, MSA 3 generally
66–58	4–3	Howiesons Poort	
77–70	5a–4	Still Bay	
105–77	5c–a	Mossel Bay	MSA II at Klasies River, MSA 2b generally, Pietersburg, Orangian
130–105	5e–d	Klasies River	MSA I at Klasies river, MSA 2a generally, Pietersburg
300–130	8–6	Early Middle Stone Age	

detail and even fewer can be located securely in time. Rising sea levels during Pleistocene interglacials obliterated traces of occupation on the coast, and most preserved sites are therefore found in the interior of South Africa, for example at Florisbad, Border Cave, Bundu Farm, Lincoln Cave, and Wonderwerk Cave (Fig. 22.1). Pinnacle Point is currently the only coastal MIS 6 site. A somewhat clearer picture emerges in MIS 5 (130–85 kyr BP) as a greater number of sites with more extensive geochronometric control is known. Research efforts historically converged on the Cape coastal region, but many of the data-rich early late Pleistocene sites occur inland, for example in the Orange River area, at the Cave of Hearths, Border Cave, Sibudu, Melikane and Rose Cottage Cave. There is an increase in archaeological visibility in glacial MIS 4 (~75–58 kyr BP). Recent insights into the Still Bay and Howiesons Poort techno-complexes come from Blombos Cave, Diepkloof, Klasies River, Umhlatuzana, Sibudu Cave, Rose Cottage Cave and Klein Kliphuis (Henshilwood, 2012). MIS 3 may have been less densely populated than MIS 5 and 4, but not significantly so (Mitchell, 2008).

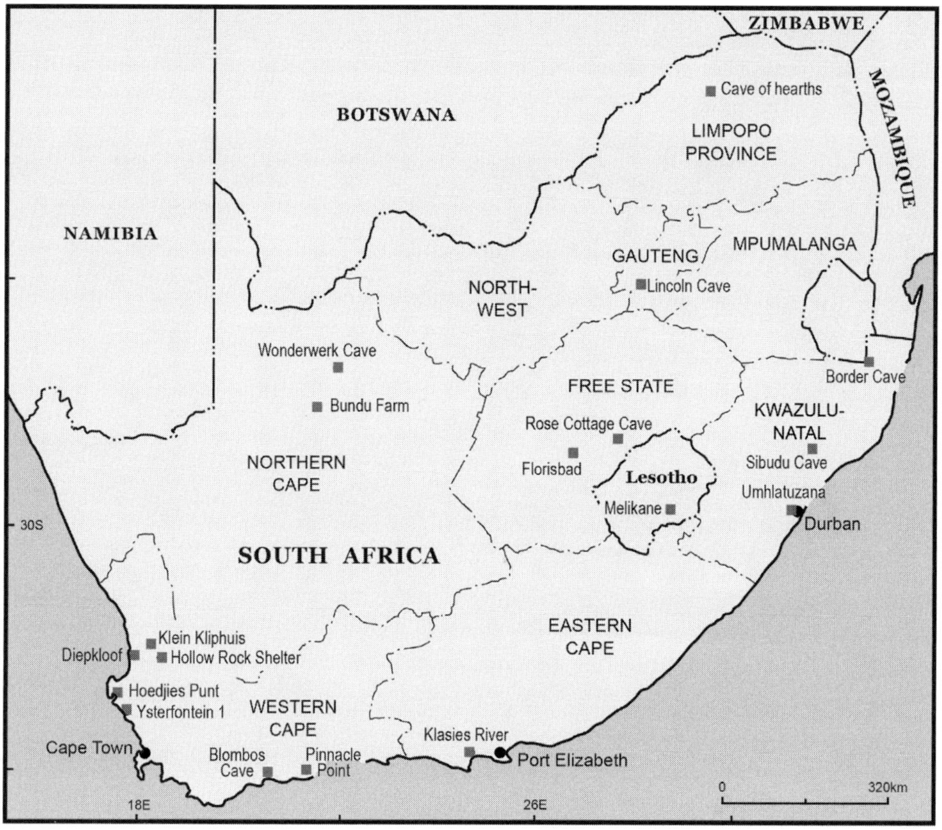

Fig. 22.1. Map of South Africa with sites mentioned in the text.

A number of research projects specifically investigating MIS 3 occupation in Lesotho, Western Cape and KwaZulu-Natal also promise a deeper understanding of this period and the beginnings of the Later Stone Age (~43–22 kyr BP). Thus, the geographical context of available MSA sites changes over time in response to climate, demography, technology and cultural practices, as well as the nature of recent research activities.

The MSA fossil record traces a slow mosaic-like development of morphologically human modern traits. The ~250 kyr Florisbad calvarium and the Hoedjiespunt fossils represent populations immediately ancestral to *Homo sapiens sapiens* that diverged ~200 kyr BP (Lombard et al., 2013). The relatively large MIS 5 fossil collection from Klasies River on the southern Cape coast (Figs. 22.2, 22.3) shows primarily modern morphological traits, with a retention of archaic characteristics, while the ~36 kyr Hofmeyr skull, a surface find from the Eastern Cape, is fully

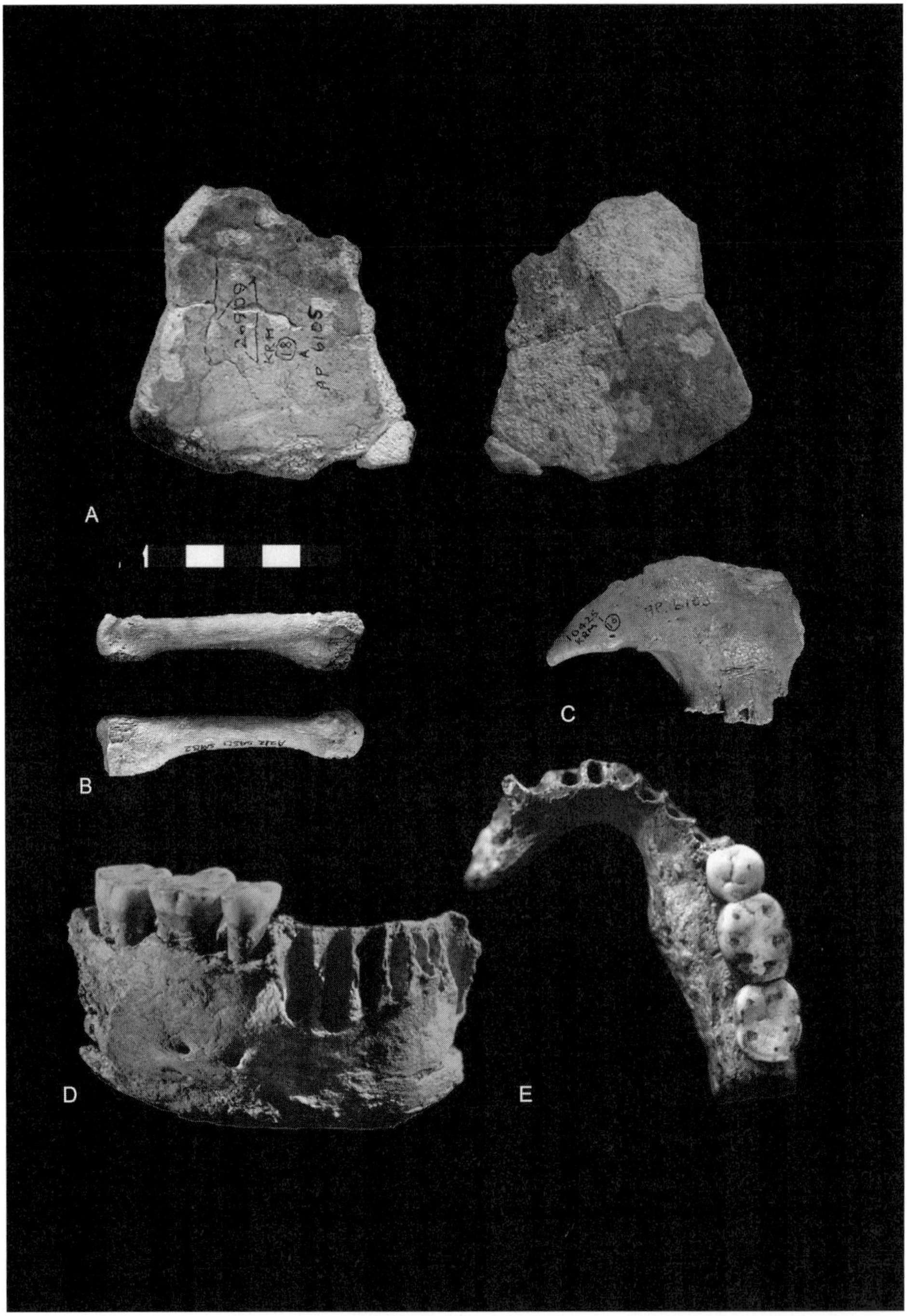

Fig. 22.2. Human fossils from Klasies River. (A) KRM26909, two views of a parietal fragment of skull, Layer 37, Cave 1; (B) metatarsal, ~ Layer SMB2 Cave 1; (C) KRM 16425, fragment of frontal bone, Layer 16, Cave 1; (D) KRM 13400, corpus of mandible, Layer 14, Cave 1; (E) KRM 13400, occlusal view (photo: Sarah Wurz).

Fig. 22.3. Klasies River main site, southern Cape, South Africa (photo: Sarah Wurz).

modern but still outside the morphological range of modern Khoe-San (Dusseldorp et al., 2013). The MSA encompasses the population history of anatomically modern *Homo sapiens* and their immediate predecessors, and one of the major issues is identifying when behavioural and cognitive 'modernity' developed. In this chapter, snapshots are provided into the aspects that most prominently influenced the formation of the South African MSA archaeological record, including technologies, subsistence and site maintenance behaviour, and how these may link to cognitive evolution.

22.2 Technological behaviour in the Middle Stone Age

The intensified interest in the MSA in recent years has resulted in new perspectives on bone, stone and fire-related technologies, and also on the production of shell beads and the utilisation of ochre. A relatively small number of formal bone tools dating to around 100 kyr BP are known from Blombos and Klasies River (Wurz, 2013). The Still Bay techno-complex at Blombos Cave (Henshilwood, 2012) and the MIS 5 to 3 layers at Sibudu yielded larger such collections (d'Errico et al., 2012b). The bone wedges, *pièces esquillées*, pressure flakers, smoothers and sequentially notched pieces from Sibudu are similar to those found in the Upper Palaeolithic. There is widespread consensus that such formally shaped bone tools are innovative and reflect cognitive complexity. There is less agreement on the cognitive implications of the deliberately shaped and designed stone tools of the MSA, even though their production is just as complex as that of bone tools. The lithic toolkit of the MSA contains flakes, blades of different proportions, unretouched and retouched points, and tool types such as denticulates, scrapers, backed and bifacials. Discoidal, prepared core (including Levallois), blade and bipolar reduction methods were used. The earliest MSA assemblages indicate that important techno-cognitive milestones had been reached. For example, the engineering constraints involved in producing the standard toolkit in the MSA necessitated mental projections of at least five to ten actions ahead, indicating modern cognitive capabilities (Hiscock, 2014). Using bonding agents to create multi-component tools denotes another complex cognitive strategy. The practice of hafting stone tools was most likely in place by the EMSA in South Africa (Wadley, 2013), perhaps even as early as 500 kyr BP (Wilkins et al., 2012; but see Rots and Plisson, 2014).

Identifying chronological and regional trends in MSA lithic technology is a major challenge. There is a strong possibility that technological conventions observable on regional levels occurred prior to 100 kyr BP, but they become more apparent after this time. A number of broadly similar assemblages from the Cape dating to MIS 5b and 5c probably denote such a convention (Wurz, 2013). The Still Bay and Howiesons Poort are markedly distinct from other MSA phases, as they occur on a subcontinental scale within bracketed time periods (but see Porraz et al., 2013). At most sites, the Still Bay dates to between ~80 and 70 kyr BP and the Howiesons Poort to between ~70 and 58 kyr BP (Wurz, 2013). There is much evidence to indicate that the Still Bay and Howiesons Poort are associated with innovative technological practices. In the Still Bay of Blombos Cave, heat-treated silcrete was pressure flaked to produce lanceolate bifacial points (Mourre et al., 2010), and in the Howiesons Poort a soft stone (Soriano et al., 2007) or perhaps a vegetal hammer was used to knap blades and bladelets which were transformed into geometric backed tools (Villa et al., 2012; Henshilwood et al., 2014). The Howiesons Poort (Figs. 22.4, 22.5) is the most intensely researched techno-complex in South Africa, and now appears to consist of several phases. Furthermore, flake production and notched tools form an important part of the wider technological repertoire at, for example, Klasies River, Diepkloof, Pinnacle Point, Sibudu (Wurz, 2013) and Klipdrift Shelter (Henshilwood et al., 2014). Finding commonalities in the lithics from different areas after ~58 kyr BP becomes problematic, but unifacial points occur in many assemblages, for example at Sibudu and Umhlatuzana.

Fig. 22.4. Artefacts from the Howiesons Poort levels at Klasies River. (A) KRM28942, silcrete backed artefact, Cave 1A, Layer 12; (B) KRM 31421, silcrete backed artefact, Cave 1A, Layer 20; (C) Silcrete notched backed artefact and blade section, Cave 1A, Layer 18; (D) Quartzite blades, Cave 2 surface; (E) Quartzite core, Cave 1A, E50 CP8; (F) KRM 26733, engraved midshaft bovid fragment, Cave 1A, Layer 19; (G) KRM 42160 bone point, Cave 1A, Layer 19 (photo: Sarah Wurz).

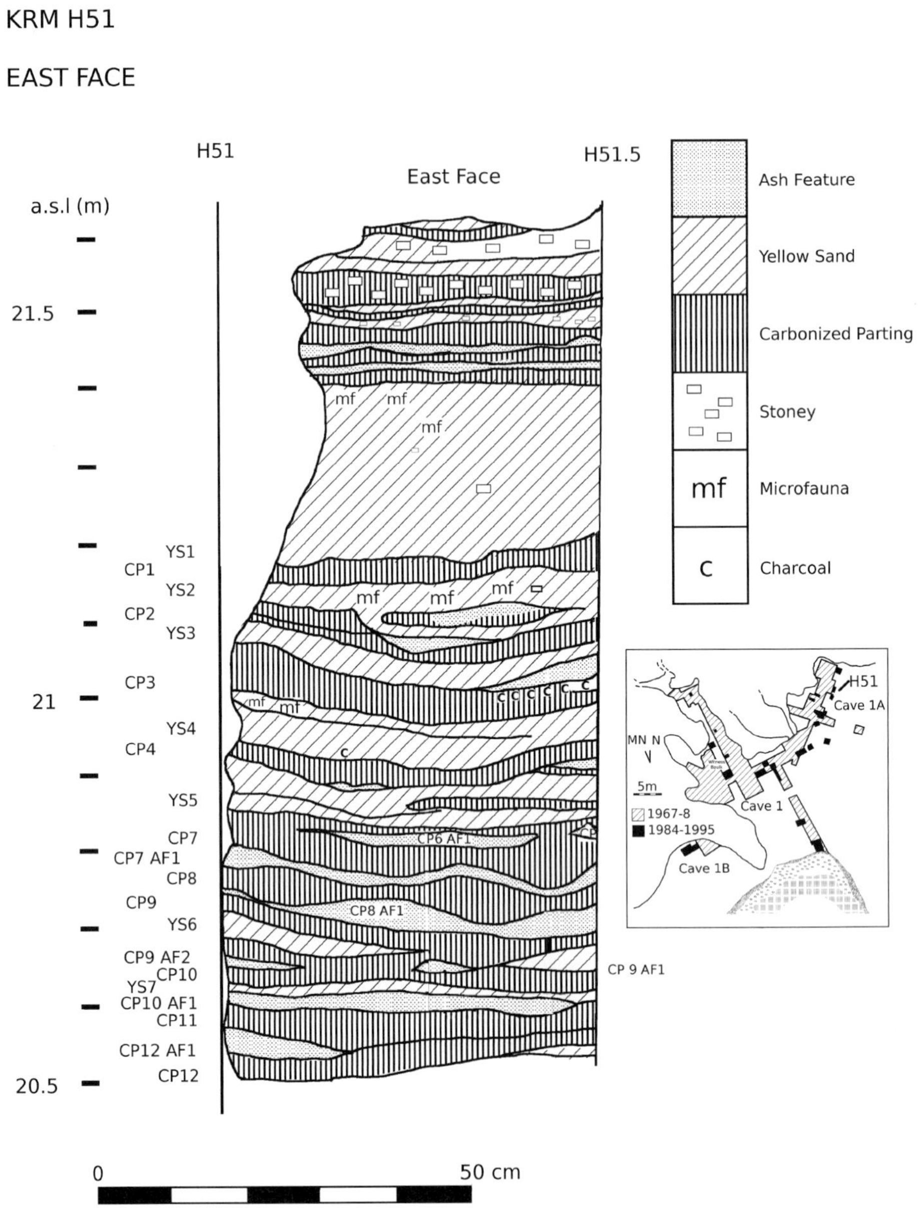

Fig. 22.5. Stratigraphic context and drawn stratigraphic log of a section of the Howiesons Poort, square H51, Cave 1A.

Typical MSA artefacts occur in various regions until 22 kyr BP, but at Border Cave, small flakes in quartz and San-like bone tools dating to 44–42 cal kyr BP are described as the earliest evidence of the Later Stone Age (LSA) in that region (d'Errico et al., 2012a; Villa et al., 2012).

Silcrete was sometimes heated to enhance flakeability (Brown et al., 2009), but the particular procedures used to accomplish this are still under debate. Field-based experiments illustrate that controlled management of the heating process, involving burying specimens in sand below a fire, is important to achieve this (Wadley and Prinsloo, 2014). A possibility is also that silcrete nodules were heated in areas surrounding open fires or directly on slightly cooled embers (Schmidt et al., 2013). Another procedure that required careful and controlled heating is the production of compound adhesives for hafting, recorded at >70 kyr BP at Sibudu and Rose Cottage (Wadley, 2013). This evidence favours the hypothesis that MSA peoples were able to intentionally control heat in different ways to achieve desired results. These technically advanced processes have been tied to modern levels of cognitive complexity (Wadley, 2013).

MSA technologies that provide more direct links to cultural ideology involve ochre and beads. Their early occurrence is key in supporting the hypothesis that by 100 kyr BP human lives were structured by objects that acted as symbols and social mediators (McBrearty and Brooks, 2000; Henshilwood and Marean, 2003). The 100 kyr-age containers at Blombos Cave were used to mix and store ochre-rich compounds and these, as well as the iconic ~77 kyr geometrically engraved ochre, are benchmarks for complex cognition in the MSA (Henshilwood et al., 2011; Wadley, 2013). Engraved ochres dating back to ~100 kyr BP at Blombos, Klasies River and Pinnacle Point (Wurz, 2013), and the MIS 4 shell beads and engraved ostrich eggshell from Diepkloof and Klipdrift Shelter, are widely regarded as indicators of modern cognition (e.g. Henshilwood, 2012; Wadley, 2013; Henshilwood et al., 2014). However, these inferences on modern cognition are under debate (see Henshilwood and Dubreuil, 2011).

22.3 Subsistence behaviour and site maintenance

For most of the MSA, both the interior and coastal areas of South Africa were productive environments with grasslands that could support large herbivores (Brink, 1999). Pleistocene assemblages contain relatively more grazers than those from the Holocene (Klein, 1983) and are often characterised by a higher ungulate richness (Faith, 2011). Grassland taxa include plains zebra (*Equus quagga*), wildebeest (*Connochaetes* spp.), giant wildebeest (*Megalotragus priscus*) and springbok (*Antidorcas* spp.), while the presence of browsers such as duiker (*Sylvicapra grimmia*), Cape grysbok/steenbok (*Raphicerus* spp.) and kudu

(*Tragelaphus strepsiceros*) indicates mosaic environments with closed vegetation in some areas and periods. Available data are too sparse to understand subsistence behaviour in the EMSA, but at Pinnacle Point Cave 13B (PP13B) in MIS 6, large and very large ungulates (size 3 and larger) such as the extinct giant buffalo (*Megalotragus priscus*), buffalo (*Syncerus antiquus*) and eland (*Tragelaphus oryx*) were selected prey. At a number of MIS 5 sites, e.g. Blombos Cave, Diepkloof Rock Shelter, Florisbad, Klasies River, Pinnacle Point Cave 13B and Ysterfontein, relatively more grazers than browsers occur. As in MIS 6, the majority of prey consists of the highest ranked species, large and very large ungulates (Clark and Kandel, 2013). It is a distinct possibility that some of the retouched and unretouched points of the MIS 6 and 5 assemblages were used as part of thrusting spears, but this hypothesis needs further investigation.

In MIS 4 a shift in faunal subsistence patterns occurs with a marked increase in the exploitation of small mammals and small (size 1) ungulates, which is an indication of widened dietary breadth (Clark and Kandel, 2013). A decline in dietary breadth occurs in MIS 3, with a return of the dominance of large and very large-bodied bovids (Clark and Kandel, 2013). Technological innovations relating to the MIS 4 dietary shift may include remote capture methods such as snares (Wadley, 2013) and bow and arrow technology at Sibudu (Lombard and Phillipson, 2010); procedures that imply complexity in cognition (Lombard and Haidle, 2012). A commonly discussed theme in the MSA, particularly regarding MIS 4, is whether there is a cause-and-effect relationship between cultural and environmental change. It is telling that, at Sibudu, changes in lithic technology from Howiesons Poort to post-Howiesons Poort precede changes in the environment as reflected by the proportional representation of habitat-sensitive species (Clark, 2013). Whether prey choice in the MSA and LSA, particularly the Holocene, reflect hunting skill is another issue of interest. Even though less dangerous species were sometimes exploited in the MSA compared to the LSA (e.g. Klein and Cruz-Uribe, 1996), similar types of ungulate prey were targeted (Faith, 2008). Resource availability and perhaps ranking of food sources rather than capabilities may have affected subsistence behaviour (Deacon, 1989). Holocene and MSA hunter-gatherers occupied different niches, as environmental changes at the Pleistocene-Holocene transition compelled an expansion of dietary breadth and more intensive exploitation of food sources such as shellfish and tortoise (Clark and Kandel, 2013).

The first evidence for shellfish in MSA diets is at 164 kyr BP at Pinnacle Point Cave 13B, although only in very low quantities. It became a more substantial, but never dominant, element of subsistence behaviour for the remainder of the MSA at coastal sites in the southern and western Cape. Proximity to the sea strongly influenced shellfish collecting strategies: these were aimed at optimising distance to the resource and their yield (Langejans et al., 2012). Fruits, seeds, nuts and geophytes must have also formed a substantial part of the MSA diet, but palaeobotanical data are limited compared to faunal and shellfish evidence. Nonetheless, at Sibudu, the analysis of plant remains such as burnt grass, sedges, charcoal, carbonised seeds, pollens, phytoliths and plant residues are the source of the most informative new perspectives on MSA behaviour (e.g. Sievers and Muasya, 2011) and climate (Bruch et al., 2012). It has, for example, been determined that sedges and other grasses were used for bedding at Sibudu at ~77 kyr BP. This is the oldest example of this practice yet identified (Wadley et al., 2011), and becomes more prevalent in the post-Howiesons Poort levels. It has also been shown that wood and leaves were used for their insecticidal properties. Aromatic leaves at Sibudu were laid on the sedge bedding, and in the post-Howiesons Poort, Tamboeti wood smoke may also have been used for this purpose (Wadley, 2013). Recent research on pyrotechnology through micromorphological and experimental methods reveals unexpected levels of behavioural sophistication (Goldberg et al., 2009; Bentsen, 2014) in the Howiesons Poort and post-Howiesons Poort. For example, hearths were maintained and altered, ash dumps formed, and bedding was burnt.

22.4 Discussion and summary

The clearest pattern in the MSA across southern Africa, in particular during MIS 4, is the apparent increase in bone tools and ornaments, and the homogeneity of stone tool technology. This may be a reflection of larger population sizes, or interactions between populations across southern Africa (Mackay et al., 2014, but see Klein and Steele, 2013). It may also be an artefact of research intensity, as the trends discussed here rely on results from relatively recent investigations from a handful of sites. It is a priority to expand the database on the MSA in order to more fully understand landscape utilisation, demographic patterns, and the geographical extent of technological conventions and innovations. Whether homogeneous developments in stone and bone tool technologies occurred on a subcontinental scale, as smaller-scale regional pulses, or as idiosyncratic expressions, will largely remain unknown until more data are available. The development of innovative behaviours in the MSA, especially in MIS 4, has been linked to climatic changes (e.g. Ziegler et al., 2013). However, neither the resolution of palaeoenvironmental (Mackay et al., 2014) nor archaeological data is sufficient to determine which occurred first.

Among the crucial evolutionary events in the early MSA are the development of anatomical modernity and complex stone tool reduction and hafting procedures. In MIS 5 the pace of cultural change as reflected in bone and stone tool, ochre and engraving technologies seems to increase, culminating in MIS 4. It is also mostly in MIS 4 that an increase in diet breadth is recorded, while faunal and shellfish exploitation patterns generally remain stable. Recent advances in experimentation and lab analysis have also provided insight into MSA cultures and practices, including weaponry, pyrotechnology, site maintenance, ochre, and plant use. These advances are deepening our understanding of the MSA and the relationship of MSA peoples to their local environment and resources, and to regional-scale Quaternary climate and environmental changes.

Acknowledgements

My thanks go to Magnus Haaland for help with the map, Petro Keene for editorial assistance and Liezl van Pletzen Vos for Figure 22.5. Two anonymous reviewers and Jasper Knight are thanked for constructive comments. This work is based on the research supported by the National Research Foundation. Any opinion, finding and conclusion or recommendation expressed in this material is that of the author and the NRF does not accept any liability in this regard.

References

Bentsen, S. E. (2014). Using Pyrotechnology: Fire-related features and activities with a focus on the African Middle Stone Age. *Journal of Archaeological Research*, 22, 141–175.

Brink, J. S. (1999). Preliminary report on a caprine from the Cape mountains, South Africa. *Archaeozoologia*, 10, 11–26.

Brown, K., Marean, C. W., Herries, A. I. R., Jacobs, Z., Tribolo, C., Braun, D., Roberts, D. L., Meyer M. C. and Bernatchez, J. (2009). Fire as an engineering tool of early modern humans. *Science*, 325, 859–862.

Bruch, A. A., Sievers, C. and Wadley, L. (2012). Quantification of climate and vegetation from southern African Middle Stone Age sites – an application using Late Pleistocene plant material from Sibudu, South Africa. *Quaternary Science Reviews*, 45, 7–17.

Clark, J. L. (2013). Exploring the relationship between climate change and the decline of the Howieson's Poort at Sibudu Cave (South Africa). In *Zooarchaeology and Modern Human Origins, Human Hunting Behavior during the Later Pleistocene*, eds. J. D. Clark and J. D. Speth. New York: Springer, pp. 9–18.

Clark, J. L. and Kandel, A. W. (2013). The Evolutionary Implications of Variation in Human Hunting Strategies and Diet Breadth during the Middle Stone Age of Southern Africa. *Current Anthropology*, 54 (S8), S269–S287.

Deacon, H. J. (1989). Late Pleistocene palaeoecology and archaeology in the southern Cape, South Africa. In *The Human Revolution, Behavioural and Biological Perspectives on the Origins of Modern Humans*, eds. P. Mellars and C. Stringer. Edinburgh: Edinburgh University Press, pp. 547–564.

d'Errico, F., Backwell, L., Villa, P., Degano, I., Lucejko, J. J., Bamford, M. K., Higham, T. F.G., Colombini, M. P. and Beaumont, P. B. (2012a). Early evidence of San material culture represented by organic artifacts from Border Cave, South Africa. *Proceedings of the National Academy of Sciences*, 109, 13214–13219.

d'Errico, F., Backwell, L. and Wadley, L. (2012b). Identifying regional variability in Middle Stone Age bone technology. The case of Sibudu Cave. *Journal of Archaeological Science*, 39, 2479–2495.

Dusseldorp, G. L., Lombard, M. and Wurz, S. (2013). Pleistocene *Homo* and the updated Stone Age sequence of South Africa. *South African Journal of Science*, 109, 46–52.

Faith, J. T. (2008). Eland, buffalo, and wild pigs, were Middle Stone Age humans ineffective hunters? *Journal of Human Evolution*, 55, 24–36.

Faith, J. T. (2011). Ungulate community richness, grazer extinctions, and human subsistence behavior in southern Africa's Cape Floral Region. *Palaeogeography, Palaeoclimatology, Palaeoecology*, 306, 219–227.

Goldberg, P., Miller, C. E., Schiegl, S., Ligouis, B., Conard, N. J. and Wadley, L. (2009). Bedding, hearths and site maintenance in the Middle Stone Age at Sibudu, KwaZulu-Natal, South Africa. *Archaeological and Anthropological Sciences*, 1, 95–122.

Henshilwood, C. S. (2012). Late Pleistocene Techno-traditions in Southern Africa, A Review of the Still Bay and Howiesons Poort, c. 75–59 ka. *Journal of World Prehistory*, 25, 205–237.

Henshilwood, C. S., d'Errico, F., van Niekerk, K. L., Coquinot, Y., Jacobs, Z., Lauritzen, S.E., Menu, M. and García-Moreno, R. (2011). A 100,000-Year-Old Ochre-Processing Workshop at Blombos Cave, South Africa. *Science*, 334, 219–222.

Henshilwood, C. S. and Dubreuil, B. (2011). The Still Bay and Howiesons Poort, 77–59 ka. *Current Anthropology*, 52, 361–400.

Henshilwood, C. S. and Marean, C. W. (2003). The origin of modern human behavior, A critique of the models and their test implications. *Current Anthropology*, 44, 627–651.

Henshilwood, C. S., van Niekerk, K. L., Wurz, S., Delagnes, A., Armitage, S., Rifkin, R., Douze, K., Keene, P., Haaland, M., Reynard, J., Discamps, E. and Mienies, S. (2014). Klipdrift Shelter, southern Cape, South Africa, preliminary report on the Howiesons Poort levels. *Journal of Archaeological Science*, 45, 284–303.

Hiscock, P. (2014). Learning in lithic landscapes, a reconsideration of the hominid 'toolmaking' niche. *Biological Theory*, 9, 27–41.

Klein, R. G. (1983). Palaeoenvironmental implications of Quaternary large mammals in the fynbos region. In *Fynbos Palaeoecology; A Preliminary Synthesis*, eds. H. J. Deacon, Q. B Hendey and J. J. N. Lambrechts. Cape Town: South African National Scientific Programmes Report, 75, pp. 116–138.

Klein, R. G. and Cruz-Uribe, K. (1996). Exploitation of large bovids and seals at Middle and Later Stone Age sites in South Africa. *Journal of Human Evolution*, 31, 315–334.

Klein, R. G. and Steele, T. E. (2013). Archaeological shellfish size and later human evolution in Africa. *Proceedings of the National Academy of Sciences*, 110, 10910–10915.

Langejans, G. H. J., van Niekerk, K. L., Dusseldorp, G. L. and Thackeray, F. (2012). Middle Stone Age shellfish exploitation, potential indications for mass collecting and resource intensification at Blombos Cave and Klasies River, South Africa. *Quaternary International*, 270, 80–94.

Lombard, M. and Haidle, M. N. (2012). Thinking a bow-and-arrow set, cognitive implications of Middle Stone Age bow and stone-tipped arrow technology. *Cambridge Archaeological Journal*, 22, 237–264.

Lombard, M. and Phillipson, L. (2010). Indications of bow and stone-tipped arrow use 64,000 years ago in KwaZulu-Natal. *Antiquity*, 84, 635–648.

Lombard, M., Schlebusch, C. and Soodyall, H. (2013). Bridging disciplines to better elucidate the evolution of early *Homo sapiens* in southern Africa. *South African Journal of Science*, 109, Art. #2013-0065, doi:10.1590/sajs.2013/20130065.

Lombard, M., Wadley, L., Deacon, J., Wurz, S., Parsons, I., Mohapi, M., Swart, J., Mitchell, P. (2012). South African and Lesotho Stone Age sequence updated (I). *South African Archaeological Bulletin*, 67, 123–144.

Mackay, A., Stewart, B. A. and Chase, B. M. (2014). Coalescence and fragmentation in the late Pleistocene archaeology of southernmost Africa. *Journal of Human Evolution*, 72, 26–51.

McBrearty, S. and Brooks, A. S. (2000). The revolution that wasn't, A new interpretation of the origin of modern human behavior. *Journal of Human Evolution*, 39, 453–563.

Mitchell, P. (2008). Developing the archaeology of Marine Isotope Stage 3. *South African Archaeological Society Goodwin Series*, 10, 52–65.

Mourre, V., Villa, P. and Henshilwood, C. S. (2010). Early use of pressure flaking on lithic artifacts at Blombos Cave, South Africa. *Science*, 330, 659–662.

Porraz, G., Parkington, J. E., Rigaud, J.-P., Miller, C. M., Poggenpoel, C., Tribolo, C., Archer, W., Cartwright, C. R., Charrié-Duhaut, A., Dayet, L., Igreja, M., Mercier, N., Schmidt, P., Verna, C. and Texier, P.-J. (2013). The MSA sequence of Diepkloof and the history of southern African Late Pleistocene populations. *Journal of Archaeological Science*, 40, 3542–3552.

Rots, V. and Plisson, H. (2014). Projectiles and the abuse of the use-wear method in a search for impact. *Journal of Archaeological Science*, 48, 154–165.

Schmidt, P., Porraz, G., Slodczyk, A., Bellot-Gurlet, L., Archer, W. and Miller, C. E. (2013). Heat treatment in the South African Middle Stone Age, temperature induced transformations of silcrete and their technological implications. *Journal of Archaeological Science*, 40, 3519–3531.

Sievers, C. and Muasya, A. M. (2011). Identification of the sedge *Cladium mariscus* subsp. jamaicense and its possible use in the Middle Stone Age at Sibudu, KwaZulu-Natal. *Southern African Humanities*, 23, 77–86.

Soriano, S., Villa, P. and Wadley, L. (2007). Blade technology and tool forms in the Middle Stone Age of South Africa: The Howiesons Poort and post-Howiesons Poort at Rose Cottage Cave. *Journal of Archaeological Science*, 34, 681–703.

Villa, P., Soriano, S., Tsanova, T., Degano, I., Higham, T. F. G., d'Errico, F., Backwell, L., Lucejko, J. J., Colombini, M. P. and Beaumont, P. B. (2012). Border Cave and the beginning of the Later Stone Age in South Africa. *Proceedings of the National Academy of Sciences*, 109, 13208–13213.

Wadley, L. (2013). Recognizing Complex Cognition through Innovative Technology in Stone Age and Palaeolithic Sites. *Cambridge Archaeological Journal*, 23, 163–183.

Wadley, L. and Prinsloo, L. C. (2014). Experimental heat treatment of silcrete implies analogical reasoning in the Middle Stone Age. *Journal of Human Evolution*, 70, 49–60.

Wadley, L., Sievers, C., Bamford, M., Goldberg, P., Berna, F. and Miller, C. (2011). Middle Stone Age bedding construction and settlement patterns at Sibudu, South Africa. *Science*, 334, 1388–1391.

Wilkins, J., Schoville, B. J., Brown, K. S. and Chazan, M. (2012). Evidence for early hafted hunting technology. *Science*, 338, 942–946.

Wurz, S. (2013). Technological trends in the Middle Stone Age of South Africa between MIS 7 and MIS 3. *Current Anthropology*, 54 (S8), S305–S319.

Ziegler, M., Simon, M. H., Hall, I. R., Barker, S., Stringer, C. and Zahn, R. (2013). Development of Middle Stone Age innovation linked to rapid climate change. *Nature Communications*, 4, 1905, doi:10.1038/ncomms2897.

23

Later Stone Age hunter-gatherers and herders

PETER MITCHELL

Abstract

No sharp divide exists between the makers of Later Stone Age artefacts and their Middle Stone Age (MSA) predecessors. When initially coined, only the former was paired with anatomically modern people and material culture resembling that of contemporary Bushmen, but both associations now extend back into the MSA. The phrase 'Later Stone Age' nevertheless remains useful as a ready reference for the hunter-gatherers (and stone-using pastoralists) who inhabited southern Africa over the past 25,000 years. This chapter provides an overview of what archaeologists have learned about these people over this period and identifies several areas of current research.

23.1 Introduction

How, when and why southern African hunter-gatherers abandoned the Middle Stone Age (MSA) toolkits that had served them for millennia is poorly understood. Almost certainly, though, there was no single or straightforward pathway. At Border Cave in KwaZulu-Natal, for example (Fig. 23.1), a quartz-dominated microlithic industry was present by 38 kyr BP (Villa et al., 2012), but in Zimbabwe's Matopo Hills prepared core techniques may have been used until almost the end of the Pleistocene (Walker, 1995). In many sequences, however, the replacement of the MSA by the Later Stone Age (LSA) appears to have been finally accomplished around the onset of Marine Isotope Stage (MIS) 2 (~25 kyr BP), which encourages us to explore links with the change to a colder and, for most of southern Africa, drier climate (Mackay and Marwick, 2011).

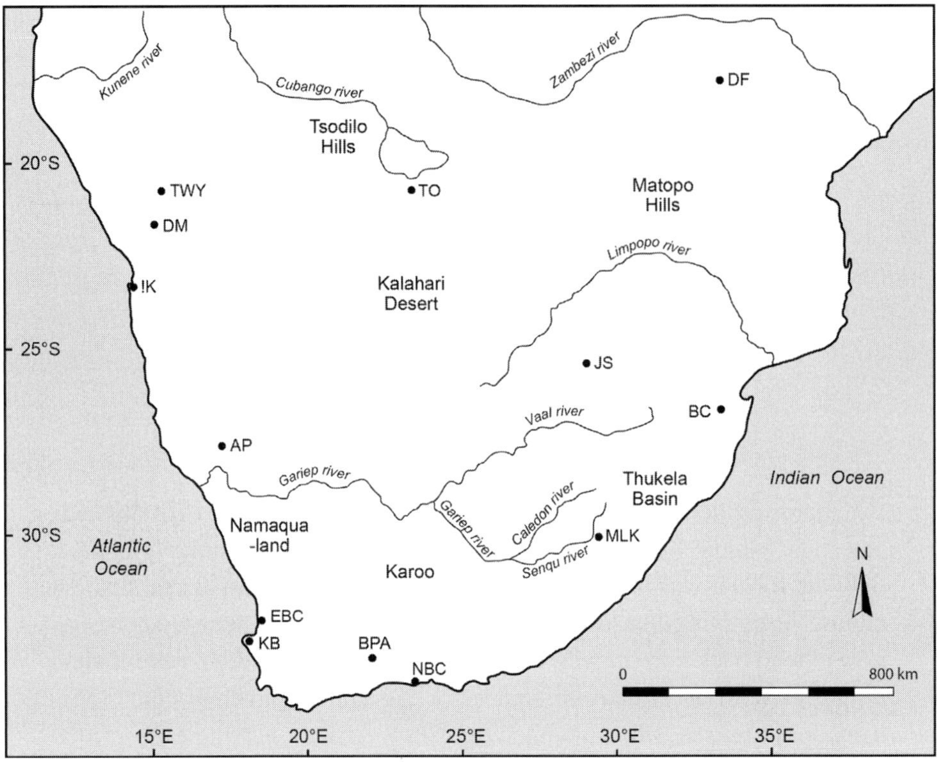

Fig. 23.1. Location of archaeological sites and places named in the text. Sites are: AP – Apollo 11 Cave; BC – Border Cave; BPA – Boomplaas Cave; DF – Duncombe Farm; DM – Dâures Massif; EBC – Elands Bay Cave; JS – Jubilee Shelter; !K – !Kuiseb Delta; KB – Kasteelberg; MLK – Melikane; NBC – Nelson Bay Cave; TO – Toteng; TWY – Twyfelfontein.

In southernmost Africa (South Africa, Lesotho and Swaziland), the earliest LSA manifestations were microlithic, made extensive use of bipolar flaking, and employed few formal tools. Most date to the period 20–25 kyr BP. Better defined is the bladelet-rich but otherwise still informal Robberg Industry, found principally in and coastward of the Cape Fold Mountain Belt and the Maloti-Drakensberg region from roughly 20–11/12 kyr BP; later survival in southeastern southern Africa is questionable as it hinges on just a few conventional radiocarbon dates. While most Robberg assemblages come from rockshelters, their recent recognition in open-air sites in the Namaqualand region extends their distribution (Orton, 2008). North of the Limpopo and Orange rivers, on the other hand, artefact assemblages were either non-microlithic (Wendt, 1972), or with fewer bladelets recorded (Campbell et al., 2010) (Fig. 23.2). This distinction may imply minimal contacts between populations in southernmost Africa and others farther north during MIS 2, consistent with genetic evidence (Pickrell et al., 2012) and the lack

Fig. 23.2. The Male Hill, Tsodilo Hills, northern Botswana. This is the largest of the three hills in the Tsodilo Hills, a major centre for archaeological research in the Kalahari that includes numerous rock art sites. Excavations here have documented the earliest evidence for freshwater fishing in southern Africa and also preserve important palaeoenvironmental archives showing that conditions have repeatedly been significantly wetter than present (photo reproduced with permission of David Nash).

of discernible connection between the Kx'a languages of the northern Kalahari and southeastern Botswana and the Tuu family to its south (Güldemann, 2008). How far these societies differed from their Holocene successors is uncertain: hunting large gregarious animals was certainly more important in the Cape region during the late Pleistocene, but this reflects ecological change from more open grassland to more closed fynbos vegetations as much as dietary choice. If bone arrowpoints and ostrich eggshell beads were used in gift-exchange (Wadley, 1987), then their rarity may indicate that at lower population densities people de-emphasised the strategies of risk management and information-sharing bound up in the gift-exchange systems observed by ethnographers.

23.2 The Pleistocene/Holocene transition

Around 12 kyr BP, Robberg toolkits were replaced by a different technocomplex, the Oakhurst, which lacked bladelets and was not microlithic. Oakhurst assemblages also occur in areas where LSA populations appear to have been largely absent during earlier millennia; they are common, for instance, at open-air locations in South Africa's Karoo and Free State, probably reflecting a genuine growth in human numbers as rainfall (and thus ecological productivity) increased in the terminal Pleistocene. Greater connectivity across southern Africa is implied by the Oakhurst's presence in Namibia and Zimbabwe, as well as in South Africa, Lesotho and Swaziland; this could have been facilitated by the expanded exchange ties perhaps signalled by increased numbers of bone points and beads (Wadley, 1987).

An additional factor in this picture of territorial expansion was the drowning of the continental shelf as sea levels rose at the end of the Pleistocene (see Compton, this volume). Its effects are evident in several archaeological sequences, perhaps most clearly at Elands Bay Cave (EBC), ~200 km north of Cape Town (Fig. 23.3). Although it is today a coastal site, at the Last Glacial Maximum EBC was located some 20 km inland. Well-preserved deposits spanning the Pleistocene/Holocene transition capture its shift from one situation to another. Like many other now-coastal sites, this is most visible in the masses of shell middens people left behind,

Fig. 23.3. Elands Bay Cave, Western Cape, South Africa. The view outwards to the Atlantic Ocean would have been dry land for much of the Pleistocene and the deposits here track the site's change from inland to coastal in location (photo reproduced with permission of G. Porraz). A black and white version of this figure will appear in some formats. For the colour version, please refer to the plate section.

but at EBC fish and rock lobsters (*Iasus lalandii*) also became important, while large mammals assumed a less prominent role. Massive increases in depositional rates, the appearance of graves, including those of women and children, and finds of items associated ethnographically with female labour (bored stone digging stick weights used to harvest underground plant foods; evidence of ostrich eggshell bead making) collectively signal an accompanying social transition from an ephemerally-used hunting location to one where entire groups of people lived for extended periods (Parkington, 1988).

23.3 The Holocene Later Stone Age: toolkits, diet, and society

The EBC sequence is interrupted in the mid-Holocene by a long period during which people did not visit the site, although they were certainly present in the broader region (Jerardino et al., 2013). During this interval, a further technological transformation saw the adoption of a new kind of microlithic technology focused on producing highly standardised formal tools, especially backed microliths and small, thumbnail-shaped scrapers. Evident as early as 10.6 kyr BP in northern Zimbabwe, this Wilton technocomplex was subsequently adopted across southern Africa along a broadly north-to-south and – within the Cape – east-to-west trajectory that was completed before 6 kyr BP (Mitchell, 2002).

In many areas (for example Lesotho), Wilton toolkits were used until the final disappearance of hunter-gatherer populations in the 19th Century AD. Changes in these toolkits included the kinds of backed microliths favoured and the size and design of scrapers. However, increased production of adzes (largely used to shape wooden artefacts, such as digging sticks) is clear from many sites within the last 4000 years. Wilton assemblages did not, however, continue everywhere, and in some areas distinctly non-microlithic industries emerged, contributing to a pattern of growing inter-regional variation over time (Table 23.1).

How people acquired food has been one major theme of southern African hunter-gatherer research. On both ecological and nutritional grounds, plants likely formed the staple resource suite, with geophytes (like *Watsonia* spp.) being key in the Fynbos and Grassland Biomes, while fruits and nuts (such as marula; *Sclerocarya birrea*) were more important in areas of savanna vegetation, and resources like !nara melons (*Acanthosicyos horridus*) were extensively used in the Namib desert. The scarcity of plants in Pleistocene deposits is almost certainly an artefact of preservation bias (Deacon, 1993), but they may genuinely have become more important in the second half of the Holocene, when increased use of small-sized animal foods (shellfish, crustaceans, fish and even rodents) is also attested in some areas, notably the Thukela Basin (Mazel, 1989) and the Fynbos Biome. There, as well as in the Forest Biome, several indicators (including stable isotope studies of

Table 23.1 *Summary of the major technocomplexes and lithic industries recognised in southern Africa over the past 25,000 years (based principally on Wendt, 1972; Walker, 1995; Campbell et al., 2010; Lombard et al., 2012).*

Technocomplex and approximate time periods	Characteristics	Distribution (with named regional variants)
Ceramic LSA ≤2 kyr BP Note that the Smithfield Industry dates to ≤0.6 kyr BP	Highly variable, but with the incorporation of pottery, itself of many different styles. Some assemblages clearly show continuity from the Wilton and emphasise backed bladelets (e.g. Swartkops). Others have few backed microliths but many larger, endstruck scrapers and a tendency towards larger flakes (Smithfield), or are relatively informal (e.g. Doornfontein)	Distributed across southern Africa. The Swartkops and Doornfontein Industries are confined to the Northern Cape Province, the Smithfield to the eastern Karoo and Free State. Walker (1995) uses the term Ceramic Matopan in the Matopo Hills of Zimbabwe. Some assemblages are associated with pastoralists, or perhaps 'hunters-with-sheep' (Sadr, 1998), rather than with hunter-gatherers
Final LSA ≤4 kyr BP	Highly variable, but without pottery. Generally microlithic, but with segments rare among backed microliths. Some assemblages are not microlithic, but flake-based and with few formal tools, other than scrapers. The Kabeljous Industry is quartzite-dominated and macrolithic and includes rare large segments and scrapers	Distributed across southern Africa; known as Amadzimba in the Matopo Hills of Zimbabwe. The Kabeljous Industry occurs in the Forest Biome of the Eastern Cape Province in both coastal and inland settings
Wilton c. 8–4 kya south of the Limpopo/Gariep Rivers, but found earlier in southern Namibia (9.4 kyr BP) and Zimbabwe (from as early as 10.5 kyr BP)	Microlithic assemblages, showing enhanced use of fine-grained rocks and proliferation of small, standardised scrapers and backed microliths, initially segments, but later also backed bladelets/points	Distributed across southern Africa; known as Nswatugi in the Matopo Hills of Zimbabwe

Oakhurst c. 12–8 kyr BP in most areas, but persisting to as late as 7/6.5 kyr BP in some parts of the South African interior.	Non-microlithic assemblages, often emphasising more coarsely grained rocks. Bladelets absent. Formal tools rare, other than large scrapers. Scrapers with adze-like lateral retouch (Woodlot scrapers) common in some areas ≤9.5 kyr BP	Distributed in the Matopo Hills of Zimbabwe (Pomongwan); central/southern Namibia, South Africa, Lesotho, Swaziland (Known as Lockshoek in the Karoo/Free State, Kuruman in the Northern Cape Province and as Albany in the southern Cape)
Late Pleistocene LSA		
Robberg c. 19–11/12 kyr BP	Microlithic assemblages with few formal tools, but rich in unmodified bladelets and bladelet cores, some of specialised kinds, preferentially made in fine-grained rocks.	South Africa and Lesotho (but confined to Namaqualand, the Cape Fold Mountain Belt, the Maloti-Drakensberg region and Free State and the Mpumalanga Escarpment)
Early LSA c. 19–12 kyr BP	Macrolithic assemblages with few if any bladelets and few formal tools save scrapers	Central/southern Namibia
Unnamed c. 19–12 kyr BP	Microlithic assemblages with some bladelets and backed microliths	Tsodilo Hills, northwestern Botswana, but note that a possibly similar assemblage occurs at Duncombe Farm, north-central Zimbabwe
Maleme c. 13–12 kyr BP	Includes some triangular points and use of prepared core techniques, but it is unclear if this is a genuine association or derives from stratigraphic mixing	Matopo Hills, Zimbabwe
Early LSA c. 20–25 kyr BP, but perhaps older. Note that at Border Cave and in the Tsodilo Hills this kind of assemblage is dated to ~38 kyr BP and ~48–31 kyr BP respectively	Microlithic, informal, quartz-dominated assemblages with extensive use of bipolar flaking	South Africa at sites in the Cape Fold Mountain Belt, the Mpumalanga Escarpment, the Magaliesberg (e.g. Jubilee Shelter), KwaZulu-Natal (e.g. Border Cave) and sites in Botswana's Tsodilo Hill

human remains) point to increasingly dense occupation by groups who (1) used artefact styles and raw material choices to distinguish themselves from adjacent groups; (2) emphasised burial at specific locations to assert claims to the land; (3) practised reduced levels of mobility; and (4) followed divergent subsistence strategies (Hall, 2000; Sealy, 2006; Dewar, 2010). How far these changes, which included storing some plant foods, can be understood as instances of hunter-gatherer intensification, remains uncertain, as does the degree of coastal sedentism (rather than movement between the seashore and areas farther inland) practised along the Atlantic coast near EBC (Jerardino et al., 2013). The absence of readily domesticable plants and animals, however, or of wild resources as productive as the acorns and salmon runs of western North America, placed an ecological constraint on the emergence of social formations that were radically different from those attested ethnographically.

Other research has pursued more avowedly social aims, using contemporary (San) Bushman ethnography to suggest how archaeologists might identify practices like community aggregation and dispersal or reciprocal gift-exchange (often termed *hxaro*, from its !Kung name) (Wadley, 1987). Ethnographic insights from the Kalahari and 19th Century sources in South Africa and Lesotho (principally those relating to the now-extinct /Xam Bushmen of the Northern Cape) have also been vital for understanding hunter-gatherer rock art. Many of the thousands of painted and engraved sites across southern Africa (Fig. 23.4) express concepts relating to the power of shamans to cure sickness, control animals, make rain and move between different tiers of the cosmos (Lewis-Williams and Pearce, 2004). Other relevant perspectives include gender, initiation, weather control and animal behaviour, but the applicability of creation mythology or ideas about the spirits of the dead remains hotly debated (Solomon, 2013). There is nevertheless broad agreement on the desirability of exploring regional variation and employing a wider range of historically contextualised ethnographies (Smith, 2010; Solomon, 2011), as well as accurately dating the art itself to identify changes within it and forge connections with other components of the archaeological record (Bonneau et al., 2011).

23.4 Later Stone Age herders

Domestic livestock began entering southern Africa in the late first millennium BC (Pleurdeau et al., 2012). Sheep reached the Fynbos Biome by the 1st/2nd Centuries AD and co-occur with cattle about that time at Toteng, northern Botswana; the latter were in Namaqualand c. AD 500, roughly the time at which dung deposits inside Boomplaas Cave suggest sheep were kept there in large numbers (Orton et al., 2013). Debate continues as to how these dispersals took place and their

Fig. 23.4. Rock painting near Maclear, Eastern Cape, South Africa (image RARI-RSA-SPR3-7D, reproduced with permission from Rock Art Research Institute, University of the Witwatersrand, South Africa). A black and white version of this figure will appear in some formats. For the colour version, please refer to the plate section.

relationship to the adoption of pottery and the spread of the Khoe language family, one variant of which was spoken by the Khoekhoen, the herders whom Europeans encountered at the Cape from the late 15th Century AD. Some hunter-gatherers may have acquired livestock without significantly altering the social and ritual aspects of their lives, becoming 'herders-with-sheep' (Sadr, 1998). However, the numbers of animals needed for long-term survival of sheep flocks and cattle herds may indicate that a fuller commitment to pastoralism was present early on, supporting the case for a herder population migrating into westernmost South Africa (Smith, 2006). Recent genetic and linguistic studies sketch a complex scenario, in which herders of East African origin first brought livestock to southern Africa's far north and then spread south, increasingly intermarrying with aboriginal hunter-gatherers; Khoe likely developed as part of this process (Güldemann, 2008; Henn et al., 2010). This model does not exclude southward dispersal of the ancestors of the historic Khoekhoen around the end of the first millennium AD (Sadr, 1998), followed by development of a more fully pastoral economy marked by significantly increased milk/beef consumption in the southern Cape (Sealy, 2010). However, beyond the extensively investigated site complex at Kasteelberg

(Smith, 2006), the many stone stock enclosures of the eastern Karoo (Sampson, 2010) and key Namibian foci such as the !Kuiseb delta and Dâures massif (Kinahan, 2001), unambiguous pastoralist signatures remain difficult to identify.

23.5 Hunter-gatherers and farmers

Pastoralists were not the only newcomers to the southern African landscape in the last 2000 years. In the mid-first millennium AD, farmers who grew cereals, kept livestock, used metal and spoke various Bantu languages settled the woodland savannas of the north and east. Their mixed economy offered greater scope for exchange with indigenous hunter-gatherers than was possible for pastoralists in western regions of southern Africa, for whom wild plant foods, game and marine resources remained extremely important. Archaeological finds from KwaZulu-Natal and Lesotho suggest that farmers traded iron, pottery and occasional livestock to foragers in return for items like ostrich eggshell beadwork and arrows (Mazel, 1989); complex interactions are also postulated for the northern Kalahari (Wilmsen and Denbow, 1990; cf. Sadr, 1997). Genetic data confirm significant intermarriage with ancestral Bantu-speaking populations, but when this happened (1st and/or 2nd millennia AD) is uncertain and it is much less evident in Mozambique and Zimbabwe (Pickrell et al., 2012); cultural and linguistic influences are also documented, especially for Nguni.

Overall, hunter-gatherers survived longest in areas farmers found unattractive because of either cold (highland Lesotho) or aridity (the Kalahari and Karoo). Elsewhere, hypergamic marriage, unequal demographic growth, loss of resources and the attractions of farmer material culture undercut the basis of hunter-gatherer social and biological reproduction (Wadley, 1996). Rock art captures something of these changes, with livestock becoming sources of ritual potency in many areas and Bushman shamans perhaps making rain for agriculturalist communities. After AD 1652, more aggressive European settlement eventually sounded the death knell for an independent forager existence in southernmost Africa: new technologies allowed farmers to expand into dryland environments, 19th Century political upheavals favoured Basotho colonisation of Lesotho's highlands and Europeans found little use for foragers or pastoralists except as a proletarianised workforce (Mitchell, 2002). Hunter-gatherers and Khoe-speaking herders survived principally in the Kalahari and Namaqualand, where their descendants continue to provide insights into southern Africa's LSA archaeological record.

23.6 Summary

Despite criticism, the phrase 'Later Stone Age' remains useful, emphasising the microlithic character of most of the stone tool industries grouped within it, the more consistent association with those industries of Bushman-like material culture and the appropriateness of cautiously using Bushman ethnography to help understand the ways of life and worldviews of their makers. Future work can be expected to seek greater integration of the excavated and rock art components of the archaeological record, to make greater use of newly available genetic data and to continue exploring past social relations using ethnographic analogies from both southern Africa and farther afield.

References

Bonneau, A., Brock, F., Higham, T. F. G., Pearce, D. G. and Pollard, A. M. (2011). An improved pretreatment protocol for radiocarbon dating black pigments in San rock art. *Radiocarbon*, 53, 419–428.

Campbell, A., Robbins, L. and Taylor, M. (eds.) (2010). *Tsodilo Hills: Copper Bracelet of the Kalahari*. Gaborone: The Botswana Society, 178pp.

Deacon, H. J. (1993). Planting an idea: an archaeology of Stone Age gatherers in South Africa. *South African Archaeological Bulletin*, 48, 86–93.

Dewar, G. (2010). Late Holocene burial cluster at Diaz Street midden, Saldanha Bay, Western Cape, South Africa. *South African Archaeological Bulletin*, 65, 26–34.

Güldemann, T. (2008). A linguist's view: Khoe-Kwadi speakers as the earliest food-producers of southern Africa. *Southern African Humanities*, 20, 93–132.

Hall, S. L. (2000). Burial sequence in the Later Stone Age of the Eastern Cape Province, South Africa. *South African Archaeological Bulletin*, 55, 137–146.

Henn, B. M., Gignoux, C., Lin, A. A., Oefner, P. J., Shen, P., Scozzari, R., Cruciani, F., Tishkoff, S. A., Mountain, J. L. and Underhill, P. A. (2010). Y-chromosomal evidence of a pastoralist migration through Tanzania to southern Africa. *Proceedings of the National Academy of Sciences*, 105, 10693–10698.

Jerardino, A., Klein, R. G., Navarro, R., Orton, J. and Horwitz, L. K. (2013). Settlement and subsistence patterns since the terminal Pleistocene in the Elands Bay and Lamberts Bay area. In *The Archaeology of the West Coast of South Africa*, eds. A. Jerardino, A. Malan and D. Braun. Oxford: Archaeopress, pp. 85–108.

Kinahan, J. (2001). *Pastoral Nomads of the Namib Desert: The People History Forgot*. Windhoek: Namibia Archaeological Trust, 167pp.

Lewis-Williams, J. D. and Pearce, D. G. (2004). *San Spirituality: Roots, Expression and Social Consequences*. Walnut Creek: AltaMira Press, 304pp.

Lombard, M., Wadley, L., Deacon, J., Wurz, S., Parsons, I., Mohapi, M., Swart, J. and Mitchell, P. J. (2012). South African and Lesotho Stone Age sequence updated (I). *South African Archaeological Bulletin*, 67, 120–144.

Mackay, A. and Marwick, B. (2011). Costs and benefits in technological decision making under variable conditions: Examples from the late Pleistocene in southern Africa. In *Keeping Your Edge: Recent Approaches to the Organisation of Stone Artefact Technology*, eds. B. Marwick and A. Mackay. Oxford: Archaeopress, pp. 119–134.

Mazel, A. D. (1989). People making history: The last ten thousand years of hunter-gatherer communities in the Thukela Basin. *Natal Museum Journal of Humanities*, 1, 1–168.

Mitchell, P. J. (2002). *The Archaeology of Southern Africa*. Cambridge: Cambridge University Press, 532pp.

Orton, J. (2008). A late Pleistocene microlithic Later Stone Age assemblage from coastal Namaqualand, South Africa. *Before Farming*, 2008/1, article 3, doi:10.3828/bfarm.2008.1.3.

Orton, J., Mitchell, P. J., Klein, R. G., Steele, T. and Horsburgh, K. A. (2013). Early dates for cattle from Namaqualand, South Africa: Implications for the origins of herding in southern Africa. *Antiquity*, 87, 108–120.

Parkington, J. E. (1988). The Pleistocene/Holocene transition in the Western Cape, South Africa: Observations from Verlorenvlei. In *Prehistoric Cultures and Environments in the Late Quaternary of Africa*, eds. J. Bower and D. Lubell. Oxford: British Archaeological Reports, pp. 349–363.

Pickrell, J. K., Patterson, N., Barbieri, C., Berthold, F., Gerlach, L., Güldemann, T., Kure, B., Mpoloka, S. W., Nakagawa, H., Naumann, C., Lipson, M., Loh, P. R., Lachance, J., Mountain, J., Bustamente, C. D., Berger, B., Tishkoff, S. A., Henn, B. M., Stoneking, M., Reich, D. and Pakendorf, B. (2012). The genetic prehistory of southern Africa. *Nature Communications*, 3, doi:10.1038/ncomms2140.

Pleurdeau, D., Imalwa, E., Détroit, F., Lesur, J., Veldman, A., Bahain, J.-J. and Marais, E. (2012). "Of sheep and men": Earliest direct evidence of caprine domestication in southern Africa at Leopard Cave (Erongo, Namibia). *PLoS ONE*, 7 (7), e40340, doi:10.1371/journal.pone.0040340.

Sadr, K. (1997). Kalahari archaeology and the Bushman debate. *Current Anthropology*, 38, 104–112.

Sadr, K. (1998). The first herders at the Cape of Good Hope. *African Archaeological Review*, 15, 101–132.

Sampson, C. G. (2010). Chronology and dynamics of Later Stone Age herders in the upper Seacow River valley, South Africa. *Journal of Arid Environments*, 74, 842–848.

Sealy, J. C. (2006). Diet, mobility, and settlement pattern among Holocene hunter-gatherers in southernmost Africa. *Current Anthropology*, 47, 569–595.

Sealy, J. C. (2010). Isotopic evidence for the antiquity of cattle-based pastoralism in southernmost Africa. *Journal of African Archaeology*, 8, 66–81.

Smith, A. B. (2006). *Excavations at Kasteelberg and the Origins of the Khoekhoen in the Western Cape, South Africa*. Oxford: Archaeopress, 101pp.

Smith, B. W. (2010). Envisioning San history: Problems in the reading of history in the rock art of the Maloti-Drakensberg mountains of South Africa. *African Studies*, 69, 345–358.

Solomon, A. C. (2011). Writing San histories: the /Xam and 'shamanism' revisited. *Journal of Southern African Studies*, 37, 99–117.

Solomon, A. C. (2013). The death of trance: Recent perspectives on San ethnographies and rock arts. *Antiquity*, 87, 1208–1213.

Villa, P., Soriano, S., Tsanova, T., Degano, I., Higham, T. F. G., d'Errico, F., Backwell, L., Lucejko, J. J., Colombini, M. P. and Beaumont, P. B. (2012). Border Cave and the beginning of the Later Stone Age in South Africa. *Proceedings of the National Academy of Sciences*, 109, 13208–13213.

Wadley, L. (1987). *Later Stone Age Hunters and Gatherers of the Southern Transvaal: Social and Ecological Interpretations*. Oxford: British Archaeological Reports, 255pp.

Wadley, L. (1996). Changes in the social relations of precolonial hunter-gatherers after agropastoral contact: An example from the Magaliesberg, South Africa. *Journal of Anthropological Archaeology*, 15, 205–217.

Walker, N. J. (1995). *Late Pleistocene and Holocene Hunter-Gatherers of the Matopos*. Uppsala: Societas Archaeologica Upsaliensis, 284pp.

Wendt, W. E. (1972). Preliminary report on an archaeological research programme in South-West Africa. *Cimbebasia B*, 2, 1–61.

Wilmsen, E. N. and Denbow, J. R. (1990). Paradigmatic history of San-speaking peoples and current attempts at revision. *Current Anthropology*, 31, 489–524.

24

Southernmost Africans, archaeology and the environment during the Holocene

MARIA H. SCHOEMAN

Abstract

Pre-colonial southernmost Africa underwent substantial changes during the Holocene. These included climate change, as well as population and economic transformations, such as shifting from a hunting and gathering economy to one dominated by herding and crop farming. These changes took place in the context of an environmentally diverse region, which both facilitated and constrained settlement. Hunter-gatherers, herders and farmers, however, developed social strategies to cope with this diversity, as well as the substantial climatic fluctuations that affected the region. This chapter explores elements of the complex relationships between pre-colonial southernmost Africans and the environment during the Holocene.

24.1 Archaeologists, environment and climate

The disciplinary context of archaeology has contributed to building long term climate chronologies, as well as exploring how people of the past engaged with their environmental situations (e.g. Mitchell, 2008; van de Noort, 2011; Davies, 2013). In southernmost Africa, archaeologists have explored the relationship between human actions and local environmental changes or constraints (e.g. Hall, 1976; Klein, 1979, 1980; Deacon et al., 1984; Feely, 1987; Huffman, 1996, 2008; Smith, 2005; Manyanga, 2007; Smith et al., 2007; Feely and Bell-Cross, 2011). A range of methods has been used in such research, including investigations of local responses to regional climate or environmental patterns (e.g. Huffman, 1996, 2008), using site-specific sequences from dendroclimatology (e.g. Hall et al., 2008, 2009), analysis of charcoal found in archaeological excavations (e.g. Esterhuysen, 1992; February, 1994; Allott, 2004), analysis of faunal remains (e.g. Plug, 1997;

Wadley et al., 2008) and historical analysis (e.g. Smith et al., 2007; Denbow et al., 2008). Sites discussed in this chapter are numbered in Figure 24.1.

Interpretations of the relationship between people and environments during the Holocene have varied from approaches suggesting environmental determinism, to agency-centred approaches (Mitchell, 1997; see Davies, 2013, for an international overview). Environmental determinist approaches were evident in the ecological adaption models that dominated southern African hunter-gatherer archaeology before the 'anthropological turn' in the 1990s (Mitchell, 1997). These earlier studies frequently linked regional demographic patterns to environmental change, and contended that hiatuses in the archaeological record, or decreases in population

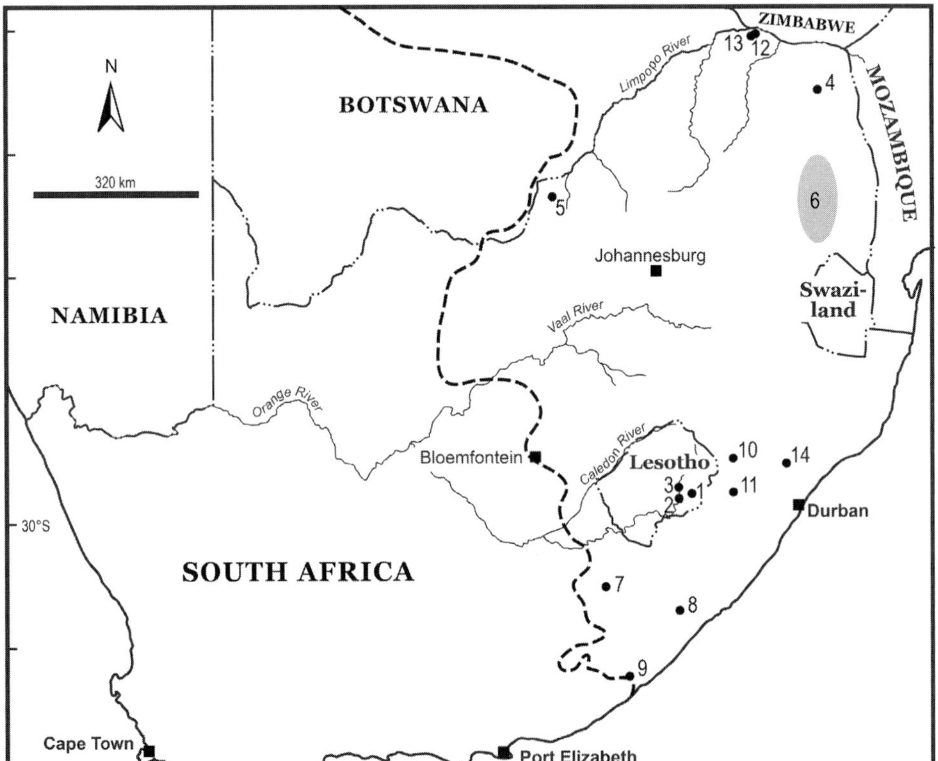

Fig. 24.1. Map of southern Africa, showing sites mentioned in the text. The stippled line demarcates the edge of the eastern extent of the summer rainfall zone that receives enough rainfall for millet cultivation (following Maggs, 1980). Numbered sites are: (1) Sehonhong, (2) Likoaeng, (3) Melikane, (4) Silver Leaves, (5) Morothodi, (6) Bokoni, (7) Oakleigh Farm, (8) Kulubele, (9) Canasta Place, (10) Mhlwazini Cave, (11) Collingham shelter, (12) Schroda, (13) K2 and Mapungubwe, (14) Karkloof forest.

numbers, were linked to climatic shifts, such as the Altithermal and Neoglacial episodes. Research from after the 'anthropological turn' has shown a much more complex pattern of occupation and abandonment that cannot be reduced to environmental factors alone (Mitchell, 1997). A shift in approach to the relationship between environment and human responses is less clear in the archaeology of southern African farming communities. Many have rejected environmental factors as drivers of change (e.g. Smith, 2005; Manyanga, 2007; Smith et al., 2007; Denbow et al., 2008; Manyanga et al., 2010). In contrast, Huffman (1996, 2008) has argued that climate change is an important driver of societal change. However, a more nuanced approach is that understanding the responses of communities to changing environmental or climatic factors should emphasise the role of societal resilience, rather than a uniform and deterministic response to climate.

24.2 Economic transformations, environment and settlement distribution

Hunter-gatherer societies in southernmost Africa underwent several transformations during the Holocene. These included the replacement of Oakhurst toolkits with Wilton toolkits, and substantial changes in social behaviour (see Mitchell, this volume). This chapter discusses only a few of the changes that occurred after 5000 BP, including a shift towards eating smaller animals and a substantial increase in numbers of archaeological sites and regions occupied. After ~5000 BP, hunter-gatherers apparently occupied, or re-occupied, most of South Africa's diverse environments (Mitchell, 1997). Some of these changes have been linked to environmental factors, specifically climate cooling between 6000–2500 BP (e.g. Holmgren et al., 2003). The occupation sequence of the area around Tsodilo Hills (Botswana) broadly reflects these climate changes, from fluctuating dry and humid periods of the early Holocene to a comparatively wet mid-Holocene (Robbins et al., 2000). Hunter-gatherer sites across Botswana become more visible after 5000 BP, and the linguistic diversity in hunter-gatherer populations has been attributed to the influx of people into the area in the mid-Holocene (Walker, 1998). While site numbers increased, access to water continued to constrain settlements in climatically marginal areas such as the Kalahari (Walker, 1998).

The occupation of the Sehonhong shelter in Lesotho (Fig. 24.1, site 1), which spans much of the Holocene, sheds light on Holocene environments here. Plug (1997), for example, suggested that the decrease in large bovids in hunter-gatherer diets in the region resulted from cooler temperatures during the Holocene, facilitating greater species diversity and creating a wider range of potential prey. At nearby Likoaeng (site 2), Mitchell et al. (2011) linked occupation phases with Holocene climatic variability, arguing that hunter-gatherers might have relied more on fish-spawning events during cooler episodes around 3000–2000 BP. Spawning, however, is a seasonal event, and occupation of sites associated with fish spawning formed part of larger regional patterns in which shelters such as Sehonhong were favoured in summer and Melikane (site 3) in winter. Many changes, however, were probably social: for example a larger population may result in increased site numbers. There also is evidence for reciprocal exchange networks (see *hxaro* below) that can lead to 'increasing identification with specific places on the landscape' (Mitchell, 1997, p398).

While environmental factors constrained hunter-gatherer settlement, the regional distributions of pre-colonial herder and farmer sites were fundamentally informed by ecological constraints. Herder sites appear to avoid areas where grazing would have been limited. In addition to the numerous sites of herder occupation located along the west coast of southern Africa (Smith and Ouzman, 2004; Smith, 2008), herder sites also occur on the fringes of the Kalahari in Botswana. Denbow (1986), however, suggested that larger parts of Botswana, with good grazing and seasonal rivers, could have played an important role in the southwards diffusion of herding practices. In addition, the distinctive rock art tradition along rivers in the interior of South Africa further suggests a preference for sites where livestock could access water (Smith and Ouzman, 2004). While there is general agreement on the distribution of these sites and the role that ecological constraints play in their distribution, the manner and timing of the introduction of herding to South Africa is disputed. Smith (2008) argued that increased rainfall in the Kalahari from 2500 to 2000 BP would have been favourable for herding, but decreased rainfall after 2000 BP would have adversely affected herders. These conditions are consistent with regional warming in southern Africa between 2500–1250 BP (Holmgren et al., 2003). Sadr (2008), however, has proposed local adoption of herding around 2000 yr BP, possibly followed by a later migration of cattle keepers at 1000 BP (see Mitchell, this volume, for more on this debate).

There is more secure evidence for the introduction of agropastoralism. Mixed farming spread into South Africa approximately 2000 years ago, with the earliest known farming community site being Silver Leaves (site 4), which dates to AD 350±50 (Vogel and Fuls, 1999). The distribution of summer rainfall required by sorghum and millet restricts the locations of farmers and their settlements (Maggs, 1980; Maggs and Whitelaw, 1991).

Southernmost African population numbers steadily increased in the 600 years following the introduction of herding and farming. In addition, economies became complex, and the region was incorporated into the Indian Ocean trade network (Huffman, 2000; Wood, 2011). This culminated in formation of the K2-Mapungubwe and Zimbabwean plateau states during the Medieval Warm Period (MWP, AD 900–1300) (Pikirayi, 2001; Huffman, 2015; Chirikure et al., 2013). The Polokwane stalagmite record shows that the period AD 880–1320 was slightly warmer and wetter than present (Holmgren et al., 1999, 2001; Lee-Thorpe et al., 2001), and may have had significant consequences for the Shashe-Limpopo Confluence Area (SLCA), where the K2-Mapungubwe state was located. At times of aridity, this area is marginal for rain-fed agriculture, but during the MWP annual rainfall in the region rose from >350 mm to <500 mm at AD 900, to >500 mm at AD 1050 (Smith, 2005; Smith et al., 2007). Increased rainfall would have made it easier for farmers to produce enough food for the growing population in the K2-Mapungubwe state.

In addition to state formation, new types of farmer sites started to appear at this time. Huffman (2002, 2004) argued that these sites were occupied by new farming communities that migrated from east Africa in two clusters. The first settled in the eastern part of southernmost Africa, developing into Nguni language speakers. The second expanded into the interior of South Africa and Botswana, developing into Sotho-Tswana speakers.

In the 15th Century AD the state centred on Great Zimbabwe was succeeded by the Khami and Mutapa states (van Waarden, 1998; Pikirayi, 2001). Increased settlement density took place across the region from the 17th Century, including the development of Tswana towns in Botswana and northwestern South Africa (Lane, 2004; Hall et al., 2006), and Bokoni towns in northeastern South Africa and Swaziland (Collett, 1982; Delius et al., 2012). The locations of Tswana towns were informed by factors relating to cultivation, political processes and defence (van Waarden, 1998; Boeyens, 2003; Lane, 2004; Hall et al., 2006). A key exception was Marathodi, which was located on the flats near the Pilanesberg in South Africa (site 5). The choice of this locality was informed by the presence of nickel sulphide pipes that were fundamental to copper and iron production (Hall et al., 2006). The availability of natural resources also informed the location of Bokoni terraced towns (Delius et al., 2012) (Fig. 24.2). Here, key factors were access to water (Collett, 1982), and farming and grazing land (Coetzee, 2015). Most of the chosen sites in Bokoni (site 6) used intensive terrace agriculture in order to mitigate against changing rainfall and temperature regimes during the last 500 years (e.g. Sjöström, 2013; Hattingh, 2014).

While distinct hunter-gathering, herding and farming economies existed in southern Africa, there were various forms of interaction and fluidity between them. Early colonial accounts from the Cape suggest that hunter-gatherer and herder communities were fundamentally entangled (Barnard, 2007), and likewise for hunter-gatherers and farmers. Walker (1998) suggested that the arrival of the first farmers, and relationships established with them, allowed hunter-gatherers in Botswana to expand into areas not previously occupied. There is also substantial evidence for extensive interaction between hunter-gatherers and farmers in the formation of the K2-Mapungubwe state (Hall and Smith, 2000; Schoeman, 2006; van Doornum, 2008), and in the northwest of South Africa and Botswana where the Tswana towns formed (e.g. Wilmsen and Denbow, 1990; Hall, 2000).

Fig. 24.2. Bokoni farming terraces in Mpumalanga, South Africa. Average homestead diameters (extent marked by the outer walls encircling the smaller livestock enclosures) are 40–55 m across (photo: Maria Schoeman).

24.3 Social strategies and resilience

While access to resources, in part, informed settlement locations, hunter-gatherers, herders and farmers used social systems to manage resource availability, and the impact of local environmental uncertainties. One of the best known hunter-gatherer social strategies was a system of aggregation and dispersal. During aggregation, social interaction would be intense, including healing and reciprocal gift-giving as observed in ethnographic contexts in the Kalahari (*hxaro*) (Wadley, 1987). These strategies were common from the Western Cape (Jerardino and Yates, 1996) to the Shashe-Limpopo Confluence Area (van Doornum, 2008) (Fig. 24.3). Herders also used social networks to offset risk, and would provide livestock in return for future favours (Viljoen, 1997). In addition, faunal remains found at herder sites show that hunting and gathering supplemented their diet (e.g. Sadr et al., 2003). During the historic period, herders in the Cape also had close relationships with hunter-gatherer communities (Barnard, 2007), which would have allowed them to manage environmental risk.

Farmer reliance on cultivation and need for grazing increased their risk from exposure due to rainfall and temperature fluctuations. This did not directly determine the choices farmers made, but rather influenced their 'lifeways' (Pikirayi, 2001). Hall (1976, p702) argued that 'traditional subsistence agriculture in southern

Fig. 24.3. Balerno Main shelter, an aggregation site in the Shashe-Limpopo Confluence Area. The shelter itself (lower left) is 16–20 m wide (photo: Maria Schoeman). A black and white version of this figure will appear in some formats. For the colour version, please refer to the plate section.

Africa may be seen as a process of evaluating risk'. Most pre-colonial farmers had strategies to offset such risks, including 'herding out cattle', establishing regional support networks with other communities, cattle transhumance, and extensive as well as intensive methods of crop farming (Hall, 1976; Pikirayi, 2001; Smith, 2005; Manyanga, 2007). These strategies were even more important for those living in areas that are marginal for rain-fed agriculture, such as the Shashe-Limpopo Confluence Area (Smith, 2007), and areas of KwaZulu-Natal (Hall, 1976) and the Eastern Cape (Prins, 1994). The following three case studies explore the relationship between farmers and environmental conditions in specific contexts: (1) pre-13th Century AD early farming communities in the Eastern Cape; (2) the development and 13th Century decline of the K2-Mapungubwe state; and (3) the period of conflict, known as the *mfecane,* in the early 19th Century AD.

24.4 Living on the edge in the Eastern Cape

As previously noted, the distribution of pre-colonial farming community settlements was limited to the summer rainfall zone, which extends into the Eastern Cape of South Africa. This region has a long hunter-gatherer record from the Middle Stone Age (e.g. Hall and Binneman, 1987; Binneman, 2001; Wurz, this volume), as well as extensive herder occupations. The earliest known date for the presence of herders in this region (1760 BP) is from Oakleigh Farm near Queenstown (Derricourt, 1977) (site 7). Mixed farming arrived in the region by AD 800, with sites in the Great Kei river valley (e.g. Kulubele, dating to AD 790–857; Binneman, 1996) (site 8). Early farmer sites generally had riverside locations, and were all located in the woodland zone with none occurring in the grasslands (Feely, 1987; Feely and Bell-Cross, 2011; see also Maggs, 1980, for site distribution in KwaZulu-Natal). In the northern part of Eastern Cape, early farmer sites were located far inland and to the limit of the woodland zone. This is not the case for southern sites such as Kulubele, which is approximately 50 km from the coast. The most southerly known site is Canasta Place near East London (site 9).

The southwestern sites do not extend to the edge of the summer rainfall zone, and consequently, Feely and Bell-Cross (2011) argued that social rather than environmental factors constrained the extent of early farming communities, including herder control of the regions to the south and west. The relationship between farmers, herders and hunter-gatherers at this time in this region, however, is poorly understood. The presence of *Pennisetum* and *Sorghum* in hunter-gatherer shelter deposits at Shongweni in KwaZulu-Natal (Davies, 1975), and the presence of early farmer pottery and sheep and cattle remains at the hunter-gatherer site Likoaeng in Lesotho, which is a substantial distance from the most westerly farmer sites (Mitchell et al., 2011), point to some form of interaction and exchange.

The distribution of early farming sites also varied through time. Prins (1994) noted a decrease in site numbers in areas that are marginally suited to rain-fed agriculture in the period AD 900–1200, suggesting this resulted from farming community responses to environmental factors, specifically lower rainfall. Decreased rainfall was identified through analysis of charcoal from Mhlwazini Cave (site 10) and Collingham Shelter (site 11) in KwaZulu-Natal (February, 1994; Prins, 1994). Prins (1994) suggested that February's (1994) rainfall data point to a local inverse relationship between rainfall and temperature during the MWP, which is dissimilar to the warmer and wetter model proposed by Tyson and Lindesay (1992).

24.5 The rise and decline of the K2-Mapungubwe state

The largest of the earliest known villages in the SLCA (that extends into Botswana, South Africa and Zimbabwe), Schroda (site 12), was occupied at the end of the 9th Century AD (Hanisch, 2002). At this time regional annual rainfall was $\leq$350 mm to $\geq$500 mm (Smith et al., 2007) which is somewhat marginal for cultivation (Smith, 2005), thus it was likely that social factors, such as participating in the Indian Ocean trade, informed population expansion in the area (Huffman, 2000). Farming communities centred on what is now Zimbabwe, expanded their control into the SLCA at the beginning of the 11th Century AD. Under their leadership, a new capital was established at K2, which was the political centre for the next 200 years. Around AD 1220 the capital shifted to the nearby Mapungubwe hill, which was occupied to the end of the 13th Century (Calabrese, 2000; Huffman, 2000, 2008; Meyer, 2000) (site 13). Annual rainfall from the 9th to 15th Centuries was $\geq$500 mm except for a brief dry spell ($<$450 mm) between AD 1200–1260 (Smith, 2005, 2007).

The rise and fall of the K2-Mapungubwe state has been linked to climate change. One of the most prominent of the environmental collapse models is Huffman's (1996) use of regional climate data (e.g. Tyson and Lindesay, 1992) to argue that favourable conditions during the MWP allowed for the growth of the K2-Mapungubwe state, and that abandonment of the area can be linked to dryness at the onset of the Little Ice Age (LIA; around AD 1500). Huffman's (1996, 2008, 2009) climate-driven model for Mapungubwe has been critiqued (e.g. Hannaford et al., 2014), and stands in contrast to Smith's (2005; Smith et al., 2007, 2010) interpretations that are based on detailed local environmental data and which argue for people's actions in terms of agency. Smith's rainfall data, as well as other environmental information used to investigate livestock and land management strategies, were generated through isotope analysis focussing on $\delta^{15}N$, $\delta^{18}O$ and $\delta^{13}C$ values, as well as $^{87}Sr/^{86}Sr$ ratios in dated faunal remains from various archaeological sites in the SLCA. The wetter and drier phases identified are broadly consistent with other regional records such as the Makapansgat stalagmite (Holmgren et al., 1999, 2001; Lee-Thorpe et al., 2001) and Tzaneen tree-ring records (Norström et al., 2005).

The relationship between internal changes in the SLCA is likely more complex than previously thought. The end of the K2-Mapungubwe state did not coincide with decreased rainfall, and socio-political and economic factors may have been contributors (Smith, 2005; Manyanga, 2006; Smith et al., 2007; Manyanga et al., 2010), such as decreased access to trade goods due to changes in trans-Indian Ocean trade networks (van Waarden, 1998; Wood, 2011).

24.6 The *mfecane*

The first part of this chapter discussed social strategies employed in pre-colonial southern Africa to manage variability. Smith (2005; Smith et al., 2007, 2010) and Manyanga (2006; Manyanga et al., 2010) argued that these were used successfully in the SLCA and Zimbabwe plateau states. These strategies, however, have not always been able to manage states of disequilibrium, such as the *mfecane* period of social strife and warfare in KwaZulu-Natal in the late 18th and early 19th Centuries AD (Hall, 1976).

A complex set of processes and environmental events resulted in the failure of resilience strategies in KwaZulu-Natal. Similar to the Eastern Cape, several areas in KwaZulu-Natal are agriculturally marginal, and during periods of decreased rainfall, social networks and strategies such as 'herding out' cattle would become important. In the mid-18th Century, sufficient rainfall allowed for increased food production and population increase, also facilitated by the introduction of maize (Hall, 1976). A sharp decline in rainfall after AD 1789 resulted in a series of droughts, which are also remembered in oral traditions. Independent evidence for drought comes from dendroclimatological data from *Podocarpus falcatus* from the Karkloof forest (site 14) (Hall, 1976), and diatom records from Lake Sibaya (Stager et al., 2013). Following so closely on favourable conditions for population expansion, the social strategies that were effective under normal circumstances failed, and food production declined sharply, resulting in famines. Simultaneously, the political and social landscape was reconfigured through a series of developments such as state formation and increasing competition over land and international trade (Hall, 1976; Eldredge, 1995). These social factors in combination with the famines culminated in the *mfecane* disruptions that plummeted large parts of southern Africa into conflict.

Huffman (1996) suggested that the fall of Tswana towns in Botswana and North West Province (South Africa) can also be attributed to the combination of drought and cultivating maize (Fig. 24.4). The application of this model to the Tswana

Fig. 24.4. The area where the Mapungubwe town once stood, Shashe-Limpopo Confluence Area (photo: Maria Schoeman).

towns, however, has been contested. Lane (2004) suggested that some of the abandonments relate to land management strategies, not environmental conditions, and that towns were regularly relocated to manage ecological factors such as declining soil fertility and drought. The abandonment of others may relate to conflict between different Tswana groups that increased due to 'population pressure and increasing competition for land and livestock, partly aggravated by an expanding Cape colonial frontier' (Boeyens, 2003, p74). Boeyens (2003) also noted that there is no historical evidence that maize cultivation had been adopted in Tswana areas.

24.7 Summary

Recent archaeological studies of the relationship between southern African communities and the environment during the Holocene recognise that ecological factors could (although not always) play a role in settlement distributions of herders and agropastoralist farmers, as can be seen in the Eastern Cape archaeological record. However, there has been a shift from earlier approaches that interpreted climate change as causing social change. Fundamental to current approaches is the recognition that the effects of climate fluctuations were offset through complex risk management strategies, which included social networks, as well as land management strategies. These strategies were used successfully in the Mapungubwe state in the SLCA and in the Tswana towns, but failed in KwaZulu-Natal in the early 19th Century.

References

Allott, L. F. (2004). Changing environments in Oxygen Isotope Stage 3: Reconstructions using archaeological charcoal from Sibudu Cave. *South African Journal of Science*, 100, 179–184.

Barnard, A. (2007). *Anthropology and the Bushman*. Oxford: Berg, 224pp.

Binneman, J. (1996). Preliminary results from investigations at Kulubele, an Early Iron Age farming settlement in the Great Kei River valley, Eastern Cape. *Southern African Field Archaeology*, 5, 28–35.

Binneman, J. (2001). An introduction to the Later Stone Age research project along the south-eastern Cape coast. *Southern African Field Archaeology*, 10, 75–87.

Boeyens, J. C. A. (2003). The Late Iron Age sequence in the Marico and early Tswana history. *South African Archaeological Bulletin*, 58, 63–78.

Calabrese, J. A. (2000). Interregional interaction in Southern Africa: Zhizo and Leopard's Kopje relations in northern South Africa, southwestern Zimbabwe, and eastern Botswana, AD 1000 to 1200. *African Archaeological Review*, 17, 183–210.

Chirikure, S., Manyanga, M., Pikirayi, I. and Pollard, M. (2013). New pathways of sociopolitical complexity in Southern Africa. *African Archaeological Review*, 30, 339–366.

Coetzee, T. J. (2015). *Mapping Bokoni: Exploring Bokoni Settlement Choices and Changes in Mpumalanga and Limpopo, South Africa using GIS Site Distribution Analysis Techniques*. MA Dissertation, Pretoria: University of Pretoria.

Collett, D. P. (1982). Excavations of the stone-walled ruin types in the Badfontein valley, Eastern Transvaal, South Africa. *South African Archaeological Bulletin*, 37, 34–43.
Davies, M. I. J. (2013). Forced moves or just good moves? Rethinking environmental decision-making among East African Intensive Cultivators. In *Humans and the Environment: New Archaeological Perspectives for the Twenty-First Century*, ed M. I. J. Davies and F. N. M'Mbogori. Oxford: Oxford University Press, pp. 57–76.
Davies, O. (1975). Excavations at Shongweni south cave: The oldest evidence to date for cultigens in Southern Africa. *Annals of the Natal Museum*, 22, 627–662.
Deacon, H. J., Deacon, J., Scholtz, A., Thackeray, J. F., Brink, J. S. and Vogel, J. C. (1984). Correlation of palaeoenvironmental data from the Late Pleistocene and Holocene deposits at Boomplaas Cave, Southern Cape. In *Late Cainozoic Palaeoclimates of the Southern Hemisphere*, ed. J. C. Vogel. Rotterdam: A.A. Balkema, pp. 330–352.
Delius, P., Maggs, T. and Schoeman, M. (2012). Bokoni: Old structures, new paradigms? Rethinking pre-colonial society from the perspective of the stone-walled sites in Mpumalanga. *Journal of Southern African Studies*, 38, 399–414.
Denbow, J. (1986). A new look at the later prehistory of the Kalahari. *The Journal of African History*, 27, 3–28.
Denbow, J., Smith, J., Ndobochani, N. M., Atwood, K. and Miller, D. (2008). Archaeological excavations at Bosutswe, Botswana: Cultural chronology, paleo-ecology and economy. *Journal of Archaeological Science*, 35, 459–480.
Derricourt, R. M. (1977). *Prehistoric Man in the Ciskei & Transkei*. Cape Town: Struik, 284pp.
Eldredge, E. A. (1995). Sources of conflict in Southern Africa c. 1800–1830. In *The Mfecane Aftermath: Reconstructive Debates in Southern African History*, ed. C. Hamilton. Johannesburg: Witwatersrand University Press and University of Natal Press, pp. 123–161.
Esterhuysen, A. B. (1992). Charcoal from Leliehoek Shelter, eastern Orange Free State, South Africa. *Palaeoecology of Africa*, 23, 193–201.
February, E. C. (1994). Rainfall reconstruction using wood charcoal from two archaeological sites in South Africa. *Quaternary Research*, 42, 100–107.
Feely, J. M. (1987). *The Early Farmers of Transkei, Southern Africa, before A.D. 1870*. Oxford: British Archaeological Reports, 142pp.
Feely, J. M. and Bell-Cross, S. M. (2011). The distribution of Early Iron Age settlement in the Eastern Cape: Some historical and ecological implications. *South African Archaeological Bulletin*, 66, 105–112.
Hall, G., Woodborne, S. and Pienaar, M. (2009). Rainfall control of the $\delta^{13}C$ ratios of Mimusops caffra from KwaZulu-Natal, South Africa. *The Holocene*, 19, 251–260.
Hall, G., Woodborne, S. and Scholes, M. (2008). Stable carbon isotope ratios from archaeological charcoal as palaeoenvironmental indicators. *Chemical Geology*, 247, 384–400.
Hall, M. (1976). Dendroclimatology, rainfall and human adaptation in the later Iron Age of Natal and Zululand. *Annals of the Natal Museum*, 22, 693–703.
Hall, S. (2000). Forager lithics and Early Moloko homesteads at Madikwe. *Natal Museum Journal of Humanities*, 12, 33–50.
Hall, S. and Binneman, J. (1987). Later Stone Age burial variability in the Cape: A social interpretation. *South African Archaeological Bulletin*, 42, 140–152.
Hall, S., Miller, D., Anderson, M. and Boeyens, J. (2006). An exploratory study of copper and iron production at Marothodi, an early 19th century Tswana town, Rustenburg District, South Africa. *Journal of African Archaeology*, 4, 3–35.

Hall, S. and Smith, B. (2000). Empowering places: Rock shelters and ritual control in farmer-forager interactions in the Northern Province. *South African Archaeological Society Goodwin Series*, 8, 30–46.

Hanisch, E. O. M. (2002). Schroda: The archaeological evidence. In *Sculptured in Clay: Iron Age Figurines from Schroda, Limpopo Province, South Africa*, ed J. A. van Schalkwyk and E. O. M. Hanisch. Pretoria: National Cultural History Museum, pp. 21–40.

Hannaford, M. J., Bigg, G. R., Jones, J. M., Phimister, I. and Staub, M. (2014). Climate variability and societal dynamics in pre-colonial southern African history (AD 900–1840): A Synthesis and Critique. *Environment and History*, 20, 411–445.

Hattingh, T. (2014). *A Phytolith Analysis of Bokoni Soils*. MSc dissertation, Johannesburg: University of the Witwatersrand, 97pp.

Holmgren, K., Karlén, W., Lauritzen, S. E., Lee-Thorp, J. A., Partridge, T. C., Piketh, S., Repinshi, P., Stevenson, C., Svanered, O. and Tyson, P. D. (1999). A 3000-year high-resolution stalagmite-based record of palaeoclimate for northeastern South Africa. *The Holocene*, 9, 295–309.

Holmgren, K. Lee-Thorp, J. A., Cooper, G. R. J., Lundblad, K., Partridge, T. C., Scott, L., Sithaldeen, R., Talma, A. S. and Tyson, P. D. (2003). Persistent millennial-scale climatic variability over the past 25,000 years in Southern Africa. *Quaternary Science Reviews*, 22, 2311–2326.

Holmgren, K., Tyson, P. D., Moberg, A. and Svanered, O. (2001). A preliminary 3000-year regional temperature reconstruction for South Africa. *South African Journal of Science*, 97, 49–51.

Huffman, T. N. (1996). Archaeological evidence for climatic change during the last 2000 years in Southern Africa. *Quaternary International*, 33, 55–60.

Huffman, T. N. (2000). Mapungubwe and the origins of the Zimbabwe culture. *South African Archaeological Society Goodwin Series*, 8, 14–29.

Huffman, T. N. (2002). Regionality in the Iron Age: the case of the Sotho-Tswana. *Southern African Humanities*, 14, 1–22.

Huffman, T. N. (2004). The archaeology of the Nguni past. *Southern African Humanities*, 16, 79–111.

Huffman, T. N. (2008). Climate change during the Iron Age in the Shashe-Limpopo Basin, southern Africa. *Journal of Archaeological Science*, 35, 2032–2047.

Huffman, T. N. (2009). A cultural proxy for drought: Ritual burning in the Iron Age of southern Africa. *Journal of Archaeological Science*, 36, 991–1005.

Huffman, T. N. (2015). Social complexity in southern Africa. *African Archaeological Review*, 32, 71–91.

Jerardino, A. and Yates, R. (1996). Preliminary results from excavations at Steenbokfontein Cave: Implications for past and future research. *South African Archaeological Bulletin*, 51, 7–16.

Klein, R. G. (1979). Palaeoenvironmental and cultural implications of late Holocene archaeological faunas from the Orange Free State and north-central Cape Province, South Africa. *South African Archaeological Bulletin*, 34, 34–49.

Klein, R. G. (1980). Environmental and ecological implications of large mammals from upper Pleistocene and Holocene sites in Southern Africa. *Annals of the South African Museum*, 81, 223–283.

Lane, P. J. (2004). Re-constructing Tswana townscapes: Toward a critical historical archaeology. In *African Historical Archaeologies*, ed A. M. Reid and P. J. Lane. London: Kluwer Academic/Plenum, pp. 269–299.

Lee-Thorp, J. A., Holmgren, K., Lauritzen, S.-E., Linge, H., Moberg, A., Partridge, T. C., Stevenson, C. and Tyson, P. D. (2001). Rapid climate shifts in the Southern African

interior throughout the mid to late Holocene. *Geophysical Research Letters*, 28, 4507–4510.
Maggs, T. (1980). The Iron Age sequence south of the Vaal and Pongola Rivers: Some historical implications. *The Journal of African History*, 21, 1–15.
Maggs, T. and Whitelaw, G. (1991). A review of recent archaeological research on food-producing communities in Southern Africa. *The Journal of African History*, 32, 3–24.
Manyanga, M. (2007). *Resilient Landscapes: Socio-Environmental Dynamics in the Shashe Limpopo Basin, Southern Zimbabwe, c. AD 800 to the Present*. Studies in Global Archaeology 11, Uppsala: Uppsala University, 235pp.
Manyanga, M., Pikirayi, M. and Chirikure, S. (2010). Conceptualising the urban mind in pre-European southern Africa: Rethinking Mapungubwe and great Zimbabwe. In *The Urban Mind: Cultural and Environmental Dynamics*, ed P. J. J. Sinclair, G. Nordquist, F. Herschend and C. Isendahl. Uppsala: Uppsala University, pp. 573–590.
Meyer, A. (2000). K2 and Mapungubwe. *South African Archaeological Society Goodwin Series*, 8, 4–13.
Mitchell, P. J. (1997). Holocene Later Stone Age hjunter-gatherers south of the Limpopo river, Ca. 10,000–2000 B.P. *Journal of World Prehistory*, 11, 359–424.
Mitchell, P. (2008). Practising archaeology at a time of climatic catastrophe. *Antiquity*, 82, 1093–1103.
Mitchell, P., Plug, I., Bailey, G., Charles, R., Esterhuysen, A., Lee Thorp, J., Parker, A. and Woodborne, S. (2011). Beyond the drip-line: A high-resolution open-air Holocene hunter-gatherer sequence from highland Lesotho. *Antiquity*, 85, 1225–1242.
Norström, E., Holmgren, K. and Mörth, C.-M. (2005). Rainfall-driven variation in $\delta^{13}C$ composition and wood anatomy of *Breonadia salicina* trees from South Africa between AD 1375 and 1995. *South African Journal of Science*, 101, 162–168.
Pikirayi, I. (2001). *The Zimbabwe Culture: Origins and Decline in Southern Zambezian States*. Walnut Creek, CA: Alta Mira, 336pp.
Plug, I. (1997). Late Pleistocene and Holocene hunter–gatherers in the Eastern Highlands of South Africa and Lesotho: A Faunal Interpretation. *Journal of Archaeological Science*, 24, 715–727.
Prins, F. E. (1994). Climate, vegetation and early agriculturist communities in Transkei and KwaZulu-Natal. *Azania*, 29–30, 179–186.
Robbins, L. H., Murphy, M. L., Brook, G. A., Ivester, A. H., Campbell, A. C., Klein, R. G., Milo, R. G., Stewart, K. M., Downey, W. S. and Stevens, N. J. (2000). Archaeology, palaeoenvironment, and chronology of the Tsodilo Hills White Paintings Rock Shelter, Northwest Kalahari Desert, Botswana. *Journal of Archaeological Science*, 27, 1085–1113.
Sadr, K. (2008). Invisible herders? The archaeology of Khoekhoe pastoralists. *Southern African Humanities*, 20, 179–203.
Sadr, K., Smith, A., Plug, I., Orton, J., and Mütti, B. (2003). Herders and Foragers on Kasteelberg: Interim Report of Excavations 1999–2002. *South African Archaeological Bulletin*, 58, 27–32.
Schoeman, M. H. (2006). Imagining rain-places: Rain-control and changing ritual landscapes in the Shashe-Limpopo Confluence Area, South Africa. *South African Archaeological Bulletin*, 61, 152–165.
Sjöström, J. (2013). *Late Holocene Palaeoenvironmental Reconstruction on a Peat Sequence from Northeastern South Africa, Using Grass Phytoliths as Main Proxy*. Masters Thesis, Stockholm: Stockholm University, 100pp.
Smith, A. B. (2008). Pastoral origins at the Cape, South Africa: Influences and arguments. *Southern African Humanities*, 20, 49–60.

Smith, B. W. and Ouzman, S. (2004). Taking stock: Identifying Khoenkoen herder rock art in Southern Africa. *Current Anthropology*, 45, 499–526.

Smith, J. M. (2005). *Climate change and agropastoral sustainability in the Shashe/Limpopo River Basin from AD 900*. Unpublished PhD Thesis, University of the Witwatersrand, 178pp.

Smith, J., Lee-Thorp, J. and Hall, S. (2007). Climate change and agropastoralist settlement in the Shashe-Limpopo River Basin, Southern Africa: AD 880 to 1700. *South African Archaeological Bulletin*, 62, 115–125.

Smith, J., Lee-Thorp, J., Prevec, S., Hall, S. and Späth, A. (2010). Pre-colonial herding strategies in the Shashe-Limpopo Basin, Southern Africa, based on strontium isotope analysis of domestic fauna. *Journal of African Archaeology*, 8, 83–98.

Stager, J. C., Ryves, D. B., King, C., Madson, J., Hazzard, M., Neumann, F. H. and Maud, R. (2013). Late Holocene precipitation variability in the summer rainfall region of South Africa. *Quaternary Science Reviews*, 67, 105–120.

Tyson, P. D. and Lindesay, J. A. (1992). The climate of the last 2000 years in southern Africa. *The Holocene*, 2, 271–278.

van de Noort, R. (2011). Conceptualising climate change archaeology. *Antiquity*, 85, 1039–1048.

van Doornum, B. (2008). Sheltered from change: Hunter-gatherer occupation of Balerno Main Shelter, Shashe-Limpopo confluence area, South Africa. *Southern African Humanities*, 20, 249–284.

van Waarden, C. (1998). The Late Iron Age. In *Ditswa Mmung: The Archaeology of Botswana*, eds P. Lane, A. Reid and A. Segobye. Gaborone: Botswana Society, pp. 115–160.

Viljoen, R. (1997). Making sense of the Khoikhoi cattle-killing of 1788: An episode of millenarianism in Khoikhoi society. *Kronos*, 24, 62–76.

Vogel, J. C. and Fuls, A. (1999). Spatial distribution of radiocarbon dates for the Iron Age in southern Africa. *South African Archaeological Bulletin*, 54, 97–101.

Wadley, L. (1987). *Later Stone Age Hunters and Gatherers of the Southern Transvaal*. Oxford: BAR, 225pp.

Wadley, L., Plug, I. and Clark, J. L. (2008). The contribution of Sibudu fauna to an understanding of KwaZulu-Natal environments at ~60 ka, ~50 ka and ~37 ka. In *People, Places and Animals of Africa: Papers in Honour of Ina Plug*, ed S. Badenhorst, P. Mitchell and J. C. Driver. Oxford: BAR International Series, 1849, pp. 34–45.

Walker, N. (1998). The Late Stone Age. In *Ditswa Mmung: The Archaeology of Botswana*, eds P. Lane, A. Reid and A. Segobye. Gaborone: Botswana Society, pp. 65–80.

Wilmsen, E. and Denbow, J. (1990). Paradigmatic history of San-speaking peoples and current attempts at revision. *Current Anthropology*, 31, 489–524.

Wood, M. (2011). *Interconnections: Glass Beads and Trade in Southern and Eastern Africa and the Indian Ocean – 7th to 16th Centuries AD*. PhD thesis, Uppsala: Uppsala University, 109pp.

25

Landscape–climate–human relations in the Quaternary of southern Africa

JASPER KNIGHT, DOMINIC STRATFORD AND STEFAN W. GRAB

Abstract

The triad of landscape–climate–human relationships provides a secure context or interpretive framework for understanding not only the macroscale landscape development of southern Africa during the Quaternary, as reflected in the geomorphological and sedimentary record, but also those patterns and processes of human physical evolution and behaviour that took place within that landscape, as reflected in the fossil, archaeological and palaeoanthropological records. In this chapter, we reflect upon the complexity of such relationships, and the limitations on our current understanding, which is based on studies that are inevitably grounded in a narrow spatial and temporal context. To remedy this situation, we propose a more explicitly integrated landscape–climate–human approach to Quaternary studies in southern Africa, which may yield a better understanding of the sensitivity of landscapes and human activity to future climate and environmental changes in the Anthropocene. This in turn should lead us to a more realistic reconstruction of the many-faceted variables of our southern African collective past.

25.1 Introduction

In southern Africa, close relationships exist between landscape-scale geomorphology, environments and associated physical processes; prevailing climatic regimes and rapid climate changes over long (10^2–10^6 years) timescales; and varied aspects of human activity including human evolution, behaviour, cultural practices, socio-political organisation, agriculture and industry over these timescales (Vrba et al., 1995). Because of these relationships, different types of geological, geochemical and biological evidence from across this region provide an opportunity to examine the interplay, forcing–response relationships and dynamical controls between landscape, climate and human factors, and the limitations of these relationships. Although these relationships have been examined individually in local case studies in different parts of southern Africa, as discussed later, the combination of the three components together is more instructive, because it is able to provide better interpretations that are relevant over the regional spatial scales and long timescales that are applicable to the evolution of the southern African landscape during the Cenozoic. As such, these regional studies can inform on large-scale landscape controls such as hemispheric climate patterns, tectonics and geologic processes (de Wit, 2007; Dirks and Berger, 2013).

The wider context of landscape–climate–human relationships in southern Africa is the role that climatic and environmental conditions during the Quaternary (last 2.6 Ma) have played on aspects of human physiological and behavioural evolution, and includes the evolution of *Homo* and the spread of the genus into different areas, exploiting different environmental resources (Mackay et al., 2014; Maslin et al., 2014). Several studies have reconstructed patterns of resource use, based upon evidence from specific, individual sites that generally represent short timeframes of activity (e.g. Copeland et al., 2011; Stewart et al., 2012; Kyriacou et al., 2014; Oestmo et al., 2014). Establishing the wider context of this relationship, however, requires information from many sites and throughout the Quaternary. Without this wider context, extra-site interpretation is more difficult and landscape–climate–human relationships may be prone to over-interpretation and lead to an over-simplified environmental determinism.

There are complex relationships between landscapes, climate and human activity, and these have interactions and feedbacks that exist over different spatial and temporal scales (Reynolds et al., 2011; Giosan et al., 2012). Measuring and understanding these relationships is usually approached through the lens of different academic disciplines, traditions and methodologies (including those in Geography, Geology, Archaeology, Ecology and Anthropology), and thus studies of these relationships have often been compartmentalised and are not truly interdisciplinary. This has been a limitation in the understanding of southern African landscape–climate–human relationships.

This chapter discusses landscape–climate–human relationships in turn, with a particular focus on how these intersect with each other in different ways and with respect to particular environmental problems, and identifies future research directions for these relationships in southern Africa, a critical region for both southern hemisphere climate records for the Quaternary, and for hominin evolution.

25.2 Landscapes and climate

Regional landscape-scale geomorphology in southern Africa has typically been viewed, over long timescales, as responding passively to cyclic variations in external forcing by climate and/or tectonics (King, 1967; Bishop, 2007; Moore et al., 2009). For example, the formation and longevity in the landscape of the Miocene to Pliocene-age African Surface and Post-African Surfaces I and II (Partridge and Maud, 1987, 2000; Partridge, 1998) have been interpreted as an equilibrium response to episodic tectonic uplift (Partridge and Maud, 1987). These surfaces are argued to be isochronous, laterally extensive (King, 1976; Partridge, 1998) and exhibiting similar geomorphic patterns, in particular with respect to river response to uplift (King, 2008). In detail, however, geochronometric studies (based on ^{10}Be and ^{26}Al cosmogenic radioisotopes) show that elements of these surfaces are not of the same age, although they can be grouped as such into broad time periods (Bierman and Caffee, 2001). Although, on the macroscale, these dates support old ideas of ancient planation surfaces, they also suggest relatively slow rates of land surface denudation (Brown et al., 2002; Tinker et al., 2008; Flowers and Schoene, 2010; Bierman et al., 2014). The high variability of ages from one sample site to another, even over short distances, however, shows that uniform and regionally consistent rates and patterns of land surface denudation cannot be inferred (Bierman and Caffee, 2001). This therefore makes it difficult to evaluate aggregated landscape responses such as climate-driven changes in weathering and sediment yield, even when averaged over long timescales (cf. Cotterill and de Wit, 2011). Important to note is that these dating studies show slow rates of landscape change, rather than high rates of landscape preservation. As such, this contrasts with outdated viewpoints of the African Surfaces which emphasise preservation and longevity (e.g. Partridge and Maud, 2000).

These macroscale hypotheses of climate forcing–landscape response can be tested when based on site-scale patterns of weathering and erosion, which are the mechanisms though which macroscale landscape responses are effected. These studies show that land surface weathering can take place through a combination of mechanical (physical), chemical and biological weathering, all of which are strongly mediated through patterns of microclimate, geology, soils, ecosystems and (increasingly) human activity (e.g. Phillips, 2005; Sumner and Nel, 2006). In turn, weathering products can be preserved as a variety of slope-driven landform types and deposits, including solifluction or boulder lobes, scree, alluvial or colluvial fans, and terraces. The stratigraphy and dating of these landforms and sediments can in many cases – but not always – be matched with significant periods of climate deterioration which promote land surface instability (e.g. Dixon et al., 2009a, b). These studies show that relationships between landscapes and climate are complex and can give rise to patterns that are not consistent, or which do not vary systematically, either over time or space. This limits the extent to which a deterministic climate forcing–landscape response paradigm can be applied to the real world.

In addition, studies now show that different landscape elements exhibit different sensitivities and responses to individual climatic events or ongoing climate change. This property, termed geomorphological sensitivity (Knight and Harrison, 2013a), is significant because it implies that not all parts of any one landscape will respond in the same way, or at the same rate, to climate forcing. The result of such spatial (and temporal) variations in geomorphological sensitivity is the formation of palimpsest landscapes (Knight, 2012). These are defined as landscapes that contain individual landforms or landform elements (e.g. summits, slopes) that are of differing ages. For example, bars within a river channel may be very young, where they are being shaped by present-day fluvial processes; but terraces located alongside the river channel may be decades to thousands of years old. Thus, the landscape contains a patchwork of elements of different ages. Clearly, our perception of whether landscapes are 'old' or 'young' depends on what is being studied or dated, thus it is no surprise that cosmogenic dating should result in a range of ages from any one landscape (Bierman et al., 2014). Figure 25.1 shows that time-evolution is an important element of landscape change.

Both geomorphological sensitivity and landscape palimpsests are important concepts in the understanding of the archaeological record (Bailey, 2007), because

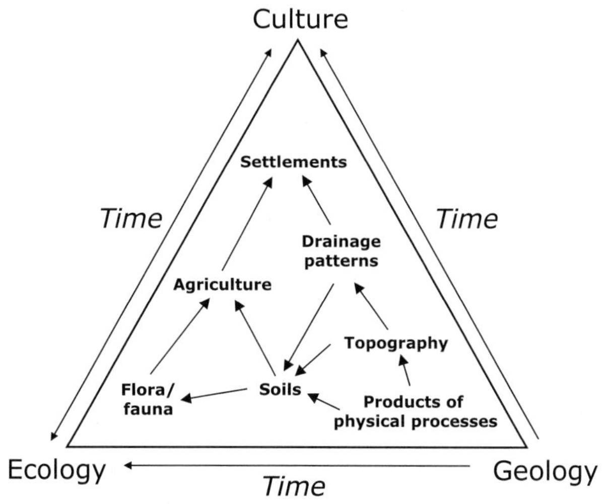

Fig. 25.1. Sketch of some of the interconnected relationships between physical and human elements of the landscape, and the role of time in landscape evolution. Time is non-dimensional in this schematic diagram.

they can help explain why the stratigraphy of archaeological sites varies from place to place, and offer caveats to the uncritical and deterministic interpretations of such site-scale evidence (cf. Reynolds et al., 2011; Stewart et al., 2012; Mackay et al., 2014).

25.3 Climate and human activity

Human impacts on Earth's physical systems are now known to have significantly altered global-scale environments and climates, in the era now commonly known as the Anthropocene (Zalasiewicz et al., 2011). Although significant anthropogenic environmental and climatic changes have taken place over the last centuries, it has recently been argued that the term 'Palaeoanthropocene' could be used to describe the smaller-scale, regional changes in ecosystems and land surface processes resulting from human subsistence and migration from ~1 Ma onwards (Foley et al., 2013), which is a much longer timeframe for human impacts on the landscape than has been previously thought (Braje and Erlandson, 2013).

Several studies from southern Africa have argued that climate played a significant role in driving human activity and innovation throughout the Quaternary (e.g. Ziegler et al., 2013; Maslin et al., 2014). For example, Huffman (2010) showed that the deliberate and ritualistic burning of Iron Age structures across southern Africa took place at the same time as El Niño-induced droughts, and that this burning was largely a cultural and socio-political response to environmental stress. Ziegler et al. (2013) argued that technological innovations in Middle Stone Age lithic use was linked to rapid changes in humidity/aridity across different parts of Africa. These direct interpretations of changing human activity to climate forcing can only be made when based upon a tightly-constrained timeframe provided in these studies by radiocarbon and U/Th dates, and by $\delta^{18}O$ wiggle matching. In the absence of such tight geochronologies, interpretations remain less certain.

More commonly, however, climate–human relationships are mediated indirectly through resource availability and use, which may be climatically limited, and it is therefore in this context of environmental resources that climate–human relationships should be framed. In order to examine these components and to establish cause-and-effect relationships, a secure chronological and stratigraphic framework is needed. However, use of a dating framework is potentially problematic for two main reasons. First is the accuracy and resolution of radiometric dating, including the clarity of the event that is being dated. Second is the meaning or interpretation of radiometric dates with respect to their environmental and stratigraphic contexts. In the more recent archaeological past (Later Stone Age, Iron Age and historical era), a wide range of dating and climate-specific isotopic evidence is available, and from different terrestrial (e.g. Norström et al., 2005, 2008; Wallington et al., 2007;

Hall et al., 2008; Scott et al., 2008) and marine settings (Dusseldorp et al., 2013). Further back in time, where limitations of radiocarbon dating exist and preserved organic matter is less abundant, different geochronometric analyses and environmental proxies may be used, such as luminescence dating (e.g. Jacobs et al., 2008), or U/Pb dating and stable isotopes from coastal cave speleothems (e.g. Bar-Matthews et al., 2010; Woodhead and Pickering, 2012). In the karstic, hominin-bearing caves of the Cradle of Humankind, no organic matter is preserved, fossil-rich deposits are time-averaged, and U/Pb dates often have low resolution. In these contexts, providing chronological control over the deposits, and therefore their palaeoenvironmental interpretations (based largely on fossil taxonomic representation, bovid ecomorphology, dental isotopes and microwear), has proven highly problematic (Partridge, 2005). Here, the stratigraphic context of both fossil-rich layers and speleothems/flowstones has been complicated by partial post-depositional reworking and dissolution/overprinting (Bruxelles et al., 2014), but it is clear that these processes affecting the integrity and meaning of the archaeological record are strongly influenced by the external (subaerial) climate outside of the cave system. At the Middle Stone age cave site of Sibudu, a range of different multidisciplinary data types has been used to reconstruct the local environment outside of the relatively sheltered confines of the cave (e.g. Wadley and Jacobs, 2006; Bruch et al., 2012).

In different archaeological settings, there remains a significant separation between the site- and landscape-scale evidence that has been used to inform the climatological context of past human activity. This is owing to several factors, including the presence and preservation of different data types, large interior areas of the country with little or no published archaeological or palaeoclimatological records, incomplete archaeological and palaeoclimatological records, archaeologically isolated sites with poor chronological control, and a general site-scale focus of much previous archaeological work. These are thus significant limitations in linking climate to different aspects of human activity over multiple spatial and temporal scales.

25.4 Landscapes and human activity

Landscapes are now identified as complex and evolutionary entities that exhibit a range of properties that change over different spatial and temporal scales (e.g. Phillips, 2009). Human interactions with landscapes include the use of their varied material and non-material resources. Material resources include soils, water, food and fibre, and different rock types that are used for different purposes. Use of these resources can result in the development of identifiable and preserved signatures or proxies of landscape disturbance, including increased downslope sediment yield (caused by land surface destabilisation by deforestation, agriculture or mining), pollen and macrofossil evidence for changed ecosystems (recorded in soils, peats, sediments), and changed local-scale geomorphology, including mining spoil heaps, hillslope terracing for agriculture and mining (e.g. Marker and Evers, 1976; Widgren, 2010). Non-material landscape resources include landscape-scale topography and relief, vistas and viewpoints, and properties such as light, colour and texture (Knight and Harrison, 2013b). The use of non-material landscape resources in an archaeological context (Anschuetz et al., 2001) has not been widely considered in southern Africa, although rock art studies make some reference to these attributes (McCall, 2010; Mazel, 2011), and ethnographic and anthropological evidence shows strong relationships between non-material landscape properties in oral traditions and through ritual and representation (e.g. Schoeman, 2006; Brunton et al., 2013). However, aspects of behaviour can be interpreted from both material and non-material evidence (e.g. McCall, 2007; Högberg and Larsson, 2011).

Understanding the interconnections between landscapes and past human activity, and thus consideration of the use and significance of both material and non-material resources, requires the integration of different types of data collected on or applicable to different scales. The richness, complexity and variation of past human cultural practices can be understood through the interpretation (or through multiple and sometimes conflicting interpretations) of a preserved material record or, in the more recent past, through relatively limited ethnographic evidence, which is projected into the past and across geographically analogous socio-cultural entities. Both the material and non-material records are subject to different sets of meanings and interpretations, and there is also discussion of what it is that constitutes 'evidence' (see Sullivan, 1978, for the context of this discussion). These are important areas of research that seek to link human behaviour and psyche to landscapes, but have not yet been fully explored within the context of archaeology in southern Africa. The most apparent or simplest relationships between landscape and human activities are revealed through lithic and zooarchaeological assemblage records. Lithic assemblages are the most commonly used cultural and chronological proxies due to their high abundance across southern Africa in all but the most recent Iron Age and historical sites. Understanding the geomorphological and stratigraphic contexts of both lithic and fossil assemblages found on the landscape or in caves requires dedicated studies (e.g. Stratford et al., 2014). For example, geomorphological processes may lead to the reworking, dispersal or concentration of assemblages, meaning that they should not be considered as wholly *in situ*. Several studies have examined these processes experimentally (e.g. Frostick and Reid, 1983; Coard, 1999; Fanning and Holdway, 2001) or reinterpreted evidence from lithic scatters or subsurface lithic/fossil accumulations within sediments, showing that gravity-driven slope movement or surface runoff can alter the state or position of lithics/fossils (Frostick and Reid, 1983; Sampson et al., 1989; Staurset and Coulson, 2014). Nonetheless, many studies uncritically look at spatial patterns of lithic assemblages and make speculations about landscape use or age that are not well supported (e.g. Forssman, 2013; Foley and Lahr, 2015). Spatial modelling techniques may be able to resolve some of these issues (e.g. Franklin et al., 2015).

A range of analytical approaches can be applied to lithic assemblages, which provide evidence of landscape use by particular cultures and at particular times, from resource exploitation, subsistence and mobility perspectives (e.g. Le Baron et al., 2010a). Similar information can also be gleaned from the zooarchaeological record with additional perspectives on seasonal exploitation practices (e.g. Klein, 1978). However, these studies offer limited insight into abstract cultural–landscape relationships, apart from interpretations seen through the cultural lenses of today's researchers. Archaeological evidence of ritual or abstract cultural significance that is found within contemporaneous landscapes is limited to the recent past (Later Stone Age, Iron Age), but can include burial, rain-making and rock art sites. This typology of sites shows that landscapes can be more than just of functional value, through their connections to memory, belief-systems, ritual, the spirit world and ancestry (e.g. Schoeman, 2006). In rock art studies, the range of non-functional uses for ritual or political is often referred to as 'Power of Place', a term that strives to integrate human agency into the landscape (Hedges, 1993; Smith, 1994; Ouzman, 1995, 1998; Lewis-Williams, 2010). Separating functional from ritual elements within a limited archaeological record is both difficult and problematic (Huffman, 2012). However, interpretations of such evidence provide the opportunity to develop theories of human activity within cultural and ritual landscapes (Anschuetz et al., 2001). Work on ritual landscapes in southern Africa has hitherto mainly focused on the practices and interactions between Later Stone Age hunter-gatherer cultures and early farming communities at specific and generally isolated ritual sites, with recognisable human modification (e.g. van der Ryst et al., 2004; van Doornum, 2008). Evidence for such interactions takes the form of painted and engraved rockshelters (e.g. Hall and Smith, 2000), engraved rock exposures (e.g. Deacon, 1988) or artefact scatters on the tops of or around hilltops (koppies) (Le Baron et al., 2010a; Brunton et al., 2013). Landscape-scale interpretations of these ritual sites, and the development of ritual landscapes more generally, require a more integrated understanding of environmental and geomorphic change, and coeval changes in human activities and/or behaviours. The fragmentary and palimpsest nature of the archaeological record, and therefore presence/absence of proxies for abstract cultural activity, makes this a challenge. However, attempts have been made to link together material and non-material evidence on a landscape scale through studies of seasonality (e.g. Parkington, 2001; Wadley, 2006) and technological innovation (Shennan, 2001). Associating intra-cultural functional evidence with coeval ritual evidence is also challenging, for reasons including the exclusive use of space, changing chronological correlations, and preservation and integrity of original physical associations.

A limitation in the interpretation of landscape–human relationships is that of scale (Jusseret, 2010). Attempts to understand site formation and assemblage integrity have led to microscale analyses of assemblage contexts within the sub-discipline of geoarchaeology (e.g. Goldberg and Berna, 2010; Karkanas and Goldberg, 2010; Le Baron et al., 2010b, 2011). Environmental proxies identified through micromorphological analyses of archaeological sites have been used to propose contemporaneous landscape processes (e.g. Karkanas and Goldberg, 2010; Le Baron et al., 2010b). Such studies undertaken on sediments within cave systems have additional depositional and interpretive complexities (Farrand, 2001; White, 2007) due to the presence of mixed environmental and geological components suggestive of both short-term or microscale climatic and chemical processes, and site-scale post-depositional modification and diagenesis (Bar-Matthews et al., 2010). Cave systems are effective sediment sinks with generally high preservation potential, so comparable records outside of the rock shelter or cave system may not be preserved, limiting the extent to which such microscale evidence can be interpreted across the region (but see Wadley et al., 2011, for attempts at Sibudu Cave).

25.5 Discussion and future research directions

25.5.1 The concerns of archaeological research

Kintigh et al. (2014) identify twenty five 'grand challenges' for the future of archaeological research, but these 'challenges' are framed as inward-looking research questions accompanied by a plea for more funding. A more helpful approach is to consider what archaeological research has been concerned with, and what is has not. Archaeological research in southern Africa, as elsewhere in the Twentieth Century, has focused on presence, materiality, evidence and continuity. This includes the faithful cataloguing and description of what is there; the belief in a simple linear 'progression' of attributes such as technology, ideas and cultures; and in the belief in the integrity and validity of 'evidence'. Although archaeologists have also been concerned with the complexities of 'progression' (if indeed there is such a thing) of material culture(s), the absence of detailed chronologies and the selectivity of preserved evidence has limited the ability to explore the intricacies of behavioural and cultural dynamics (Parsons and Lombard, 2011). Even technology, established by early archaeologists like Augustus Pitt Rivers and Flinders Petrie as the ultimate proxy of cultural evolution and sophistication, is understood to be influenced by many dimensions of cultural identity, tradition, environment, and individual identity and behaviour. Current ideas in historical archaeology and in the genetics of hominin evolution challenge these paradigms because these ideas show evidence for multiple and parallel histories, the ambiguity and complexity of 'evidence', and suggest periods of rapid cultural and/or evolutionary change separated by periods of quasi-stability (Scharf, 2014). Also, research into the archaeological record has been less concerned with identifying and interpreting absence, transience, non-materiality, plurality and palimpsest (e.g. Bille et al., 2010; Mackay et al., 2014). These components are important, however, because they allow for a more nuanced, more critical, and less deterministic interpretation of archaeological and palaeoanthropological evidence, and enable linkages to cognate areas of interest in cultural heritage, memorialisation and memory, and of promoting dissonant voices. The concerns of archaeological research have therefore become more complex and multidimensional in recent decades, and with more cultural interpretive power than hitherto appreciated.

25.5.2 Climate and environmental change and the landscape record

A wide range of different types of climatic and environmental proxies is now available from individual sites in different climatic or environmental contexts across southern Africa, and these can provide high resolution data that are usually constrained by a geochronological framework. These proxies can also inform on different aspects of climate and environmental change during the Quaternary, thus, when taken together, can provide a more holistic and integrated understanding of these changes. The highest resolution direct climate records come from layered biological (tree-ring) or mineralogical (speleothem) sources. Tree-ring thickness and $\delta^{13}C$ and $\delta^{18}O$ data from climatically sensitive (e.g. *Podocarpus* spp., *Mimusops caffra*) (e.g. February and Stock, 1998; Hall et al., 2009) or long-lived species (e.g. *Adansonia digitata*, *Breonadia salicina*) (e.g. Robertson et al., 2006; Norström et al., 2008) can resolve annual climate records and have potential to allow for regional climate comparison. Data on $\delta^{13}C$ and $\delta^{18}O$ values within cave speleothems can likewise be used to reconstruct precipitation and air temperature, and surface ecosystem composition (in particular C_3/C_4 relative abundance) (Holmgren et al., 1999; Bar-Matthews et al., 2010). Global-scale climate records such as from Antarctic ice cores can also be used. The problem with such records is that they are relatively sparse in number and their temporal resolution, although high, cannot be readily matched with other lower-resolution records from the same locations. Biological records, in particular from lowland wetlands, mires, lakes or offshore basins, can extend over tens of thousands of years, and may be analysed for a range of properties including pollen, charcoal, phytoliths, testate amoebae, stable isotopes, *n*-alkanes, esters and lipid biomarkers (e.g. Norström et al., 2009; Kristen et al., 2010; Neumann et al., 2010; Scott et al., 2012). In addition, a range of different data analysis approaches can be used, most commonly including principal components analysis (PCA) (e.g. Scott et al., 2012), but also neural network and Bayesian analysis and ecosystem modelling. These latter approaches have not been commonly used in southern Africa but are important ways forward.

Specific links between reconstructed patterns of past climate and land surface processes are not well formulated. For example, speleothems commonly contain aeolian dust, but this is hardly ever described even though its geochemistry can pinpoint dust source regions. Peat cores commonly contain sedimentary interbeds of fluvial, aeolian or slope origin but as these are usually devoid of pollen (although macro plant remains can be present), these interbeds are usually ignored. There is thus a need for greater integration of data on different land surface processes and properties *sensu lato* that are commonly contained within the same cores. Marine records from around southern Africa indicate how this can be done effectively (e.g. Gasse et al., 2008; Dupont et al., 2011).

25.5.3 Resolving landscapes and human activity: geoarchaeology and its limitations

Geoarchaeology as a subdiscipline of archaeology is critical in developing an understanding of landscape, climate and human activities through the archaeological record of southern Africa. In contrast to the large-scale site excavations undertaken by archaeologists such as Pitt Rivers and Mortimer Wheeler, total excavation of a site is now generally impossible and seen as scientifically unethical, and thus low-disturbance geophysical and microscale survey and analytical techniques are now viewed as valid alternatives. This research approach, using methods that are more commonly applied in geology, geography and geomorphology, is the basis of geoarchaeology. Although geoarchaeological studies have been useful in archaeological research, an increasing focus on the microscale record of individual sites does not favour development of a landscape-scale approach, and may tend towards over-interpretation of the limited evidence available. Geoarchaeological studies can also be applied to landscape archaeology, but this has not been commonly done in southern Africa due to the general absence of a high site density that can allow for spatial comparison of coeval patterns of occupation of landscape use (Kent and Scholtz, 2003; Oestmo et al., 2014). Reconciling the disparity of scales between high-resolution archaeological excavations and landscape archaeology requires a better understanding of the site and its landscape context, regardless of the presence of archaeological evidence (e.g. Blumenschine et al., 2012). Archaeological sampling theory provides the framework for such studies (e.g. Krakker et al., 1983; Read, 1986).

As human activities, civilisations and cultures become increasingly dissociated from the constraints of the physical landscape, two future themes may emerge in landscape–human research, and which have a bearing on development of geoarchaeological research approaches. One is emphasis on the research of landscape–human activities of the past, which is increasingly focused on *Homo*'s power to deliver global change, as seen by discussion of the origins of the Anthropocene (Erlandson and Braje, 2013). Research on hominin evolution in southern Africa can also inform on development of the southern African landscape and the integrity of the archaeological record. For example, Balter et al. (2012), based on isotopic data from tooth dynamics, showed that dietary changes in early *Homo* and *Paranthropus* probably reflect changes in ecosystem composition and habitat use. Human use of landscape resources is multidimensional and may include food-resource utilisation, agriculture, stone-tool/lithic manufacture, water utilisation; and using landscape topography as vantage points, spiritual sites, hunting sites, refuge/shelter sites, etc. Oestmo et al. (2014) described the landscape context of Middle Stone Age open-air and cave/rock-shelter sites along the Eastern Cape coast, and showed that their occupants interacted with landscape resources in different ways. Studies that draw on multidisciplinary and multi-scalar data to identify patterns between lithic/fossil distributions, technologies, DNA, isotopic analyses and ethnography within established geochronological frameworks may be able to facilitate understanding of broader human movements and behaviours in relation to local to regional environmental changes (e.g. Lombard et al., 2013; Breton et al., 2014). These studies are at the moment applied to behavioural records of *Homo sapiens,* but are limited by the availability of contexts with suitable archaeological, fossil and sediment records. These considerations have been examined in a variety of settings in southern Africa such as the Brandberg, Namibia (Lenssen-Erz, 2004), Tsodilo Hills, Botswana (Segadika, 2006), and the Limpopo Valley, South Africa (Le Baron et al., 2010a). Human activities through time have also had physical impacts on landscapes, ranging from the movement of mined earth materials since at least the Middle Stone Age (Wadley et al., 2004), to microscale alterations on rock surfaces (engravings), which in some cases suggest a ritual significance for that particular site (e.g. Deacon, 1988). The second theme is that of the impacts of present-day climate and environmental change on the ongoing narrative of the human species with the Earth. Human activities today may be preserved as the archaeology of the future. Consequently, our understanding of landscape–human relationships should also consider the multiple, diverse and contradictory ways in which humans engage with the physical landscape; the non-materiality of cultural histories, communities and traditions; and the tensions between absence, presence and preservation. The ways in which geologists, geomorphologists and archaeologists have studied climate and environmental changes of the past may be inappropriate in the context of the Anthropocene (Knight and Harrison, 2014). A new sub-discipline of Physical Geography, anthropogeomorphology, may provide the context for developing such an understanding (Thornbush, 2012).

25.6 Summary

Southern Africa offers a rich resource for the understanding of landscape–climate–human relationships during the Quaternary, because of the interplay of these three components over long (10^6-year) timescales, and in response to Quaternary glacial–interglacial cycles. Moreover, the record of these relationships is relatively continuous in a few key sites with high sediment preservation potential, such as caves. However, these key sites are relatively isolated and it is difficult to extrapolate their records to the wider landscape, as has been done more commonly in Europe and North America. This is a challenge for future research in southern Africa. Developments in the fields of landscape morphodynamics, climate change, and hominin evolution and geoarchaeology, offer new datasets that may permit a greater understanding of landscape–climate–human relationships. These datasets may include dating of land surface processes, sediments, archaeological sites and bones; isotopic data reflecting changing precipitation patterns or ecosystem composition; remote sensing data on land surface properties including geology, soil geochemistry, and digital elevation modelling; groundwater and sediment flux modelling inland and in the coastal zone; ecosystem modelling and species' range analysis; and palaeoanthropological interpretations of artefactual data including rock art. These varied data types illustrate the potential complex interconnections between the disciplines of geology, geography, archaeology and anthropology that should be the concern of future research in southern Africa.

Acknowledgements

We thank Lyn Wadley, Marlize Lombard and Garth Sampson for very useful comments on a draft of this chapter.

References

Anschuetz, K. F., Wilshusen, R. H. and Scheick, C. L. (2001). An Archaeology of landscapes: Perspectives and directions. *Journal of Archaeological Research*, 9, 157–211.

Bailey, G. (2007). Time perspectives, palimpsests and the archaeology of time. *Journal of Anthropological Archaeology*, 26, 198–223.
Balter, V., Braga, J., Télouk, P. and Thackeray, F. (2012). Evidence for dietary change but not landscape use in South African early hominins. *Nature*, 489, 558–560.
Bar-Matthews, M., Marean, C. W., Jacobs, Z., Karkanas, P., Fisher, E., Herries, A. I. R., Brown, K., Williams, H. M., Bernatchez, J., Ayalon, A. and Nilssen, P. J. (2010). A high resolution and continuous isotopic speleothem record of paleoclimate and paleoenvironment from 90 to 53 ka from Pinnacle Point on the south coast of South Africa. *Quaternary Science Reviews*, 29, 2131–2145.
Bierman, P. R. and Caffee, M. (2001). Slow rates of rock surface erosion and sediment production across the Namib Desert and escarpment, southern Africa. *American Journal of Science*, 301, 326–358.
Bierman, P. R., Coppersmith, R., Hanson, K., Neveling, J., Portenga, E. W. and Rood, D. H. (2014). A cosmogenic view of erosion, relief generation, and the age of faulting in southern Africa. *GSA Today*, 24 (9), 4–11.
Bille, M., Hastrup, F. and Sørensen, T. F. (eds) (2010). *An Anthropology of Absence. Materializations of Transcendence and Loss*. Berlin: Springer, 221pp.
Bishop, P. (2007). Long-term landscape evolution: Linking tectonics and surface processes. *Earth Surface Processes and Landforms*, 32. 329–365.
Blumenschine, R., Stanistreet, I. and Masao, F. T. (2012). Olduvai Gorge and the Olduvai landscape palaeoanthropology project. *Journal of Human Evolution*, 63, 247–250.
Braje, T. J. and Erlandson, J. M. (2013). Looking forward, looking back: Humans, anthropogenic change, and the Anthropocene. *Anthropocene*, 4, 116–121.
Breton, G., Schlebusch, C. M., Lombard, M., Sjödin, P., Soodyall, H. and Jakobsson, M. (2014). Lactase persistence alleles reveal partial East African ancestry of southern African Khoe pastoralists. *Current Biology*, 24, 852–858.
Brown, R. W., Summerfield, M. A. and Gleadow, A. J. W. (2002). Denudational history along a transect across the Drakensberg Escarpment of southern Africa derived from apatite fission track thermochronology. *Journal of Geophysical Research: Solid Earth*, 107, 74–77.
Bruch, A. A., Sievers, C. and Wadley, L. (2012). Quantification of climate and vegetation from southern African Middle Stone Age sites – an application using Late Pleistocene plant material from Sibudu, South Africa. *Quaternary Science Reviews*, 45, 7–17.
Brunton, S., Badenhorst, S. and Schoeman, M. H. (2013). Ritual fauna from Ratho Kroonkop: A second millennium AD rain control site in the Shashe-Limpopo Confluence area of South Africa. *Azania-Archaeological Research in Africa*, 48, 111–132.
Bruxelles, L., Clarke, R. J., Maire, R., Ortega, R. and Stratford, D. (2014). Stratigraphic analysis of the Sterkfontein StW 573 *Australopithecus* skeleton and implications for its age. *Journal of Human Evolution*, 70, 36–48.
Coard, R. (1999). One Bone, Two Bone, Wet Bones, Dry Bones: Transport potentials under experimental conditions. *Journal of Archaeological Science*, 26, 1369–1375.
Copeland, S. R., Sponheimer, M., De Ruiter, D. J., Lee-Thorp, J. A., Cordon, D., le Roux, P. J., Grimes, V. and Richards, M. P. (2011). Strontium isotope evidence for landscape use by early hominins. *Nature*, 747, 76–79.
Cotterill, F. P. D. and de Wit, M. J. (2011). Geoecodynamics and the Kalahari epeirogeny: Linking its genomic record, tree of life and palimpsest into a unified narrative of landscape evolution. *South African Journal of Geology*, 114, 489–514.
Deacon, J. (1988). The power of a place in understanding southern San rock engravings. *World Archaeology*, 20, 129–140.

de Wit, M. (2007). The Kalahari Epierogeny and climate change: Differentiating cause and effect from core to space. *South African Journal of Geology*, 110, 367–392.

Dirks, P. H. G. M. and Berger, L. R. (2013). Hominin-bearing caves and landscape dynamics in the Cradle of Humankind, South Africa. *Journal of African Earth Sciences*, 78, 109–131.

Dixon, J. L., Heimsath, A. M. and Amundson, R. (2009a). The critical role of climate and saprolite weathering in landscape evolution. *Earth Surface Processes and Landforms*, 34, 1507–1521.

Dixon, J. L., Heimsath, A. M., Kaste, J. and Amundson, R. (2009b). Climate-driven processes of hillslope weathering. *Geology*, 37, 975–978.

Dupont, L. M., Caley, T., Kim, J.-H., Castañeda, I., Malaizé, B. and Giraudeau, J. (2011). Glacial-interglacial vegetation dynamics in South Eastern Africa coupled to sea surface temperature variations in the Western Indian Ocean. *Climate of the Past*, 7, 1209–1224.

Dusseldorp, G., Lombard, M. and Wurz, S. (2013). Pleistocene Homo and the updated Stone Age sequence of South Africa. *South African Journal of Science*, 109, Art. #0042, 7pp, http://dx.doi.org/10.1590/sajs.2013/20120042

Erlandson, J. M. and Braje, T. J. (2013). Archeology and the Anthropocene. *Anthropocene*, 4, 1–7.

Fanning, P. and Holdaway, S. (2001). Stone artifact scatter in Western NSW, Australia: Geomorphic controls on artifact size and distribution. *Geoarchaeology*, 16, 667–686.

Farrand, W. R. (2001). Sediments and stratigraphy in rockshelters and caves: A personal perspective on principles and pragmatics. *Geoarchaeology*, 16, 537–557.

February, E. C. and Stock, W. D. (1998). An assessment of the dendrochronological potential of two *Podocarpus* species. *The Holocene*, 8, 747–750.

Flowers, R. M. and Schoene, B. (2010). (U-Th)/He thermochronometry constraints on unroofing of the eastern Kaapvaal craton and significance for uplift of the southern African plateau. *Geology*, 38, 827–830.

Foley, R. A. and Lahr, M. M. (2015). Lithic landscapes: Early human impact from stone tool production on the Central Saharan environment. *PLoS ONE*, 10 (3), e0116482, doi:10.1371/journal.pone.0116482.

Foley, S. F., Gronenborn, D., Andreae, M. O., Kadereit, J. W., Esper, J., Scholz, D., Pöschl, U., Jacob, D. E., Schöne, B. R., Schreg, R., Vött, A., Jordan, D., Lelieveld, J., Weller, C. G., Alt, K. W., Gaudzinski-Windheuser, S., Bruhn, K.-C., Tost, H., Sirocko, F. and Crutzen, P. J. (2013). The Palaeoanthropocene – The beginnings of anthropogenic environmental change. *Anthropocene*, 3, 83–88.

Forssman, T. (2013). Missing pieces: Later Stone Age surface assemblages on the greater Mapungubwe landscape, South Africa. *Southern African Humanities*, 25, 65–85.

Franklin, J., Potts, A. J., Fisher, E. C., Cowling, R. M. and Marean, C. W. (2015). Paleodistribution modeling in archaeology and paleoanthropology. *Quaternary Science Reviews*, 110, 1–14.

Frostick, L. and Reid, I. (1983). Taphonomic significance of sub-aerial transport of vertebrate fossils on steep semi-arid slopes. *Lethaia*, 16, 157–164.

Gasse, F., Chalié, F., Vincens, A., Williams, M. A. J. and Williamson, D. (2008). Climatic patterns in equatorial and southern Africa from 30,000 to 10,000 years ago reconstructed from terrestrial and near-shore proxy data. *Quaternary Science Reviews*, 27, 2316–2340.

Giosan, L., Fuller, D. Q., Nicoll, K., Flad, R. K. and Clift, P. D. (eds) (2012). *Climates, Landscapes, and Civilizations*. Washington, D.C.: American Geophysical Union, Geophysical Monograph 198, 226pp.

Goldberg, P. and Berna, F. (2010). Micromorphology and context. *Quaternary International*, 214, 56–62.

Hall, G., Woodborne, S. and Pienaar, M. (2009). Rainfall control of the $\delta^{13}C$ ratios of Mimusops caffra from KwaZulu-Natal, South Africa. *The Holocene*, 19, 251–260.

Hall, G., Woodborne, S. and Scholes, M. (2008). Stable carbon isotope ratios from archaeological charcoal as palaeoenvironmental indicators. *Chemical Geology*, 247, 384–400.

Hall, S. L. and Smith, B. (2000). Empowering places: Rock shelters and ritual control in farmer-forager interactions in the Northern Province. *The South African Archaeological Society Goodwin Series*, 8, 30–46.

Hedges, K. (1993). Places to see and places to hear: Rock art and features of the sacred landscape. In *Time and Space: Dating and Spatial Considerations in Rock Art Research*, ed. J. Steinbring, A. Watchman, P. Faulstitch and P. S. C. Taçon. Melbourne: Occasional AURA Publication 8, pp. 121–127.

Högberg, A. and Larsson, L. (2011). Lithic technology and behavioural modernity: New results from the Still Bay site, Hollow Rock Shelter, Western Cape Province, South Africa. *Journal of Human Evolution*, 61, 133–155.

Holmgren, K., Karlén, W., Lauritzen, S. E., Lee-Thorp, J. A., Partridge, T. C., Piketh, S., Repinski, P., Stevenson, C., Svanered, O. and Tyson, P. D. (1999). A 3000-year high-resolution stalagmite-based record of palaeoclimate for northeastern South Africa. *The Holocene*, 9, 295–309.

Huffman, T. N. (2010). Intensive El Niño and the Iron Age of south-eastern Africa. *Journal of Archaeological Science*, 37, 2572–2586.

Huffman, T. (2012). Ritual space in Pre-Colonial Farming Societies in Southern Africa. *Ethnoarchaeology*, 4, 119–146.

Jacobs, Z., Roberts, R. G., Galbraith, R. F. Deacon, H. J., Grün, R., Mackay, A., Mitchell, P., Vogelsang, R. and Wadley, L. (2008). Ages for the Middle Stone Age of southern Africa: implications for human behavior and dispersal. *Science*, 322, 733–735.

Jusseret, S. (2010). Socializing geoarchaeology: Insights from Bourdieu's theory of practice applied to Neolithic and Bronze Age Crete. *Geoarchaeology*, 25, 675–708.

Karkanas, P. and Goldberg, P. (2010). Site formation processes at Pinnacle Point Cave 13B (Mossel Bay, Western Cape Province, South Africa): Resolving stratigraphic and depositional complexities with micromorphology. *Journal of Human Evolution*, 59, 256–273.

Kent, S. and Scholtz, N. (2003). Perspectives on the geology of an open-air Middle Stone Age site, eastern Free State, South Africa. *South African Journal of Science*, 99, 422–424.

King, L. C. (1967). *South African Scenery: A Textbook of Geomorphology* (3rd Ed). Edinburgh: Oliver & Boyd, 308pp.

King, L. C. (1976). Planation remnants upon highlands. *Zeitschrift für Geomorphologie*, 20, 133–148.

King, R. B. (2008). South African pediments and interfluves. In *Landscape Evolution: Denudation, Climate and Tectonics Over Different Time and Space Scales*, ed. K. Gallagher, S. J. Jones and J. Wainwright. Geological Society of London: Special Publications, 296, pp. 167–181.

Kintigh, K. W., Altschul, J. H., Beaudry, M. C., Drennan, R. D., Kinzig, A. P., Kohler, T. A., Limp, W. F., Maschner, H. D. G., Michener, W. K., Pauketat, T. R., Peregrine, P., Sabloff, J. A., Wilkinson, T. J., Wright, H. T. and Zeder, M. A. (2014). Grand challenges for archaeology. *American Antiquity*, 79, 5–24.

Klein, R. G. (1978). Stone Age predation on large African bovids. *Journal of Archaeological Science*, 5, 195–217.

Knight, J. (2012). Development of Palimpsest Landscapes. *Vignettes – Key Concepts in Geomorphology*. Available at http://serc.carleton.edu/vignettes/collection/67822.html.

Knight, J. and Harrison, S. (2013a). The impacts of climate change on terrestrial Earth surface systems. *Nature Climate Change*, 3, 24–29.

Knight, J. and Harrison, S. (2013b). 'A land history of men': The intersection of geomorphology, culture and heritage in Cornwall, southwest England. *Applied Geography*, 42, 186–194.

Knight, J. and Harrison, S. (2014). Limitations of uniformitarianism in the Anthropocene. *Anthropocene*, 5, 71–75.

Krakker, J. J., Shott, M. J. and Welch, P. D. (1983). Design and evaluation of shovel-test sampling in regional archaeological survey. *Journal of Field Archaeology*, 10, 469–480.

Kristen, I., Wilkes, H., Vieth, A., Zink, K.-G., Plessen, B., Thorpe, J., Partridge, T. C. and Oberhänsli, H. (2010). Biomarker and stable carbon isotope analyses of sedimentary organic matter from Lake Tswaing: evidence for deglacial wetness and early Holocene drought from South Africa. *Journal of Paleolimnology*, 44, 143–160.

Kyriacou, K., Parkington, J. E., Marais, A. D. and Braun, D. R. (2014). Nutrition, modernity and the archaeological record: Coastal resources and nutrition along Middle Stone Age hunter-gatherers on the western Cape coast of South Africa. *Journal of Human Evolution*, 77, 64–73.

Le Baron, J. C., Kuman, K. and Grab, S. W. (2010a). The landscape distribution of stone-age artefacts on the Hackthorne Plateau, Limpopo River Valley, South Africa. *South African Archaeological Bulletin*, 65, 123–131.

Le Baron, J. C., Grab, S. W. and Kuman, K. (2010b). Grain-size characteristics of surface sands and sandstones: indicators of potential sediment sources for the southern Tuli Basin, South Africa. *South African Geographical Journal*, 92, 7–21.

Le Baron, J. C., Grab, S. W. and Kuman, K. (2011). Surface texture analysis of southern Tuli Basin sediments: Implications for Limpopo Valley geoarchaeological contexts. *Journal of African Earth Sciences*, 59, 384–394.

Lenssen-Erz, T. (2004). The landscape setting of rock-painted sites in the Brandberg (Namibia): infrastructure, Gestaltung, use and meaning. In *The Figured Landscapes of Rock-art: Looking at Pictures in Place*, ed. G. Nash and C. Chippindale. Cambridge: Cambridge University Press, pp. 131–150.

Lewis-Williams, J. D. (2010). The imagistic web of San myth, art and landscape. *Southern African Humanities*, 22, 1–18.

Lombard, M., Schlebusch, C. and Soodyall, H. (2013). Bridging disciplines to better elucidate the evolution of early *Homo sapiens* in southern Africa. *South African Journal of Science*, 109 (11/12), Art. #2013-0065, 8pp, http://dx.doi.org/10.1590/sajs.2013/20130065

Mackay, A., Stewart, B. A. and Chase, B. M. (2014). Coalescence and fragmentation in the late Pleistocene archaeology of southernmost Africa. *Journal of Human Evolution*, 72, 26–51.

Marker, M. E. and Evers, T. M. (1976). Iron Age Settlement and Soil Erosion in the Eastern Transvaal, South Africa. *South African Archaeological Bulletin*, 31, 153–165.

Maslin, M. A., Brierley, C. M., Milner, A. M., Shultz, S., Trauth, M. H. and Wilson, K. E. (2014). East African climate pulses and early human evolution. *Quaternary Science Reviews*, 101, 1–17.

Mazel, A. (2011). Time, color, and sound: Revisiting the rock art of Didima Gorge, South Africa. *Time & Mind – The Journal of Archaeology Consciousness and Culture*, 4, 283–296.

McCall, G. S. (2007). Behavioural ecological models of lithic technological change during the later Middle Stone Age of South Africa. *Journal of Archaeological Science*, 34, 1738–1751.

McCall, G. S. (2010). Changing views of Drakensberg San rock art: Examining landscape use, ritual activity, and contact through multivariate content-based spatial analysis. *American Antiquity*, 75, 773–791.

Moore, A., Blenkinsop, T. and Cotterill, F. (2009). Southern African topography and erosion history: Plumes or plate tectonics? *Terra Nova*, 21, 310–315.

Neumann, F. H., Scott, L., Bousman, C. B. and van As, L. (2010). A Holocene sequence of vegetation change at Lake Eteza, coastal KwaZulu-Natal, South Africa. *Review of Palaeobotany and Palynology*, 162, 39–53.

Norström, E., Holmgren, K. and Mörth, C. (2005). Rainfall-driven variations in $\delta^{13}C$ composition and wood anatomy of *Breonadia salicina* trees from South Africa between AD 1375 and 1995. *South African Journal of Science*, 101, 162–168.

Norström, E., Holmgren, K. and Mörth, C. (2008). A 600 year $\delta^{18}O$ record from cellulose of *Breonadia salicina* trees, South Africa. *Dendochronologia*, 26, 21–33.

Norström, E., Scott, L., Partridge, T. C., Risberg, J. and Homgren, K. (2009). Reconstruction of environmental and climate changes at Braamhoek wetland, eastern escarpment South Africa, during the last 16,000 years with emphasis on the Pleistocene–Holocene transition. *Palaeogeography, Palaeoclimatology, Palaeoecology*, 271, 240–258.

Oestmo, S., Schoville, B. J., Wilkins, J. and Marean, C. W. (2014). A Middle Stone Age Paleoscape near the Pinnacle Point caves, Vleesbaai, South Africa. *Quaternary International*, 350, 147–168.

Ouzman, S. (1995). Spiritual and political uses of a rock engraving site and its imagery by San and Tswana-speakers. *The South African Archaeological Bulletin*, 50, 55–67.

Ouzman, S. (1998). Toward a mindscape of landscape: Rock-art as expression of world-understanding. In *The Archaeology of Rock-Art*, ed. C. Chippindale and P. S. C. Taçon. Cambridge: Cambridge University Press, pp. 30–41.

Parkington, J. E. (2001). Milestones: The impact of the systematic exploitation of marine foods on human evolution. In *Humanity from African Naissance to Coming Millennia: Colloquia in Human Biology and Palaeoanthropology*, ed. P. V. Tobias, M. A. Raath, J. Moggi-Cecchi and G. A. Doyle. Florence: Florence University Press, pp. 327–336.

Parsons, I. and Lombard, M. (2011). The multi-disciplinary landscape of mid-late Upper Pleistocene archaeology. *South African Archaeological Bulletin*, 66, 178–185.

Partridge, T. C. (1998). Of diamonds, dinosaurs and diastrophism: 150 million years of landscape evolution in southern Africa. *South African Journal of Geology*, 101, 167–184.

Partridge, T. C. (2005). Dating of the Sterkfontein hominids: Progress and possibilities. *Transactions of the Royal Society of South Africa*, 60, 107–109.

Partridge, T. C. and Maud, R. R. (1987). Geomorphic evolution of southern Africa since the Mesozoic. *South African Journal of Geology*, 90, 179–208.

Partridge, T. C. and Maud, R. R. (eds) (2000). *The Cenozoic of Southern Africa*. Oxford: Oxford University Press, 406pp.

Phillips, J. D. (2005). Weathering instability and landscape evolution. *Geomorphology*, 67, 255–272.

Phillips, J. D. (2009). Landscape evolution space and the relative importance of geomorphic processes and controls. *Geomorphology*, 109, 79–85.

Read, D. W. (1986). Sampling procedures for regional surveys: A problem of representativeness and effectiveness. *Journal of Field Archaeology*, 13, 477–491.
Reynolds, S. C., Bailey, G. N. and King, G. C. P. (2011). Landscapes and their relation to hominin habitats: Case studies from Australopithecus sites in eastern and southern Africa. *Journal of Human Evolution*, 60, 281–298.
Robertson, I., Loader, N. J., Froyd, C. A., Zambatis, N., Whyte, I. and Woodborne, S. (2006). The potential of the baobab (*Adansonia digitata* L.) as a proxy climate archive. *Applied Geochemistry*, 21, 1674–1680.
Sampson, C. G., Hart, T. J. G., Wallsmith, D. L. and Blagg, J. D. (1989). The ceramic sequence in the Upper Seacow Valley: Problems and implications. *The South African Archaeological Bulletin*, 44, pp. 3–16.
Scharf, E. A. (2014). Deep time: The emerging role of archaeology in landscape ecology. *Landscape Ecology*, 29, 563–569.
Schoeman, M. H. (2006). Imagining rain-places: Rain-control and changing ritual landscapes in the Shashe-Limpopo Confluence Area, South Africa. *South African Archaeological Bulletin*, 61, 152–165.
Scott, L., Holmgren, K. and Partridge, T. C. (2008). Reconciliation of vegetation and climatic interpretations of pollen profiles and other regional records from the last 60 thousand years in the Savanna Biome of Southern Africa. *Palaeogeography, Palaeoclimatology, Palaeoecology*, 257, 198–206.
Scott, L., Neumann, F. H., Brook, G. A., Bousman, C. B., Norström, E. and Metwally, A. A. (2012). Terrestrial fossil-pollen evidence of climate change during the last 26 thousand years in Southern Africa. *Quaternary Science Reviews*, 32, 100–118.
Segadika, P. (2006). Managing intangible heritage at Tsodilo. *Museum International*, 58, 31–40.
Shennan, S. (2001). Demography and cultural innovation: A model and its implications for the emergence of modern human culture. *Cambridge Archaeological Journal*, 11, 5–16.
Smith, A. B. (1994). Metaphors of space: Rock art and territoriality in southern Africa. In *Contested Images: Diversity in Southern African Rock Art Research*, ed. T. A. Dowson and J. D. Lewis-Williams. Johannesburg: Witwatersrand University Press, pp. 373–384.
Staurset, S. and Coulson, S. (2014). Sub-surface movement of stone artefacts at White Paintings Shelter, Tsodilo Hills, Botswana: Implications for the Middle Stone Age chronology of central southern Africa. *Journal of Human Evolution*, 75, 153–165.
Stewart, B. A., Dewar, G. I., Morley, M. W., Inglis, R. H., Wheeler, M., Jacobs, Z. and Roberts, R. G. (2012). Afromontane foragers of the Late Pleistocene: Site formation, chronology and occupational pulsing at Melikane Rockshelter, Lesotho. *Quaternary International*, 270, 40–60.
Stratford, D., Grab, S. and Pickering, T. R. (2014). The stratigraphy and formation history of fossil- and artefact-bearing sediments in the Milner Hall, Sterkfontein Cave, South Africa: New interpretations and implications for palaeoanthropology and archaeology. *Journal of African Earth Sciences*, 96, 155–167.
Sullivan, A. P. (1978). Inference and evidence in archaeology: A discussion of the conceptual problems. *Advances in Archaeological Method and Theory*, 1, 183–222.
Sumner, P. D. and Nel, W. (2006). Surface-climate attributes at Injisuthi Outpost, Drakensberg, and possible ramifications for weathering. *Earth Surface Processes and Landforms*, 31, 1445–1451.
Thornbush, M. J. (2012). Archaeogeomorphology as an application in physical geography. *Applied Geography*, 34, 325–330.

Tinker, J., de Wit, M. and Brown, R. (2008). Mesozoic exhumation of the Southern Cape, South Africa, quantified using apatite fission track thermochronology. *Tectonophysics*, 455, 77–93.

van der Ryst, M., Lombard, M. and Biemond, W. (2004). Rocks of potency: Engravings and cupules from the Doverdale Ward, Southern Tuli block, Botswana. *South African Archaeological Bulletin*, 59, 1–11.

van Doornum, B. L. (2008). Sheltered from change: Hunter-gatherer occupation of Balerno Main Shelter, Shashe-Limpopo confluence area, South Africa. *Southern African Humanities*, 20, 249–284.

Vrba, E. S., Denton, G. H., Partridge, T. C. and Burckle, L. H. (eds) 1995. *Paleoclimate and Evolution, with Emphasis on Human Origins*. New Haven: Yale University Press, 547pp.

Wadley, L. (2006). The use of space in the Late Middle Stone Age of Rose Cottage Cave, South Africa. In *Transitions before the Transitions: Evolution and Stability in the Middle Paleolithic and Middle Stone Age*, ed. E. Hovers and S. L. Kuhn. New York: Springer, pp. 279–294.

Wadley, L. and Jacobs, Z. (2006). Sibudu Cave: Background to the excavations, stratigraphy and dating. *Southern African Humanities*, 18, 1–26.

Wadley, L., Sievers, C., Bamford, M., Goldberg, P., Berna, F. and Miller, C. (2011). Middle Stone Age Bedding Construction and Settlement Patterns at Sibudu, South Africa. *Science*, 224, 1388–1391.

Wadley, L, Williamson, B. and Lombard, M. (2004). Ochre in hafting in Middle Stone Age southern Africa: A practical role. *Antiquity*, 78, 661–675.

Wallington, B. P., McKechnie, A. E., Owen-Smith, N. and Woodborne, S. (2007). Stable carbon isotope analysis of eland (*Taurotragus oryx*) diet in the Suikerbosrand Nature Reserve. *South African Journal of Wildlife Research*, 37, 127–131.

White, W. B. (2007). Cave sediments and Paleoclimate. *Journal of Cave and Karst Science*, 69, 76–93.

Widgren, M. (2010). Besieged Palaeonegritics or innovative Farmers: Historical Ecology of Intensive and Terraced Agriculture in West Africa and Sudan. *African Studies*, 69, 323–343.

Woodhead, J. and Pickering, R. (2012). Beyond 500 ka: Progress and prospects in the U–Pd chronology of speleothems, and their application to studies in palaeoclimate, human evolution, biodiversity and tectonics. *Chemical Geology*, 322–323, 290–299.

Zalasiewicz, J., Williams, M., Haywood, A. and Ellis, M. (2011). The Anthropocene: A new epoch of geological time? *Philosophical Transactions of the Royal Society, Series A*, 369, 835–841.

Ziegler, M., Simon, M. H., Hall, I. R., Barker, S., Stringer, C. and Zahn, R. (2013). Development of Middle Stone Age innovation linked to rapid climate change. *Nature Communications*, 4, 1905, doi:10.1038/ncomms2897.

Index